AN INTRODUCTION TO THE STUDY OF STELLAR STRUCTURE

By S. CHANDRASEKHAR

Yerkes Observatory

DOVER PUBLICATIONS, INC.

Published in Canada by General Publishing Company, Ltd., 30 Lesmill Road, Don Mills, Toronto, Ontario.
Published in the United Kingdom by Constable and Company, Ltd., 10 Orange Street, London WC 2.

This Dover edition, first published in 1958, is an unabridged and corrected republication of the work originally published in 1939 by the University of Chicago Press. It is republished by special arrangement with the University of Chicago Press.

Standard Book Number: 486-60413-6

Library of Congress Catalog Card Number: 58-162

Manufactured in the United States of America
Dover Publications, Inc.
180 Varick Street
New York, N. Y. 10014

TABLE OF CONTENTS

PREFACE

The present volume forms the second in the series of the "Astrophysical Monographs." The plan and scope of this book are set forth in the introductory chapter, and there remains only the pleasant task of thanking those who have helped me. I am under obligation to Mr. B. Strömgren for many valuable discussions during the writing of the monograph and also for allowing me to incorporate in chapters vi and vii some of his unpublished investigations. In the same way, Mr. G. P. Kuiper has allowed me to use the results of his study on the empirical mass-luminosity relation before publication. I am also deeply grateful to Mr. W. W. Morgan for reading the whole book both in manuscript and in proof. It is also a pleasure to record the generous encouragement I have received from Mr. O. Struve.

Finally, I wish to express my very grateful appreciation of the unfailing courtesy and consideration which the officials of the University of Chicago Press have shown me during the printing of this volume.

S. C.

YERKES OBSERVATORY
December 1938

INTRODUCTION

In this monograph an attempt is made to develop the theory of stellar structure from a consistent point of view and, as far as possible, rigorously. This and considerations of space have placed a somewhat severe restriction on the problems that are to come under review, while requiring at the same time a detailed treatment of other aspects of the subject. Thus, on the physical side, questions requiring the application of relatively advanced methods of statistical mechanics have to be avoided, while, on the astronomical side, questions concerned with problems of the type of stellar rotation and stellar variability or stability have had to be entirely omitted. This may seem a drawback, but, on the other hand, there is more space to develop the fundamentals with which the reader should be thoroughly familiar. In this introduction we shall make some comments on the type of problems with which we shall be mainly concerned and then outline the plan and scope of the monograph.

As we have already indicated, we shall restrict ourselves to the consideration of stars which are in equilibrium and which are in a steady state. Such an equilibrium configuration can be characterized by three parameters: its mass, M; its radius, R; and its luminosity, L (L being defined as the amount of radiant energy, expressed in ergs, radiated by the star per second to the space outside). It is beyond the scope of the monograph to discuss how the values of these parameters for individual stars are determined in practice. We shall assume, however, that we do have sets of values of these quantities for a number of stars. Stellar structure deals with these results of observational astronomy.

Our first problem, then, is to present the observational material in some form suitable for further discussion. There are two plots which we shall find useful: (*a*) the mass-luminosity diagram, and (*b*) the mass-radius diagram. In diagram (*a*) it is customary to plot $\log M$ (M expressed in solar units) against, essentially, $2.5 \log L$ (in practice, the absolute bolometric magnitude). In diagram (*b*) we plot log M against log R (R expressed in solar units). In Figures 1 and 2 we have collected together the results of observations; the material presented has been provided by Dr. G. P. Kuiper.

The ultimate objects of studies in stellar structure are the following:

1. To derive the complete march of the physical variables (the density, ρ; the temperature, T; etc.), on the one hand, and the variation of the chemical composition (the relative abundances of the different elements), on the other, throughout the entire configuration.

2. To describe quantitatively the kind of steady state (radiative, convective, etc.) that exists, eventually as a function of the radius vector r.

3. To specify the fundamental physical processes that are responsible for the setting-up of the steady states described under (2).

4. To evaluate quantitatively the irreversible processes that must be taking place which should be responsible for the continual loss of energy at the rate L by a star.

It is clear that complete and entirely satisfactory answers to all the foregoing problems require detailed information about physical phenomena which we do not have at the present time; even if we possessed this information, we should be faced with a mathematical problem of a very high order of complexity. From one point of view the most serious lack of information (at least until recently) concerns the nature of the physical processes involved under (4) above.

The question now arises as to how we can formulate, at least provisionally, the fundamental problem of stellar structure the solution of which will not only be of value but will also enable us to make substantial progress toward the solution of the complete problem. In other words, we need to formulate a somewhat restricted problem of stellar structure. The problem we shall consider is: Can we establish some relation between all three parameters, L, M, and R? That we can hope to make some progress toward the solution of this problem can be seen in the following way. When we observe a star, we see that in a prescribed spherical volume of radius R an amount of material of total mass M is inclosed; we also know that through this mass there occurs a continual streaming-out of a cer-

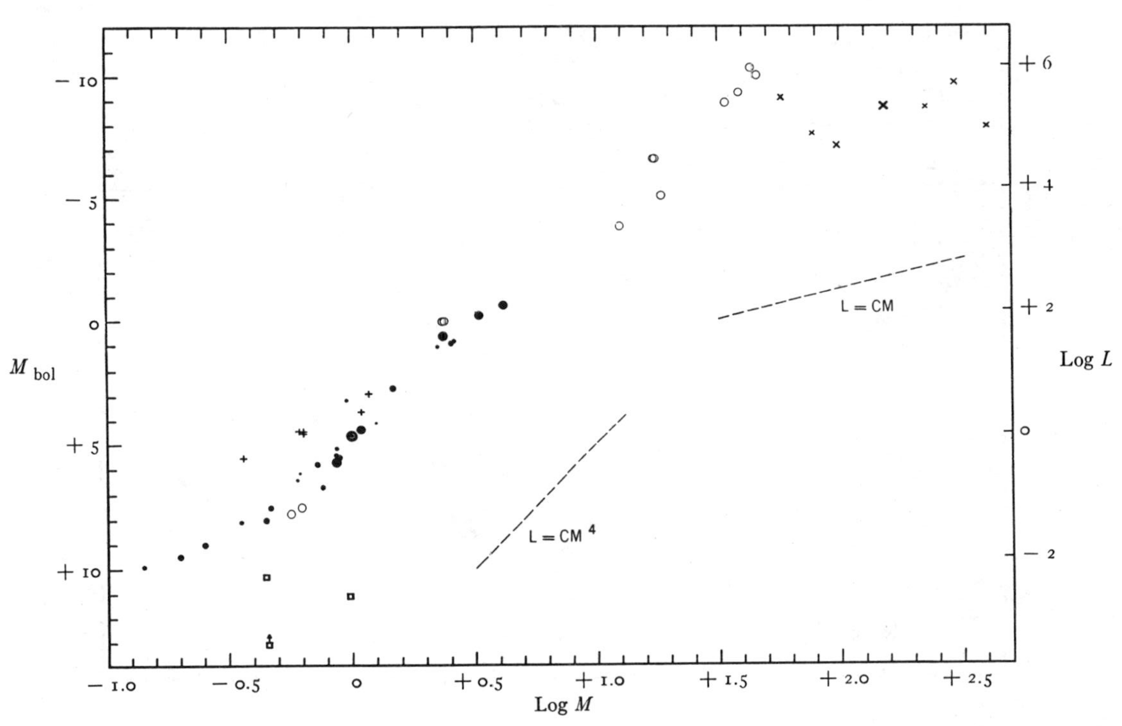

FIG. 1.—The mass-luminosity diagram. Solid dots: visual binaries; open circles: spectroscopic binaries; vertical crosses: the stars in the Hyades, diagonal crosses: the Trumpler stars, squares: white dwarfs.

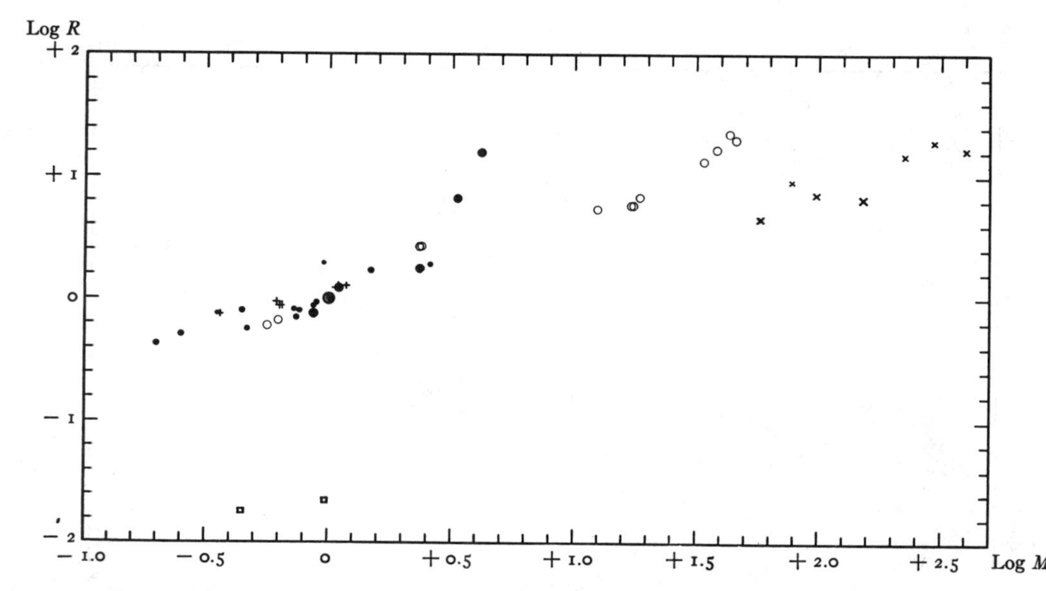

FIG. 2.—The mass-radius diagram. Solid dots: visual binaries; open circles: spectroscopic binaries; vertical crosses: the stars in the Hyades; diagonal crosses: the Trumpler stars, squares: white dwarfs.

tain mean flux of radiant energy specified by the luminosity, L. By hypothesis the star is in a steady state. The question we can then ask is: "How is it that a certain specified march of the net flux of radiant energy is able to support (against the gravitational attraction) an amount of mass equal to M inside a spherical volume of precisely the radius, R?" It will be noticed that some uncertainty has already been introduced. The luminosity, L, specifies the net flux of energy given by $L/4\pi R^2$ at the boundary of the star; we can, of course, take this as an index of a certain average flux that exists in the interior, but the solution of the mathematical problem of equilibrium would require a knowledge of the complete march of the function $L(r)$ and not merely a certain unspecified average depending on L. It is precisely for this reason that progress toward the solution of the restricted problem is made by means of the study of stellar models.

From the observed L and M we infer that each gram of the stellar material liberates on the average an amount of energy, $\bar{\epsilon} = L/M$. It may be safely assumed that $\epsilon(r)$—the rate of liberation of energy per gram of the material at the point r—is zero in the outer parts of the star where the physical conditions are relatively "mild," so that $L(r) = L$ in the outer parts of the star (these parts constitute the stellar envelope studied in chapter viii). Presumably, $L(r)$ decreases inward in such a way that $\epsilon(r) = L(r)/M(r)$ tends to some finite value as $r \to 0$, with or without a maximum for $r > 0$ (in the latter case $\epsilon(0)$ will be the maximum value of the function $\epsilon(r)$). Two obvious limiting cases suggest themselves: (*a*) $\epsilon(r) = \bar{\epsilon} =$ Constant, and (*b*) $\epsilon(r) = 0$, $r \neq 0$. The former case corresponds to a uniform distribution of the energy sources, while the latter leads us to the case where all the energy sources are concentrated at the center (this is the "point-source model").

We can investigate these two limiting cases as well as other, "intermediate," stellar models. After studying such models we attempt to abstract from the ensemble of the results thus obtained features which can be regarded as common to all the models. It would be safe to conclude that such common features must have some counterpart in nature; this is the manner in which progress has been made. One rather unexpected feature introduces an essential simplification. We shall discuss the origin of this simplification later (chaps. ii, vii, and viii); but it may be stated here that it follows from very general considerations that the majority of the normal stars, such as the sun and Capella, are gaseous and that radiation pressure as a factor in the equation of the hydrostatic equilibrium can be neglected (though, of course, it is important in determining the temperature gradient set up). This last circumstance in turn reveals another unexpected feature: the form of the relation between L, M, and R is independent of the stellar model considered. We shall not go farther into the consideration of these matters, but enough has been said to show that progress toward the solution of the restricted problem is in fact possible.

There is one other matter of importance to which we shall draw attention: we cannot assume beforehand that the chemical composition of all the stars is the same. Actually, under stellar conditions matter is generally so highly ionized that, as we shall see in greater detail in chapter vii, the uncertainty in the chemical composition is essentially due to the uncertainty in the abundance of the two lightest elements, namely, hydrogen and helium. The abundance of the lightest elements has then to be considered as a fresh parameter in the discussion. We can thus summarize by saying that our fundamental problem is to seek a theoretical relation of the kind

$$F[L, M, R, \text{abundance of hydrogen and helium}] = 0 . \qquad \text{(I)}$$

Our main object, then, is to describe the theory and the methods that have been developed toward this end.

We shall now proceed to outline the general plan:

The monograph divides itself into two distinct parts: the "classical" (chaps. i–iv) and the "modern" (chaps. v–xii). Furthermore, of the twelve chapters, two (chaps. i and x) deal essentially with physical theories (the laws of thermodynamics and quantum statistics, respectively), and chapter v deals also with a physical theory (the formal theory of radiation) presented, however, from an astrophysical angle. The last, chapter xii, on stellar energy, is on a plane different from the rest in that it summarizes the recent work on some of the most thorny problems of the subject. In greater detail the contents of each of the chapters are as follows:

Chapter i.—The laws of thermodynamics are here presented following Carathéodory's axiomatic standpoint. The reasons for including this chapter are twofold: first, there exists no treatise in English which gives Carathéodory's theory; and second, in the writer's view Carathéodory's theory is not merely an alternative, but elegant, approach to thermodynamics but is the only physically correct approach to the second law. Incidentally, the logical rigor and the beauty of Carathéodory's theory may be regarded as an example of the standard of perfection which should be demanded eventually of any physical theory, including the theory of stellar structure.

Chapter ii.—In this chapter we consider a number of physical theorems, the adiabatic and the polytropic laws, the virial theorem, homologous transformations, etc., and some immediate applications to the general theory of stellar structure.

Chapter iii.—Here we attempt to go as far as possible with our problem without any special assumptions except that the stars are in hydrostatic equilibrium. It is made clear in this chapter that no special assumptions are required to derive the orders of magnitude of the most important physical quantities which describe the structure of a star.

Chapter iv.—This chapter presents what is perhaps the most important contribution which "stellar structure" has made to applied mathematics. It represents, largely, the work of the great pioneers—Ritter, Emden, and Kelvin. As Schwarzschild has said, the theory of polytropes is a beautiful example of the flowering of a complete mathematical theory out of a physical problem. The bibliographical note for this chapter has been made rather extensive, as there appears to be, at the present time, a great deal of confusion with regard to the historical developments of the subject. It may be stated further that no fundamentally new contribution has been made to the subject since the publication of Emden's book (1907).

Chapter v.—Here the formal theory of radiation is presented and the equations of radiative equilibrium are derived. The number of final results obtained is small, but the amount of formal developments required is rather considerable; for accuracy and precision they cannot, however, be avoided.

Chapter vi.—In this chapter "gaseous stars" are considered. Some general theorems for stars in radiative equilibrium are obtained, and a fundamental formula—the luminosity formula—is derived. This last is made the starting-point of the whole discussion.

Chapter vii.—Here the general theory (which leads to a definite relation of the type I) is used to derive from the observational material the abundances of the lightest elements for individual stars, and an attempt is made to draw some general conclusions. The emphasis thus laid on Strömgren's work on the varying abundance of the lightest elements from star to star is one of the more important features of the monograph. In discussing this theory it is important to realize that the theory of the stellar absorption coefficient and of the mean molecular weight (which are also described in this chapter) have been developed as accurately as is necessary for the purposes at hand.

Chapter viii.—In this chapter the theory of stellar envelopes is made an independent starting-point for the theory of gaseous stars. This serves partly to confirm the results described in chapter vii and partly to go beyond the range of that theory.

Chapter ix.—Some further stellar models are considered in this chapter which partly confirm the results of chapter vi and partly extend them.

Chapter x.—A rather detailed account of the Gibbs statistical mechanics (the quantum mechanical version) is given in this chapter. In view of the astrophysical applications the theory is developed to take account of the relativistic effects from the outset.

Chapter xi.—The theory of degeneracy developed in chapter x is here applied to elucidate the structure of the white dwarfs. For the white dwarfs the structure depends, to a good approximation, only on M, R, and the abundance of the lightest elements. This result arises essentially from the circumstance of the white dwarfs being highly "underluminous."

Chapter xii.—In this chapter some general trends in the current investigations on the problem of stellar energy are outlined. There are, as yet, no very definitive results to report, but some general ideas which are likely to prove fruitful in the future developments of the subject are considered.

The foregoing brief comments on the contents of each of the chapters may be supplemented by their introductory paragraphs, where more specific statements of the problems considered are made.

In concluding this brief outline of the monograph, it should be emphasized again that the particular arrangement of the subject matter has arisen in the attempt to present the subject from a unified standpoint. This, in turn, has required a somewhat detailed treatment of certain aspects of the subject which may not appear to deserve that prominence. However, in the opinion of the writer the general standpoint taken appears to be the only fruitful one under the present limitations of our knowledge.

Finally, in the actual developments an attempt has been made to give the full details, both of mathematical derivations and of physical theories, as far as this has proved feasible. This method may involve the disadvantage that there is a danger of the reader losing the general perspective in the details of the solution of a mathematical problem or in the arguments of a physical theory. It will therefore be advantageous—even though it may not be strictly necessary—if the reader acquires during the study of the monograph some familiarity with the general results. For an attractive account, which in several respects runs parallel to the more detailed treatment of the monograph, reference may be made to B. Strömgren, "Die Theorie des Sterninnern und die Entwicklung der Sterne," *Ergebnisse der Exakten Naturwissenschaften*, **16**, 465, 1937.

We shall now consider two "technical" matters concerning the monograph.

1. *Bibliographical notes.*—With regard to references to the literature, it was decided to resist the temptation of making it a rule to give running references in the text; actually only a very few references are made. This has resulted in a more continuous arrangement of the arguments than would otherwise have been possible. However, at the end of each chapter bibliographical notes are appended

in which specific references to each particular section are made. It should, moreover, be stated that it has not been the intention even to attempt to give a complete list of references; only those investigations are quoted which have been incorporated in the text or the results of which further amplify the points concerned.

2. *The numbering of the sections and the equations.*—The equations and the sections in the different chapters have been numbered separately. References to equations or sections (§) in the same chapter are made by giving the appropriate numbers, e.g., equation (37) or § 10; references to equations or sections in a different chapter are distinguished by giving the chapter number as a Roman numeral, e.g., equation (37), v (or more simply as (37), v), or § 10, v.

CHAPTER I

THE LAWS OF THERMODYNAMICS

In this chapter we shall be concerned mainly with the first and the second laws of thermodynamics. In our presentation of the fundamental principles of thermodynamics we shall follow Carathéodory's axiomatic point of view. This axiomatic presentation of the laws of thermodynamics has the advantage of reducing the number of new undefinables to a minimum and achieves at the same time the maximum logical simplicity. Since a proper appreciation of the meaning and content of the laws of thermodynamics is necessary for the developments in the succeeding chapters, we shall accordingly develop the fundamental ideas *ab initio*.

1. We shall consider only the simplest of thermodynamical systems, namely, those composed of chemically noninteracting mixtures of gases and liquids. We shall assume that the elementary notions concerning mass, force, pressure, work, and volume are familiar; we shall, however, define accurately the purely thermal notions, such as "temperature," "quantity of heat," etc.

In the purely mechanical discussions of the equilibrium of a body —as, for instance, in hydrodynamics—the inner state of a fluid of known mass is determined when we know its specific volume, V, the volume per unit mass of the fluid. But this is not generally true, as we can alter the pressure exerted by a gas without altering its specific volume, V. For this purpose it is necessary to consider physical processes which are associated with "heating." In thermodynamics such physical situations are realized, and we introduce both the pressure, p, and the volume, V, as independent variables. Thus, V and p specify completely the inner state of a system.

We assume that individual systems can be isolated from the outside world by means of inclosures, or that two parts of a given system can be separated by walls. Though we shall not include these inclosures or walls as a part of the thermodynamical system, we shall yet have to make certain specific ideal requirements for these partitions. We shall have to consider two types of such partitions.

a) *Adiabatic inclosures.*—If a body is inclosed in an adiabatic inclosure and if it is in equilibrium, then, in the absence of external fields of forces, the only way in which we can change the inner state of the body is by means of actual displacements of at least some finite part of the walls of the inclosure. If we assume the notion of heat, this means that the only way in which we can change the inner state of a body in an adiabatic inclosure is by doing external work, and that, furthermore, the walls of the inclosure are opaque to the communication of heat.

b) *Diathermic partitions.*—If two bodies are inclosed in an adiabatic inclosure but are mutually separated by a diathermic wall, then a certain definite relation between the four parameters p_1, V_1; p_2, V_2 (defining the state of the two bodies, respectively) must exist in order that there may be equilibrium; the relation depends on the nature of the two bodies only. Thus, we must have

$$F(p_1, V_1, p_2, V_2) = 0 . \tag{1}$$

We shall say that two bodies are in "thermal contact" if they are both inclosed in the same adiabatic inclosure but are separated by a diathermic wall. Equation (1) then expresses the condition for thermal equilibrium.

Thus, it is empirically found that, if two perfect gases are in thermal contact, we always have

$$p_1 V_1 - p_2 V_2 = 0 .$$

2. *Empirical temperature.*—Experience shows the following characteristic of thermal equilibrium. If (p_1, V_1), (p_2, V_2), $(\bar{p}_1, \overline{V}_1)$, and $(\bar{p}_2, \overline{V}_2)$ define two distinct states of two different systems (not necessarily those of two different bodies) and if both (p_1, V_1) and (p_2, V_2) are in thermal equilibrium with $(\bar{p}_1, \overline{V}_1)$, and if, further, (p_1, V_1) is in thermal equilibrium with $(\bar{p}_2, \overline{V}_2)$, then it is always true that (p_2, V_2) will be in thermal equilibrium with $(\bar{p}_2, \overline{V}_2)$. This simply means that, if two bodies are separately in thermal equilibrium with a third body, then the two original bodies, if brought into thermal contact, would also be in thermal equilibrium. By equation (1),

which specifies the condition for thermal equilibrium, the foregoing means that the equations

$$\left.\begin{array}{l} F(p_1, V_1, \bar{p}_1, \overline{V}_1) = 0 , \qquad F(p_2, V_2, \bar{p}_1, \overline{V}_1) = 0 , \\ F(p_1, V_1, \bar{p}_2, \overline{V}_2) = 0 , \end{array}\right\} \tag{2}$$

imply the validity of

$$F(p_2, V_2, \bar{p}_2, \overline{V}_2) = 0 . \tag{3}$$

But this is then, and only then, possible if the relation $F(p, V, \bar{p}, \overline{V}) = 0$ has the form

$$t(p, V) - \bar{t}(\bar{p}, \overline{V}) = 0 . \tag{4}$$

In (4) t and $\bar{t}$ are not uniquely determined, for the condition of equilibrium, (4), can also be written as

$$T[t(p, V)] = T[\bar{t}(\bar{p}, \overline{V})] , \tag{4'}$$

where $T(x)$ can be any arbitrary function in x.

Of all the possible forms which the condition of equilibrium can take, let us choose arbitrarily one particular form and write it in the form (4). The values $t(p, V)$ and $\bar{t}(\bar{p}, \overline{V})$ define on an arbitrary scale the *empirical temperature* of the two bodies; if the two bodies are in thermal contact and are in equilibrium, then we should always have the equality of the empirical temperatures. If

$$t = t(p, V), \qquad \bar{t} = \bar{t}(\bar{p}, \overline{V}) , \tag{5}$$

then in equilibrium

$$t = \bar{t} . \tag{6}$$

The equations (5) define in the (p, V) and in the $(\bar{p}, \overline{V})$ planes, respectively, a one-parametric family of curves which are called "isothermals." The equations (5) are called the "equations of state."

If the empirical temperature scale is once selected and defined, then we can always choose any two of the three variables p, V, and t as the independent variables defining the state of a system. In the same way two arbitrary functions of the physical variables p, V, and t would also suffice to specify a state of the system.

3. *The First Law of Thermodynamics.*—The experiments of Joule establish the following circumstance:

In order to bring a body (or a system of bodies) from a prescribed initial state to another prescribed final state adiabatically, then the same constant amount of mechanical work (or an equivalent electrical work), which is independent of how the change is carried out and which depends only on the prescribed initial and final states, has to be done.

Let the initial state be specified by p_0, V_0, ... , and the final state by p_1, V_1, ... Let the work done to carry out the change of state adiabatically be W. Then, according to the first law, if we keep the initial state fixed, W depends only on the final state. We can therefore write

$$W = U - U_0 \, , \tag{7}$$

where U is a function of the parameters determining the state of the system—p and V, if there is only one body—and U_0 is its value in the initial state. U, thus defined, is called the "internal energy" of the system.

If we define our unit of heat as the mechanical work (expressed in ergs) required to change the (empirical) temperature, t, of water of unit volume (at constant volume) between two definite values, then we obtain the so called "mechanical equivalent of heat."

4. *Quantity of heat.*—Suppose that we know the internal energy as a function of the physical parameters from a series of calorimetric experiments, as, for instance, Joule's experiments. Suppose, now, that in some given arbitrary nonadiabatic process the internal energy of a system changes by $(U - U_0)$; further, let W be the amount of work done on the system. Then we say that a quantity Q of heat, where

$$Q = (U - U_0) - W \, , \tag{8}$$

has been supplied to the system.

We see that the notion of the quantity of heat has no independent meaning apart from the First Law of Thermodynamics. $(U - U_0)$

is a physical quantity which can be determined experimentally, while the notion of Q is a derived one.

5. *The internal energy of a system of bodies.*—If two or more bodies are isolated from each other adiabatically, then by definition the energy of the system is equal to the sum of the energies of the individual bodies:

$$U = U_1 + U_2 . \tag{9}$$

In general, when the two bodies are brought into contact, the energy is not additive; it is easy to see, however, that the deviation must be proportional to the common surface area of the bodies, and hence, for large volumes the deviations from the additive law can be neglected.

6. *Stationary and quasi-statical processes.*—In the formulation of the first law we assumed that the work done can in principle be measured. But to evaluate the work done during a given process we need an apparatus to register continuously the forces exerted on, and the displacements of, the walls of the inclosure, for the work done is simply the integral over the product of the force and the displacement. In practice this limits us to only two essentially distinct procedures for which we can measure the work done. These are:

a) *Stationary processes.*—For example, as in Joule's experiments, there is a stirrer which rotates in the fluid at a constant rate. This would give rise to a stationary system of currents in which the stirrer experiences a constant friction. If we neglect the relatively small acceleration in the beginning and the end of the interval during which the stirrer rotates, then the work done is simply the product of the torque times the rate of working of the stirrer.

b) *Quasi-statical processes.*—We conduct the process infinitely slowly, so that we can regard the state of the system at any given moment as one of equilibrium. We refer to such processes as "quasi-statical processes." They are generally referred to as "reversible processes" because, in general, quasi-statical processes can be conducted in the reverse sense. We shall refer to a process as "nonstatical" if it is not quasi-statical.

7. *Infinitesimal quasi-statical adiabatic changes.*—If we have a body inclosed in an adiabatic inclosure, and if we do an infinitesimal amount of mechanical work, dW (by displacing the walls of the in-

closure), carried out quasi-statically, then we say that we have carried out an "infinitesimal quasi-statical adiabatic change." If during such an infinitesimal quasi-statical adiabatic change the change in volume amounts to dV, then clearly

$$dW = -p dV , \tag{10}$$

where p is the equilibrium pressure. Then, according to the first law,

$$dQ = dU + p dV = 0 . \tag{11}$$

For a system of two bodies which are both inclosed in the same adiabatic inclosure but which are separated from one another by means of a diathermic wall, we have, since both Q and U are additive,

$$\begin{aligned} dQ &= dQ_1 + dQ_2 , \\ &= dU_1 + dU_2 + p_1 dV_1 + p_2 dV_2 = 0 . \end{aligned} \tag{12}$$

Finite quasi-statical adiabatic changes are simply continuous sequences of equilibrium states and therefore are curves in the phase-space (i.e., the p, V plane for a single body) which satisfy at each point equations of the form (11) or (12). Equations (11) and (12) are called the "equations of the adiabatics."

If we consider U as a function of V and t, then

$$dU = \left(\frac{\partial U}{\partial V}\right) dV + \left(\frac{\partial U}{\partial t}\right) dt . \tag{13}$$

Hence (11) takes the form

$$dQ = \left(\frac{\partial U}{\partial V} + p\right) dV + \frac{\partial U}{\partial t} dt = 0 . \tag{14}$$

Equation (12) has interest only when the two bodies are in thermal contact. The system then can be described by three independent variables, V_1, V_2, and t, the common empirical temperature:

$$t(p_1, V_1) = \bar{t}(p_2, V_2) = t . \tag{15}$$

Equation (12) can then be written as

$$\left.\begin{aligned} dQ = \left(\frac{\partial U_1}{\partial V_1} + p_1\right) dV_1 + \left(\frac{\partial U_2}{\partial V_2} + p_2\right) dV_2 \\ + \left(\frac{\partial U_1}{\partial t} + \frac{\partial U_2}{\partial t}\right) dt = 0\,. \end{aligned}\right\} \quad (16)$$

Equations (14) and (16) are the equations of the adiabatics. Equations of the form (14) and (16) are called "Pfaffian differential equations." We must now study some mathematical properties of these differential equations.

8. *Mathematical theorems on Pfaffian differential equations.*—We shall consider first a Pfaffian differential expression in two variables x and y:

$$dQ = X(x, y)dx + Y(x, y)dy\,, \quad (17)$$

which has the same form as equation (14). The integral of dQ between two points 1 and 2 depends in general on the path of the integration. Hence $\int_1^2 dQ$ cannot in general be written as $Q(x_2, y_2) - Q(x_1, y_1)$, which means that dQ is not "integrable." This in turn means that dQ in general is not a perfect differential of the function $Q(x, y)$. If dQ were a perfect differential, we should have $dQ = d\sigma$, where σ is a function of x and y; we should have further

$$d\sigma = \frac{\partial \sigma}{\partial x} dx + \frac{\partial \sigma}{\partial y} dy\,. \quad (18)$$

Comparing (17) and (18), we have

$$X(x, y) = \frac{\partial \sigma}{\partial x}\,; \qquad Y(x, y) = \frac{\partial \sigma}{\partial y}\,, \quad (19)$$

or

$$\frac{\partial X}{\partial y} = \frac{\partial^2 \sigma}{\partial x \partial y} = \frac{\partial Y}{\partial x}\,. \quad (20)$$

Condition (20) between the coefficients in the Pfaffian expression need not, of course, be true.

Corresponding to (17), the Pfaffian equation in two variables is

$$dQ = Xdx + Ydy = 0 , \tag{21}$$

or

$$\frac{dy}{dx} = -\frac{X}{Y} . \tag{22}$$

The right-hand side of equation (22) is a known function of x and y, and hence the Pfaffian equation (21) defines a definite direction at each point in the (x, y) plane. The solving of the equation simply consists of drawing a system of curves in the (x, y) plane such that at any point the tangent to the curve (at that point) has the same direction as that specified by (21). Hence, the solution of the equation (21) defines a one-parametric family of curves in the (x, y) plane. The solution can therefore be written as $\sigma(x, y) = c =$ constant. Then

$$\frac{\partial \sigma}{\partial x} + \frac{\partial \sigma}{\partial y}\frac{dy}{dx} = 0 . \tag{23}$$

From (22) and (23) we easily find, that

$$Y\frac{\partial \sigma}{\partial x} = X\frac{\partial \sigma}{\partial y} = \frac{XY}{\tau} , \tag{24}$$

where $\tau(x, y)$ is a factor depending on x and y. Equation (24) can also be written as

$$X = \tau\frac{\partial \sigma}{\partial x} ; \qquad Y = \tau\frac{\partial \sigma}{\partial y} . \tag{25}$$

Inserting (25) into (17), we have

$$dQ = \tau\left(\frac{\partial \sigma}{\partial x}dx + \frac{\partial \sigma}{\partial y}dy\right) = \tau d\sigma , \tag{26}$$

or

$$\frac{dQ}{\tau} = d\sigma ; \tag{27}$$

i.e., if we divide the Pfaffian expression (17) by τ, we obtain a perfect differential. A factor, τ, which has this property is called an "integrating denominator." A Pfaffian differential expression, then, in two variables always admits of an integrating denominator.

If we replace σ by another function of σ, say $S[\sigma(x, y)]$, then $S = c =$ constant will again represent the solutions of the differential equation. In that case

$$dS = \frac{dS}{d\sigma}\, d\sigma = \frac{dS}{d\sigma}\frac{dQ}{\tau}\,, \tag{28}$$

$$= \frac{1}{T(x, y)}\, dQ\,, \tag{29}$$

where

$$T(x, y) = \tau(x, y)\,\frac{d\sigma}{dS}\,. \tag{30}$$

Therefore, T is also an integrating denominator. Hence, if a Pfaffian expression admits of one integrating denominator, it must admit of an infinity of them. This result is easily seen to be true for a Pfaffian expression in any number of variables.

We shall now proceed to consider a Pfaffian expression in three variables. (The generalization to more than three variables is immediate.) Consider the Pfaffian expression

$$dQ = X dx + Y dy + Z dz\,, \tag{31}$$

where X, Y, and Z are functions of the variables x, y, and z. Our thermodynamical equation (16) is of this form. The ratio $dx{:}dy{:}dz$ defines a definite direction in the (x, y, z) space. The equation $dQ = 0$, corresponding to (31), specifies that dx, dy, and dz must satisfy a linear equation at each point in the space, and hence specifies a certain tangential plane at each point in the (x, y, z) space. A solution of a Pfaffian equation, $dQ = 0$, passing through a given point, (x, y, z), must lie in the tangential plane through that point; but its direction in the tangential plane is arbitrary.

Now, dQ in general will not be a perfect differential. If it were, $dQ = d\sigma$, where σ is some function of x, y, z, so that

$$dQ = d\sigma(x, y, z) = \frac{\partial\sigma}{\partial x}\, dx + \frac{\partial\sigma}{\partial y}\, dy + \frac{\partial\sigma}{\partial z}\, dz\,.$$

Hence, by comparison with (31),

$$X = \frac{\partial \sigma}{\partial x}; \qquad Y = \frac{\partial \sigma}{\partial y}; \qquad Z = \frac{\partial \sigma}{\partial z}, \tag{32}$$

or

$$\frac{\partial Y}{\partial z} = \frac{\partial Z}{\partial y}; \qquad \frac{\partial Z}{\partial x} = \frac{\partial X}{\partial z}; \qquad \frac{\partial X}{\partial y} = \frac{\partial Y}{\partial x}. \tag{33}$$

The relations (33) need not be valid for arbitrary functions X, Y, Z.

But we can ask: Does the Pfaffian expression admit of an integrating denominator? In other words, can we determine a function, τ, of x, y, and z such that

$$\frac{dQ}{\tau(x, y, z)} = d\sigma = \frac{\partial \sigma}{\partial x} dx + \frac{\partial \sigma}{\partial y} dy + \frac{\partial \sigma}{\partial z} dz \quad ? \tag{34}$$

If we can determine an integrating denominator $\tau(x, y, z)$, then every solution of the differential equation $dQ = 0$ would also be a solution of $d\sigma = 0$; or the solution can be written in the form $\sigma(x, y, z) = c =$ constant; i.e., the solutions can be any arbitrary curve lying on any one of the one-parametric family of surfaces $\sigma(x, y, z) = c$. It is, however, important to realize that we cannot, in general, find integrating denominators for Pfaffian expressions in more than two variables. This can be verified by the following example. Consider the equation

$$dQ = -ydx + xdy + kdz = 0, \tag{35}$$

where k is a constant. If the Pfaffian expression (35) admitted of an integrating denominator τ, then

$$\frac{dQ}{\tau} = -\frac{y}{\tau} dx + \frac{x}{\tau} dy + \frac{k}{\tau} dz = d\sigma \tag{36}$$

is a perfect differential. Hence, we should have

$$\frac{\partial \sigma}{\partial x} = -\frac{y}{\tau}; \qquad \frac{\partial \sigma}{\partial y} = \frac{x}{\tau}; \qquad \frac{\partial \sigma}{\partial z} = \frac{k}{\tau}. \tag{37}$$

We have

$$\frac{\partial}{\partial y}\left(-\frac{y}{\tau}\right) = -\frac{1}{\tau} + \frac{y}{\tau^2}\frac{\partial \tau}{\partial y} = \frac{\partial}{\partial x}\left(\frac{x}{\tau}\right) = \frac{1}{\tau} - \frac{x}{\tau^2}\frac{\partial \tau}{\partial x}, \tag{38}$$

or

$$2\tau = x\frac{\partial \tau}{\partial x} + y\frac{\partial \tau}{\partial y}. \tag{39}$$

Again

$$\frac{\partial}{\partial z}\left(-\frac{y}{\tau}\right) = \frac{y}{\tau^2}\frac{\partial \tau}{\partial z} = \frac{\partial}{\partial x}\left(\frac{k}{\tau}\right) = -\frac{k}{\tau^2}\frac{\partial \tau}{\partial x}, \tag{40}$$

or

$$\frac{\partial \tau}{\partial x} = -\frac{y}{k}\frac{\partial \tau}{\partial z}. \tag{41}$$

Similarly,

$$\frac{\partial}{\partial y}\left(\frac{k}{\tau}\right) = -\frac{k}{\tau^2}\frac{\partial \tau}{\partial y} = \frac{\partial}{\partial z}\left(\frac{x}{\tau}\right) = -\frac{x}{\tau^2}\frac{\partial \tau}{\partial z}, \tag{42}$$

or

$$\frac{\partial \tau}{\partial y} = \frac{x}{k}\frac{\partial \tau}{\partial z}. \tag{43}$$

From (39), (41), and (43) we have $\tau \equiv 0$, thus leading to a contradiction.

By means of such examples we realize that Pfaffian expressions in three (or more) variables will not in general admit of integrating denominators except under very special circumstances. It is necessary to appreciate this, for precisely such special circumstances obtain in thermodynamics.

We have seen that the Pfaffian differential expressions fall into two classes, those which admit of integrating denominators and those which do not. We must look for a less abstract characteristic of this difference. Consider a Pfaffian equation in two variables. Then through every point in the (x, y) plane there passes just one curve of the family $\sigma(x, y) = c$. Hence from any given point in the plane we cannot certainly reach all the neighboring points by means of curves which satisfy the Pfaffian equation. We shall refer to this circumstance by the statement that not all the neighboring points are *accessible* from a given point.

Now consider a Pfaffian expression in three variables. If it admits of an integrating denominator, the situation is the same as in the plane; all the solutions lie on one or other of the family of surfaces $\sigma(x, y, z) = c$, so that we cannot reach all points in the neighborhood of a given point. Only those points will be accessible which are on the surface belonging to the family $\sigma(x, y, z) = c$, which passes through the point under consideration.

We now ask the converse question: If in the neighborhood of a point (however near) there are points which are inaccessible to it along curves which are solutions of the Pfaffian equation, then does the Pfaffian expression admit of an integrating denominator? Carathéodory has shown that the answer to the foregoing question is in the affirmative. The proof is as follows:

All those points which are accessible to a given point, P_0 (accessible along curves which are solutions of the Pfaffian equation), and which are in its immediate neighborhood, must form, together with P_0, a continuous domain of points; hence we have three possibilities: all the accessible points in the immediate neighborhood of P_0 either fill a certain volume element containing P_0, or a surface element containing P_0, or a line element passing through P_0. The first possibility is excluded because all points in a sufficiently close neighborhood of P_0 would then be accessible to P_0; this contradicts our hypothesis that in the neighborhood of a point, however near, there are always points inaccessible to it. Again, the last possibility is also excluded because $dQ = 0 = Xdx + Ydy + Zdz$ already defines an infinitesimal surface element containing only points accessible to P_0. Hence, the points which are accessible to P_0 and which are in its neighborhood must form a surface element, dF_0. If we now consider the boundary points P' of dF_0, we can again define surface elements dF' containing all the points accessible to the points P' on the boundary of dF_0. These surface elements dF' must overlap dF_0; at the same time the elements dF' cannot form surface elements lying above or below dF_0, for then along paths going from P_0 to a point P' on the boundary of dF_0, and thence from P' along a curve lying in an appropriate element dF', we should be able to reach all the points in an immediate spatial neighborhood of P_0; this would again contradict our hypothesis. Thus, the element dF_0, together with the elements dF', must form a continuous set of surface elements. By this process of continuation, only points lying on a definite surface passing through P_0 are obtained, and hence all the points accessible to P_0 must lie on a definite surface F_0. If we now start at a point P_1 not on F_0, we must obtain in the same way another surface F_1 which cannot either intersect or touch the surface F_0. In this way we can construct a whole family of nonintersecting surfaces F_0, F_1, F_2, ... , continuously filling the whole (x, y, z) space, such that only points on any given surface are accessible to points on the surface itself. These surfaces then form a one-parametric family of surfaces, $\sigma(x, y, z) =$ constant, such that $d\sigma = 0$ implies $dQ = 0$. Hence, we must have

$$dQ = \tau(x, y, z)d\sigma(x, y, z) , \tag{44}$$

where

$$\tau = \frac{X}{\frac{\partial \sigma}{\partial x}} = \frac{Y}{\frac{\partial \sigma}{\partial y}} = \frac{Z}{\frac{\partial \sigma}{\partial z}} . \tag{45}$$

We have thus proved Carathéodory's theorem:

If a Pfaffian expression

$$dQ = Xdx + Ydy + Zdz$$

has the property that in every arbitrarily close neighborhood of a point P *there are inaccessible points, i.e., points which cannot be connected to* P *along curves which satisfy the equation* dQ = 0, *then the Pfaffian expression must admit of an integrating denominator.*

It is easily seen that the foregoing theorem must also be true for Pfaffian expressions in more than three variables. Further, it is clear that, if a Pfaffian expression admits of one integrating denominator, it must admit of infinitely many integrating denominators.

For the family of surfaces, $\sigma(x, y, z) =$ constant can also be written as $S[\sigma(x, y, z) =]$ constant, where $S(\sigma)$ is an arbitrary function in σ. Then we have

$$dS = \frac{dS}{d\sigma} d\sigma = \frac{dS}{d\sigma} \frac{dQ}{\tau} , \tag{46}$$

or

$$dQ = T(x, y, z)dS , \tag{47}$$

where

$$T = \tau \frac{d\sigma}{dS} = \frac{X}{\frac{\partial S}{\partial x}} = \frac{Y}{\frac{\partial S}{\partial y}} = \frac{Z}{\frac{\partial S}{\partial z}} . \tag{48}$$

Carathéodory's theorem, which expresses the mathematical equivalence of the inaccessibility along curves $dQ = 0$ with the existence of an integrating denominator $\tau(x, y, z)$ to Q, contains, as we shall see, the essence of the Second Law of Thermodynamics.

9. *The Second Law of Thermodynamics.*—The physical basis for the second law is the realization that certain processes are not physically realizable. The most sweeping statement of this character is that without "compensation" it is not possible to transfer heat from a colder to a hotter body; more precisely, the law is included in Kelvin's principle, which states: *In a cycle of processes it is impossible to transfer heat from a heat reservoir and convert it all into work, without at the same time transferring a certain amount of heat from a hotter to a colder body.* The second law is sometimes also stated in the form: *It is impossible that, at the end of a cycle of changes, heat has been transferred from a colder to a hotter body without at the same time converting a certain amount of work into heat.* This latter statement of the second law is due to Clausius. However, the essential point of Carathéodory's theory is that it formulates the facts of experience in a very much more general way, enabling us at the same time to obtain all the mathematical consequences of the second law without any further physical discussion. In fact, in order to obtain the full mathematical content of the second law, it is sufficient that there exist certain processes that are not physically realizable. Carathéodory states his principle in the following form: *Arbitrarily near to any given state there exist states which cannot be reached from an initial state by means of adiabatic processes.*

From Carathéodory's principle it follows in particular that there exist states neighboring a given one which cannot be reached by means of quasi-static adiabatic processes.

In the first instance we shall only apply Carathéodory's principle to quasi-static adiabatic processes. Later (§ 10), we shall have occasion to usc the principle in its wider form, namely, that there exist states neighboring a given one which are inaccessible to it along nonstatic adiabatic processes.

From the restricted form of Carathéodory's principle, it follows that there are states neighboring a given one which cannot be reached along adiabatics (Eqs. [14] and [16]); hence, by Carathéodory's theorem the Pfaffian differential expression for dQ must admit of an integrating denominator:

$$dQ = \tau d\sigma . \tag{49}$$

For one single substance whose state is characterized by the two parameters V and t, Carathéodory's principle does not lead to anything new, because a Pfaffian expression in two variables always admits of an integrating denominator.

When, however, we consider a system composed of two bodies adiabatically inclosed and in thermal contact, Carathéodory's principle asserts something new in so far as we can now assert that $dQ = dQ_1 + dQ_2$ can always be written in the form

$$dQ = dQ_1 + dQ_2 = \tau(V_1, V_2, t)d\sigma(V_1, V_2, t) . \tag{50}$$

On the other hand, we have for each of the two bodies

$$dQ_1 = \tau_1(V_1, t_1)d\sigma_1(V_1, t_1) , \tag{51}$$

$$dQ_2 = \tau_2(V_2, t_2)d\sigma_2(V_2, t_1) . \tag{52}$$

If the two bodies are in thermal contact, we have

$$t_1 = t_2 = t . \tag{53}$$

Hence,

$$\tau d\sigma = \tau_1 d\sigma_1 + \tau_2 d\sigma_2 . \tag{54}$$

If we now choose σ_1, σ_2 and t as the independent variables, instead of V_1, V_2, and t, we can regard τ and σ as functions of σ_1, σ_2 and t; from (54) we then have

$$\frac{\partial\sigma}{\partial\sigma_1} = \frac{\tau_1(\sigma_1, t)}{\tau(\sigma_1, \sigma_2, t)} ; \qquad \frac{\partial\sigma}{\partial\sigma_2} = \frac{\tau_2(\sigma_2, t)}{\tau(\sigma_1, \sigma_2, t)} ; \qquad \frac{\partial\sigma}{\partial t} = 0 . \tag{55}$$

From the last equation it follows that σ is independent of t; hence, σ depends only on σ_1 and σ_2, or

$$\sigma = \sigma(\sigma_1, \sigma_2) \ . \tag{56}$$

From the first two equations in (55) it follows that τ_1/τ and τ_2/τ are also functions independent of t. Hence,

$$\frac{\partial}{\partial t}\left(\frac{\tau_1}{\tau}\right) = 0 \ ; \qquad \frac{\partial}{\partial t}\left(\frac{\tau_2}{\tau}\right) = 0 \ , \tag{57}$$

or

$$\frac{1}{\tau_1}\frac{\partial \tau_1}{\partial t} = \frac{1}{\tau_2}\frac{\partial \tau_2}{\partial t} = \frac{1}{\tau}\frac{\partial \tau}{\partial t} \ . \tag{58}$$

Now τ_1 is a function only of σ_1 and t, and τ_2 is a function only of σ_2 and t. Hence, the first equality in (58) can be valid only if the two quantities are functions of t only. We can therefore write (58) as

$$\frac{\partial \log \tau_1}{\partial t} = \frac{\partial \log \tau_2}{\partial t} = \frac{\partial \log \tau}{\partial t} = g(t) \ , \tag{59}$$

where $g(t)$ must be a universal function, because it has the same value for two arbitrary systems and also for the "combined" system. We are thus led to a universal function of the empirical temperature, t.

From (59) we have, on integration,

$$\log \tau = \int g(t)dt + \log \Sigma(\sigma_1, \sigma_2) \ , \tag{60}$$

$$\log \tau_i = \int g(t)dt + \log \Sigma_i(\sigma_i) \ , \qquad (i = 1, 2) \ , \tag{61}$$

where the constants of integration Σ and Σ_i are independent of t and are functions only of the other physical variables characterizing the system. Equations (60) and (61) can also be written as

$$\tau = \Sigma(\sigma_1, \sigma_2) \cdot e^{\int g(t)dt} \ ; \qquad \tau_i = \Sigma_i(\sigma_i) \cdot e^{\int g(t)dt} \ . \tag{62}$$

Thus, for any thermodynamical system the integrating denominator consists of two factors, one factor which depends on the temperature (and which is the same for all substances) and another factor which depends on the remaining variables characterizing the system. We therefore introduce the *absolute temperature*, T, defined by

$$T = Ce^{\int g(t)dt}, \tag{63}$$

where C is an arbitrary constant (instead of which we can also introduce an arbitrary lower limit to the integral in the exponent in [63]), and which is determined in such a way that two fixed points (e.g., the freezing- and the boiling-point of water) differ by 100 on the absolute scale. It should be noticed that T does not contain any additive constant—in other words, the zero of the absolute scale of temperature is physically determined. From (49), (62), and (63) we have

$$dQ = \tau d\sigma = T\frac{\Sigma}{C}d\sigma\,, \qquad dQ_i = \tau_i d\sigma_i = T\frac{\Sigma_i}{C}d\sigma_i\,. \tag{64}$$

If we are dealing with a single homogeneous body the state of which is defined by the independent variables t and σ_1, then Σ_1 depends only on σ_1, so that we can introduce the function S_1, which is defined as

$$S_1 = \frac{1}{C}\int \Sigma_1(\sigma_1)d\sigma_1 + \text{constant}\,. \tag{65}$$

The function S_1 depends only on σ_1 and is determined apart from an arbitrary additive constant. Furthermore, S_1 is constant along an adiabatic. The function S_1, so defined, is called the "entropy." One can now write

$$dQ_1 = TdS_1\,. \tag{66}$$

If we now consider a system composed of two bodies in thermal contact, we have for the two bodies separately

$$dQ_1 = \tau_1 d\sigma_1 = T\frac{\Sigma_1(\sigma_1)}{C}d\sigma_1 = TdS_1\,, \tag{67}$$

$$dQ_2 = \tau_2 d\sigma_2 = T\frac{\Sigma_2(\sigma_2)}{C}d\sigma_2 = TdS_2\,, \tag{68}$$

and for the combined system

$$dQ = \tau d\sigma = T\,\frac{\Sigma(\sigma_1, \sigma_2)}{C}\,d\sigma(\sigma_1, \sigma_2)\,, \tag{69}$$

$$= dQ_1 + dQ_2 = T\,\frac{\Sigma_1(\sigma_1)}{C}\,d\sigma_1 + T\,\frac{\Sigma_2(\sigma_2)}{C}\,d\sigma_2\,. \tag{69'}$$

Hence,

$$\Sigma(\sigma_1, \sigma_2)d\sigma = \Sigma_1(\sigma_1)d\sigma_1 + \Sigma_2(\sigma_2)d\sigma_2\,. \tag{70}$$

From (70) it follows that

$$\Sigma(\sigma_1, \sigma_2)\,\frac{\partial\sigma}{\partial\sigma_1} = \Sigma_1(\sigma_1)\,; \qquad \Sigma(\sigma_1, \sigma_2)\,\frac{\partial\sigma}{\partial\sigma_2} = \Sigma_2(\sigma_2)\,. \tag{71}$$

Hence,

$$\frac{\partial\Sigma_1}{\partial\sigma_2} = \frac{\partial\Sigma}{\partial\sigma_2}\,\frac{\partial\sigma}{\partial\sigma_1} + \Sigma\,\frac{\partial^2\sigma}{\partial\sigma_1\partial\sigma_2} = 0\,, \tag{72}$$

$$\frac{\partial\Sigma_2}{\partial\sigma_1} = \frac{\partial\Sigma}{\partial\sigma_1}\,\frac{\partial\sigma}{\partial\sigma_2} + \Sigma\,\frac{\partial^2\sigma}{\partial\sigma_1\partial\sigma_2} = 0\,. \tag{73}$$

From (72) and (73) it follows that the functional determinant

$$\frac{\partial\Sigma}{\partial\sigma_1}\,\frac{\partial\sigma}{\partial\sigma_2} - \frac{\partial\Sigma}{\partial\sigma_2}\,\frac{\partial\sigma}{\partial\sigma_1} = \frac{\partial(\Sigma, \sigma)}{\partial(\sigma_1, \sigma_2)} \tag{74}$$

is zero, and consequently $\Sigma(\sigma_1, \sigma_2)$ contains the variables σ_1 and σ_2 only in the combination $\sigma(\sigma_1, \sigma_2)$. We can therefore write

$$\Sigma(\sigma_1, \sigma_2) = \Sigma(\sigma)\,. \tag{75}$$

Equation (69) can be written as

$$dQ = \tau d\sigma = TdS\,, \tag{76}$$

where

$$dS = \frac{\Sigma(\sigma)}{C}\,d\sigma\,, \tag{77}$$

or

$$S = \frac{1}{C}\int\Sigma(\sigma)d\sigma + \text{constant}\,, \tag{78}$$

where S is now the "total" entropy of the system. From (67), (68), and (76) we further have that

$$dS = dS_1 + dS_2 = d(S_1 + S_2) , \tag{79}$$

or, in words: the change of entropy of a system composed of two bodies in thermal contact, during a quasi-statical process, is the sum of the entropy changes in the two bodies separately.

By a suitable choice of the additive constant entering into our definition of entropy we can arrange so that

$$S = S_1 + S_2 , \tag{80}$$

or: the entropy of a system is the sum of the entropies of its different parts.

Equation (76) contains the mathematical statement of the Second Law of Thermodynamics, which follows as a purely mathematical consequence of the Carathéodory principle: *The differential of the heat,* dQ, *for an infinitesimal quasi-statical change, when divided by the absolute temperature* T, *is a perfect differential,* dS, *of the entropy function.*

The essential differences between (47) and (76) should be noted. In (47) T and S (and τ and σ) are functions of *all* the physical variables; while in (76), τ and T depend only on the empirical temperature, t, which is the same for the different parts of the system; furthermore, σ and S depend only on the variables (σ_1 and σ_2) which do not alter their values for adiabatic changes; finally, T is a universal function of t, and S is a function only of $\sigma(\sigma_1, \sigma_2)$.

We shall now show that the gas-thermometer scale, $pV = t$, defines a temperature scale proportional to the absolute temperature. It should be emphasized that the usual assumption that $pV = t$ defines, apart from a constant factor, the absolute temperature scale is logically unsound. To assume beforehand that the absolute temperature scale should be precisely $pV = t$ and not any other monotonic function, $t = f(pV)$, is to beg the question. We shall see that we cannot identify $pV \propto T$ without an appeal to the Second Law of Thermodynamics. To do this logically, we need to know the internal energy, U, as a function of the state of the gas. The experimental basis is the idealized Joule-Kelvin experiment, which shows that, when a gas expands adiabatically without doing any external work, the product pV (i.e., the gas temperature, $t = f[pV]$) does not change. (It should be noticed that an appeal is made here to an irreversible process. As Carathéodory has pointed out, it is necessary at some stage to appeal to an irreversible process to fix the zero-point of the absolute temperature scale.) It follows, then, from the Joule-Kelvin experiment that U is independent of V. Hence, we can write

$$U = U(t) \; ; \qquad pV = F(t) \; , \tag{81}$$

where t is the empirical temperature. For the differential of the heat for a quasi-statical change, we have

$$\begin{aligned} dQ = dU + pdV &= \frac{dU}{dt}\, dt + F(t)\, \frac{dV}{V} \\ &= F(t)\left[\frac{1}{F(t)}\, \frac{dU}{dt}\, dt + d \log V\right] . \end{aligned} \tag{82}$$

Define a quantity, χ, by the equation

$$\log \chi = \int \frac{1}{F(t)}\, \frac{dU}{dt}\, dt + \text{constant} \, . \tag{83}$$

Equation (82) can be re-written as

$$dQ = F(t) d \log \chi V \, . \tag{84}$$

Hence, we can choose $F(t)$ as the integrating denominator

$$\tau = F(t) \; ; \qquad \sigma = \log \chi V \, . \tag{85}$$

Equation (84) now takes the standard form

$$dQ = \tau d\sigma \, . \tag{86}$$

We can, of course, choose the integrating factor in many other ways. If

$$\sigma^* = \sigma^*(\sigma) \; ; \qquad \tau^* = F(t)\, \frac{d\sigma}{d\sigma^*} \, , \tag{85'}$$

equation (86) can be written as

$$dQ = \tau^* d\sigma^* . \tag{86'}$$

Hence, there is no a priori reason to choose $\tau = F(t) = pV$ as the integrating denominator. But we have shown that

$$g(t) = \frac{\partial \log \tau}{\partial t} \tag{87}$$

is a universal function which is the same in whatever way we may choose to define the integrating denominator. $g(t)$, defined by (87), is invariant to the transformations (85'). From our definition of the absolute temperature (Eq. [63]) we have

$$T = Ce^{\int g(t)dt} = CF(t) = CpV . \tag{88}$$

Thus the absolute temperature scale agrees with the temperature on the gas-thermometer scale.

From $dQ = TdS$, we find that

$$dS = \frac{1}{C} d \log \chi V , \tag{89}$$

or

$$S = \frac{1}{C} \log \chi V + \text{constant} . \tag{90}$$

If we write $U = c_V T$ and consider c_V as a constant, and further define $R = 1/C$, we have

$$\log \chi = \int \frac{c_V}{RT} dT = \frac{c_V}{R} \log T + \text{constant} . \tag{91}$$

Hence, finally,

$$S = S_0 + c_V \log T + R \log V , \tag{92}$$

where S_0 is a constant.

10. *The principle of the increase of entropy.*—So far we have considered only quasi-statical changes of state, though at one point

(§ 9) we had to consider a nonstatical process when we appealed to an idealized Joule-Kelvin experiment. We shall now discuss nonstatical processes more generally.

We shall consider, as we have done so far, an adiabatically inclosed system composed of two bodies in thermal contact. The equilibrium state of such a system can be characterized by three independent variables, such as V_1, V_2, t (the variables we have used so far). We shall now choose V_1, V_2, and S as the independent variables. Let V_1^0, V_2^0, and S^0 be the values of the physical variables in an initial state and V_1, V_2, and S in a final state. We now assert that S is either always greater than S^0 or always less than S^0.

To show this, we consider the final state as being reached in two steps:

a) We alter the volumes V_1^0 and V_2^0 by means of a quasi-statical and adiabatical process such that the volumes at the end are V_1 and V_2. In this way we keep the entropy constant and equal to S^0.

b) We then alter the state of the system, keeping the volumes fixed, but change the entropy by means of adiabatical but nonstatical processes (such as stirring, rubbing, etc., in which $dQ = 0$ but $dQ \neq TdS$) such that the entropy changes from S^0 to S.

If, now, S were greater than S^0 in some processes and less than S^0 in others, then it should be possible to reach every close neighboring state, (V_1, V_2, S), of the initial state, (V_1^0, V_2^0, S^0), by means of adiabatic processes. (After reaching the state (V_1, V_2, S), we can reach all the states, (V_1', V_2', S), by means of processes [*a*]). This contradicts Carathéodory's principle in its more general form, which postulates that in any arbitrarily near neighborhood of a state, (V_1^0, V_2^0, S), there exist adiabatically inaccessible states even when we allow nonstatical processes. Consequently, by means of the processes (*b*), and therefore also by means of the processes (*a*) and (*b*), the entropy S^0 of the system can either only increase or only decrease. Since this is true for every initial state, we see that, because of the continuity of the impossibility of "increase" or "decrease," the entropy of the system we have considered must either never increase or never decrease. The same must also be true for two independent systems because of the additive nature of entropy. We have thus proved: *For all the possible changes (quasi-statical or otherwise) that an adiabatically inclosed system can undergo, the entropy*, S, *must either never increase or never decrease.*

Whether the entropy decreases or increases depends in the first instance on the sign of C introduced in our definition of entropy (78). This is naturally chosen in such a way that the absolute temperature is positive. Then one single experiment is sufficient to determine the sign of the entropy change. By the expansion of an ideal gas, G, into a vacuum, the entropy S_G of the gas increases, as can be seen from equation (92) (V increases and T remains the same). We now consider a system composed of the gas, G, and of another body, K. If we consider such changes of state in which the entropy S_K of the body remains constant and S_G changes, then $S = S_G + S_K$ must increase (since, as we have just seen, S_G always increases); consequently, S can never decrease. Hence, if we consider processes in which the entropy of the gas remains constant, it is clear that, as S can only increase, S_K can only increase; this is true also when K and G are adiabatically separated. Hence, in general we have proved the following important result:

For an adiabatically inclosed system the entropy can never decrease:

$$\left.\begin{aligned} S &> S^0\,, &&\text{(nonstatical process)}\,, \\ S &= S^0 &&\text{(statical process)}\,. \end{aligned}\right\} \tag{93}$$

It follows that if in any change of state of an adiabatically inclosed system the entropy becomes different, then no adiabatic change can be realized which will change the system from the final to the initial state. In this sense, therefore, every change of state in which the entropy changes must be irreversible. This can also be stated as follows: For an adiabatically inclosed system the entropy must tend to a maximum.

Still another formulation of the foregoing is

$$\oint \frac{dQ}{T} \leqslant 0\,, \tag{94}$$

where the integral is taken over a closed cycle of changes, it being assumed that during the cycle the system can be characterized at each instant by a unique value for T. To prove this let us consider a cycle of changes in which the working substance is carried through states A and B, and in which, further, the part of the cycle from A to B is carried out adiabatically (but not necessarily statically) while the part of the cycle from B to A is carried out reversibly. For this cycle of changes

$$\oint \frac{dQ}{T} = \int_A^B \frac{dQ}{T} + \int_B^A \frac{dQ}{T} .$$

Since the part of the cycle from A to B has been carried out adiabatically, we have

$$\oint \frac{dQ}{T} = \int_B^A \frac{dQ}{T} = S_A - S_B , \tag{94'}$$

which, according to (93), must be zero or negative. We have thus proved (94) for the special cycle of changes considered. The arguments can be extended to prove (94) quite generally.

We thus see that the full mathematical content of the second law can be deduced from Carathéodory's principle. But the question still remains whether Carathéodory's principle can lead us to Kelvin's formulation of the second law. To answer this, we must supplement Carathéodory's principle with some additional axioms before we can derive Kelvin's or Clausius' formulation of the second law. The arguments necessary to establish this involve some rather delicate considerations, and these go beyond the scope of our present chapter. The interested reader may refer to an illuminating discussion by T. Ehrenfest Afanassjewa quoted in the bibliographical note at the end of the chapter.

11. *The free energy and the thermodynamical potential.*—We have shown in § 10 that

$$\oint \frac{dQ}{T} \leqslant 0 , \tag{95}$$

where the integral is taken over a closed cycle of changes. Let us suppose that the closed cycle of changes carries the working substance through states A and B, and that, further, the part of the cycle for B to A is along a reversible path. Then

$$\oint \frac{dQ}{T} = \int_A^B \frac{dQ}{T} + \int_B^A \frac{dQ}{T}, \tag{96}$$

or, since the path from B to A is reversible, we have, according to (95) and (96),

$$\int_A^B \frac{dQ}{T} \leqslant S_B - S_A. \tag{97}$$

Equation (97) is, of course, equivalent to (95).

Let us now consider an isothermal change. Then (97) can be written as

$$\int_A^B dQ \leqslant T(S_B - S_A), \tag{98}$$

where T denotes the constant temperature. By the First Law of Thermodynamics we now have.

$$U_B - U_A + W_{AB} \leqslant T(S_B - S_A), \tag{99}$$

where W_{AB} is the work done by the system. Equation (99) can be written alternatively in the form

$$F_B - F_A + W_{AB} \leqslant 0, \tag{100}$$

where

$$F \equiv U - TS. \tag{101}$$

The function F, thus introduced, is called the "free energy" of the system. From (100) it follows that for an isothermal change in which no work is done the free energy cannot increase.

Another function of importance is the thermodynamical potential, defined by

$$G = F + pV = U + pV - TS. \tag{102}$$

It is clear that if the temperature and the external forces are kept constant G cannot increase.

12. *Some thermodynamical formulae.*—So far we have concerned ourselves only with general principles. We shall conclude this chapter with the derivation of some thermodynamical formulae which are of considerable practical importance.

Let us consider a homogeneous isotropic medium. Then for a quasi-statical change (in Eq. [14] we shall now use the absolute temperature, T, instead of the empirical temperature, t)

$$dQ = \left[\left(\frac{\partial U}{\partial V}\right)_T + p\right] dV + \left(\frac{\partial U}{\partial T}\right)_V dT \,. \qquad (103)$$

Since dQ/T is a perfect differential, we should have

$$\frac{\partial}{\partial T}\left[\frac{1}{T}\left(\frac{\partial U}{\partial V} + p\right)\right] = \frac{\partial}{\partial V}\left(\frac{1}{T}\frac{\partial U}{\partial T}\right), \qquad (104)$$

or, carrying out the differentiations,

$$-\frac{1}{T^2}\left[\left(\frac{\partial U}{\partial V}\right)_T + p\right] + \frac{1}{T}\left[\frac{\partial^2 U}{\partial T \partial V} + \left(\frac{\partial p}{\partial T}\right)_V\right] = \frac{1}{T}\frac{\partial^2 U}{\partial V \partial T}, \qquad (105)$$

or

$$\left(\frac{\partial U}{\partial V}\right)_T = T\left(\frac{\partial p}{\partial T}\right)_V - p \,. \qquad (106)$$

Let us next consider the free energy. By definition (Eq. [101])

$$dF = dU - TdS - SdT \,, \qquad (107)$$

or, since

$$dQ = TdS = dU + pdV \,, \qquad (108)$$

we have

$$dF = -SdT - pdV \,. \qquad (109)$$

dF is, however, a perfect differential. Hence, we should have

$$\left(\frac{\partial F}{\partial T}\right)_V = -S \,; \qquad \left(\frac{\partial F}{\partial V}\right)_T = -p \,. \qquad (110)$$

Finally, let us consider the thermodynamical potential, G. We have

$$dG = dF + pdV + Vdp\,, \tag{111}$$

or, using (109),

$$dG = -SdT + Vdp\,. \tag{112}$$

Hence, we should have

$$\left(\frac{\partial G}{\partial T}\right)_p = -S\,; \qquad \left(\frac{\partial G}{\partial p}\right)_T = V\,. \tag{113}$$

We shall have occasion later to use (106), (110), and (113).

BIBLIOGRAPHICAL NOTES

C. CARATHÉODORY's paper appeared in *Math. Ann.*, **67**, 355, 1909. Further developments are contained in C. CARATHÉODORY, *Sitz. Ber. d. Prü. Akad.*, p. 39, Berlin, 1925; T. EHRENFEST AFANASSJEWA, *Zs. f. Phys.*, **33**, 933, 1925.

General expositions have been given by M. BORN, *Phys. Zs.*, **22**, 218, 249, 282, 1921, and by A. LANDÉ, *Handb. d. Phys.*, **9**, chap. iv, Berlin, 1926. Our account is based largely on Born's and Landé's expositions.

CHAPTER II

PHYSICAL PRINCIPLES

In this chapter we shall be concerned with some miscellaneous problems which form the background to the study of stellar structure. The main topics for consideration are: the thermodynamics of a perfect gas, uniform expansion (or contraction) of gaseous configurations, the virial theorem, and the thermodynamics of black-body radiation.

1. *The specific heats of a perfect gas.*—We shall consider a perfect gas for which, according to the results of the last chapter,

$$pV = RT \; ; \qquad U = U(T) \; , \tag{1}$$

where T is the absolute temperature and where the constant of proportionality, R, so introduced, is called the "gas constant."

For an infinitesimal quasi-statical change of state we have

$$dQ = dU + pdV \; , \tag{2}$$

or, according to (1),

$$dQ = \frac{dU}{dT}\, dT + pdV \; . \tag{3}$$

Let a be a function of the physical variables. Then the specific heat, c_a, at constant a is defined by

$$c_a = \left(\frac{dQ}{dT}\right)_{a\,=\,\text{constant}} . \tag{4}$$

The right-hand side of (4) is to be determined from (3) in such a way that a remains constant. Thus, the specific heat, c_V, at constant volume is given by

$$c_V = \frac{dU}{dT} \; . \tag{5}$$

To determine the specific heat, c_p, at constant pressure, we proceed as follows: From the equation of state we have

$$pdV + Vdp = RdT \; . \tag{6}$$

From (3) and (6) we have

$$dQ = \left(\frac{dU}{dT} + R\right)dT - Vdp\ , \tag{7}$$

from which it follows that

$$c_p = \frac{dU}{dT} + R\ . \tag{8}$$

Combining (5) and (8), we have the important result

$$c_p - c_V = R\ . \tag{9}$$

The ratio of the specific heats, denoted by γ, is defined as c_p/c_V.

In further work, we shall assume that c_V is independent of T. This is a consequence of the kinetic theory of gases, to which we shall return in chapter x. From (5) we have, then, for the internal energy U:

$$U = c_V T\ . \tag{10}$$

2. *Adiabatic changes.* Using (5), we can write for the differential dQ for a quasi-statical change

$$dQ = c_V dT + p dV\ , \tag{11}$$

or, using the equation of the state,

$$dQ = c_V dT + \frac{RT}{V}\, dV\ . \tag{12}$$

For a quasi-statical adiabatic change, therefore,

$$c_V dT + \frac{RT}{V}\, dV = 0\ , \tag{13}$$

or, using (9),

$$c_V \frac{dT}{T} + (c_p - c_V)\frac{dV}{V} = 0\ , \tag{14}$$

from which we obtain

$$c_V \log T + (c_p - c_V) \log V = \text{constant}\ . \tag{15}$$

In terms of the ratio of the specific heats we can re-write (15) as

$$TV^{\gamma-1} = \text{constant} . \tag{16}$$

Using $pV = RT$, we can eliminate T in (16) and obtain

$$pV^{\gamma} = \text{constant} . \tag{17}$$

Similarly, by eliminating V between (16) and (17), we have

$$p^{1-\gamma}T^{\gamma} = \text{constant} . \tag{18}$$

Hence, along an adiabatic we have

$$pV^{\gamma} = \text{constant} ; \quad p^{1-\gamma}T^{\gamma} = \text{constant} ; \quad TV^{\gamma-1} = \text{constant} . \tag{19}$$

The foregoing equations (19) are due to Poisson. The derivation in the form given above is due to Lord Kelvin.

3. *Polytropic changes.*—A polytropic change is a quasi-statical change of state carried out in such a way that the specific heat remains constant (at some prescribed value) during the entire process. Thus,

$$\frac{dQ}{dT} = c = \text{constant} . \tag{20}$$

An adiabatic, then, is a polytropic of zero specific heat, and an isothermal a polytropic of infinite heat capacity. It is also clear that quasi-statical changes in which the pressure and the volume are kept constant are polytropics of specific heats c_p and c_V, respectively. Polytropic changes were first introduced in thermodynamics by G. Zeuner and have been used extensively by Helmholtz and especially by Emden.

From (11) and (20) we have, for an infinitesimal polytropic change,

$$(c_V - c)dT + pdV = 0 . \tag{21}$$

Equation (21) is the equation of a polytropic. From (21) we have

$$(c_V - c)\frac{dT}{T} + (c_p - c_V)\frac{dV}{V} = 0 , \tag{22}$$

or, integrating, we have

$$T^{(c_V - c)}\, V^{(c_p - c_V)} = \text{constant} . \qquad (23)$$

We shall define the polytropic exponent, γ', by

$$\gamma' = \frac{c_p - c}{c_V - c} . \qquad (24)$$

We have

$$\gamma' - 1 = \frac{c_p - c_V}{c_V - c} . \qquad (25)$$

Equation (23) can then be written as

$$T V^{\gamma' - 1} = \text{constant} , \qquad (26)$$

which is of the same form as (16) except that the polytropic exponent, γ', replaces the ratio of the specific heats, γ. Hence, quite similarly, as in the last section, we have that along a polytropic

$$p V^{\gamma'} = \text{constant} ; \quad p^{1-\gamma'} T^{\gamma'} = \text{constant} ; \quad T V^{\gamma'-1} = \text{constant} . \qquad (27)$$

4. *A theorem due to Emden.*—Let AB and CD be two polytropics of heat capacity c_1 and exponent γ_1; further, let AD and BC be two other polytropics of heat capacity c_2 and exponent γ_2. Let these four polytropics intersect at the points A, B, C, and D. Let p_A, V_A, and T_A be the values of the physical variables at A; p_B, V_B, and T_B, the values at B; and so on.

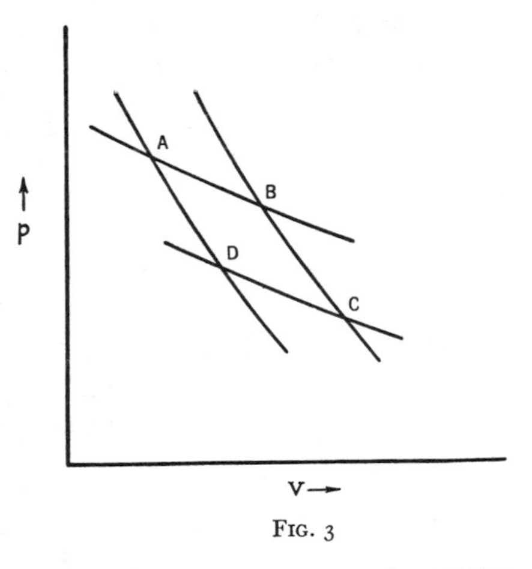

FIG. 3

Consider the case in which the gas goes through the cycle $ABCD$ quasi-statically. Since dQ/T is a perfect differential,

$$\oint \frac{dQ}{T} \qquad (28)$$

over a closed cycle must be zero. We shall evaluate the foregoing integral for the cycle under consideration. Since $dQ = c_1 dT$ along AB and CD, and $dQ = c_2 dT$ along AD and BC,

$$\oint \frac{dQ}{T} = c_1 \int_A^B \frac{dT}{T} + c_2 \int_B^C \frac{dT}{T} + c_1 \int_C^D \frac{dT}{T} + c_2 \int_D^A \frac{dT}{T} = 0 , \quad (29)$$

or

$$c_1 \log \frac{T_B}{T_A} + c_2 \log \frac{T_C}{T_B} + c_1 \log \frac{T_D}{T_C} + c_2 \log \frac{T_A}{T_D} = 0 , \quad (30)$$

or, again,

$$(c_1 - c_2) \log \frac{T_B T_D}{T_A T_C} = 0 . \quad (31)$$

Since $c_1 \neq c_2$, we have

$$T_B T_D = T_A T_C , \quad (32)$$

or

$$\frac{T_A}{T_B} = \frac{T_D}{T_C} ; \qquad \frac{T_A}{T_D} = \frac{T_B}{T_C} . \quad (33)$$

Since along the polytropics AB and CD we have the relations (27) with $\gamma' = \gamma_1$, we have

$$\frac{T_A}{T_B} = \frac{V_B^{\gamma_1 - 1}}{V_A^{\gamma_1 - 1}} . \quad (34)$$

Similarly,

$$\frac{T_D}{T_C} = \frac{V_C^{\gamma_1 - 1}}{V_D^{\gamma_1 - 1}} . \quad (35)$$

Combining (33), (34), and (35), we have

$$\frac{V_A}{V_B} = \frac{V_D}{V_C} ; \qquad \frac{V_A}{V_D} = \frac{V_B}{V_C} . \quad (36)$$

Similarly, we can show that

$$\frac{p_A}{p_B} = \frac{p_D}{p_C} ; \qquad \frac{p_A}{p_D} = \frac{p_B}{p_C} . \quad (37)$$

Thus we have proved: *If a pair of polytropics belonging to a given class (i.e., a given exponent) is intersected by another polytropic belong-*

ing to a different class, then the ratio of the physical variables (p, V, *or* T) *at the points of intersection are the same whatever polytropic belonging to the second class we may choose.*

We can state the foregoing theorem, due to Emden, in the following somewhat different way: A polytropic AB of exponent γ_1 is cut at the point A by another polytropic AD belonging to another class, of exponent γ_2 (γ_2 arbitrary but different from γ_1). Along AD we consider the point D such that p_A/p_D (or T_A/T_D or V_A/V_D) is some fixed constant. We now allow AD to be *any* polytropic belonging to class γ_2. The locus of D is then another polytropic, belonging to class γ_1. We shall see that, stated in the foregoing form, Emden's theorem has an important application to the theory of gaseous configurations.

5. *Polytropic temperature and the Emden variables.*—As we saw in § 3, along polytropics belonging to the class γ' we have

$$pV^{\gamma'} = \text{constant} \; ; \quad TV^{\gamma'-1} = \text{constant} \; ; \quad p^{1-\gamma'}T^{\gamma'} = \text{constant} \, . \tag{38}$$

In the (p, V) plane, the polytropics belonging to a given exponent γ' form a one-parametric family of curves, the parameter being the "constant" occurring in the first of the foregoing formulae. This family of curves can be classified by labeling each curve by what Emden calls the appropriate "polytropic temperature"; the latter is defined as the temperature along the given polytropic where the specific volume, V (and therefore also the density), has the value unity. We shall use $\Theta_{\gamma'}$ to denote the polytropic temperature. Then,

$$TV^{\gamma'-1} = \Theta_{\gamma'} \cdot 1 = \Theta_{\gamma'} \, . \tag{39}$$

Since the isothermal is a polytropic of infinite heat capacity, $\gamma' = 1$, and we have

$$T = \Theta_{\infty} \, , \tag{40}$$

i.e., the polytropic temperatures for the isothermals agree with actual temperatures labeling the isothermals.

In terms of the polytropic temperature we can represent the physical variables very conveniently. Let us write

$$\rho = \lambda\theta^n \; ; \qquad n = \frac{1}{\gamma' - 1} \, , \tag{41}$$

where λ is some constant factor to allow for a change in the scale on which density is measured; n introduced as above is called the "polytropic index." Since the density, ρ, is the reciprocal of the specific volume, V, we have, from (39), that

$$T = \Theta_{\gamma'}\rho^{\gamma'-1} = \Theta_{\gamma'}\rho^{1/n} = \lambda^{1/n}\Theta_{\gamma'}\theta \, . \tag{42}$$

If we choose λ to be unity, we see from (42) that θ is the temperature in a scale in which the polytropic temperature is unity. We further have

$$p = R\rho T = R\lambda^{(n+1)/n}\Theta_{\gamma'}\theta^{n+1} \, . \tag{43}$$

If we consider polytropics with zero specific heat, then we have adiabatics as a special case. Then $\gamma' = \gamma$, and we have definitions for the adiabatic temperature and adiabatic index.

6. *Entropy changes.*—We have

$$dS = \frac{dQ}{T} = c_V \frac{dT}{T} + \frac{p}{T} dV \, , \tag{44}$$

or, using the equation of state,

$$dS = c_V \frac{dT}{T} - (c_p - c_V) \frac{d\rho}{\rho} \, . \tag{45}$$

Now

$$T = \Theta_{\gamma'}\rho^{\gamma'-1} \, . \tag{46}$$

By differentiating (46),

$$\frac{dT}{T} = \frac{d\Theta_{\gamma'}}{\Theta_{\gamma'}} + (\gamma' - 1) \frac{d\rho}{\rho} \, . \tag{47}$$

Inserting (47) in (45), we have

$$dS = c_V \left[\frac{d\Theta_{\gamma'}}{\Theta_{\gamma'}} + (\gamma' - \gamma) \frac{d\rho}{\rho} \right] , \tag{48}$$

or

$$S = S_0 + c_V[\log \Theta_{\gamma'} + (\gamma' - \gamma) \log \rho] \, . \tag{49}$$

For an adiabatic $\gamma' = \gamma$, and

$$S = S_0 + c_V \log \Theta_{\gamma} \, . \tag{50}$$

Again, for (48) we have

$$dQ = TdS = c_V T\left[\frac{d\Theta_{\gamma'}}{\Theta_{\gamma'}} + (\gamma' - \gamma)\frac{d\rho}{\rho}\right], \tag{51}$$

from which we obtain

$$(dQ)_{ad} = c_V T \frac{d\Theta_\gamma}{\Theta_\gamma}. \tag{52}$$

Hence, the withdrawal of heat lowers the adiabatic temperature, while the supply of heat increases the adiabatic temperature.

If we consider changes along a given polytropic, then $\Theta_{\gamma'}$ does not change; and hence, by (47) and (48), along a polytropic,

$$dS = c_V(\gamma' - \gamma)\frac{d\rho}{\rho}; \qquad \frac{dT}{T} = (\gamma' - 1)\frac{d\rho}{\rho}, \tag{53}$$

or

$$dS = c_V \frac{(\gamma' - \gamma)}{\gamma' - 1}\frac{dT}{T}. \tag{54}$$

Equation (54) could have been derived directly from the definition $dQ = cdT$. Along an adiabatic dS is, of course, zero.

7. *Uniform expansion and contraction of gaseous configurations; cosmogenetic changes.*—Consider a perfect gas configuration in gravitational equilibrium. Then

$$\frac{dp}{dr} = -\frac{GM(r)}{r^2}\rho, \tag{55}$$[1]

where r denotes the radius vector with the center of the configuration as origin. Furthermore, $M(r)$ is the mass inclosed inside a spherical surface of radius r, and $p = R\rho T$. The foregoing equation is an elementary consequence of hydrostatic equilibrium. (The meaning of [55] is further commented upon in chap. iii.) We shall refer to a perfect gas configuration satisfying (55) as a "gas sphere."

An expansion or contraction of a spherical distribution of matter is said to be uniform if the distance between any two points is altered in the same way as the radius of configuration.

[1] In writing this equation we have neglected radiation pressure (see chaps. iii and vi).

Let the radii[2] of the initial and the final configuration be R_0 and R_1, and, further, let

$$R_1 = yR_0 \, . \tag{56}$$

Then, if r_1 and r_0 are the distances of any specified element of matter from the center before and after the expansion,[3]

$$r_1 = yr_0 \tag{57}$$

if the expansion has been carried out uniformly. More generally, if an element has an extension ds_0 in the initial configuration, then it will have an extension yds_0 after the uniform expansion:

$$ds_1 = yds_0 \, . \tag{58}$$

In particular,

$$dr_1 = ydr_0 \, . \tag{58'}$$

Let ρ_0, p_0, T_0 and ρ_1, p_1, T_1 be the density, pressure and temperature at "corresponding points" (i.e., at a distance r_0 in the initial configuration and at a distance $r_1 = yr_0$ in configuration after expansion). It is clear that

$$\rho_1 = y^{-3}\rho_0 \, , \tag{59}$$

since the corresponding volume elements in the two configurations are in the ratio y^3, while the mass inclosed in either is the same.

We shall now consider a uniform expansion of a gas sphere. Then, we should have

$$\frac{dp_0}{dr_0} = -\frac{GM(r_0)}{r_0^2}\rho_0 \, , \tag{60}$$

$$\frac{dp_1}{dr_1} = -\frac{GM(r_1)}{r_1^2}\rho_1 \, . \tag{61}$$

Since, however, $M(r_0) = M(r_1)$, we have, according to equations (58′), (59), and (61),

$$dp_1 = -\frac{GM(r_0)}{y^2 r_0^2} \cdot y^{-3}\rho_0 \cdot ydr_0 = -y^{-4}\frac{GM(r_0)}{r_0^2}\rho_0 dr_0 \, . \tag{62}$$

[2] At the boundary of the configuration, p, ρ, and T are all zero. The vanishing of p, ρ, and T defines the radius of a gas sphere.

[3] For brevity we shall explicitly refer only to "expansion" and not repeat each time "expansion or contraction."

By (60), then,

$$dp_1 = y^{-4} dp_0 \,, \tag{63}$$

from which it readily follows that

$$p_1 = y^{-4} p_0 \,. \tag{64}$$

Since $p = R\rho T$, we have, from (59) and (64),

$$p_1 = R\rho_1 T_1 = R y^{-3} \rho_0 T_1 = y^{-4} p_0 = y^{-4} R \rho_0 T_0 \,, \tag{65}$$

or

$$T_1 = y^{-1} T_0 \,. \tag{66}$$

Equations (56), (59), (64), and (66) can be written as

$$\frac{\rho_1}{\rho_0} = \frac{V_0}{V_1} = \left(\frac{R_0}{R_1}\right)^3 ; \qquad \frac{p_1}{p_0} = \left(\frac{R_0}{R_1}\right)^4 ; \qquad \frac{T_1}{T_0} = \frac{R_0}{R_1} \,. \tag{67}$$

We have thus proved the following theorem: *By a uniform expansion (or contraction) of a gas sphere, the density, pressure, and temperature at every point alter according to the inverse third, fourth, and unit power, respectively, of the ratio of the initial to the final radius.* The theorem in this general form is due to P. Rudzki (1902), though in a less general form it was known to Homer Lane (1869) and also to A. Ritter (1878). We shall refer to the foregoing theorem as "Lane's theorem."

Since the heat energy is proportional to $c_V T$, a further consequence of Lane's theorem is that the total heat energy in a gas sphere varies inversely as the radius during the process of uniform expansion.

For an infinitesimal uniform expansion we clearly have

$$\frac{dT}{T} = \frac{1}{3}\frac{d\rho}{\rho} = \frac{1}{4}\frac{dp}{p} = -\frac{dR_0}{R_0} \,. \tag{68}$$

From (67) it follows that

$$\frac{p_1}{p_0} = \left(\frac{\rho_1}{\rho_0}\right)^{4/3} = \left(\frac{V_0}{V_1}\right)^{4/3} = \left(\frac{T_1}{T_0}\right)^4 . \tag{69}$$

Alternatively,

$$p_1 V_1^{4/3} = p_0 V_0^{4/3} \,; \qquad T_1 V_1^{1/3} = T_0 V_0^{1/3} \,; \qquad T_1 p_1^{-1/4} = T_0 p_0^{-1/4} \,. \tag{70}$$

Thus, if a gas sphere expands (or contracts) uniformly through a sequence of equilibrium configurations, then the matter at every point undergoes a polytropic change belonging to the exponent $\gamma' = 4/3$, *or* n = 3. This result is due to Ritter, who was thus the first to recognize the special "cosmological" importance of polytropic changes of exponent 4/3. For this reason, he called polytropic changes of index 3 "cosmogenetic changes."

Since, according to Ritter's theorem, the physical variables change along a cosmogenetic during a uniform expansion, we can apply the results of § 6 to calculate the corresponding change in entropy. The appropriate formula to use is (54) with $\gamma' = 4/3$. Hence, for an infinitesimal expansion, the change in entropy, dS, is given by

$$dS = c_V(4 - 3\gamma)\frac{dT}{T}, \tag{71}$$

or, by (68),

$$dS = -c_V(4 - 3\gamma)\frac{dR_0}{R_0}. \tag{72}$$

Further, we have

$$dQ = TdS = -c_V T(4 - 3\gamma)\frac{dR_0}{R_0}. \tag{73}$$

8. *Uniform expansion (or contraction) of polytropic gas spheres.*— If, in a gas sphere, the pressure and density are related according to equations (41) and (43) with some definite value for $\Theta_{\gamma'}$, then the gas sphere is said to be a "polytropic gas sphere of index n," or more simply as a "polytrope of index n." This means that, if we plot the pressures at the different points in the gas sphere against the specific volumes at the respective points, then the points must all lie along a definite polytropic of index n and exponent γ'. Let us fix our attention on one definite point, A_0, on this polytropic. Through this point A_0 draw a polytropic of index 3. By Ritter's theorem, the effect of a uniform expansion (or contraction) is to displace the point A_0 along the polytropic of index 3 to another point, A_1, such that the temperature at A_1 bears a constant ratio (R_0/R_1) to the temperature at A_0. Now this happens to *all* the points, A_0, on the original polytropic as a result of the uniform expansion. By Emden's theorem (§ 4), we obtain in this way another polytropic of index n with a polytropic temperature different from the polytropic temperature of the original gas sphere. We thus see that the values of p and V at the points in the new configuration again lie on a polytropic of index n. Thus, *the configuration resulting from the uniform expansion of a polytropic gas sphere is again another polytropic gas sphere belonging to the same index.*

As we have just seen, the polytropic temperature of a polytrope is altered as a result of a uniform expansion. Let $\Theta_{\gamma'}(0)$ and $\Theta_{\gamma'}(1)$ be the polytropic temperature before and after the expansion. Then by (42) and (67),

$$\frac{\Theta_{\gamma'}(1)}{\Theta_{\gamma'}(0)} = \frac{T_1 V_1^{\gamma'-1}}{T_0 V_0^{\gamma'-1}} = \left(\frac{R_0}{R_1}\right)^{1-3(\gamma'-1)} = y^{-(4-3\gamma')} , \tag{74}$$

or

$$\Theta_{\gamma'}(0) R_0^{4-3\gamma'} = \Theta_{\gamma'}(1) R_1^{4-3\gamma'} . \tag{75}$$

Hence, for an infinitesimal expansion,

$$\frac{d\Theta_{\gamma'}}{\Theta_{\gamma'}} = -(4 - 3\gamma') \frac{dR_0}{R_0} . \tag{76}$$

From (48) we can now calculate the change in entropy for an infinitesimal expansion of a polytrope. Since

$$dS = c_V \left[\frac{d\Theta_{\gamma'}}{\Theta_{\gamma'}} + (\gamma' - \gamma) \frac{d\rho}{\rho} \right] , \tag{77}$$

we have, according to (76) and (68),

$$dS = -c_V[(4 - 3\gamma') + 3(\gamma' - \gamma)] \frac{dR_0}{R_0} = -c_V(4 - 3\gamma) \frac{dR_0}{R_0} , \tag{78}$$

thus recovering our earlier result (72).

9. *The virial theorem.*—We shall now consider the general motion of a cloud of particles. The "particles" may be gaseous molecules, dust particles, or even stars.

Let m denote the mass of a particle; x, y, and z its co-ordinates; and X, Y, and Z the components of the force acting on it. Then, by Newton's laws of motion,

$$m \frac{d^2x}{dt^2} = X ; \qquad m \frac{d^2y}{dt^2} = Y ; \qquad m \frac{d^2z}{dt^2} = Z . \tag{79}$$

We have

$$\frac{1}{2}\frac{d^2}{dt^2}(mx^2) = m\frac{d}{dt}\left(x\frac{dx}{dt}\right) = mx\frac{d^2x}{dt^2} + m\left(\frac{dx}{dt}\right)^2, \tag{80}$$

or, using the first of the equations (79),

$$\frac{1}{2}\frac{d^2}{dt^2}(mx^2) = m\left(\frac{dx}{dt}\right)^2 + xX. \tag{81}$$

Similarly,

$$\frac{1}{2}\frac{d^2}{dt^2}(my^2) = m\left(\frac{dy}{dt}\right)^2 + yY, \tag{82}$$

$$\frac{1}{2}\frac{d^2}{dt^2}(mz^2) = m\left(\frac{dz}{dt}\right)^2 + zZ. \tag{83}$$

Adding the foregoing equations, (81), (82), and (83), we obtain

$$\frac{1}{2}\frac{d^2}{dt^2}(mr^2) = m\left[\left(\frac{dx}{dt}\right)^2 + \left(\frac{dy}{dt}\right)^2 + \left(\frac{dz}{dt}\right)^2\right] + (xX + yY + zZ). \tag{84}$$

The first term on the right-hand side is simply twice the kinetic energy of the particle. Hence, summing the foregoing equation over all the particles, we have

$$\frac{1}{2}\frac{d^2I}{dt^2} = 2T + \Sigma(xX + yY + zZ), \tag{85}$$

where I is the moment of inertia about the origin defined by

$$I = \Sigma(mr^2), \tag{86}$$

and T is the kinetic energy of motion of the particles forming the cloud. The second term occurring on the right-hand side of (85) is called the "virial of Clausius."

To evaluate the virial, we fix our attention on two specific particles of masses m_1 and m_2 at the points (x_1, y_1, z_1) and (x_2, y_2, z_2). Let the force exerted by the second on the first have components A, B, and C, so that the force exerted by the first on the second will have the components $-A$, $-B$, and $-C$. The contribution of this pair of forces to the virial is given by

$$A(x_1 - x_2) + B(y_1 - y_2) + C(z_1 - z_2), \tag{87}$$

and hence

$$\text{Virial} = \Sigma\Sigma[A(x_1 - x_2) + B(y_1 - y_2) + C(z_1 - z_2)] , \tag{88}$$

where the summation is extended over all the pairs of particles. For a cloud of density so low that the ideal gas laws may be assumed to hold, all forces except the gravitational forces may be neglected. Thus we may take for A, B, and C the components of the force Gm_1m_2/r_{12}^2 (where G is the constant of gravitation) directed from 2 to 1. Hence, the components are

$$-G\frac{m_1m_2}{r_{12}^2} \times \begin{cases} \dfrac{x_1 - x_2}{r_{12}} , & \text{along the } X\text{-axis} \\ \dfrac{y_1 - y_2}{r_{12}} , & \text{along the } Y\text{-axis} \\ \dfrac{z_1 - z_2}{r_{12}} , & \text{along the } Z\text{-axis} \end{cases} \tag{89}$$

and

$$\text{Virial} = -\sum\sum\frac{Gm_1m_2}{r_{12}} . \tag{90}$$

Now each term inside the summation sign is simply the work done in separating the pair of particles to infinity against the gravitational attraction. Thus the virial is seen to be the potential energy, Ω, of the cloud of particles under consideration. Hence, we have

$$\frac{1}{2}\frac{d^2I}{dt^2} = 2T + \Omega , \tag{91}$$

an equation derived by Poincaré and Eddington. If the system is in a steady state, I is constant, and consequently we have

$$2T + \Omega = 0 . \tag{92}$$

Equation (92) expresses what is generally called the "virial theorem."

10. *An application of the virial theorem.*—Let us apply the virial theorem to a perfect gas configuration in gravitational equilibrium. Consider an element of mass dm at temperature T. From the kinetic theory of gases (see chap. x) the mean kinetic energy of a single

molecule in this element is $\frac{3}{2}kT$, where k is the Boltzmann constant. Let there be dN molecules in the element of mass under consideration. The contribution to the kinetic energy (of molecular motion) due to this element of mass is given by

$$dT = \tfrac{3}{2}kTdN = \tfrac{3}{2}RTdm = \tfrac{3}{2}(c_p - c_V)Tdm\,. \qquad (93)$$

But the internal energy, dU, of the element of mass is given by (Eq. [10])

$$dU = c_V T dm\,. \qquad (94)$$

Hence,

$$dT = \tfrac{3}{2}(\gamma - 1)dU\,, \qquad (95)$$

or, for the whole configuration,

$$T = \tfrac{3}{2}(\gamma - 1)U\,. \qquad (96)$$

By the virial theorem, then,

$$3(\gamma - 1)U + \Omega = 0\,. \qquad (97)$$

Let E be the total energy. Then

$$U + \Omega = E\,. \qquad (98)$$

From (97) and (98) we easily obtain

$$E = -(3\gamma - 4)U = \frac{3\gamma - 4}{3(\gamma - 1)}\Omega\,. \qquad (99)$$

The foregoing equation has the following consequences:

a) For a mass of gas for which $\gamma = 4/3$, we see that $E = 0$ in a steady state (independent of the radius of the configuration). A small radial expansion of the mass is, accordingly, possible, the mass changing from one equilibrium configuration to an adjacent configuration of equilibrium without change of energy. It follows that, if we consider a sequence of equilibrium configurations in which γ varies continuously, then at $\gamma = 4/3$ a change from stability to instability (for radial oscillations) must set in.[4] On the other hand, we

[4] This is intuitively obvious, but for a general discussion see J. H. Jeans, *Problems of Cosmogony and Stellar Dynamics*, pp. 20–23, Cambridge, 1919.

see that for $\gamma = 1$, $\Omega = 0$ for any prescribed E; i.e., for $\gamma = 1$ no stable configuration is possible. Hence, it follows that we have "stable" gas spheres only for $\gamma > 4/3$.

This result is originally due to Ritter and Emden; our proof, however, is due to Poincaré.

b) For $\gamma > 4/3$, equation (99) shows that E must be negative; or, in other words, in a steady state the energy is less than in a state of diffusion at infinity. Suppose, now, that the configuration contracts so that the potential energy changes by an amount $\Delta\Omega$. If ΔE and ΔU are corresponding changes in the total energy, E, and the internal energy, U, then by (99)

$$\Delta E = -(3\gamma - 4)\Delta U = \frac{3\gamma - 4}{3(\gamma - 1)}\,\Delta\Omega\,. \tag{100}$$

Hence, the amount of energy lost by radiation is $-\Delta E$:

$$-\Delta E = -\frac{3\gamma - 4}{3(\gamma - 1)}\,\Delta\Omega\,, \tag{101}$$

which is positive for a contraction of the configuration. At the same time, the internal energy increases by an amount

$$\Delta U = -\frac{1}{3(\gamma - 1)}\,\Delta\Omega\,, \tag{102}$$

which is again positive for a contraction. The reason for the increase in the internal energy consequent to a contraction of the configuration is that of the work $|\Delta\Omega|$ done by contraction, only the fraction $[(3\gamma - 4)/3(\gamma - 1)]$ is lost in radiation to space outside, and the remaining fraction $[1 - (3\gamma - 4)/3(\gamma - 1)] = [1/(3\gamma - 1)]$ is used in raising the temperature of the mass.

11. *The Stefan-Boltzmann law.*—We shall now consider the application of thermodynamics to inclosures containing radiation. Consider a perfectly black body, M, contained in an inclosure with perfectly reflecting walls. The inclosure will be traversed in all directions by radiation. Let the temperature of the black body contained in the inclosure be T. In a steady state, the inclosure is traversed by "black-body radiation" at temperature T. We shall assume that quasistatical processes can be carried out with the radiation. We shall suppose, further, that the radiation is the same throughout the inclosure. Let the energy of radiation per unit volume be u, so that the internal energy U will be uV:

$$U = uV . \tag{103}$$

There is a certain analogy between radiation and a perfect gas. The energy of both depends on temperature, and both exert pressure. According to the electromagnetic theory of light, radiation exerts the pressure

$$p = \tfrac{1}{3}u . \tag{104}$$ [5]

Let us allow the inclosure to expand quasi-statically while the temperature is maintained constant. Let the volume, V, increase by an amount dV while u and p remain unaltered. Consequently, the internal energy, U, increases by an amount udV, so that

$$\left(\frac{\partial U}{\partial V}\right)_T = u . \tag{105}$$

We shall now use the thermodynamical formula established in the last chapter (Eq. [106], i):

$$\left(\frac{\partial U}{\partial V}\right)_T = T\left(\frac{\partial p}{\partial T}\right)_V - p . \tag{106}$$

In our present case p depends only on T; and hence, according to (104) and (105), we can write (106) as

$$u = \tfrac{1}{3}T\frac{du}{dT} - \tfrac{1}{3}u , \tag{107}$$

or

$$T\frac{du}{dT} = 4u , \tag{108}$$

or again,

$$u = aT^4 ; \qquad p = \tfrac{1}{3}aT^4 . \tag{109}$$

[5] This is proved in chapter v, which deals with radiation problems in greater detail. Here we are only concerned with one straightforward application of thermodynamics to inclosures containing radiation.

Thus, the energy of black-body radiation per unit volume is proportional to the fourth power of the temperature. This is the statement of Stefan's law. The constant a introduced in (109) is called the "Stefan-Boltzmann constant." (Stefan empirically discovered the law in 1879 and Boltzmann gave the proof [essentially the one given here] in 1884.)

12. *Adiabatic changes in an inclosure containing matter and radiation.*—(*a*) We shall first consider the case of an inclosure containing radiation only. For a quasi-statical change,

$$dQ = dU + pdV \,, \tag{110}$$

or, by (103) and (104),

$$dQ = d(uV) + \tfrac{1}{3}udV = Vdu + \tfrac{4}{3}udV \,. \tag{110$'$}$$

For a quasi-statical adiabatic change, then,

$$Vdu + \tfrac{4}{3}udV = 0 \,, \tag{111}$$

or

$$uV^{4/3} = \text{constant} \,, \tag{112}$$

or, since $u = aT^4$,

$$TV^{1/3} = \text{constant} \,. \tag{113}$$

From (109) and (112) we have

$$pV^{4/3} = \text{constant} \,. \tag{114}$$

Thus radiation, in this respect, behaves like a perfect gas with a ratio of the specific heats $\gamma = 4/3$.

b) Let us now consider an inclosure containing both matter and radiation. We shall only consider the case where the matter is a perfect gas. The internal energy of such a system is, according to (103), (109), and (10),

$$U = aVT^4 + c_V T \,. \tag{115}$$

Let p_r denote the radiation pressure and p_g the gas pressure. Then the total pressure, P, is, accordingly,

$$P = p_r + p_g = \tfrac{1}{3}aT^4 + \frac{R}{V}T \,. \tag{116}$$

For a quasi-statical change

$$dQ = \left(\frac{\partial U}{\partial T}\right)_V dT + \left(\frac{\partial U}{\partial V}\right)_T dV + PdV\,. \tag{117}$$

By (115) and (116)

$$\left(\frac{\partial U}{\partial T}\right)_T = 4aVT^3 + c_V = \frac{V}{T}\left(12p_r + \frac{c_V}{R}\,p_g\right), \tag{118}$$

$$\left(\frac{\partial U}{\partial V}\right)_V = aT^4 = 3p_r\,. \tag{119}$$

Inserting (118) and (119) in (117), we have

$$dQ = \frac{V}{T}\left(12p_r + \frac{c_V}{c_p - c_V}\,p_g\right)dT + (4p_r + p_g)dV\,. \tag{120}$$

Let γ be the ratio of the specific heats ($\gamma = c_p/c_V$). Then

$$dQ = \frac{V}{T}\left(12p_r + \frac{1}{\gamma - 1}\,p_g\right)dT + (4p_r + p_g)dV\,. \tag{121}$$

For an adiabatic change,

$$\left(12p_r + \frac{1}{\gamma - 1}\,p_g\right)\frac{dT}{T} + (4p_r + p_g)\,\frac{dV}{V} = 0\,. \tag{122}$$

We define the adiabatic exponents Γ_1, Γ_2, and Γ_3 by the relations

$$\frac{dP}{P} + \Gamma_1\,\frac{dV}{V} = 0\,, \tag{123}$$

$$\frac{dP}{P} + \frac{\Gamma_2}{1 - \Gamma_2}\,\frac{dT}{T} = 0\,, \tag{124}$$

$$\frac{dT}{T} + (\Gamma_3 - 1)\,\frac{dV}{V} = 0\,. \tag{125}$$

Now

$$dP = d(p_r + p_g) = (4p_r + p_g)\,\frac{dT}{T} - p_g\,\frac{dV}{V}\,. \tag{126}$$

Hence, (123) is the same as

$$(4p_r + p_g)\frac{dT}{T} + [\Gamma_1(p_r + p_g) - p_g]\frac{dV}{V} = 0\,. \tag{127}$$

From (122) and (127) we find that

$$\frac{12p_r + \dfrac{1}{\gamma - 1}p_g}{4p_r + p_g} = \frac{4p_r + p_g}{\Gamma_1(p_r + p_g) - p_g}\,. \tag{128}$$

Let us now define the quantity β as follows:

$$\beta P = p_g\,; \qquad (1 - \beta)P = p_r\,. \tag{129}$$

Equation (128) can now be written as

$$\frac{12(\gamma - 1)(1 - \beta) + \beta}{(\gamma - 1)(4 - 3\beta)} = \frac{4 - 3\beta}{\Gamma_1 - \beta}\,. \tag{130}$$

Solving for Γ_1, we find that

$$\Gamma_1 = \beta + \frac{(4 - 3\beta)^2(\gamma - 1)}{\beta + 12(\gamma - 1)(1 - \beta)}\,. \tag{131}$$

From (131) we see that $\Gamma_1 = \gamma$ when $\beta = 1$, and $\Gamma_1 = 4/3$ when $\beta = 0$.

Again, from (124) and (126) we have

$$\left[4p_r + p_g + \frac{\Gamma_2}{1 - \Gamma_2}(p_r + p_g)\right]\frac{dT}{T} - p_g\frac{dV}{V} = 0\,. \tag{132}$$

From (122), (132), and (129) we now have

$$\frac{12(1 - \beta) + \dfrac{1}{\gamma - 1}\beta}{(4 - 3\beta) + \dfrac{\Gamma_2}{1 - \Gamma_2}} = -\frac{4 - 3\beta}{\beta}\,. \tag{133}$$

Solving for Γ_2, we find that

$$\Gamma_2 = 1 + \frac{(4 - 3\beta)(\gamma - 1)}{\beta^2 + 3(\gamma - 1)(1 - \beta)(4 + \beta)}\,. \tag{134}$$

From (127) and (132) we have

$$1 + \frac{1}{4 - 3\beta} \frac{\Gamma_2}{1 - \Gamma_2} = -\frac{\beta}{\Gamma_1 - \beta}, \tag{135}$$

or

$$\frac{\Gamma_2}{\Gamma_2 - 1} = (4 - 3\beta) \frac{\Gamma_1}{\Gamma_1 - \beta}. \tag{136}$$

Finally, we obtain the following equation, expressing Γ_2 in terms of Γ_1:

$$\Gamma_2 = \frac{(4 - 3\beta)\Gamma_1}{3(1 - \beta)\Gamma_1 + \beta}. \tag{137}$$

We see that when $\beta = 1$, $\Gamma_1 = \Gamma_2 = \gamma$; and when $\beta = 0$, $\Gamma_1 = \Gamma_2 = 4/3$.

To determine Γ_3 we proceed as follows: Eliminating dP/P between (123) and (124), we have

$$\frac{dT}{T} + \frac{(\Gamma_2 - 1)\Gamma_1}{\Gamma_2} \frac{dV}{V} = 0. \tag{138}$$

Comparing this with (125), we have

$$\Gamma_3 - 1 = \frac{(\Gamma_2 - 1)\Gamma_1}{\Gamma_2}. \tag{139}$$

By (136) and (139) we find

$$\Gamma_3 = 1 + \frac{\Gamma_1 - \beta}{4 - 3\beta}. \tag{140}$$

From (131) and (140) we finally have that

$$\Gamma_3 = 1 + \frac{(4 - 3\beta)(\gamma - 1)}{\beta + 12(\gamma - 1)(1 - \beta)}. \tag{141}$$

Γ_3 has the same limiting values for $\beta = 1$ and $\beta = 0$ as Γ_1 and Γ_2. Table 1 gives the values of the exponents Γ_1, Γ_2, and Γ_3 for different values of β for a monatomic gas ($\gamma = 5/3$).

TABLE 1

$1-\beta$	Γ_1	Γ_2	Γ_3	$1-\beta$	Γ_1	Γ_2	Γ_3
0.	1.667	1.667	1.667	0.6	1.405	1.343	1.359
0.1	1.563	1.484	1.510	0.7	1.386	1.338	1.350
0.2	1.511	1.417	1.444	0.8	1.368	1.335	1.344
0.3	1.476	1.383	1.408	0.9	1.350	1.333	1.338
0.4	1.449	1.363	1.386	1.0	1.333	1.333	1.333
0.5	1.426	1.351	1.370				

Equations (121) and (126) enable us to determine the specific heats at constant volume and pressure for an inclosure containing matter and radiation. Thus from (121) we have

$$C_V = \left(\frac{dQ}{dT}\right)_{dV=0} = \frac{V}{T}\left(12p_r + \frac{1}{\gamma - 1}p_g\right), \tag{142}$$

or, in terms of β,

$$C_V = \frac{c_V}{\beta}\left[\beta + 12(\gamma - 1)(1 - \beta)\right]. \tag{143}$$

Using equation (130), we have alternatively

$$C_V = c_V(4 - 3\beta)^2 \frac{\gamma - 1}{\beta(\Gamma_1 - \beta)}. \tag{144}$$

Similarly, eliminating dV between (121) and (126) and putting $dP = 0$, we find that

$$C_P = \frac{c_V}{\beta^2}\left[\beta^2 + (\gamma - 1)(4 - 3\beta)^2 + 12(\gamma - 1)\beta(1 - \beta)\right], \tag{145}$$

or, using (131),

$$C_P = \frac{c_V}{\beta^2}\Gamma_1\left[\beta + 12(\gamma - 1)(1 - \beta)\right]. \tag{146}$$

Equations (130) and (136) enable us to express C_P in the following alternative forms:

$$C_P = c_V \frac{(\gamma - 1)(4 - 3\beta)^2\Gamma_1}{\beta^2(\Gamma_1 - \beta)} = c_V \frac{(\gamma - 1)(4 - 3\beta)\Gamma_2}{\beta^2(\Gamma_2 - 1)}. \tag{147}$$

From (143) and (146) we find that

$$\frac{C_P}{C_V} = \frac{\Gamma_1}{\beta}. \tag{148}$$

BIBLIOGRAPHICAL NOTES

Polytropic changes were first considered by G. Zeuner (*Technische Thermodynamik*, **1**, Leipzig, 1887). The systematic use of a "polytropic" as a fundamental thermodynamical notion is, however, due to Emden. The results contained in §§ 4, 5, 6, and 8 are due to Emden. Reference should be made here to his classical treatise *Gaskugeln*, Leipzig, 1907.

The uniform expansion and contraction of gaseous configurations were first considered by A. Ritter (*Wiedemann Annalen*, **5**, 543, 1878). The treatment of the problem given in the text (§ 7) follows, more or less closely, Ritter's original treatment. Historical remarks concerning the association of a part of the results of § 7 with the name of Homer Lane are made in the bibliographical note for chapter iv.

The virial theorem proved in § 9 is due to H. Poincaré (*Leçons sur les hypothèses cosmogoniques*, § 74, Paris, 1911) and A. S. Eddington (*M.N.*, **76**, 528, 1916). The applications in the form given in § 10 are due to Poincaré.

The expression Γ_1 (Eq. [131]), for the adiabatic exponent for inclosures containing matter and radiation, is due to Eddington (*M.N.*, **79**, 2, 1918); it has, however, been generally overlooked that there are two other equally possible definitions for the adiabatic exponent, namely, Γ_2 and Γ_3. The expressions for C_P and C_V (Eqs. [143] and [146]) are given here for the first time.

CHAPTER III

INTEGRAL THEOREMS ON THE EQUILIBRIUM OF A STAR

As was emphasized in the Introduction, the structure of a star depends on a multitude of variables, and an approach toward a detailed theory is made only by introducing assumptions and approximations of various kinds with a view toward discriminating between the relevant and the less relevant aspects of the physical situation. It is therefore necessary to introduce one assumption at a time and investigate how far we can proceed with one assumption before we feel the need to make another. In this chapter we shall be mainly concerned with an attempt to discover how far we can proceed with the assumption that a star is in a steady state in gravitational equilibrium. We shall supplement this further by the assumption that the density distribution is such that the mean density $\bar{\rho}(r)$, interior to given point r inside the star, does not increase outward from the center. We shall see that these two assumptions already enable us to determine the order of magnitude of some of the more important physical variables describing a star. The method consists in finding inequalities for quantities like the central pressure, mean pressure, the potential energy, the mean value of gravity, etc. Before proceeding to establish the inequalities, however, we shall obtain the equations of equilibrium and some general formulae.

1. *Equations of gravitational equilibrium.*—We shall be concerned only with spherically symmetrical distributions of matter. Let r denote the radius vector, measured from the center of the configuration. Since we have a spherically symmetrical distribution of matter, the total pressure P, the density ρ, and the other physical variables will all be functions of r only. Let $M(r)$ be the mass inclosed inside r. Then

$$M(r) = \int_0^r 4\pi r^2 \rho dr \ ; \qquad dM(r) = 4\pi r^2 \rho dr \ . \tag{1}$$

We shall denote by $\bar{\rho}(r)$ the mean density inside r, and by $\bar{\rho}$ the mean density for the whole configuration:

$$\bar{\rho}(r) = \frac{M(r)}{\frac{4}{3}\pi r^3}\,; \qquad \bar{\rho} = \frac{M}{\frac{4}{3}\pi R^3}\,, \tag{2}$$

where M is the mass of the configuration and R defines the radius of the configuration at which ρ and P vanish.

Consider an infinitesimal cylinder at distance r from the center of height dr, and of unit cross-section at right angles to r (see Fig. 4).

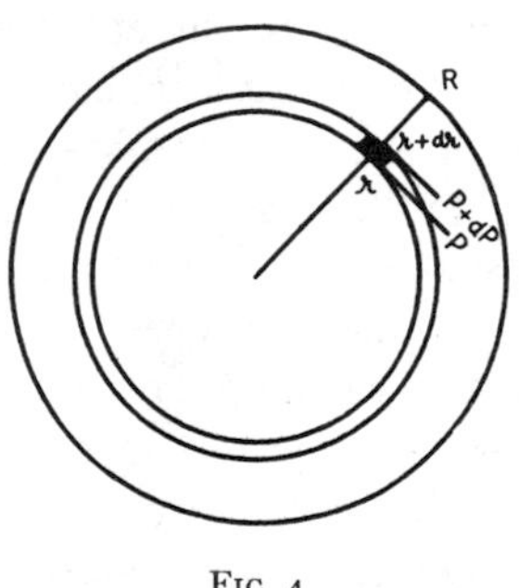

FIG. 4

Let P be the pressure at r and let the increment in P as we go from r to $r + dr$ be dP. The difference in pressure dP represents a force, $-dP$, acting on the element of mass considered, in the direction of increasing r. This must be counteracted by the gravitational attraction to which the element of mass is subjected. The mass of the infinitesimal cylinder considered is ρdr. The force of attraction between $M(r)$ and ρdr is, according to elementary potential theory, the same as between a mass $M(r)$ at the center and ρdr at r. By Newton's law this attractive force is given by $GM(r)\rho dr/r^2$, where G is the constant of gravitation. Further, the attraction due to the material outside r is zero. Hence, for equilibrium we should have

$$-dP = \frac{GM(r)\rho dr}{r^2}\,, \tag{3}$$

or

$$\frac{dP}{dr} = -\frac{GM(r)}{r^2}\,\rho\,. \tag{4}$$

It should be noticed that we have used P to denote the total pressure; thus, if we are considering a gaseous star, P is the sum of the gas kinetic pressure and the radiation pressure (according to the Stefan-Boltzmann law). We shall then write

$$P = \frac{k}{\mu H}\,\rho T + \tfrac{1}{3}aT^4\,, \tag{5}$$

where k is the Boltzmann constant, μ the mean molecular weight, H the mass of the proton, and a the Stefan-Boltzmann constant (chap. ii, § 11). In (5) we have used $(k/\mu H)$ in place of the gas constant "R," as hitherto. This is more convenient, and in the future we shall adopt this definition consistently.

Finally from (4) and (1) we have our fundamental equation of equilibrium:

$$\frac{1}{r^2}\frac{d}{dr}\left(\frac{r^2}{\rho}\frac{dP}{dr}\right) = -4\pi G\rho . \tag{6}$$

2. *The potential and the potential energy.*—The gravitational potential V is defined as the function the derivative of which in a given direction represents the gravitational attractive force in that direction acting on unit mass. For a spherically symmetrical distribution of matter, V must be clearly such that

$$\frac{dV}{dr} = \frac{GM(r)}{r^2} . \tag{7}$$

Equation (4) can now be written as

$$\frac{1}{\rho}\frac{dP}{dr} = -\frac{dV}{dr} . \tag{8}$$

If $r \geqslant R$, $M(r) = M =$ constant; we can therefore integrate (7) and obtain

$$V = -\frac{GM}{r} \qquad (r \geqslant R) . \tag{9}$$

(Equation [9] is so normalized that $V = 0$ as $r \to \infty$.) In particular, the potential V_1 at the boundary is given by

$$V_1 = -\frac{GM}{R} . \tag{10}$$

The potential energy of a given distribution of matter is defined as the work done (on the system) to bring the matter "diffused" to infinity into the given distribution. We shall denote the potential energy by Ω. For a spherically symmetrical distribution of matter, Ω can be calculated as follows: Suppose that we have already "brought" from infinity an amount of material $M(r)$. The work done to bring an additional amount of matter $dM(r)$ (as a spherical shell of thickness dr) is

$$-GM(r)dM(r)\int_r^\infty \frac{dr}{r^2} = -\frac{GM(r)dM(r)}{r}. \tag{11}$$

Hence, the potential energy Ω of the configuration is given by

$$-\Omega = G\int_0^R \frac{M(r)dM(r)}{r}. \tag{12}$$

Equation (12) is perfectly general, and is independent of the equation of hydrostatic equilibrium. For the case of hydrostatic equilibrium, equation (12) can be further transformed as follows:

$$-\Omega = \tfrac{1}{2}G\int_0^R \frac{d}{dM(r)}\,[M^2(r)]\,\frac{1}{r}\,dM(r) \tag{13}$$

$$= \tfrac{1}{2}G\,\frac{M^2}{R} + \tfrac{1}{2}G\int_0^R \frac{M^2(r)}{r^2}\,dr, \tag{14}$$

or by (7)

$$-\Omega = \frac{1}{2}\,\frac{GM^2}{R} + \frac{1}{2}\int_0^R \frac{dV}{dr}\,M(r)dr. \tag{15}$$

Again integrating by parts and using (10) for the value of V at $r = R$, we find that

$$\Omega = \frac{1}{2}\int_0^R V dM(r); \tag{16}$$

the important point to notice is the appearance of the factor $1/2$ in (16).

We shall now proceed to establish a number of integral theorems for configurations in gravitational equilibrium. The first three theorems (due to Milne) are true for any equilibrium configuration;

even the assumption "$\bar{\rho}(r)$ does not increase outward" is not introduced (the assumption is first made in Theorem 6).

3. THEOREM 1.—*In any equilibrium configuration the function*

$$P + \frac{GM^2(r)}{8\pi r^4} \tag{17}$$

decreases outward.

Proof: Equations (1) and (4) can be combined into

$$\frac{dP}{dr} = -\frac{GM(r)}{4\pi r^4}\frac{dM(r)}{dr}. \tag{18}$$

Now,

$$\frac{d}{dr}\left[P + \frac{GM^2(r)}{8\pi r^4}\right] = \frac{dP}{dr} + \frac{GM(r)}{4\pi r^4}\frac{dM(r)}{dr} - \frac{GM^2(r)}{2\pi r^5}. \tag{19}$$

By (18), then,

$$\frac{d}{dr}\left[P + \frac{GM^2(r)}{8\pi r^4}\right] = -\frac{GM^2(r)}{2\pi r^5} < 0, \tag{20}$$

from which the theorem follows.

Corollary: If P_c denotes the central pressure, then we should have

$$P_c > P + \frac{GM^2(r)}{8\pi r^4} > \frac{GM^2}{8\pi R^4}. \tag{21}$$

The outer members of the foregoing inequality give

$$P_c > \frac{GM^2}{8\pi R^4}, \tag{22}$$

or, inserting numerical values,

$$P_c > 4.44 \times 10^{14}\left(\frac{M}{\odot}\right)^2\left(\frac{R_\odot}{R}\right)^4 \text{ dynes cm}^{-2}, \tag{23}$$

or

$$P_c > 4.50 \times 10^{8}\left(\frac{M}{\odot}\right)^2\left(\frac{R_\odot}{R}\right)^4 \text{ atmospheres}, \tag{24}$$

where $\odot$ and $R_\odot$ refer to the mass and the radius of the sun.

4. THEOREM 2.—*For any equilibrium configuration*

$$I_\nu = \int_0^R \frac{GM(r)dM(r)}{r^\nu} = 4\pi(4-\nu)\int_0^R Pr^{3-\nu}\,dr \tag{25}$$

if

$$\nu < 4 . \tag{26}$$ [1]

From (18) and the definition of I_ν we have

$$I_\nu = -4\pi\int_0^R \frac{dP}{dr} r^{4-\nu}\,dr , \tag{27}$$

or, integrating by parts and remembering that $\nu < 4$, we have

$$I_\nu = 4\pi(4-\nu)\int_0^R Pr^{3-\nu}\,dr , \tag{28}$$

which proves the theorem.

We notice that when $\nu = 4$, (27) can be integrated, and we have

$$I_4 = 4\pi P_c . \tag{29}$$

Again, by (12)

$$I_1 = \int_0^R \frac{GM(r)dM(r)}{r} = -\Omega ; \tag{30}$$

and hence, by the theorem

$$-\Omega = 12\pi\int_0^R Pr^2dr , \tag{31}$$ [2]

or

$$-\Omega = 3\int_0^R PdV , \tag{32}$$

where dV stands for the volume element.

[1] Actually, we shall see (§ 11) that under "normal" circumstances the integral I_ν converges for $\nu < 6$.

[2] Equation (31) was known to A. Ritter (*Wiedemann Annalen*, **8**, 160, 1879).

Now the value of the gravity g at r is clearly $GM(r)/r^2$. Hence, if we denote by $\bar{g}$ the mean value of gravity defined by

$$M\bar{g} = \int_0^R g dM(r) = \int_0^R \frac{GM(r)dM(r)}{r^2} = I_2 \, , \tag{33}$$

then

$$M\bar{g} = 8\pi \int_0^R P r dr \, . \tag{34}$$

5. THEOREM 3.—*For any equilibrium configuration*

$$\nu\pi P_c R^{4-\nu} + \frac{4-\nu}{8}\frac{GM^2}{R^\nu} > I_\nu > \frac{GM^2}{2R^\nu} \tag{35}$$

if

$$\nu < 4 \, . \tag{36}$$

Proof: By Theorem 2

$$I_\nu = 4\pi(4-\nu)\int_0^R P r^{3-\nu} dr \, . \tag{37}$$

But by Theorem 1

$$\frac{GM^2}{8\pi R^4} - \frac{GM^2(r)}{8\pi r^4} < P < P_c - \frac{GM^2(r)}{8\pi r^4} \, . \tag{38}$$

By (37) and (38) we have

$$\left. \begin{aligned} 4\pi(4-\nu)\int_0^R \left[P_c - \frac{GM^2(r)}{8\pi r^4}\right] r^{3-\nu} dr > I_\nu \\ > 4\pi(4-\nu)\int_0^R \left[\frac{GM^2}{8\pi R^4} - \frac{GM^2(r)}{8\pi r^4}\right] r^{3-\nu} dr \, , \end{aligned} \right\} \tag{39}$$

or

$$4\pi P_c R^{4-\nu} > I_\nu + \frac{4-\nu}{2}\int_0^R \frac{GM^2(r)dr}{r^{1+\nu}} > \frac{GM^2}{2R^\nu} \, . \tag{40}$$

Now,

$$\int_0^R \frac{GM^2(r)}{r^{1+\nu}} dr = -\frac{1}{\nu}\int_0^R GM^2(r)\frac{d}{dr}\left(\frac{1}{r^\nu}\right) dr \, , \tag{41}$$

or, after an integration by parts,

$$\int_0^R \frac{GM^2(r)}{r^{1+\nu}}\,dr = -\frac{GM^2}{\nu R^\nu} + \frac{2}{\nu}\int_0^R \frac{GM(r)dM(r)}{r^\nu} \tag{42}$$

$$= -\frac{GM^2}{\nu R^\nu} + \frac{2}{\nu} I_\nu\,. \tag{43}$$

Inserting (43) in (40), we have

$$4\pi P_c R^{4-\nu} > I_\nu + \frac{4-\nu}{\nu} I_\nu - \frac{4-\nu}{2\nu}\frac{GM^2}{R^\nu} > \frac{GM^2}{2R^\nu}\,, \tag{44}$$

or, simplifying, we have

$$\nu\pi P_c R^{4-\nu} + \frac{4-\nu}{8}\frac{GM^2}{R^\nu} > I_\nu > \frac{GM^2}{2R^\nu}\,, \tag{45}$$

which proves the theorem.

Corollary 1: If $\nu = 1$, $I_1 = -\Omega$, we have

$$\pi P_c R^3 + \frac{3}{8}\frac{GM^2}{R} > -\Omega > \frac{GM^2}{2R}\,. \tag{46}$$

That $GM^2/2R$ sets the absolute minimum to $-\Omega$ was first proved by Ritter (*Wiedemann Annalen*, **16,** 183, 1882).

Corollary 2: If $\nu = 2$, $I_2 = M\bar{g}$, and we have

$$2\pi P_c R^2 + \frac{1}{4}\frac{GM^2}{R^2} > M\bar{g} > \frac{GM^2}{2R^2}\,. \tag{47}$$

The following theorem is due to Ritter.

6. THEOREM 4.—*In a gaseous configuration in equilibrium in which the radiation pressure is negligible,*

$$\bar{T} > \frac{1}{6}\frac{\mu H}{k}\frac{GM}{R}\,, \tag{48}$$

where the mean temperature $\bar{T}$ *is defined by*

$$M\bar{T} = \int_0^R T\,dM(r)\,; \tag{49}$$

μ *is further assumed to be constant in the configuration.*

Proof: If the radiation pressure can be neglected,

$$P = \frac{k}{\mu H} \rho T \,, \qquad \text{or} \qquad T = \frac{\mu H}{k} \frac{P}{\rho} \,. \tag{50}$$

Hence,

$$M\bar{T} = \int_0^R T dM(r) = \frac{\mu H}{k} \int_0^R \frac{P}{\rho} dM(r) \tag{51}$$

$$= \frac{\mu H}{k} \int_0^R P dV \,. \tag{52}$$

By (32), then,

$$M\bar{T} = -\frac{1}{3} \frac{\mu H}{k} \Omega \,. \tag{53}$$

By Theorem 3, Corollary 1, we have

$$\bar{T} > \frac{1}{6} \frac{\mu H}{k} \frac{GM}{R} \,, \tag{54}$$

thus proving the theorem.

Inserting numerical values in (54), we find that

$$\bar{T} > 3.84 \times 10^6 \mu \frac{M}{\odot} \frac{R_\odot}{R} \,; \tag{54'}$$

in other words, we may expect the temperature to be of the order of a few million degrees in stellar interiors.

Equation (53), derived above, has an important physical meaning. If we are considering a gaseous configuration (and if we neglect radiation pressure), then the internal energy is given by

$$U = c_V \int T dM(r) = c_V M\bar{T}$$

$$= -\tfrac{1}{3} c_V \frac{\mu H}{k} \Omega = -\frac{1}{3} \frac{c_V}{c_p - c_V} \Omega \,,$$

or, finally,

$$U = -\frac{1}{3(\gamma - 1)} \Omega \,, \tag{55}$$

a formula which was derived independently by A. Ritter and J. Perry.[3] (We shall refer to [55] as "Ritter's relation.") We have already derived (55) from the virial theorem (chap. ii, § 10).

7. THEOREM 5.—*If* $\overline{P}(r)$ *is the mean pressure interior to* r, *defined by*

$$M(r)\overline{P}(r) = \int_0^r P dM(r) , \tag{56}$$

then in any equilibrium configuration

$$\overline{P}(r) - P(r) > \frac{1}{12\pi}\frac{GM^2(r)}{r^4} \quad (r > 0) . \tag{57}$$

Proof: Integrating by parts the integral defining $\overline{P}(r)$, we have

$$M(r)[\overline{P}(r) - P(r)] = -\int_0^r M(r) dP , \tag{58}$$

or, using (18),

$$M(r)[\overline{P}(r) - P(r)] = \frac{G}{12\pi}\int_0^r \frac{d[M^3(r)]}{r^4} , \tag{59}$$

or, again integrating by parts,

$$M(r)[\overline{P}(r) - P(r)] = \frac{G}{12\pi}\frac{M^3(r)}{r^4} + \frac{G}{3\pi}\int_0^r \frac{M^3(r)}{r^5} dr . \tag{60}$$

Since the second term on the right-hand side of (60) is positive, we have the inequality stated.

Corollary: If we put $r = R$ in (57), we have for the mean pressure $\overline{P}$ defined for the whole configuration the inequality

$$\overline{P} > \frac{1}{12\pi}\frac{GM^2}{R^4} . \tag{61}$$

[3] A. Ritter, *op. cit.*, pp. 160–162; J. Perry, *Nature*, **60**, 247, 1899. Lord Kelvin, in his work (referred to in greater detail in chap. iv), refers to (55) as the "Ritter-Perry theorem."

8. THEOREM 6.—*In any equilibrium configuration in which the mean density* $\bar{\rho}(r)$ *interior to* r *does not increase outward, we have*

$$\tfrac{1}{2}G(\tfrac{4}{3}\pi)^{1/3}\bar{\rho}^{4/3}(r)M^{2/3}(r) \leqslant P_c - P \leqslant \tfrac{1}{2}G(\tfrac{4}{3}\pi)^{1/3}\rho_c^{4/3}M^{2/3}(r) ; \qquad (62)$$

where ρ_c *is the central density.*

Proof: From (18) we have

$$P_c - P = \frac{G}{4\pi}\int_0^r \frac{M(r)dM(r)}{r^4} . \qquad (63)$$

From the definition of the mean density $\bar{\rho}(r)$ (Eq. [2]), we have

$$r^4 = \left[\frac{M(r)}{\frac{4}{3}\pi\bar{\rho}(r)}\right]^{4/3} . \qquad (64)$$

Inserting (64) in (63), we have

$$P_c - P = \frac{1}{4\pi}(\tfrac{4}{3}\pi)^{4/3}G\int_0^r \bar{\rho}^{4/3}(r)M^{-1/3}(r)dM(r) . \qquad (65)$$

Since by hypothesis $\bar{\rho}(r)$ does not increase outward, wc havc from (65) that

$$P_c - P \geqslant \frac{1}{4\pi}(\tfrac{4}{3}\pi)^{4/3}G\bar{\rho}^{4/3}(r)\int_0^r M^{-1/3}(r)dM(r) . \qquad (66)$$

The integral on the right-hand side of (66) can be evaluated, and we have

$$P_c - P \geqslant \tfrac{1}{2}(\tfrac{4}{3}\pi)^{1/3}G\bar{\rho}^{4/3}(r)M^{2/3}(r) . \qquad (67)$$

Again, from (65), according to our hypothesis,

$$P_c - P \leqslant \frac{1}{4\pi}(\tfrac{4}{3}\pi)^{4/3}G\rho_c^{4/3}\int_0^r M^{-1/3}(r)dM(r) , \qquad (68)$$

or

$$P_c - P \leqslant \tfrac{1}{2}(\tfrac{4}{3}\pi)^{1/3}G\rho_c^{4/3}M^{2/3}(r) . \qquad (69)$$

Combining (67) and (69), we have the required inequality.

Corollary: If we put $r = R$ in the inequality of Theorem 6, we obtain

$$\tfrac{1}{2}G(\tfrac{4}{3}\pi)^{1/3}\bar{\rho}^{4/3}M^{2/3} \leqslant P_c \leqslant \tfrac{1}{2}G(\tfrac{4}{3}\pi)^{1/3}\rho_c^{4/3}M^{2/3} . \tag{70}$$

In the left-hand side of the inequality we can substitute for $\bar{\rho}$ its expression $M/\frac{4}{3}\pi R^3$. We then find

$$\frac{3}{8\pi}\frac{GM^2}{R^4} \leqslant P_c \leqslant \tfrac{1}{2}G(\tfrac{4}{3}\pi)^{1/3}\rho_c^{4/3}M^{2/3} . \tag{71}$$

We see that the additional restriction imposed on the density distribution, namely, that $\bar{\rho}(r)$ does not increase outward, enables us to improve the inequality obtained for P_c in Theorem 1. Numerically we now have that

$$P_c \geqslant 1.35 \times 10^9 \left(\frac{M}{\odot}\right)^2 \left(\frac{R_\odot}{R}\right)^4 \text{ atmospheres} . \tag{72}$$

Equation (71) was first given by Eddington,[4] but the complete theorem and the proof given are due to Chandrasekhar. Milne has given the following instructive alternative proof for the inequality

$$P_c \geqslant \frac{3}{8\pi}\frac{GM^2}{R^4} . \tag{73}$$

Consider the expression

$$P + a\frac{GM^2(r)}{8\pi r^4} , \tag{74}$$

where a is, for the present, an arbitrary number. Now,

$$\frac{d}{dr}\left[P + a\frac{GM^2(r)}{8\pi r^4}\right] = \frac{dP}{dr} + a\frac{GM(r)}{4\pi r^4}\frac{dM(r)}{dr} - a\frac{GM^2(r)}{2\pi r^5} , \tag{75}$$

or by (18)

$$\frac{d}{dr}\left[P + a\frac{GM^2(r)}{8\pi r^4}\right] = -(a - 1)\frac{dP}{dr} - a\frac{GM^2(r)}{2\pi r^5} , \tag{76}$$

[4] Eddington stated the result only for ρ (not $\bar{\rho}(r)$) decreasing outward.

or, again, by (4)

$$\frac{d}{dr}\left[P + \alpha\frac{GM^2(r)}{8\pi r^4}\right] = \frac{GM(r)}{r^2}\left[(\alpha - 1)\rho - \alpha\frac{M(r)}{2\pi r^3}\right], \qquad (77)$$

or by (2)

$$\frac{d}{dr}\left[P + \alpha\frac{GM^2(r)}{8\pi r^4}\right] = \frac{GM(r)}{r^2}\left[(\alpha - 1)\rho - \frac{2\alpha}{3}\bar{\rho}(r)\right]. \qquad (78)$$

If the mean density decreases outward, it is clear that $\bar{\rho}(r) \geqslant \rho$. Hence, if we choose $\alpha = 3$, we have

$$\frac{d}{dr}\left[P + 3\frac{GM^2(r)}{8\pi r^4}\right] = -2\frac{GM(r)}{r^2}[\bar{\rho}(r) - \rho] \leqslant 0. \qquad (79)$$

Hence, the expression (74) considered with $\alpha = 3$ is a decreasing function of r:

$$P_c \geqslant P + \frac{3}{8\pi}\frac{GM^2(r)}{r^4} \geqslant \frac{3}{8\pi}\frac{GM^2}{R^4}, \qquad (80)$$

thus establishing the inequality. Milne's proof cannot, however, be extended to give the complete Theorem 6.

9. THEOREM 7.—*The ratio* $(1 - \beta_c)$ *of the radiation pressure to the total pressure at the center of a wholly gaseous configuration in equilibrium in which* $\bar{\rho}(r)$ *does not increase outward, satisfies the inequality*

$$1 - \beta_c \leqslant 1 - \beta^*, \qquad (81)$$

where β^* *satisfies the quartic equation*

$$M = \left(\frac{6}{\pi}\right)^{1/2}\left[\left(\frac{k}{\mu_c H}\right)^4 \frac{3}{a}\frac{1 - \beta^*}{\beta^{*4}}\right]^{1/2}\frac{1}{G^{3/2}}; \qquad (82)$$

μ_c *is the mean molecular weight at the center.*

Proof: Now, according to (5), the total pressure P is given by

$$P = \frac{k}{\mu H}\rho T + \tfrac{1}{3}aT^4. \qquad (83)$$

Define the quantity $(1 - \beta)$ by

$$(1 - \beta)P = \tfrac{1}{3}aT^4 ; \qquad \beta P = \frac{k}{\mu H} \rho T . \tag{84}$$

By (84)

$$\frac{1}{1 - \beta} \frac{a}{3} T^4 = \frac{1}{\beta} \frac{k}{\mu H} \rho T , \tag{85}$$

or

$$T = \left[\frac{k}{\mu H} \frac{3}{a} \frac{1 - \beta}{\beta} \right]^{1/3} \rho^{1/3} . \tag{86}$$

Again,

$$P = \frac{1}{\beta} \frac{k}{\mu H} \rho T = \left[\left(\frac{k}{\mu H} \right)^4 \frac{3}{a} \frac{1 - \beta}{\beta^4} \right]^{1/3} \rho^{4/3} . \tag{87}$$

Hence, at the center of the configuration

$$P_c = \left[\left(\frac{k}{\mu_c H} \right)^4 \frac{3}{a} \frac{1 - \beta_c}{\beta_c^4} \right]^{1/3} \rho_c^{4/3} . \tag{88}$$

By Theorem 6, on the other hand,

$$P_c \leqslant \tfrac{1}{2} G (\tfrac{4}{3}\pi)^{1/3} M^{2/3} \rho_c^{4/3} . \tag{89}$$

Comparing (88) and (89), we have

$$\left[\left(\frac{k}{\mu_c H} \right)^4 \frac{3}{a} \frac{1 - \beta_c}{\beta_c^4} \right]^{1/3} \leqslant \left(\frac{\pi}{6} \right)^{1/3} G M^{2/3} , \tag{90}$$

or

$$M \geqslant \left(\frac{6}{\pi} \right)^{1/2} \left[\left(\frac{k}{\mu_c H} \right)^4 \frac{3}{a} \frac{1 - \beta_c}{\beta_c^4} \right]^{1/2} \frac{1}{G^{3/2}} . \tag{91}$$

Defining $(1 - \beta^*)$ as in equation (82), we have

$$\frac{1 - \beta^*}{\beta^{*4}} \geqslant \frac{1 - \beta_c}{\beta_c^4} . \tag{92}$$

But $(1 - \beta)/\beta^4$ is a monotonic increasing function of $(1 - \beta)$. Hence, we should have

$$1 - \beta^* \geqslant 1 - \beta_c , \tag{93}$$

which proves the theorem.

10. Theorem 7 (which is due to Chandrasekhar) shows that for a gaseous star the value of $(1 - \beta)$ at the center cannot exceed an amount depending on the mass of the star only. Table 2 gives the value of $(1 - \beta^*)$ for different values of the mass M.

TABLE 2

VALUES OF $(1-\beta^*)$

$1 - \beta^*$	$\left(\frac{M}{\odot}\right)\mu_c^2$	$1 - \beta^*$	$\left(\frac{M}{\odot}\right)\mu_c^{-2}$
0.025	0.908	0.5	15.432
.05	1.352	0.6	26.41
.1	2.130	0.7	50.72
.2	3.812	0.8	122.0
.3	6.099	0.9	517.6
0.4	9.585	1.0	∞

As an example of the application of Table 2, we see that for the sun $(1 - \beta_c) < 0.03$ while for Capella $(M = 4.18\odot)$, $(1 - \beta_c) < 0.22$, assuming in both the cases that $\mu_c = 1$.

11. THEOREM 8.—*For* I_ν, *defined as in Theorem 2, and under the conditions of Theorem 6, we have*

$$\frac{3}{6-\nu}\frac{GM^2}{R^\nu} \leqslant I_\nu \leqslant \frac{3}{6-\nu}\frac{GM^2}{r_c^\nu} \qquad (\nu < 6), \quad (94)$$

where r_c *is defined by*

$$\tfrac{4}{3}\pi r_c^3 \rho_c = M \; . \qquad (95)$$

Proof: Now,

$$r^\nu = \left[\frac{M(r)}{\frac{4}{3}\pi\bar{\rho}(r)}\right]^{\nu/3} . \qquad (96)$$

Substituting the foregoing in the integral defining I_ν, we have

$$I_\nu = G(\tfrac{4}{3}\pi)^{\nu/3}\int_0^R \bar{\rho}^{\nu/3}(r) M^{(3-\nu)/3}(r) dM(r) \; . \qquad (97)$$

Since $\bar{\rho}(r)$ does not increase outward, the minimum value for the integral on the right-hand side is obtained by replacing $\bar{\rho}(r)$ by $\bar{\rho}$, and taking it outside the integral sign. Similarly, the maximum value is obtained by replacing $\bar{\rho}(r)$ by ρ_c, and taking it outside the integral sign. In this way we find that

$$\frac{3}{6-\nu} G(\tfrac{4}{3}\pi\rho_c)^{\nu/3}M^{(6-\nu)/3} \geqslant I_\nu \geqslant \frac{3}{6-\nu} G(\tfrac{4}{3}\pi\bar{\rho})^{\nu/3}M^{(6-\nu)/3} . \quad (98)$$

But by definition,

$$\tfrac{4}{3}\pi r_c^3\rho_c = M = \tfrac{4}{3}rR^3\bar{\rho} . \quad (99)$$

Using (99), (98) is found to reduce to

$$\frac{3}{6-\nu}\frac{GM^2}{R^\nu} \leqslant I_\nu \leqslant \frac{3}{6-\nu}\frac{GM^2}{r_c^\nu} , \quad (100)$$

which proves the theorem.

Incidentally, we have also proved that the integral defining I_ν *converges for* $\nu < 6$ *if the mean density decreases outward and if further* ρ_c *is finite.*

Corollary 1: If $\nu = 1$, $I_1 = -\Omega$, and we have

$$\frac{3}{5}\frac{GM^2}{R} \leqslant -\Omega \leqslant \frac{3}{5}\frac{GM^2}{r_c} . \quad (101)$$

Corollary 2: If $\nu = 2$, $I_2 = M\bar{g}$, and we have

$$\frac{3}{4}\frac{GM}{R^2} \leqslant \bar{g} \leqslant \frac{3}{4}\frac{GM}{r_c^2} . \quad (101')$$

Corollary 3: If we put $\nu = 6$ on the right hand side of (97) and extend the range of integration from $r = r$ to $r = R$, we find

$$\frac{GM^2}{R^6}\log\frac{M}{M(r)} \leqslant \int_r^R \frac{GM(r)dM(r)}{r^6} \leqslant \frac{GM^2}{r_c{}^6}\log\frac{M}{M(r)} . \quad (102)$$

12. THEOREM 9.—*In a gaseous configuration in equilibrium in which the radiation pressure is negligible and in which, further,* $\bar{\rho}$(r) *does not increase outward,*

$$\frac{1}{5}\frac{\mu H}{k}\frac{GM}{r_c} \geqslant \bar{T} \geqslant \frac{1}{5}\frac{\mu H}{k}\frac{GM}{R} , \quad (103)$$

where μ (the molecular weight) is assumed to be constant in the configuration, and $\overline{T}$ is the mean temperature defined as in Theorem 4.

Proof: By (53) we have for the case considered

$$M\overline{T} = -\frac{1}{3}\frac{\mu H}{k}\Omega . \qquad (104)$$

By Corollary 1 of the last theorem we have

$$\frac{3}{5}\frac{GM^2}{R} \leqslant -\Omega \leqslant \frac{3}{5}\frac{GM^2}{r_c} . \qquad (105)$$

Combining (104) and (105), we have the required inequality.

We are thus able to replace the "1/6" that occurred in Theorem 4 by "1/5" because of our additional hypothesis concerning $\bar{\rho}(r)$. Numerically (103) reduces to

$$\overline{T} \geqslant 4.61 \times 10^6 \mu \frac{M}{\odot}\frac{R_\odot}{R} . \qquad (106)$$

13. THEOREM 10.—*If $I_{\sigma,\nu}$ is the integral defined by*

$$I_{\sigma,\nu} = \int_0^R \frac{GM^\sigma(r)dM(r)}{r^\nu} \qquad [3(\sigma+1) > \nu] , \qquad (107)$$

then under the conditions of Theorem 6

$$\frac{3}{3\sigma+3-\nu}\frac{GM^{\sigma+1}}{R^\nu} \leqslant I_{\sigma,\nu} \leqslant \frac{3}{3\sigma+3-\nu}\frac{GM^{\sigma+1}}{r_c^\nu} , \qquad (108)$$

where r_c is defined as in Theorem 8.

Proof: Since

$$r^\nu = \left[\frac{M(r)}{\frac{4}{3}\pi\bar{\rho}(r)}\right]^{\nu/3} , \qquad (109)$$

we have from (107) that

$$I_{\sigma,\nu} = G(\tfrac{4}{3}\pi)^{\nu/3}\int_0^R \bar{\rho}^{\nu/3}(r)M^{(3\sigma-\nu)/3}(r)dM(r) . \qquad (110)$$

Arguing as in Theorem 8, we easily obtain the inequality (108).

14. THEOREM 11.—*If* $\bar{P}$ *is the mean pressure defined by*

$$M\bar{P} = \int_0^R P dM(r) , \tag{111}$$

then, under the conditions of Theorem 6,

$$\frac{3}{20\pi}\frac{GM^2}{r_c^4} \geqslant \bar{P} \geqslant \frac{3}{20\pi}\frac{GM^2}{R^4} . \tag{112}$$

Proof: By definition

$$M\bar{P} = \int_0^R P dM(r) , \tag{113}$$

or, integrating by parts,

$$M\bar{P} = -\int_0^R M(r) dP . \tag{114}$$

Since (Eq. [18])

$$dP = -\frac{GM(r)}{4\pi r^4} dM(r) , \tag{115}$$

we can re-write (114) as

$$M\bar{P} = \frac{G}{4\pi}\int_0^R \frac{M^2(r)dM(r)}{r^4} , \tag{116}$$

or, in terms of the integral $I_{\sigma, \nu}$, introduced in Theorem 10, we have

$$M\bar{P} = \frac{1}{4\pi} I_{2, 4} . \tag{117}$$

By Theorem 10 we then have

$$\frac{3}{20\pi}\frac{GM^2}{r_c^4} \geqslant \bar{P} \geqslant \frac{3}{20\pi}\frac{GM^2}{R^4} , \tag{118}$$

which proves the theorem.

Numerically we have

$$\bar{P} \geqslant 5.4 \times 10^8 \left(\frac{M}{\odot}\right)^2 \left(\frac{R_\odot}{R}\right)^4 \text{atmospheres} . \tag{119}$$

15. The physical content of Theorems 6, 8, 9, 10, and 11 is the following: We are given a certain equilibrium configuration of mass M and radius R with some arbitrary density distribution, arbitrary except for the condition that the density $\bar{\rho}(r)$ does not increase outward. From the given configuration we can construct two other configurations of uniform density—one with a constant density equal to $\bar{\rho}$, and the other with a constant density equal to ρ_c (see Fig. 5). The radii of these two configurations are clearly R and r_c,

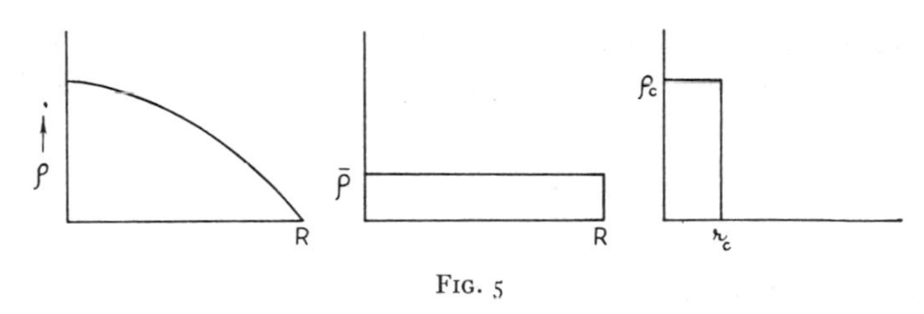

FIG. 5

respectively. Theorems 6, 8, 9, 10, and 11 simply state that the physical variables characterizing the given equilibrium configuration, namely, P_c, $-\Omega$, $\bar{g}$, $\overline{T}$ (for the case of negligible radiation pressure), and $\overline{P}$, have values respectively less than those for the configuration of uniform density with $\rho = \rho_c$, and respectively greater than those for the configuration of uniform density with $\rho = \bar{\rho}$. Thus, the given configuration is, in this sense, intermediate between the two configurations of uniform density with $\rho = \rho_c$ and $\rho = \bar{\rho}$, respectively.

16. THEOREM 12.—*Under the conditions of Theorem 6 we have*

$$\frac{P_c}{\rho_c^{\nu/3}} \leqslant \frac{1}{6-\nu} \left(\tfrac{4}{3}\pi\right)^{(\nu-3)/3} G R^{\nu-4} M^{(6-\nu)/3} , \tag{120}$$

provided

$$6 > \nu \geqslant 4 , \tag{121}$$

where (120) is a strict inequality for $\nu > 4$.

Proof: Consider the integral I_ν:

$$I_\nu = \int_0^R \frac{GM(r)dM(r)}{r^\nu} . \tag{122}$$

By (18) we can transform this into

$$I_\nu = -4\pi \int_0^R \frac{dP}{r^{\nu-4}} . \tag{123}$$

Since we have assumed that $\nu \geqslant 4$, we clearly have

$$I_\nu \geqslant -\frac{4\pi}{R^{\nu-4}} \int_0^R dP = \frac{4\pi}{R^{\nu-4}} P_c . \tag{124}$$

In (124) we have the equality sign for the case $\nu = 4$. For $\nu > 4$ we have a strict inequality. On the other hand, by Theorem 8 (Eq. [98]) we have

$$I_\nu \leqslant \frac{3}{6-\nu} (\tfrac{4}{3}\pi\rho_c)^{\nu/3} G M^{(6-\nu)/3} \qquad (\nu < 6) . \tag{125}$$

Combining (124) and (125), we have

$$\frac{4\pi}{R^{\nu-4}} P_c \leqslant I_\nu \leqslant \frac{3}{6-\nu} (\tfrac{4}{3}\pi\rho_c)^{\nu/3} G M^{(6-\nu)/3} , \tag{126}$$

or

$$\frac{P_c}{\rho_c^{\nu/3}} \leqslant \frac{1}{6-\nu} (\tfrac{4}{3}\pi)^{(\nu-3)/3} G R^{\nu-4} M^{(6-\nu)/3} . \tag{127}$$

Again, (127) is a strict inequality for $\nu > 4$. This proves the theorem.

If we write

$$\nu = 3\left(1 + \frac{1}{n}\right) , \tag{128}$$

equation (127) reduces to

$$\frac{P_c}{\rho_c^{(n+1)/n}} \leqslant S_n G R^{(3-n)/n} M^{(n-1)/n} \tag{129}$$

if

$$1 < n \leqslant 3 . \tag{130}$$

Further, S_n introduced in (122) stands for the numerical coefficient

$$S_n = (\tfrac{4}{3}\pi)^{1/n} \frac{n}{3(n-1)} . \tag{131}$$

Finally, (129) is a strict inequality for $n < 3$. Equations (129) and (131) bring out clearly the critical nature of $n = 1$ and $n = 3$, a circumstance we shall again encounter in the future. Table 3 gives the numerical values of S_n for different values of n.

TABLE 3

VALUES OF S_n

n	S_n	n	S_n
3.0	0.806	2.0	1.364
2.5	0.985	1.5	2.599

17. *Homologous transformations.—A general homologous transformation is one in which the density and the linear dimensions at each point are multiplied by constant factors to obtain another equilibrium configuration.*

The general homologous transformation is best considered as being "built" up of two elementary homologous transformations: (a) the transformation in which the radial dimensions are kept unaltered while the density at each point is multiplied by a constant factor x; (b) the transformation in which the configuration is submitted to a uniform expansion or contraction (in the sense already defined in chap. ii, § 7) when the radial dimensions are altered in the ratio $1:y$.

We shall prove the following theorem:

THEOREM 13.—*If the radiation pressure is a fraction* $(1 - \beta_0)$ *of the total pressure at a given point in an equilibrium configuration, and if it is a fraction* $(1 - \beta_1)$ *at the corresponding point in a homologously transformed configuration, then*

$$\frac{1 - \beta_1}{\mu_1^4 \beta_1^4} = \frac{1 - \beta_0}{\mu_0^4 \beta_0^4} \left(\frac{M_1}{M_0} \right)^2 , \tag{132}$$

where M_0 *and* M_1 *refer to the mass of the configuration before and after the homologous transformation.*

Proof: We shall consider the homologous transformation as built up of two elementary homologous transformations, as already explained.

First let us consider the homologous transformation in which the radial dimensions are unaltered. Then the density, ρ, and the mass interior to r, $M(r)$, are each multiplied by x. From the equation

$$\frac{dP}{dr} = -\frac{GM(r)}{r^2}\rho \tag{133}$$

we see that P gets multiplied by x^2. But according to equation (87),

$$P = \left[\left(\frac{k}{\mu H}\right)^4 \frac{3}{a} \frac{1-\beta}{\beta^4}\right]^{1/3} \rho^{4/3} . \tag{134}$$

Hence, the left-hand side of the foregoing equation gets multiplied by x^2 while $\rho^{4/3}$ on the right-hand side gets multiplied by $x^{4/3}$. Hence, the term involving β must get multiplied by $x^{2/3}$. In other words,

$$\frac{1-\beta_1}{\mu_1^4\beta_1^4} = \frac{1-\beta_0}{\mu_0^4\beta_0^4} x^2 = \frac{1-\beta_0}{\mu_0^4\beta_0^4}\left(\frac{M_1}{M_0}\right)^2 . \tag{135}$$

Now, consider a uniform expansion, in which the linear dimensions are increased in the ratio $1:y$. As shown in chapter ii, § 7, the effect of this transformation is to multiply P and ρ by y^{-4} and y^{-3}, respectively. From (134) we now see that for this transformation the left-hand side gets multiplied by y^{-4}, while $\rho^{4/3}$ on the right-hand side also gets multiplied by y^{-4}. Hence, $(1 - \beta)/\beta^4\mu^4$ is invariant to this transformation. Hence, (135) is true for a general homologous transformation.

The foregoing theorem is of importance in the theory of gaseous stars in so far as it shows that, if we consider a sequence of homologous gaseous configurations in equilibrium, then the relative importance of the radiation pressure—as measured by $1-\beta$—increases in the direction of increasing mass along the sequence.

BIBLIOGRAPHICAL NOTES

Integral theorems have been considered by Ritter, Eddington, Milne, and Chandrasekhar.

A. Ritter, *Widemann Annalen*, **16**, 183, 1882. The inequality for Ω (Theorem 3, Corollary 1) and Theorem 4 are proved here.

A. S. Eddington, *The Internal Constitution of the Stars*, pp. 90–94, Cambridge, England. The inequality (73) is stated (and proved by means of physical arguments). Theorem 9 is also considered, though the proof given in the text is different from Eddington's.

E. A. MILNE, *M.N.*, **89**, 739, 1929; *ibid.*, **96**, 179, 1936. Theorems 1, 2, and 3 are proved here, while parts of Theorems 6 and 8 are proved by different methods.

S. CHANDRASEKHAR, *M.N.*, **96**, 644, 1936; *Ap. J.*, **85**, 372, 1937. Theorems 6, 7, 8, 10, 11, and 12 are proved in these papers. Theorem 5 is proved here for the first time.

CHAPTER IV

POLYTROPIC AND ISOTHERMAL GAS SPHERES

In the last chapter we considered the most general properties of equilibrium configurations. In this chapter we shall be concerned with the detailed study of a class of equilibrium configurations resulting from a special kind of relation between P and ρ. Formally, the fundamental problem is the study of equilibrium configurations in which P and ρ are connected by a relation of the kind

$$P = K\rho^{(n+1)/n} , \tag{1}$$

where K and n are constants. This problem, toward the solution of which fundamental contributions have been made by Lane, Ritter, Kelvin, Emden, and Fowler, is also of considerable physical interest. We shall, therefore, first consider the physical circumstances which led initially to the study of the equilibrium configurations with an underlying "equation of state" of the kind (1).

1. *Convective and polytropic equilibrium.*—The physical notion of convective equilibrium was first introduced by Lord Kelvin in 1862 in connection with some of his considerations relating to the temperature of the earth's atmosphere.[1] Kelvin defined convective equilibrium in the following terms:

> Any fluid under the influence of gravity is said to be in convective equilibrium if the density and the temperature are so distributed throughout the whole fluid mass that the surfaces of equal density and of equal temperature remain unchanged when currents are produced in it by any disturbing influence so gentle that changes of pressure due to inertia of motions are negligible.[2]

Kelvin further comments that

> the essence of convective equilibrium is that if a small spherical or cubic portion of the fluid in any position, P, is ideally enclosed in a sheath impermeable to

[1] Sir W. Thomson (Lord Kelvin), *Mathematical and Physical Papers*, **3**, 255–260, Cambridge, 1911.

[2] *Ibid.*, **5**, 254–283. The quotation is from p. 256.

heat and expanded or contracted to the density of the fluid at any other place P', its temperature will be altered, by the expansion or contraction, from the temperature which it had at P to the actual temperature of the fluid at P'.

It is clear that the process considered is a quasi-statical adiabatic change, and consequently the equations to be used are (Eq. [19], ii)

$$p = \text{constant} \cdot \rho^{\gamma} \; ; \qquad p^{1-\gamma}T^{\gamma} = \text{constant} \; ; \qquad T\rho^{1-\gamma} = \text{constant} \; , \qquad (2)$$

where γ is the ratio of the specific heats. It is seen that the relation connecting p and ρ is of the form (1).

The gravitational equilibrium of a gaseous configuration in which p and ρ are related as in (2) was first considered by Lane[3] (1870), but the same problem was independently considered by Ritter[4] (1878) and also by Kelvin[5] (1887).

In applying the equations (2) of adiabatic expansion or contraction to a spherical mass of gas in convective equilibrium, Kelvin[6] makes the following interesting remarks:

If a gas is enclosed in a rigid spherical shell impermeable to heat and left to itself for a sufficiently long time, it settles into the condition of gross-thermal equilibrium by "conduction of heat" till the temperature becomes uniform throughout. But if it were stirred artificially all through its volume, currents not considerably disturbing the static distribution of pressure and density will bring it approximately to what I have called convective equilibrium of temperature. The *natural stirring* produced in a great free fluid mass like the Sun's by the cooling at the surface, must, I believe, maintain a somewhat close approximation to convective equilibrium throughout the whole mass.

It follows from Kelvin's remarks that we are entitled to use the equations (2) for an adiabatic expansion or contraction provided that during the process of "stirring" the appropriate $dQ = 0$. But this need not in general be the case. Indeed, in his very first application of the idea of convective equilibrium (to the earth's atmosphere, with a view to calculate the fall of temperature with height), Kelvin had to consider the case where the "stirring" led to a physical process in which $dQ \neq 0$. The difficulty arises from the circumstance

[3] J. Homer Lane, *Amer. J. Sci.*, 2d ser., **50**, 57, 1870.

[4] A. Ritter, *Wiedemann Annalen*, **6**, 135, 1878.

[5] W. Thomson, *Phil. Mag.*, **22**, 287, 1887.

[6] *Ibid.* Also Thomson's *Collected Papers*, **5**, 184–190. The quotation appears on p. 186.

that if we consider the "natural stirring" of a moist atmosphere the condensation of vapor in the upward currents of air is of considerable importance. This latter problem was also considered by Kelvin (at Joule's suggestion) and is, of course, of fundamental importance in meteorology. In modern versions[7] of Kelvin's work such changes are generally considered to be represented by the equation $dQ = cdT$, where c is taken to be approximately constant. More generally, if, during the process of stirring, the quantity of heat, dQ, supplied is proportional to the instantaneous change of temperature, dT, then $dQ = cdT$; this is the definition of a polytropic change. We then have

$$p = \text{constant} \cdot \rho^{\gamma'} ; \qquad \gamma' = \frac{c_p - c}{c_V - c} . \tag{3}$$

Hence, the consideration of polytropic changes is more general than the consideration of adiabatic changes; the latter is obtained as a special case when we put $c = 0$. For this reason Emden considered polytropic-convective equilibrium.[8]

If we use the variables introduced in chapter ii, § 5, we can write

$$\rho = \lambda\theta^n ; \qquad p = \frac{k}{\mu H}\Theta_{\gamma'}\lambda^{(n+1)/n}\theta^{n+1} ; \qquad n = \frac{1}{\gamma' - 1} , \tag{4}$$

where $\Theta_{\gamma'}$ is the polytropic temperature, which is, of course, the same for all parts of the gaseous sphere. For the adiabatic-convective case $\gamma' = \gamma$ and Θ_γ is the adiabatic temperature. Since in all these considerations radiation pressure has been neglected, we can write

$$P = K\rho^{1+(1/n)} ; \qquad K = \frac{k}{\mu H}\Theta_{\gamma'} . \tag{5}$$

We are thus led to consider the mathematical problem of determining the structure of an equilibrium configuration in which P and ρ are related according to equation (1); when we wish to con-

[7] See L. Weickmann, "Mechanik und Thermodynamik der Atmosphäre," in *Lehrbuch der Geophysik* herausgegeben von B. Guttenberg, pp. 797–965 (Berlin, 1929).

[8] K. Schwarzschild, *Vierteljahrsschrift der astronomischen Gesellschaft*, **43**, 26, 1908.

sider convective (either an adiabatic or a polytropic) equilibrium, the appropriate value of K is defined by (5). In treating the problem in this manner, i.e., in regarding (1) as an a priori relation, we gain the further advantage of being able to apply the analysis to cases where we have a relation of the type, (1), without any appeal to considerations of the kind which lead us to (4) and (5). It is necessary to keep in mind this possibility, and, indeed, when we come to consider radiative equilibrium in chapter vi we are led to equations of the type (1) from quite a different viewpoint. Further, in a given region inside a star we can often approximate the relation between P and ρ by a mononomial relation and the study of the equilibrium of such spherical shells will again lead us to essentially the same mathematical problems. In the first instance, however, we shall only be concerned with those equilibrium configurations in which a relation of the kind (1) is valid throughout the entire mass with a given constant value for K.

2. *The equations of equilibrium.*—The equations governing the equilibrium are (Eqs. [1], [4], [6], iii)

$$\frac{dP}{dr} = -\frac{GM(r)}{r^2}\rho \; ; \qquad \frac{dM(r)}{dr} = 4\pi r^2 \rho \, , \tag{6}$$

$$\frac{1}{r^2}\frac{d}{dr}\left(\frac{r^2}{\rho}\frac{dP}{dr}\right) = -4\pi G\rho \, , \tag{7}$$

where, according to our assumptions, we can write

$$\rho = \lambda\theta^n \; ; \qquad P = K\rho^{1+(1/n)} = K\lambda^{1+(1/n)}\theta^{n+1} \, , \tag{8}$$

where λ is, for the present, an arbitrary constant. Substituting (8) in (7), we find

$$\left[\frac{(n+1)K}{4\pi G}\lambda^{(1/n)-1}\right]\frac{1}{r^2}\frac{d}{dr}\left(r^2\frac{d\theta}{dr}\right) = -\theta^n \, . \tag{9}$$

We now introduce the dimensionless variable ξ, which is defined by

$$r = \alpha\xi \; ; \qquad \alpha = \left[\frac{(n+1)K}{4\pi G}\lambda^{(1/n)-1}\right]^{1/2} . \tag{10}$$

Equation (9) becomes

$$\frac{1}{\xi^2}\frac{d}{d\xi}\left(\xi^2\frac{d\theta}{d\xi}\right) = -\theta^n . \tag{11}$$

We shall refer to (11) as the "Lane-Emden equation of index n."

Equation (11) must govern the density distribution in any region where the relation (8) is valid;[9] the region of validity of (8) need not, of course, extend throughout the entire mass. We shall consider first, however, only *complete polytropes*, i.e., equilibrium configurations in which a relation of the kind (8) is valid for the entire mass. In that case we can choose λ to be equal to the central density ρ_c; we must then seek a solution of (11) which takes the value unity at the origin. Further, it is clear that $d\theta/d\xi$ must vanish at the origin; this easily follows from the first of the equations (6) and (8). Thus, with the "normalization" $\lambda = \rho_c$ we must seek a solution of (11) which satisfies the boundary conditions

$$\theta = 1 ; \qquad \frac{d\theta}{d\xi} = 0 \qquad \text{at} \qquad \xi = 0 . \tag{12}$$

We shall refer to the solution of (11) which satisfies the boundary conditions (12) as the "Lane-Emden function of index n," and denote it by θ_n. It is interesting to recall that Lord Kelvin referred to the function θ_n as "Homer Lane's function," "because he [Lane] first used it and expressed in terms of it all the features of a wholly gaseous spherical nebula in convective equilibrium and calculated it for the cases" $n = 1.5$ and $n = 2.5$.[10]

The problem, then, is to solve the Lane-Emden equation, and in particular to find the Lane-Emden functions (for different values of the index, n) which satisfy the boundary conditions (12). We shall first consider the various transformations of the Lane-Emden equation which are useful in the discussion of the general solution.

[9] Eq. (8) may be written in a somewhat more general form:

$$\rho = \lambda\theta^n ; \qquad P = K\lambda^{(n+1)/n}\theta^{n+1} + D , \tag{8'}$$

where D is a constant. With (8′) we still have (9), (10), and (11).

[10] See W. Thomson, *Collected Papers*, **5**, 254–283. The quotation is taken from p. 266. Kelvin used κ instead of n to denote the index.

3. *Transformations of the Lane-Emden equation.*—
a) Put

$$\theta = \frac{\chi}{\xi} . \tag{13}$$

The equation (11) easily reduces to

$$\frac{d^2\chi}{d\xi^2} = -\frac{\chi^n}{\xi^{n-1}} . \tag{14}$$

b) *Kelvin's transformation.*—Instead of ξ, introduce the new variable x defined by

$$x = \frac{1}{\xi} ; \qquad \frac{d}{d\xi} = -x^2 \frac{d}{dx} . \tag{15}$$

The Lane-Emden equation now transforms into

$$x^4 \frac{d^2\theta}{dx^2} = -\theta^n . \tag{16}$$

c) *The singular solution for* n > 3.—We first ascertain whether (16) has a solution of the form

$$\theta = a x^{\bar{\omega}} \tag{17}$$

for a suitably chosen a and $\bar{\omega}$. Substituting (17) in (16), we have

$$a\bar{\omega}(\bar{\omega} - 1)x^{\bar{\omega}+2} = -a^n x^{n\bar{\omega}} , \tag{18}$$

an equation which must be valid for all values of x. Hence, we should have

$$\bar{\omega} + 2 = n\bar{\omega} ; \qquad a^{n-1} = \bar{\omega}(1 - \bar{\omega}) , \tag{19}$$

or

$$\bar{\omega} = \frac{2}{n-1} ; \qquad a = \left[\frac{2(n-3)}{(n-1)^2}\right]^{1/(n-1)} . \tag{20}$$

For $n > 3$, and $\bar{\omega} < 1$ we therefore have the singular solution

$$\theta_s = \left[\frac{2(n-3)}{(n-1)^2}\right]^{1/(n-1)} x^{2/(n-1)} , \tag{21}$$

or, in terms of ξ,

$$\theta_s = \left[\frac{2(n-3)}{(n-1)^2}\right]^{1/(n-1)} \frac{1}{\xi^{2/(n-1)}} . \tag{22}$$

For $n \leqslant 3$ we have no proper singular solution of the type (17).

d) Emden's transformations.—Since (17) defines a solution of (16) (for $n > 3$), we make the substitution

$$\theta = A x^{\bar{\omega}} z ; \qquad \bar{\omega} = \frac{2}{n-1} , \tag{23}$$

where A is, for the present, an arbitrary constant, which we shall, specify later. From (23) we obtain

$$\frac{d^2\theta}{dx^2} = A\left[x^{\bar{\omega}} \frac{d^2 z}{dx^2} + 2\bar{\omega} x^{\bar{\omega}-1} \frac{dz}{dx} + \bar{\omega}(\bar{\omega}-1) x^{\bar{\omega}-2} z\right] . \tag{24}$$

Substituting (24) in (16) and using the relation $\bar{\omega} + 2 = n\bar{\omega}$, we have

$$x^2 \frac{d^2 z}{dx^2} + 2\bar{\omega} x \frac{dz}{dx} + \bar{\omega}(\bar{\omega}-1) z + A^{n-1} z^n = 0 . \tag{25}$$

We can eliminate x from the foregoing equation by making the further substitution

$$x = \frac{1}{\xi} = e^t ; \qquad t = \log x = -\log \xi . \tag{26}$$

From (26) we easily find that

$$\frac{dz}{dx} = e^{-t} \frac{dz}{dt} ; \qquad \frac{d^2 z}{dx^2} = e^{-2t}\left[\frac{d^2 z}{dt^2} - \frac{dz}{dt}\right] . \tag{27}$$

Substituting (27) in (25), we obtain

$$\frac{d^2 z}{dt^2} + (2\bar{\omega} - 1) \frac{dz}{dt} + \bar{\omega}(\bar{\omega}-1) z + A^{n-1} z^n = 0 . \tag{28}$$

We shall consider two forms of (28):

Case i: $n > 3$.—For $n > 3$ the singular solution (21) is proper, and we shall therefore choose $A = a$. By (19),

$$A^{n-1} = a^{n-1} = \bar{\omega}(1 - \bar{\omega}) \qquad (n > 3, \bar{\omega} < 1) . \tag{29}$$

Equation (28) now takes the form

$$\frac{d^2z}{dt^2} + (2\bar{\omega} - 1)\frac{dz}{dt} - \bar{\omega}(1 - \bar{\omega})z(1 - z^{n-1}) = 0\,, \tag{30}$$

or, since $\bar{\omega} = 2/(n-1)$, we have

$$\frac{d^2z}{dt^2} + \frac{5-n}{n-1}\frac{dz}{dt} - \frac{2(n-3)}{(n-1)^2}z(1 - z^{n-1}) = 0\,. \tag{31}$$

The singular solution (21) is defined by $z = 1$.

Case ii.—We choose $A = 1$. Our equation then is

$$\frac{d^2z}{dt^2} + \frac{5-n}{n-1}\frac{dz}{dt} + \frac{2(3-n)}{(n-1)^2}z + z^n = 0\,. \tag{32}$$

4. *The Lane-Emden functions for* n = 0, 1, *and* 5.—We shall consider these three cases separately.

Case i: n = 0.—The Lane-Emden equation is

$$\frac{1}{\xi^2}\frac{d}{d\xi}\left(\xi^2\frac{d\theta}{d\xi}\right) = -1\,, \tag{33}$$

which, after a first integration, yields

$$\xi^2\frac{d\theta}{d\xi} = -\tfrac{1}{3}\xi^3 - C\,, \tag{34}$$

where $-C$ is an integration constant. A second integration now yields

$$\theta = D + \frac{C}{\xi} - \tfrac{1}{6}\xi^2\,, \tag{35}$$

where D is a second integration constant.

We see that the general solution of (33) has a singularity at the origin, and that

$$\theta \sim \frac{C}{\xi} \qquad (\xi \to 0)\,. \tag{36}$$

If, however, we restrict ourselves to solutions which are finite at the origin, then $C = 0$ and

$$\theta = D - \tfrac{1}{6}\xi^2\,. \tag{37}$$

The Lane-Emden function is characterized by $\theta = 1$ at the origin, and hence the Lane-Emden function θ_0 is given by

$$\theta_0 = 1 - \tfrac{1}{6}\xi^2 . \tag{38}$$

The function θ_0 has its first zero at $\xi = \xi_1$, where

$$\xi_1 = \sqrt{6} ; \qquad \theta(\xi_1) = 0 . \tag{39}$$

Case ii: n = 1.—Consider the Lane-Emden equation in the form (14). Then, for $n = 1$

$$\frac{d^2\chi}{d\xi^2} = -\chi . \tag{40}$$

The general solution of (40) is

$$\chi = C \sin (\xi - \delta) , \tag{41}$$

where C and δ are constants of integration. By (13),

$$\theta = \frac{C \sin (\xi - \delta)}{\xi} . \tag{42}$$

If $\delta \neq 0$, the general solution has a singularity at the origin:

$$\theta \sim \frac{\text{constant}}{\xi} \qquad (\xi \to 0) . \tag{43}$$

Again, if we restrict ourselves to solutions which are finite at the origin, $\delta = 0$ and

$$\theta = \frac{C \sin \xi}{\xi} . \tag{44}$$

The solutions in the foregoing forms were first given by Ritter. The Lane-Emden function θ_1 is given by

$$\theta_1 = \frac{\sin \xi}{\xi} . \tag{45}$$

The foregoing function has its first zero at $\xi = \pi$ and is monotonically decreasing in the interval $(0, \pi)$.

Case iii: n = 5.—We shall consider the equation in the form (31) with $n = 5$. We have

$$\frac{d^2z}{dt^2} = \frac{1}{4}z(1 - z^4) , \tag{46}$$

where our variables z and t are, according to equations (22), (23), and (26),

$$\frac{1}{x} = \xi = e^{-t} ; \qquad \theta = \left(\frac{x}{2}\right)^{1/2} z = (\tfrac{1}{2}e^t)^{1/2} z . \tag{47}$$

Multiplying both sides of (46) by dz/dt, we have

$$\frac{1}{2}\frac{d}{dt}\left[\left(\frac{dz}{dt}\right)^2\right] = \frac{1}{4}z(1 - z^4)\frac{dz}{dt} , \tag{48}$$

which can be integrated as it stands:

$$\frac{1}{2}\left(\frac{dz}{dt}\right)^2 = \frac{1}{8}z^2 - \frac{1}{24}z^6 + D , \tag{49}$$

where D is a constant of integration. If $z \to \pm\infty$, then according to (49), $(dz/dt)^2 \to -\infty$, and this is impossible since dz/dt is real. We can therefore write

$$\frac{dz}{\pm[2D + \frac{1}{4}z^2 - \frac{1}{12}z^6]^{1/2}} = dt , \tag{50}$$

and z can at most oscillate to and fro between the greatest and the least roots of

$$2D + \tfrac{1}{4}z^2 - \tfrac{1}{12}z^6 = 0 . \tag{51}$$

The integration of (50) for nonzero D is complicated and involves elliptic integrals. The case of interest is, however, when $D = 0$. Then,

$$\frac{dz}{z(1 - \frac{1}{3}z^4)^{1/2}} = -\tfrac{1}{2}dt , \tag{52}$$

where the ambiguous sign has been so chosen that $t \to \infty$. Make the substitution

$$\tfrac{1}{3}z^4 = \sin^2\zeta , \tag{53}$$

from which we have

$$4\frac{dz}{z} = 2\frac{\cos\zeta}{\sin\zeta}d\zeta\,, \tag{54}$$

and (52) becomes

$$\operatorname{cosec}\zeta d\zeta = -dt\,, \tag{55}$$

which can be integrated. We obtain

$$\tan\tfrac{1}{2}\zeta = Ce^{-t}\,, \tag{56}$$

where C is a constant of integration. From (53) we now have

$$\tfrac{1}{3}z^4 = \frac{4\tan^2\frac{1}{2}\zeta}{(1+\tan^2\frac{1}{2}\zeta)^2}\,, \tag{57}$$

or from (56)

$$z = \pm\left[\frac{12C^2e^{-2t}}{(1+C^2e^{-2t})^2}\right]^{1/4}. \tag{58}$$

By (47), then,

$$\theta = \left[\frac{3C^2}{(1+C^2\xi^2)^2}\right]^{1/4}. \tag{59}$$

The Lane-Emden function θ_5 is therefore given by

$$\theta_5 = \frac{1}{(1+\frac{1}{3}\xi^2)^{1/2}}\,, \tag{60}$$

which was independently discovered by Schuster and Emden. We see that θ_5 is a decreasing function and tends to zero only as $\xi \to \infty$, which means that the corresponding equilibrium configuration extends to infinity.

5. *The Lane-Emden functions for general* n.—We have seen that the Lane-Emden function can be explicitly given for $n = 0, 1$, and 5. Such explicit expressions for other values of n do not seem to exist, and recourse must be had to numerical methods. A method of constructing the Lane-Emden function would be to start with a series expansion near the origin. We assume a series of the form

$$\theta = 1 + c\xi^2 + d\xi^4 + \ldots\ldots \tag{61}$$

The series is thus chosen in order that the boundary conditions (12) be satisfied; there can clearly be no term in ξ, since $d\theta/d\xi$ has to vanish at the origin and consequently the series can contain only terms of even powers in ξ. By substituting the foregoing series in the Lane-Emden equation and equating the coefficients of like powers in ξ, we can successively determine the coefficients $c, d, \ldots\ldots$ Thus, the series including the first three terms is found to be

$$\theta = 1 - \tfrac{1}{6}\xi^2 + \frac{n}{120}\xi^4 - \ldots\ldots \tag{62}$$

By taking a sufficient number of terms in such a series we can calculate the values of θ for $\xi < 1$ to any required degree of accuracy. For $\xi > 1$ the solution can then be continued by means of standard numerical methods.

The solution so constructed monotonically decreases from the center, and for $n < 5$ has a zero for some finite $\xi = \xi_1$ (say). At $\xi = \xi_1$, θ has its first zero, and thus the configuration has a definite boundary. As we have already seen for $n = 5$, the configuration extends to infinity; the same is true for $n > 5$, as we shall see in § 20.

Tables of the Lane-Emden functions, θ_n, are given in Emden's book (1907) for the values of $n = 0.5$, 1, 1.5, 2, 2.5, 3, 4, 4.5, 4.9, and 6. Tables of these functions were also computed by G. Green (1908) for $n = 1.5$, 2.5, 3, and 4; these tables formed an appendix to a paper by Lord Kelvin. Recently these functions have been computed very accurately for $n = 1.5$, 2, 2.5, 3, 3.5, 4, and 4.5 by D. H. Sadler and J. C. P. Miller.

Table 4 gives the values of ξ_1 and of certain other functions (involving $d\theta/d\xi$ and ξ) at ξ_1 which are of interest.

6. *Physical characteristics.*—We thus see that if the Lane-Emden function is known, then we can construct for a fixed value of K (i.e., for a given polytropic temperature if we are considering convective-polytropic equilibrium) a one-parametric family of configurations by allowing λ to vary continuously. Before we proceed to show how the Lane-Emden functions are to be used in practice, we shall first derive some necessary formulae.

TABLE 4

THE CONSTANTS OF THE LANE-EMDEN FUNCTIONS*

n	ξ_1	$-\xi_1^2\left(\frac{d\theta_n}{d\xi}\right)_{\xi=\xi_1}$	$\rho_c/\bar{\rho}$	${}_0\omega_n = -\xi_1^{\frac{n+1}{n-1}}\left(\frac{d\theta_n}{d\xi}\right)_{\xi=\xi_1}$	N_n	W_n	$-\frac{1}{(n+1)\xi_1\left(\frac{d\theta_n}{d\xi}\right)_{\xi=\xi_1}}$
0	2.4494	4.8988	1.0000	0.33333	....	0.119366	0.5
0.5	2.7528	3.7871	1.8361	0.02156	2.270	0.26227	0.53847
1.0	3.14159	3.14159	3.28987	....	0.63662	0.392699	0.5
1.5	3.65375	2.71406	5.99071	132.3843	0.42422	0.770140	0.53849
2.0	4.35287	2.41105	11.40254	10.4950	0.36475	1.63818	0.60180
2.5	5.35528	2.18720	23.40646	3.82662	0.35150	3.90906	0.69956
3.0	6.89685	2.01824	54.1825	2.01824	0.36394	11.05066	0.85432
3.25	8.01894	1.94980	88.153	1.54716	0.37898	20.365	0.96769
3.5	9.53581	1.89056	152.884	1.20426	0.40104	40.9098	1.12087
4.0	14.97155	1.79723	622.408	0.729202	0.47720	247.558	1.66606
4.5	31.83646	1.73780	6189.47	0.394356	0.65798	4922.125	3.33100
4.9	169.47	1.7355	934800	0.14239	1.340	3.693×10^6	16.550
5.0	∞	1.73205	∞	0	∞	∞	∞

* The values for $n = 0.5$ and 4.9 are computed from Emden's integrations of θ_n; for $n = 3.25$ an unpublished integration by Chandrasekhar has been used. $n = 5$ corresponds to the Schuster-Emden integral. For the other values of n the *British Association Tables*, Vol. II, has been used.

a) *Radius.*—The radius R of the star is given by (cf. Eq[10]):

$$R = a\xi_1 = \left[\frac{(n+1)K}{4\pi G}\right]^{1/2} \lambda^{(1-n)/2n} \xi_1 , \tag{63}$$

where ξ_1 defines the first zero of θ_n.

The value $n = 1$ is a critical case, for if $n = 1$, $\xi_1 = \pi$, and we have

$$R = \left[\frac{K}{2\pi G}\right]^{1/2} \pi \qquad (n = 1) , \tag{64}$$

which is independent of λ. Hence, the radius of a polytrope of index 1 depends only on K, and is independent of the central density λ. If we are considering a configuration in convective-polytropic equilibrium, the result shows that the radius of a polytrope of index 1 depends only on its polytropic temperature.

Further, it is clear that for $n = 5$, $R = \infty$ for all finite values of λ.

b) *The mass relation.*—The mass $M(\xi)$ interior to ξ is given by

$$M(\xi) = \int_0^{a\xi} 4\pi\rho r^2 dr = 4\pi a^3 \lambda \int_0^{\xi} \xi^2 \theta^n d\xi , \tag{65}$$

or, using (11),

$$M(\xi) = -4\pi a^3 \lambda \int_0^{\xi} \frac{d}{d\xi}\left(\xi^2 \frac{d\theta}{d\xi}\right) d\xi , \tag{66}$$

or

$$M(\xi) = -4\pi a^3 \lambda \, \xi^2 \frac{d\theta}{d\xi} ; \tag{67}$$

substituting for a (Eq. [10]), we have

$$M(\xi) = -4\pi \left[\frac{(n+1)K}{4\pi G}\right]^{3/2} \lambda^{(3-n)/2n} \left(\xi^2 \frac{d\theta}{d\xi}\right) . \tag{68}$$

The total mass, M, of the configuration is given by

$$M = -4\pi \left[\frac{(n+1)K}{4\pi G}\right]^{3/2} \lambda^{(3-n)/2n} \left(\xi^2 \frac{d\theta_n}{d\xi}\right)_{\xi=\xi_1} . \tag{69}$$

The value $n = 3$ is also a critical case, for, when $n = 3$,

$$M = -4\pi \left[\frac{K}{\pi G}\right]^{3/2} \left(\xi^2 \frac{d\theta_3}{d\xi}\right)_{\xi=\xi_1} \qquad (n = 3) . \tag{70}$$

We thus see that the mass of the configuration depends only on K and is independent of λ. For the convective-polytropic case this shows that the mass of a polytrope of index 3 depends only on its polytropic temperature.

We notice, further, that when $n = 5$ the mass is finite, though the configuration extends to infinity, for, according to the Schuster-Emden expression for θ_5, we find that

$$\lim_{\xi \to \infty} \left(-\xi^2 \frac{d\theta_5}{d\xi} \right) = \sqrt{3} \, . \tag{71}$$

The values of $[-\xi^2 d\theta_n/d\xi]_1$ for different values of n are given in Table 4.

c) The mass-radius relation.—Eliminating λ between (63) and (69), we have

$$GM^{(n-1)/n} R^{(3-n)/n} = \frac{(n+1)K}{(4\pi)^{1/n}} \left[-\xi^{(n+1)/(n-1)} \frac{d\theta_n}{d\xi} \right]^{(n-1)/n}_{\xi=\xi_1} . \tag{72}$$

We shall denote by ${}_0\omega_n$ the quantity

$${}_0\omega_n = -\xi_1^{(n+1)/(n-1)} \left(\frac{d\theta_n}{d\xi} \right)_{\xi=\xi_1} . \tag{73}$$

We can re-write (72) as

$$K = N_n G M^{(n-1)/n} R^{(3-n)/n} , \tag{74}$$

where N_n stands for the numerical coefficient

$$N_n = \frac{1}{n+1} \left[\frac{4\pi}{{}_0\omega_n^{n-1}} \right]^{1/n} . \tag{75}$$

The coefficients N_n are tabulated in Table 4. For the convective polytropic case, equation (74) is used to evaluate the polytropic temperature for a configuration known to be a polytrope of a specified index n of given mass, M, and radius, R.

d) The ratio of the mean to the central density.—Let $\bar{\rho}(\xi)$ denote the mean density of matter interior to $r = \alpha\xi$. Then,

$$\bar{\rho}(\xi) = \frac{M(\xi)}{\frac{4}{3}\pi \alpha^3 \xi^3} , \tag{76}$$

or, by (67),

$$\bar{\rho}(\xi) = -\frac{3}{\xi}\left(\frac{d\theta}{d\xi}\right)\lambda , \tag{77}$$

or, since λ is the central density, we have

$$\rho_c = \lambda = -\left[\frac{\xi}{3}\frac{1}{\dfrac{d\theta_n}{d\xi}}\right]_{\xi=\xi_1}\bar{\rho} . \tag{78}$$

Relation (78) shows that for a polytrope of a given index, n, the central density is a definite multiple of the mean density. The factor by which we have to multiply the mean density to obtain the central density for different values of n are given in Table 4, column 4. This column, incidentally, brings out a very important feature of the polytropes as a class. The comparatively small range of n (where $0 \leqslant n \leqslant 5$) includes a variety of density-distributions, including the two limiting cases of the uniform distribution of density and the infinite concentration of the mass toward the center.

e) The central pressure.—Since $\theta_n = 1$ at $\xi = 0$, we have

$$P_c = K\lambda^{(n+1)/n} . \tag{79}$$

Substituting (74) and (78) for K and λ, respectively, we obtain, after some minor transformations,

$$P_c = W_n \frac{GM^2}{R^4} , \tag{80}$$

where W_n stands for the quantity

$$W_n = \frac{1}{4\pi(n+1)\left[\left(\dfrac{d\theta_n}{d\xi}\right)_{\xi=\xi_1}\right]^2} . \tag{81}$$

The values of W_n are given in Table 4.

We are now in a position to see how a knowledge of the Lane-Emden functions enables us to determine the complete march of P, ρ, etc., in an equilibrium configuration of a given mass, M, and radius, R, and known to be a polytrope of a specified index, n. From our definitions it is clear that θ^n and θ^{n+1} give the density and the pressure in the scales in which ρ_c and P_c are regarded as units. For a given M and R we can calculate the mean density, $\bar{\rho}$, and (78) enables us to calculate ρ_c. In the same way, equation (80) enables us to calculate P_c. Finally, equation (74) determines K. Equations (68) and (77) then describe further features of the configuration.

7. *The potential energy.*—Let V be the potential. Then, according to equation (8), chapter iii,

$$\frac{1}{\rho}\frac{dP}{dr} = -\frac{dV}{dr}. \tag{82}$$

From $P = K\rho^{(n+1)/n}$ we easily find that

$$\frac{1}{\rho}\frac{dP}{dr} = (n+1)\frac{d}{dr}\left(\frac{P}{\rho}\right). \tag{83}$$

By (82) and (83), we have

$$(n+1)\frac{P}{\rho} = -V + V_1, \tag{84}$$

where V_1 is the potential at the boundary. By equation (10), chapter ii, equation (84) can also be written as

$$-V = (n+1)\frac{P}{\rho} + \frac{GM}{R}. \tag{85}$$

Again, by equation (16), chapter ii, the potential energy, Ω, is given by

$$\Omega = \frac{1}{2}\int_0^R V dM(r), \tag{86}$$

or, substituting for V its value (85), we have

$$-\Omega = \tfrac{1}{2}(n+1)\int_0^R \frac{P}{\rho}\,dM(r) + \frac{1}{2}\frac{GM}{R}\int_0^R dM(r), \tag{87}$$

or, if dV is the volume element,

$$-\Omega = \tfrac{1}{2}(n+1)\int_0^R P dV + \frac{1}{2}\frac{GM^2}{R}. \tag{88}$$

Again, by equation (32), chapter iii, we have

$$-\Omega = -\tfrac{1}{6}(n + 1)\Omega + \frac{1}{2}\frac{GM^2}{R} , \tag{89}$$

or, finally,

$$-\Omega = \frac{3}{5 - n}\frac{GM^2}{R} , \tag{90}$$

a formula due to Betti and Ritter. We see that $-\Omega$ is "infinite" for $n = 5$; the reason for this is that the polytrope of index $n = 5$ of finite radius is infinitely "concentrated" toward the center.

If we consider a gaseous equilibrium configuration (in which the radiation pressure is negligible), then according to Ritter's relation, equation (55), chapter iii, the internal energy U is given by

$$U = -\frac{1}{3(\gamma - 1)}\Omega . \tag{91}$$

If the configuration is, further, a polytrope of index n, according (90) and (91), we have

$$U = \frac{1}{(5 - n)(\gamma - 1)}\frac{GM^2}{R} . \tag{92}$$

Finally, if we consider the case of convective-adiabatic equilibrium, then

$$n = \frac{1}{\gamma - 1} , \tag{93}$$

and we have

$$-\Omega = \frac{3(\gamma - 1)}{(5\gamma - 6)}\frac{GM^2}{R} , \tag{94}$$

$$U = \frac{1}{(5\gamma - 6)}\frac{GM^2}{R} . \tag{95}$$

The relations (94) and (95) are due to Betti and Ritter.

8. *The homology theorem and homology invariant functions.*—We shall first prove the following theorem: *If* $\theta(\xi)$ *is a solution of the Lane-Emden equation of index* n, *then* $A^{2/(n-1)}\theta(A\xi)$ *is also a solution of the equation, where* A *is an arbitrary real number.*

Let $\theta = f(\xi)$ be a solution of the Lane-Emden equation. We have to show that $A^{2/(n-1)}f(A\xi)$ is also a solution. To prove this, write

$$\eta = A\xi\ ; \qquad \theta = A^{2/(n-1)}\,\phi \tag{96}$$

in the Lane-Emden equation. We then find

$$\frac{1}{\eta^2}\frac{d}{d\eta}\left(\eta^2\frac{d\phi}{d\eta}\right) = -\phi^n\ , \tag{97}$$

which is identical in form with the original equation in the θ, ξ variables. We can therefore choose for ϕ the solution $f(\eta)$. According to (96), this solution in θ, ξ variables is

$$\theta^*(\xi) = A^{2/(n-1)}f(\eta) = A^{2/(n-1)}f(A\xi)\ , \tag{98}$$

which proves the theorem.

Whenever a differential equation has the property that from one given solution a whole class of solutions can be derived by a simple change of the "scale," as above, then the differential equation is said to "admit of a homology transformation," and the constant (such as A, above) used in such a transformation is called a "homology constant."

Suppose we choose for $f(\xi)$ the Lane-Emden function $\theta_n(\xi)$. Then from this one function we can construct a whole class of solutions which we shall denote by $\theta_E(\xi)$; these are obtained from the Lane-Emden function by means of the transformation

$$\theta_E(\xi) = A^{\bar{\omega}}\theta_n(A\xi) \qquad \left(\bar{\omega} = \frac{2}{n-1}\right), \tag{99}$$

or

$$\theta_E(\xi/A) = A^{\bar{\omega}}\theta_n(\xi)\ , \tag{100}$$

where A is an arbitrary constant. In words, the value of $\theta_E(\xi)$ at ξ is $A^{\bar{\omega}}$ times the value which $\theta_n(\xi)$ takes at the point $A\xi$. In particular, if ξ_1 defines the zero of θ_n, then ξ_1/A defines the zero of θ_E defined according to (99).

We thus see that from one solution of the Lane-Emden equation a whole continuous family of solutions can be derived. We shall denote by $\{\theta(\xi)\}$ all solutions which can be transformed, one into the other, by means of the homologous transformation $\theta(\xi) \to A^{\bar{\omega}}\theta(A\xi)$. $\{\theta(\xi)\}$ is then said to define a "homologous family" of solutions. Thus, $\{\theta_n(\xi)\}$ defines a homologous family of solutions which are all finite at the origin and which further have $d\theta/d\xi = 0$ at the origin. We may say that the one boundary condition $d\theta/d\xi = 0$ at $\xi = 0$ defines the family of solutions $\{\theta_n(\xi)\}$. We shall see presently (§ 9) that the condition that θ shall be finite at $\xi = 0$ already defines the family $\{\theta_n(\xi)\}$. We shall refer to solutions belonging to the family $\{\theta_n(\xi)\}$ as "*E*-solutions."

Now, the Lane-Emden equation is a differential equation of the second order, and consequently the general solution must be characterized by two integration constants. But, as we have seen above, one of the constants must be "trivial" in the sense that it merely defines the scale-factor A. It is clear, then, that we should be able to reduce the equation to one of the first order.

Thus the variables used in § 3(*d*) already enable us to reduce the Lane-Emden equation to one of the first order.

The variables chosen are (Eq. [23] with $A = 1$, and Eq. [26]):

$$\xi = e^{-t}\,; \qquad z = \xi^{\bar{\omega}}\theta\,; \qquad \bar{\omega} = \frac{2}{n-1}\,. \tag{101}$$

z then satisfies the differential equation (Eq. [28] with $A = 1$)

$$\frac{d^2z}{dt^2} + (2\bar{\omega} - 1)\frac{dz}{dt} + \bar{\omega}(\bar{\omega} - 1)z + z^n = 0\,. \tag{102}$$

We now introduce the new variable, y, defined by

$$y = \frac{dz}{dt}\,. \tag{103}$$

Then,

$$\frac{d^2z}{dt^2} = \frac{dy}{dt} = \frac{dy}{dz}\frac{dz}{dt} = y\frac{dy}{dz}\,. \tag{104}$$

Equation (102) then becomes

$$y\frac{dy}{dz} + (2\bar{\omega} - 1)y + \bar{\omega}(\bar{\omega} - 1)z + z^n = 0\,, \tag{105}$$

which is an equation of the first order. The reason for this reduction of the order of the equation is that the functions y and z, defined by

$$z = \xi^{\bar{\omega}}\theta , \tag{106}$$

and by

$$y = \frac{dz}{dt} = -\xi \frac{dz}{d\xi} = -\xi^{\bar{\omega}}\left(\bar{\omega}\theta + \xi \frac{d\theta}{d\xi}\right), \tag{107}$$

or

$$y = -\xi^{\bar{\omega}+1}\frac{d\theta}{d\xi} - \bar{\omega}z , \tag{108}$$

are both invariant to homologous transformations. To show this, let $\theta(\xi)$ be a solution of the Lane-Emden equation and let $z(\xi)$ be the corresponding function defined as in (106). Let further, $\theta^*(\xi)$ be obtained by applying a homologous transformation to $\theta(\xi)$, so that

$$\theta^*(\xi) = A^{\bar{\omega}}\theta(A\xi) . \tag{109}$$

Let $z^*(\xi)$ be the corresponding function defined as in (106). Consider the corresponding points ξ and ξ/A on the solution-curves θ and θ^*, respectively. Then

$$z^*(\xi/A) = (\xi/A)^{\bar{\omega}}\theta^*(\xi/A) , \tag{110}$$

or by (109)

$$z^*(\xi/A) = (\xi/A)^{\bar{\omega}}A^{\bar{\omega}}\theta(\xi) = \xi^{\bar{\omega}}\theta(\xi) = z(\xi) . \tag{111}$$

In words, there is a one-to-one correspondence and an equality between the set of values which z takes along a given solution and the set of values which it takes along a solution homologous to the original one. This proves that z is a homology-invariant function. To show that y is also homology invariant, it is sufficient to show that $\xi^{\bar{\omega}+1}d\theta/d\xi$ is homology invariant. As before, if we consider the corresponding points ξ and ξ/A along θ and θ^*, respectively, we have

$$\theta^*(\xi/A) = A^{\bar{\omega}}\theta(\xi) ; \qquad \left(\frac{d\theta^*}{d\xi}\right)_{\xi=\xi/A} = A^{\bar{\omega}+1}\left(\frac{d\theta}{d\xi}\right)_{\xi=\xi} . \tag{112}$$

Hence,

$$(\xi/A)^{\bar{\omega}+1}\left(\frac{d\theta^*}{d\xi}\right)_{\xi=\xi/A} = (\xi/A)^{\bar{\omega}+1} \cdot A^{\bar{\omega}+1}\left(\frac{d\theta}{d\xi}\right)_{\xi=\xi} , \tag{113}$$

or

$$(\xi/A)^{\bar{\omega}+1}\left(\frac{d\theta^*}{d\xi}\right)_{\xi=\xi/A} = \xi^{\bar{\omega}+1}\left(\frac{d\theta}{d\xi}\right)_{\xi=\xi}, \tag{114}$$

which proves that $\xi^{\bar{\omega}+1}d\theta/d\xi$ is homology invariant. Hence, as y and z are both homology-invariant functions, we have a first-order differential equation between them.

It is possible, of course, to construct other homology-invariant functions, and we can therefore derive an arbitrarily large number of first-order differential equations all equivalent to the Lane-Emden equation. As another example of such a first-order equation, we shall consider the following two functions, u and v, defined as

$$u = -\frac{\xi\theta^n}{\theta'}; \qquad v = -\frac{\xi\theta'}{\theta}, \tag{115}$$

where we have used θ' to denote $d\theta/d\xi$.

We can show that u and v are homology invariant by the same kind of reasoning that we adopted to prove the homology invariance of y and z. The first-order equation between u and v can be obtained as follows:

We have

$$\frac{1}{u}\frac{du}{d\xi} = \frac{1}{\xi} + \frac{n}{\theta}\theta' - \frac{\theta''}{\theta'}. \tag{116}$$

Since, according to the Lane-Emden equation,

$$\theta'' = -\theta^n - \frac{2}{\xi}\theta', \tag{117}$$

we can re-write (116) as

$$\frac{1}{u}\frac{du}{d\xi} = \frac{1}{\xi}\left[3 + n\frac{\xi\theta'}{\theta} + \frac{\xi\theta^n}{\theta'}\right], \tag{118}$$

or,

$$\frac{1}{u}\frac{du}{d\xi} = \frac{1}{\xi}(3 - nv - u). \tag{119}$$

Now we have

$$\frac{1}{v}\frac{dv}{d\xi} = \frac{1}{\xi} - \frac{\theta'}{\theta} + \frac{\theta''}{\theta'}, \tag{120}$$

or, again by (117), we find

$$\frac{1}{v}\frac{dv}{d\xi}=\frac{1}{\xi}(-1+u+v)\,. \tag{121}$$

From (119) and (121) we have

$$\frac{u}{v}\frac{dv}{du}=-\frac{u+v-1}{u+nv-3}\,, \tag{122}$$

which is the required first-order differential equation. We shall return to the foregoing equation in § 21.

We pass on now to a general discussion of the Lane-Emden equation. The fundamental problem is the following: If we prescribe the value θ_0 and its derivative θ_0' at a given point ξ_0, then the Lane-Emden equation specifies uniquely a solution-curve passing through the point (ξ_0, θ_0) and in the given direction prescribed by θ_0'. The problem is: What is the nature of such a solution for all the points (ξ_0, θ_0) in the (ξ, θ) plane and for all possible starting-slopes? In other words, what is the arrangement of the solutions of the Lane-Emden equation? We shall only be concerned with values of $n > 1$. For $n = 1$ the solution can be given explicitly, while for $n < 1$ there seem to be formal difficulties of a far deeper character than those encountered for $n > 1$. The solution for $n = 0$, however, is explicitly known.

9. *The* E-*solutions.*—We shall prove that *solutions of the Lane-Emden equation which are finite at the origin necessarily have* $d\theta/d\xi = 0$ *at* $\xi = 0$, and that, consequently, the homologous family $\{\theta_n(\xi)\}$ includes all the solutions which are finite at the origin.

Consider the Lane-Emden equation in the form (Eqs. [13] and [14])

$$\frac{d^2\chi}{d\xi^2}=-\frac{\chi^n}{\xi^{n-1}} \qquad (\chi=\theta\xi)\,. \tag{123}$$

Solutions which are finite at the origin in the (ξ, θ) plane correspond to solutions passing through the origin in the (χ, ξ) plane. We have

$$\frac{d\theta}{d\xi}=\frac{1}{\xi}\frac{d\chi}{d\xi}-\frac{\chi}{\xi^2}\,. \tag{124}$$

Hence,

$$\left(\frac{d\theta}{d\xi}\right)_{\xi=0} = \lim_{\xi=0}\left[\frac{\xi\dfrac{d\chi}{d\xi} - \chi}{\xi^2}\right]. \tag{125}$$

Since we are considering solutions passing through $\chi = 0$, $\xi = 0$, we can write

$$\chi(\xi) = \xi\left(\frac{d\chi}{d\xi}\right)_{\xi=0} + \frac{\xi^2}{2}\left(\frac{d^2\chi}{d\xi^2}\right)_{\xi=0} + \ldots, \tag{126}$$

$$\frac{d\chi}{d\xi} = \left(\frac{d\chi}{d\xi}\right)_{\xi=0} + \xi\left(\frac{d^2\chi}{d\xi^2}\right)_{\xi=0} + \ldots. \tag{127}$$

By (125), (126), and (127) we have

$$\left(\frac{d\theta}{d\xi}\right)_{\xi=0} = \frac{1}{2}\left(\frac{d^2\chi}{d\xi^2}\right)_{\xi=0}, \tag{128}$$

or, according to (123),

$$\left(\frac{d\theta}{d\xi}\right)_{\xi=0} = -\tfrac{1}{2}\lim_{\xi=0}\left[\left(\frac{\chi}{\xi}\right)^{n-1}\chi\right] = 0, \tag{129}$$

since $\chi/\xi = \theta$ is finite at the origin, and, further, $\chi = 0$ at $\xi = 0$ for the solutions considered. This proves the theorem.

10. *The* (y, z) *plane.*—We shall discuss the solution-curves in the (y, z) plane. The functions y and z, as we have shown, are homology-invariant functions, and consequently each solution-curve in the (y, z) plane corresponds to a complete homologous family of solutions in the (ξ, θ) plane. In particular, there is just one curve in the (y, z) plane which corresponds to the E-solutions which are included in the homologous family $\{\theta_n(\xi)\}$. We shall call the curve which corresponds to the family $\{\theta_n(\xi)\}$ the "Emden-curve," or the "E-curve," and denote it by $y_E(z)$.

To repeat, our equations are

$$y\frac{dy}{dz} + (2\bar{\omega} - 1)y + \bar{\omega}(\bar{\omega} - 1)z + z^n = 0, \tag{130}$$

where

$$z = \xi^{\bar{\omega}}\theta, \tag{131}$$

and

$$y = -\xi^{\bar{\omega}}(\bar{\omega}\theta + \xi\theta') = -\xi^{\bar{\omega}+1}\theta' - \bar{\omega}z\,. \tag{132}$$

Further,

$$y = \frac{dz}{dt}\,; \qquad \xi = e^{-t}\,. \tag{133}$$

We notice that, according to these transformations, different directions through a fixed point in the (ξ, θ) plane correspond to different points on a definite line parallel to the y-axis.

From the foregoing equations we can derive the following formulae, which we shall need.

From (132) we have

$$\theta' = -\xi^{-\bar{\omega}-1}(y + \bar{\omega}z)\,. \tag{134}$$

If we denote dy/dz by y', we have, according to (130),

$$y' = -\frac{(2\bar{\omega} - 1)y + \bar{\omega}(\bar{\omega} - 1)z + z^n}{y}\,, \tag{135}$$

or, substituting for z and y according to (131) and (132), we have

$$y' = -\frac{(2\bar{\omega} - 1)\xi\theta' + \bar{\omega}^2\theta - \xi^2\theta^n}{\xi\theta' + \bar{\omega}\theta}\,. \tag{136}$$

From (136), solving for θ', we have

$$\theta' = \frac{\xi^2\theta^n - \bar{\omega}^2\theta - \bar{\omega}\theta y'}{\xi(y' + 2\bar{\omega} - 1)}\,. \tag{137}$$

We see that the origin, $y = z = 0$ (which we shall call O_1), is a singular point of the equation (130), since, when $y = 0$ and $z = 0$, dy/dz is indeterminate. In the same way, if $n > 3$, $\bar{\omega} < 1$, the solution has another singular point:

$$y = 0\,, \qquad z = z_s = [\bar{\omega}(1 - \bar{\omega})]^{1/(n-1)} \tag{138}$$

We shall call this singular point O_2. The existence of this second singular point O_2 corresponds to the existence of the proper singular solutions that exist for $\bar{\omega} < 1$; for if $\bar{\omega} < 1$, then, as we have already seen (Eq. [22]),

$$\theta_s = [\bar{\omega}(1 - \bar{\omega})]^{\bar{\omega}/2}\xi^{-\bar{\omega}} \tag{139}$$

satisfies the Lane-Emden equation. Equation (139) is equivalent to

$$z_s = [\bar{\omega}(1 - \bar{\omega})]^{\bar{\omega}/2} \qquad (140)$$

in the (y, z) plane, which is identical with (138).

Finally, we have the following correspondence between the (y, z) and the (ξ, θ) planes. From (135) we see that along the y-axis $(z = 0)$

$$y' = -(2\bar{\omega} - 1) . \qquad (141)$$

But from (136), y' takes the value $-(2\bar{\omega} - 1)$ for $\theta = 0$. Hence, the y-axis corresponds to the ξ-axis.

11. *Behavior near the singular point* y = 0, z = 0.—It is clear that the E-curve which is characterized by $\theta' = 0$ at $\xi = 0$ in the (ξ, θ) plane must pass through the origin (see Eqs. [131] and [132]); and we have, according to (136),

$$(y'_E)_{O_1} = -\bar{\omega} = -\frac{2}{n - 1} . \qquad (142)$$

Hence, the E-curve touches the line $y + \bar{\omega}z = 0$ at the origin. On the other hand, we can show that there cannot be two solution-curves which are both tangential to $y + \bar{\omega}z = 0$ at the origin. To prove this, suppose y and y^* are two different solutions such that

$$y \sim -\bar{\omega}z ; \qquad y^* \sim -\bar{\omega}z \qquad (z \to 0) . \qquad (143)$$

We may, without loss of generality, assume that $y < y^*$ near the origin. Then we should have

$$\Delta = y^* - y > 0 ; \qquad \lim_{\substack{z=0 \\ \Delta=0}} \frac{\Delta}{z} = 0 . \qquad (144)$$

From the differential equation (130) we have

$$\frac{d\Delta}{dz} = [\bar{\omega}(\bar{\omega} - 1)z + z^n] \frac{\Delta}{yy^*} . \qquad (145)$$

Since $yy^* \sim \bar{\omega}^2 z^2$, we have from the foregoing that

$$\lim_{\substack{z=0 \\ \Delta=0}} \left(\frac{d \log \Delta}{d \log z}\right) = \lim_{z=0} \left[\frac{\bar{\omega}(\bar{\omega} - 1)z^2 + z^{n+1}}{yy^*}\right] , \qquad (146)$$

or

$$\lim_{\substack{z=0 \\ \Delta=0}} \left(\frac{d \log \Delta}{d \log z} \right) = \frac{\bar{\omega}(\bar{\omega} - 1)}{\bar{\omega}^2} . \tag{147}$$

But

$$\lim_{\substack{z=0 \\ \Delta=0}} \frac{d \log \Delta}{d \log z} = \lim_{\substack{z=0 \\ \Delta=0}} \frac{\log \Delta}{\log z} . \tag{148}$$

Combining (147) and (148), we have

$$\lim_{\substack{z=0 \\ \Delta=0}} \frac{\log \Delta}{\log z} = 1 - \frac{1}{\bar{\omega}} , \tag{149}$$

or

$$\lim_{\substack{z=0 \\ \Delta=0}} \frac{\log \dfrac{\Delta}{z}}{\log z} = -\frac{1}{\bar{\omega}} , \tag{150}$$

which leads to a contradiction since, according to (144), we should have a non-negative limit for the left-hand side of (150). This proves that *the E-curve is the only solution-curve which is tangential to* $y + \bar{\omega}z = 0$ *at the origin.*

The line $y + \bar{\omega}z = 0$ has further significance. By (134), along this line $\theta' = 0$ and $\theta' < 0$ only above the line $y + \bar{\omega}z = 0$. We shall refer to the direction $y + \bar{\omega}z = 0$ as the "*Y*-direction."

Now the origin O_1, as we have already seen, is a singular point of the differential equation; we shall now investigate whether the differential equation characterizes directions other than the *Y*-direction along which solutions can start at the origin.

From (135)

$$(y')_{y,\, z=0} = -\lim_{y,\, z=0} \left[(2\bar{\omega} - 1) + \bar{\omega}(\bar{\omega} - 1) \frac{z}{y} + \frac{z^n}{y} \right] , \tag{151}$$

which (since $n > 1$) is easily seen to be equivalent to

$$(y')_{y,\, z=0} = -(2\bar{\omega} - 1) - \frac{\bar{\omega}(\bar{\omega} - 1)}{(y')_{y,\, z=0}} , \tag{152}$$

or $(y')_{y,\, z=0}$ satisfies the equation

$$y'^2 + (2\bar{\omega} - 1)y' + \bar{\omega}(\bar{\omega} - 1) = 0 . \tag{153}$$

Hence,

$$y' = -\bar{\omega} \quad \text{or} \quad y' = -(\bar{\omega} - 1) . \tag{154}$$

Thus there is a second direction defined by $y = -(\bar{\omega} - 1)z$ to which solutions can be tangential at the origin. We shall refer to this second direction, $y + (\bar{\omega} - 1)z = 0$, as the "$X$-direction."

If we substitute $y' = -(\bar{\omega} - 1)$ in (137), we have, correspondingly,

$$\theta'_X = \left(\frac{\xi^2\theta^n - \bar{\omega}\theta}{\bar{\omega}\xi}\right)_{z=0} . \tag{155}$$

Since $z = \xi^{\bar{\omega}}\theta$, $z = 0$ implies $\xi^{\bar{\omega}}\theta = 0$ or since $\bar{\omega} = 2/(n - 1)$, $z = 0$ implies $\xi^2\theta^{n-1} = 0$. Hence, according to (155),

$$\theta'_X = -\frac{\theta}{\xi} . \tag{156}$$

If $z = 0$ corresponds to $\xi = 0$, then, from (156) it follows that if θ remains positive, $\theta' \to -\infty$ as $\xi \to 0$. On the other hand, from (124) it follows that $\theta' \to -\infty$, $\xi \to 0$, implies that $\chi(= \theta\xi)$ is finite at the origin or $\theta \to \infty$ as $\xi \to 0$. But this is true only if $z = 0$ when approached along the X-direction corresponds to $t \to +\infty$ or $\xi \to 0$. We shall see, however (§§ 19, 20), that under certain circumstances ($n \geqslant 5$) the origin O_1 approached along the X-direction corresponds to $t \to -\infty$ or $\xi \to \infty$.

12. *The case* $\bar{\omega} \neq 1$.—We shall now consider in greater detail the behavior of the solutions in the immediate neighborhood of the origin.

If we are in a sufficiently close neighborhood to the origin and if, further, $\bar{\omega} \neq 1$ (i.e., $n \neq 3$ but $n > 1$), we can write equation (130) as

$$y\frac{dy}{dz} + (2\bar{\omega} - 1)y + \bar{\omega}(\bar{\omega} - 1)z = 0 , \tag{157}$$

or, since $y = dz/dt$, we can re-write the foregoing as

$$\frac{d^2z}{dt^2} + (2\bar{\omega} - 1)\frac{dz}{dt} + \bar{\omega}(\bar{\omega} - 1)z = 0 . \tag{158}$$

The general solution of (158) is seen to be

$$z = Ae^{-\bar{\omega}t} + Be^{-(\bar{\omega}-1)t} , \tag{159}$$

where A and B are two integration constants. Since $y = dz/dt$, we have

$$y = -\bar{\omega}Ae^{-\bar{\omega}t} - (\bar{\omega} - 1)Be^{-(\bar{\omega}-1)t} . \tag{160}$$

From (159) and (160), we obtain

$$y + \bar{\omega}z = Be^{-(\bar{\omega}-1)t} , \tag{161}$$

$$y + (\bar{\omega} - 1)z = -Ae^{-\bar{\omega}t} . \tag{162}$$

From the foregoing, it follows that

$$[y + (\bar{\omega} - 1)z]^{\bar{\omega}-1} = C[y + \bar{\omega}z]^{\bar{\omega}} , \tag{163}$$

where C is a constant.

We shall choose the X and the Y directions as defined, respectively, by $y + (\bar{\omega} - 1)z = 0$ and $y + \bar{\omega}z = 0$, as defining a new frame of oblique system of axes. Let ϑ_X and ϑ_Y be the angles which the X and the Y directions make with the z-axis. Then

$$\tan \vartheta_X = -(\bar{\omega} - 1) ; \qquad \tan \vartheta_Y = -\bar{\omega} . \tag{164}$$

Let X and Y be the co-ordinates of a point with respect to the new system of axes. Then we have

$$z = X \cos \vartheta_X + Y \cos \vartheta_Y , \tag{165}$$

$$y = X \sin \vartheta_X + Y \sin \vartheta_Y . \tag{166}$$

From the foregoing, we find

$$X = \frac{y \cos \vartheta_Y - z \sin \vartheta_Y}{\sin (\vartheta_X - \vartheta_Y)} = \frac{\cos \vartheta_Y}{\sin (\vartheta_X - \vartheta_Y)} (y - z \tan \vartheta_Y) , \tag{167}$$

or by (164)

$$X = \frac{\cos \vartheta_Y}{\sin (\vartheta_X - \vartheta_Y)} (y + \bar{\omega}z) . \tag{168}$$

Similarly,

$$Y = -\frac{\cos \vartheta_X}{\sin (\vartheta_X - \vartheta_Y)} [y + (\bar{\omega} - 1)z] . \tag{169}$$

Hence, the solution (163) can be written as

$$Y^{\bar{\omega}-1} = CX^{\bar{\omega}} , \tag{170}$$

or

$$Y = CX^{\bar{\omega}/(\bar{\omega}-1)} . \tag{171}$$

Since $\bar{\omega} = 2/(n - 1)$, equation (171) can also be written as

$$Y = CX^{2/(3-n)} . \tag{172}$$

From (172) it follows that we have to distinguish again between the cases $\bar{\omega} > 1$ and $\bar{\omega} < 1$, i.e., $n < 3$ and $n > 3$; the case $n = 3$ requires special treatment.

13. *The case* $\bar{\omega} \geqslant 1$, $n \leqslant 3$.—The first part of the discussion for $\bar{\omega} > 1$ is valid also for the case $\bar{\omega} = 1$.

The differential equation can be written as

$$\dot{y} = \frac{dy}{dt} = -(2\bar{\omega} - 1)y - \bar{\omega}(\bar{\omega} - 1)z - z^n , \tag{173}$$

$$\dot{z} = \frac{dz}{dt} = y , \tag{174}$$

$$\frac{dy}{dz} = -(2\bar{\omega} - 1) - \frac{\bar{\omega}(\bar{\omega} - 1)z + z^n}{y} , \tag{175}$$

with

$$z = \xi^{\bar{\omega}}\theta = e^{-\bar{\omega}t}\theta ; \qquad y = -\xi^{\bar{\omega}+1}\frac{d\theta}{d\xi} - \bar{\omega}z . \tag{176}$$

Since we need to consider only $\theta \geqslant 0$, $z \geqslant 0$, we shall therefore restrict ourselves to a discussion of the solution-curves in the half-plane in which z is positive. Further, if $\bar{\omega} \geqslant 1$, then, in the half-plane considered there is only one singular point, namely, O_1. It is this last circumstance which makes the discussion relatively simple, for the solution-curves of (175) must form a one-parametric family of curves at all points in the half-plane considered, except at the singular point O_1.

From (173) and (174) it follows that along the z-axis, $\dot{z} = 0$ (or that the solution-curves, if they intersect the z-axis, must do so vertically) and that the locus of points at which $\dot{y}$ vanishes is given by

$$(2\bar{\omega} - 1)y = -\bar{\omega}(\bar{\omega} - 1)z - z^n , \qquad (177)$$

a curve which (since $\bar{\omega} \geqslant 1$) lies entirely in the lower quadrant.

In the three regions marked *I*, *II*, and *III* we have the signs of $\dot{y}$ and $\dot{z}$ as shown in Figure 6.

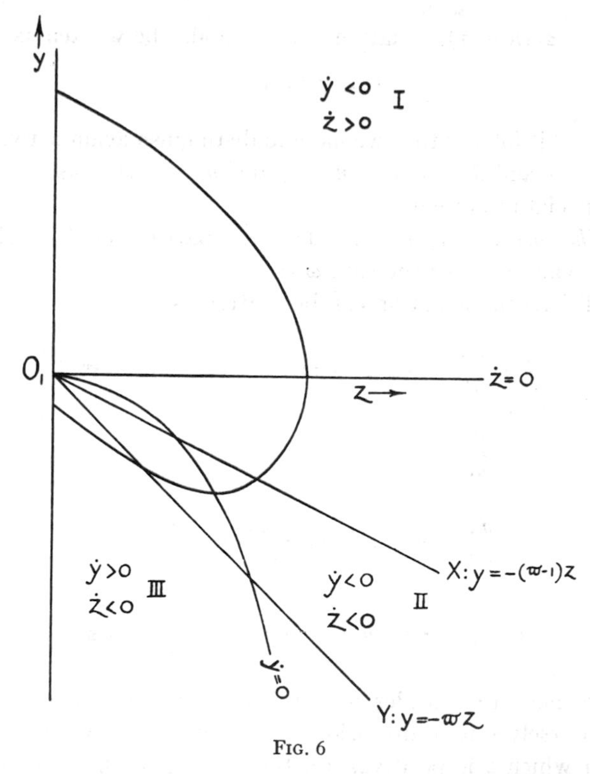

FIG. 6

Finally, the locus of the inflex is obtained by differentiating the differential equation (130) and setting $d^2y/dz^2 = 0$. We obtain in this way,

$$\left(\frac{dy}{dz}\right)^2 + (2\bar{\omega} - 1)\frac{dy}{dz} + \bar{\omega}(\bar{\omega} - 1) + nz^{n-1} = 0 . \qquad (178)$$

Eliminating dy/dz between (175) and (178), we obtain, after some elementary transformations, for the locus of the inflex

$$y = -\frac{1}{2}\frac{\bar{\omega}(\bar{\omega}-1)z + z^n}{\bar{\omega}(\bar{\omega}-1) + nz^{n-1}}\left[(2\bar{\omega}-1) \pm \sqrt{1 - 4nz^{n-1}}\right] . \quad (179)$$

As $z \to 0$, we obtain from the foregoing for $\bar{\omega} \neq 1$

$$y = -\tfrac{1}{2}z[(2\bar{\omega}-1) \pm 1] \qquad (\bar{\omega} \neq 1) . \quad (180)$$

Hence,

$$y = -\bar{\omega}z \qquad \text{or} \qquad y = -(\bar{\omega}-1)z \quad (\bar{\omega} \neq 1) . \quad (181)$$

or, at the origin the locus of the inflex (for $\bar{\omega} \neq 1$) touches the X and the Y directions as defined in § 11.

We shall now prove the following lemmas, due to E. Hopf.

LEMMA 1.—Any solution-curve $y(t)$, $z(t)$ starting at a point on the positive y-axis falls monotonically with y decreasing and z increasing and intersects the z-axis vertically at a finite point, after which the solution continues to fall monotonically with both y and z decreasing until it intersects horizontally, at a finite point, the curve $\dot{y} = 0$; after this the solution rises monotonically with y increasing and z decreasing and either reaches a point on the negative y-axis for a finite value of t or tends toward the origin as $t \to \infty$.

This is intuitively obvious from Figure 6. To prove it, let $z(t_0) = 0$ and $y(t_0) = y_0 > 0$. As long as $\dot{y} > 0$, $z(t)$ increases, while $y(t)$ decreases.

Now, since $\bar{\omega} \geqslant 1$, we have, according to (173),

$$\dot{y} < -z^n . \quad (182)$$

Let $t_1 > t_0$ be sufficiently near t_0. Then $y(t_1) < y_0$ and $z(t_1) = z_1 > 0$. By (182), $\dot{y} < -z_1^n$, for $t > t_1$, as z must increase in the positive quadrant. Hence,

$$y \leqslant y(t_1) - z_1^n(t - t_1) < y_0 - z_1^n(t - t_1) . \quad (183)$$

Therefore, the curve must cross the z-axis for $t < t_1 + y_0 z_1^{-n}$. After crossing the z-axis vertically it is clear from Figure 6 that, so long as we are in region II, y decreases and z also decreases. Clearly the solution-curve cannot avoid the curve $\dot{y} = 0$. After crossing this curve horizontally, it is clear that y has to increase while z continues to decrease. There are two possibilities: either the curve remains in the III-region for all large values of t or it remains there for only a finite interval in t. In either case the curve must tend to a limit point, $z^* \geqslant 0$, $y^* \leqslant 0$. In the first case the curve must necessarily approach the singular point, O_1. In the second case $z^* = 0$ and $y^* < 0$.

LEMMA 2.—Every solution-curve must be of the form described in Lemma 1.

We have to show that any solution-curve must for t decreasing reach a point on the positive y-axis.

We shall consider the most unfavorable case, namely, a solution starting in the III-region. In this region $\dot{y} > 0$, and hence, by equation (177),

$$y < -\frac{\bar{\omega}(\bar{\omega} - 1)z + z^n}{(2\bar{\omega} - 1)} . \tag{184}$$

On the other hand, if y is negative, we have from (175) that

$$\frac{dy}{dz} > -(2\bar{\omega} - 1) , \tag{185}$$

or

$$y \geqslant y_1 - (2\bar{\omega} - 1)(z - z_1) , \tag{186}$$

where y_1 and z_1 define the initial point. For large z, (184) and (186) are contradictory, and consequently z cannot tend to infinity for the solution-curve in region III. Hence, the solution-curve must enter the II-region for some finite z. In this region, z continues to increase (as t decreases), while y begins to increase. If $y_2 < y$, then so long as y is negative, $|y_2| > |y|$. By (175),

$$\frac{dy}{dz} > -(2\bar{\omega} - 1) + \frac{\bar{\omega}(\bar{\omega} - 1)z + z^n}{|y_2|} , \tag{187}$$

where we can choose for y_2 the value of y at the intersection of the solution-curve and the curve $\dot{y} = 0$. From (187) we derive

$$y > -(2\bar{\omega} - 1)z + \frac{\frac{1}{2}\bar{\omega}(\bar{\omega} - 1)z^2 + \frac{1}{n+1} z^{n+1}}{|y_2|} + C , \tag{188}$$

where C is a constant. From (188) it follows that y must become zero for some finite z and hence must intersect the z-axis (vertically). After crossing the z-axis, y increases and z decreases (for t continuing to decrease); since dy/dz is bounded, it is clear that the solution can be continued to a point on the positive y-axis. This proves our second lemma.

Consider now the solution $y(z; y_0)$, which intersects the negative y-axis at $y_0 < 0$. Lemma 2 has described the character of such a solution. Then, since the $y(z)$ curves form a one-parametric family (except at the singular point), it is clear that for sufficiently small z (in the lower quadrant) we should have

$$y(z; y_0) < y(z; \check{y}_0) \tag{189}$$

if

$$y_0 < \check{y}_0 \tag{190}$$

Let us now consider the limit function

$$\lim_{y_0 \to -0} y(z; y_0) . \tag{191}$$

We shall show that this is the E-curve, $y_E(z)$.

To show this we compare y with another function w which satisfies the differential equation

$$\frac{dw}{dz} = -(2\bar{\omega} - 1) - [\bar{\omega}(\bar{\omega} - 1) + \epsilon] \frac{z}{w} , \tag{192}$$

where $\epsilon(<\frac{1}{4})$ is a constant. Equation (192) can also be written as

$$w \frac{dw}{dz} + (2\bar{\omega} - 1)w + [\bar{\omega}(\bar{\omega} - 1) + \epsilon]z = 0 . \tag{193}$$

The foregoing equation is of exactly the same form as equation (157), which we have already considered. Analogous to the solution (163) of (157), we now have

$$\left(\frac{w + q_2 z}{w_0}\right)^{q_2} = \left(\frac{w + q_1 z}{w_0}\right)^{q_1} , \tag{194}$$

where $w = w_0$ for $z = 0$ and $-q_1$ and $-q_2$ are the roots of the equation

$$q^2 + (2\bar{\omega} - 1)q + [\bar{\omega}(\bar{\omega} - 1) + \epsilon] = 0 . \tag{195}$$

We write

$$q_1 = \bar{\omega} - \tfrac{1}{2} + \tfrac{1}{2}\sqrt{1 - 4\epsilon} , \tag{196}$$

$$q_2 = \bar{\omega} - \tfrac{1}{2} - \tfrac{1}{2}\sqrt{1 - 4\epsilon} . \tag{197}$$

The quantities q_1 and q_2 are real if $\epsilon < \frac{1}{4}$. We have, accordingly, assumed $\epsilon < \frac{1}{4}$. We shall consider the special solution of (193) which is obtained from (194) by making $w_0 \to -0$. We have

$$w = -q_1 z . \tag{198}$$

Since

$$\frac{dy}{dz} = -(2\bar{\omega} - 1) - [\bar{\omega}(\bar{\omega} - 1) + z^{n-1}]\frac{z}{y} , \tag{199}$$

we have

$$\frac{dy}{dz} < -(2\bar{\omega} - 1) - [\bar{\omega}(\bar{\omega} - 1) + \epsilon]\frac{z}{y} \tag{200}$$

if

$$z^{n-1} < \epsilon \qquad \text{and} \qquad y < 0 . \tag{201}$$

Subtracting (200) from (192), we have

$$\frac{d}{dz}(w - y) > [\bar{\omega}(1 - \bar{\omega}) + \epsilon]z\,\frac{w - y}{yw} \qquad [z < (\epsilon)^{1/(n-1)};\ y < 0] . \tag{202}$$

We write the foregoing in the form

$$\frac{d\Delta}{dz} > [\bar{\omega}(1 - \bar{\omega}) + \epsilon]z\,\frac{\Delta}{yw} \qquad [z < (\epsilon)^{1/(n-1)};\ y < 0] , \tag{203}$$

where

$$\Delta = w - y = -q_1 z - y(z;\, y_0) \qquad (y_0 < 0) , \tag{204}$$

according to our choice of (198) as the appropriate solution of (192). From (204) it follows that

$$\Delta_{z=0} = -y_0 > 0 . \tag{205}$$

From (203) and (205) we conclude that, under the circumstances specified, Δ is positive and increases. Hence,

$$-q_1 z - y(z;\, y_0) > 0 \qquad [y_0 < 0;\, z < (\epsilon)^{1/(n-1)}] . \quad (206)$$

Now, as $\epsilon < \frac{1}{4}$, we have

$$\sqrt{1 - 4\epsilon} > 1 - 4\epsilon . \quad (207)$$

By (196), therefore,

$$q_1 > \bar{\omega} - 2\epsilon > 0 \qquad [\bar{\omega} > 1;\, \epsilon < \tfrac{1}{4}] . \quad (208)$$

Hence, by (206) and (208),

$$y(z;\, y_0) < -q_1 z < -(\bar{\omega} - 2\epsilon) z . \quad (209)$$

Equation (209) is valid so long as $y < 0$, $z < (\epsilon)^{1/(n-1)}$, $\epsilon < \frac{1}{4}$, and hence so long as $z < (\epsilon)^{1/(n-1)}$, $\epsilon < \frac{1}{4}$ without the restriction $y < 0$. The inequality $y_0 < 0$ is, of course, essential.

For a given positive value of $z < (\frac{1}{4})^{n-1}$ we can choose ϵ to be $(z)^{n-1}$. Hence, by (209) we have

$$y(z;\, y_0) < -(\bar{\omega} - 2z^{n-1}) z \qquad [0 < z < (\tfrac{1}{4})^{n-1}] , \quad (210)$$

or

$$y(z;\, y_0) < -\bar{\omega} z + 2z^n , \quad (211)$$

which is an inequality due to E. Hopf. Consider the limit function $y_E(z)$ as $y_0 \to -0$ (we anticipate by our notation that the limit function is, in fact, the *E*-curve).

From (211) it now follows that

$$y_E(z) \leqslant -\bar{\omega} z + 2z^n . \quad (212)$$

We will now show that

$$y_E(z) \geqslant -\bar{\omega} z \qquad (z > 0) . \quad (213)$$

To prove (213) we proceed as follows: From (199) we have in particular

$$\frac{dy_E}{dz} > -(2\bar{\omega} - 1) - \bar{\omega}(\bar{\omega} - 1) \frac{z}{y_E} \qquad (y_E < 0) . \quad (214)$$

In equation (192) choose $\epsilon = 0$ (we are now considering the equation [157]). We now have

$$\frac{d}{dz}(y_E - w) > \frac{\bar{\omega}(\bar{\omega} - 1)z}{y_E w}(y_E - w) . \qquad (215)$$

Choose for w a solution such that, for $z = 0$, $w = w_0 < 0$. Since $y_E = 0$ at $z = 0$, $(y_E - w) = -w_0 > 0$ at $z = 0$. Equation (215) now shows that, so long as $y_E < 0$ and $w < 0$, $(y_E - w)$ is positive and increasing. Now let $w_0 \to -0$, in which case we obtain the solution (198) with $\epsilon = 0$, i.e., the solution $w = -\bar{\omega}z$. Hence, we obtain

$$y_E(z) \geqslant -\bar{\omega}z \qquad (y_E < 0) . \quad (216)$$

From our lemma it now follows that the inequality (216) is always true.

From (212) and (213) it now follows that, as $z \to 0$, y_E must become tangential to the line $y + \bar{\omega}z = 0$. We have already seen in § 11 that the E-curve is tangential to the line $y + \bar{\omega}z = 0$ at the origin and that there can only be one solution with this property. Hence, the function defined as the limit $y(z;\, y_0)$, $y_0 \to -0$ is, in fact, the E-curve.

We have now shown that the E-curve passes through the origin and is tangential to the direction $y + \bar{\omega}z = 0$ at the origin. Draw the complete E-curve. Let this curve cut the y-axis at $y_0(E)$. It is clear that a solution starting at a point y_0 of the positive y-axis with a value for $y_0 < y_0(E)$ must necessarily remain entirely in the region bounded by the E-curve and the part of the y-axis $0 \leqslant y \leqslant y_0(E)$, and, according to our Lemma 1, must tend to the origin as $t \to \infty$ (or $\xi \to 0$). We shall refer to such solutions as "M-solutions." (As we shall see, along an M-solution, $\theta \to \infty$ as $\xi \to 0$, monotonically.) On the other hand, a solution-curve starting at a point $y_0 > y_0(E)$ must remain outside the region bounded by the E-curve and the part of the y-axis $0 \leqslant y \leqslant y_0(E)$, and hence must reach a point $z = 0$, $y_1 < 0$ of the negative y-axis for a finite value of t, according to the definition of E-curve and to Lemma 1. We shall refer to the solutions outside the E-curve as the "F-solutions." Hence, the whole family of the solution-curves is divided into two regions by the E-curve, the region of the M-solutions, and the region of the F-solutions [cf. Figs. 7 and 8].

14. *The case* $\bar{\omega} > 1$.—So far our discussion is valid for $\bar{\omega} = 1$ as well. We shall now exclude the case $\bar{\omega} = 1$, $n = 3$ and consider the case $\bar{\omega} > 1$, $n < 3$. According to our analysis of § 12, the asymptotic form of the solutions near the origin is given by

$$[y + (\bar{\omega} - 1)z]^{\bar{\omega}-1} = C[y + \bar{\omega}z]^{\bar{\omega}} ; \tag{217}$$

or, in the oblique system of co-ordinates defined by the directions X and Y, we have (cf. Eq. [172])

$$Y = CX^{2/(3-n)} , \tag{218}$$

or

$$\left(\frac{dY}{dX}\right)_0 = 0 , \qquad C \text{ finite} . \tag{219}$$

From (219) it follows that (if $n < 3$) all the solutions must touch the X-axis, or, in other words, the solutions must all be tangential to the line $y + (\bar{\omega} - 1)z = 0$ at the origin except the one $y + \bar{\omega}z = 0$, obtained from (217) when $C = \infty$; this last case corresponds to the E-curve. Hence, all solutions other than the E-solution passing through the origin must touch the X-direction at the origin.

From (159) we have for the corresponding behavior of θ near the origin:

$$\theta = z\xi^{-\bar{\omega}} = ze^{\bar{\omega}t} = A + Be^{t} \quad (t \to +\infty) , \tag{220}$$

or

$$\theta = A + \frac{B}{\xi} \qquad (\xi \to 0) . \tag{221}$$

If $B = 0$, we get the solution finite at the origin, and hence a solution belonging to the homologous family $\{\theta_n(\xi)\}$. For $B \neq 0$ and positive we have the behavior near the origin of the M-solutions which are seen to tend to ∞ monotonically as $\xi \to 0$.

15. *The case* $\bar{\omega} = 1$.—The analysis of § 12 does not apply to the case $\bar{\omega} = 1$, $n = 3$, and hence the arguments of the last section cannot be used for this case. The differential equation for $n = 3$ is

$$y\frac{dy}{dz} + y + z^3 = 0 , \tag{222}$$

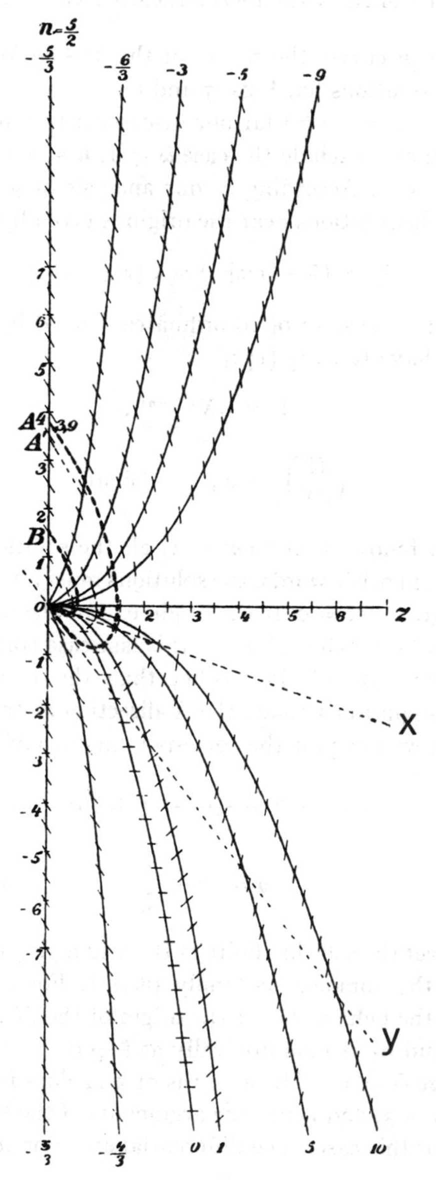

FIG. 7.—The system of isoclinical curves in the (y, z) plane for $n = 2.5$. (The diagram is reproduced from Emden's *Gaskugeln*.)

and equation (157) no longer gives the behavior of the solutions at the origin. However, the discussion of the E-curve (with regard to its existence and uniqueness as developed in § 13) is unaltered. In particular, we have the result that the E-curve is tangential to $y + z = 0$ at the origin.

Now, according to a theorem[11] due to Hardy, *any rational function,* $H(x, y, y')$, *is ultimately monotonic along a solution* $y(x)$ *of an algebraic differential equation of the form* $f(x, y, y') \equiv \Sigma A x^m y^n y'^p = 0$. If we apply Hardy's theorem to the ratio X_i/X_j of any two terms of the differential equation itself, it follows that, since X_i/X_j has to be ultimately monotonic, we should have one of the relations

$$\frac{X_i}{X_j} \to 0 \; ; \qquad \frac{X_i}{X_j} \to -1 \; ; \qquad \frac{X_i}{X_j} \to \infty \; . \tag{223}$$ [12]

Equation (222) is an algebraic differential equation, and Hardy's theorem is applicable. We should ultimately have one of the relations

$$\frac{dy}{dz} \to -1 \; ; \qquad \frac{dy}{dz} \to -\infty \; ; \qquad \frac{dy}{dz} \to 0 \; . \tag{224}$$

The first of the possibilities leads clearly to the E-curve; the second is impossible since the M-solutions are all above the curve $y + z = 0$. Hence, the only remaining possibility is that, as $t \to \infty$, $dy/dz \to 0$. Hence, according to (222),

$$y = \frac{dz}{dt} \sim -z^3 \, , \tag{225}$$

or

$$z \sim \left[\frac{1}{2(t - c)}\right]^{1/2} , \tag{226}$$

where c is a constant of integration. Remembering that $\xi = e^{-t}$, we have

$$\theta = z\xi^{-\bar{\omega}} \sim \frac{1}{\xi}\left[\frac{1}{2 \log \dfrac{C}{\xi}}\right]^{1/2} , \tag{227}$$

[11] For a proof of Hardy's theorem see G. H. Hardy, *Orders of Infinity* ("Cambridge Tracts in Mathematics and Mathematical Physics," No. 12), pp. 57–60, 1924.

[12] In exceptional cases, more than two terms being of equal highest order, we may have $X_i \sim$ constant X_j instead of $X_i \sim -X_j$.

where $C = e^{-c}$ is a constant. Equation (227) then gives the behavior of the M-solutions. It should be noticed that these solutions in the (y, z) plane touch the z-axis, which is for this case the X-direction as well. Hence, the behavior of the solutions near the origin is still qualitatively the same as for the case $\bar{\omega} > 1$; the proof given in § 14 is not, however, valid in the present case. The results proved in this section are due to Fowler.

16. *Fowler's theorem.*—We can now proceed to derive the fundamental theorem (due to Fowler) concerning the arrangement of the solutions of the Lane-Emden equation in the (ξ, θ) plane:

To a given set of initial values

$$\xi = \xi_0 ; \qquad \theta = \theta_0 \geqslant 0 ; \qquad \left(\frac{d\theta}{d\xi}\right)_{\xi=\xi_0} = \theta_0' , \tag{228}$$

there corresponds a definite point $z_0 \geqslant 0$, y_0 in the (y, z) plane:

$$z_0 = \xi_0^{\bar{\omega}} \theta_0 ; \qquad y_0 + \bar{\omega} z_0 = -\xi_0^{\bar{\omega}+1} \theta_0' . \tag{229}$$

Let us specify a certain point (ξ_0, θ_0) in the (ξ, θ)-plane and let θ_0' correspond to all possible starting-slopes at (ξ_0, θ_0). Then the corresponding point in the (y, z) plane describes a vertical line through $(z_0, 0)$. Let $z = z_0(E)$ be the point where the E-curve intersects the z-axis.

From the arrangement of the different types of solutions in the (y, z) plane already described at the end of § 13 we can easily derive the following (see Fig. 8):

If $z_0 > z_0(E)$, then the solution passing through (z_0, y_0), where y_0 is arbitrary, is necessarily an F-solution.

If $z_0 = z_0(E)$, then the solution passing through $[z_0(E), y_0 = 0]$ is an E-solution, while the solution passing through $[z_0(E), y_0 > 0]$ is an F-solution.

If $z_0 < z_0(E)$, then for a given $z_0 < z_0(E)$ there are two points $[z_0, y_0^{(1)}(E)]$ and $[z_0, y_0^{(2)}(E)]$ which lie on the E-curve. Of these two points one must be in the positive quadrant and the other must be in the lower quadrant. Let $y_0^{(1)}(E) > 0$. Then a solution passing through the point $[z_0, y_0^{(1)}(E) > y_0 > y_0^{(2)}(E)]$ is clearly an M-solution, while solutions passing through $[z_0, y_0 > y_0^{(1)}(E)]$ and $[z_0,$

$y_0 < y_0^{(2)}(E)$] are F-solutions. Finally, the solution passing through [z_0, $y_0^{(1)}(E)$] or [z_0, $y_0^{(2)}(E)$] is an E-solution.

Now a characteristic of an F-solution is that $z = 0$ for two finite values of t, or the corresponding solution[12] $\theta(\xi)$ is such that it has at least two zeros and that, further, it must be characterized by having

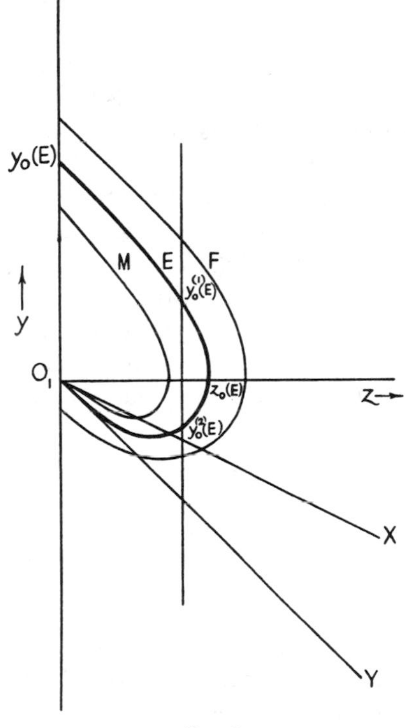

FIG. 8

a maximum in the interval in which it is positive. The characteristic of an M-solution is that the corresponding $\theta(\xi)$ solutions tend monotonically to infinity as $\xi \to 0$. Near the origin it has definite asymptotic forms, according as $\bar{\omega} > 1$ or $\bar{\omega} = 1$ (see Eqs. [221] and [227]).

[12] $\theta(\xi)$ is only one of a homologous family which can be derived from a given solution-curve in the (y, z) plane.

We can translate the foregoing results in the (ξ, θ) plane as follows:

a) Starting-points (ξ_0, θ_0) are divided into two classes by a critical curve $\theta = z_0(E)\xi^{-\bar{\omega}}$ ($z_0(E)$ being determined by $\bar{\omega}$). This curve is the envelope of all the solutions belonging to the family $\{\theta_n(\xi)\}$.

b) Any solution $\theta(\xi)$ starting at a point (ξ_0, θ_0) above the critical curve is an F-solution with two zeros ξ_1 and ξ_2 such that $\xi_1 \leqslant \xi_0 \leqslant \xi_2$.

c) If the starting-point lies on the critical curve, all solutions are again F-solutions except the one which is tangential to the critical curve at the starting point; the latter is an E-solution belonging to the family $\{\theta_n(\xi)\}$.

d) If the point (ξ_0, θ_0) lies below the critical curve, there exist two starting-slopes $\theta_0' = \theta_0'^{(1)}(E)$ and $\theta_0' = \theta_0'^{(2)}(E)$, and $\theta_0'^{(2)}(E) < \theta_0'^{(1)}(E) < 0$. All solutions corresponding to $\theta_0' > \theta_0'^{(1)}(E)$ or $\theta_0' < \theta_0'^{(2)}(E)$ are F-solutions. Slopes between $\theta_0'^{(1)}(E)$ and $\theta_0'^{(2)}(E)$, i.e., $\theta_0'^{(2)}(E) < \theta_0' < \theta_0'^{(1)}(E)$, correspond to M-solutions which become infinite as $\xi \to 0$. For $\theta_0' = \theta_0'^{(1)}(E)$ or $\theta_0' = \theta_0'^{(2)}(E)$ we have E-solutions.

e) The asymptotic forms of the M-solutions are

$$\left.\begin{aligned} \theta &\sim A + \frac{B}{\xi} && (\bar{\omega} > 1)\,; \\ \theta &\sim \frac{1}{\xi}\left[\frac{1}{2\log\dfrac{C}{\xi}}\right]^{1/2} && (\bar{\omega} = 1)\,. \end{aligned}\right\} \tag{230}$$

Fowler refers to the circumstance summarized in (*a*)–(*d*) above by the very convenient statement that "the E-solutions form a grid for use in analysing the other solutions."

If we apply the theorem to the special case where the starting-point (ξ_0, θ_0) is on the ξ-axis, we have

$$z_0 = 0\,; \quad y_0 = -\xi_0^{\bar{\omega}+1}\theta_0'\,. \tag{231}$$

As a special case of Fowler's theorem, or directly from Figure 7 or 8, it is now clear that if $y = y_0(E)$, or

$$-\theta_0'(E) = \frac{y_0(E)}{\xi_0^{\bar{\omega}+1}}\,, \tag{232}$$

then we have an E-solution. If $y_0 < y_0(E)$, or

$$-\theta_0' < \frac{y_0(E)}{\xi_0^{\bar{\omega}+1}} = -\theta_0'(E) \, , \tag{233}$$

we have an M-solution; finally, if

$$-\theta_0' > \frac{y_0(E)}{\xi_0^{\bar{\omega}+1}} = -\theta_0'(E) \tag{234}$$

we have an F-solution. We have an F-solution also when y_0 is negative.

17. *The case* $\bar{\omega} > \frac{1}{2}$, $1 < n < 5$.—The general discussion of this case presents somewhat greater difficulties, since in the positive half-plane we are considering there can be two singular points, depending upon whether $\bar{\omega} \geqslant 1$ or $\frac{1}{2} < \bar{\omega} < 1$. We therefore do not expect the discussion of § 13 to be valid in the present, more general case. In particular, Lemma 1 is not true for $\frac{1}{2} < \bar{\omega} < 1$, but we can prove the following:

LEMMA.—Any solution-curve $[y(t), z(t)]$ starting at a point $y_1 < 0$ on the negative y-axis falls monotonically, y decreasing and z increasing as t decreases, until it cuts horizontally, for a finite z, the curve $\dot{y} = 0$; after this the curve monotonically rises, both y and z increasing (for t continuing to decrease) until it cuts the z-axis at a regular point (i.e., for $\bar{\omega} < 1$ the curve cuts the z-axis at a point $z > z_s = [\bar{\omega}(1 - \bar{\omega})]^{\bar{\omega}/2}$). For further decrease of t, the curve continues to rise, with z now decreasing until the curve cuts the y-axis again at a point on the positive y-axis.

We have already proved the lemma for $\bar{\omega} \geqslant 1$, and we shall therefore restrict ourselves here to the case $1 > \bar{\omega} > \frac{1}{2}$. In this case there is a singular point on the z-axis (denoted by O_2). The foregoing lemma shows that solutions which cut the negative y-axis ($y_1 < 0$) do not approach any of the singular points in the half-plane considered, and in all cases for which $\bar{\omega} > \frac{1}{2}$ these solutions avoid the singular points.

It should be noticed that the locus of points at which $dy/dt = 0$, specified by (cf. Eq. [177])

$$(2\bar{\omega} - 1)y = \bar{\omega}(1 - \bar{\omega})z - z^n \, , \tag{235}$$

is, for $\bar{\omega} < 1$, no longer a monotonically decreasing curve lying entirely in the lower negative quadrant. The locus (235) lies in the positive upper quadrant for $0 \leqslant z \leqslant z_s$, and crosses the z-axis at the second singular point (see Fig. 9). We shall now prove the lemma.

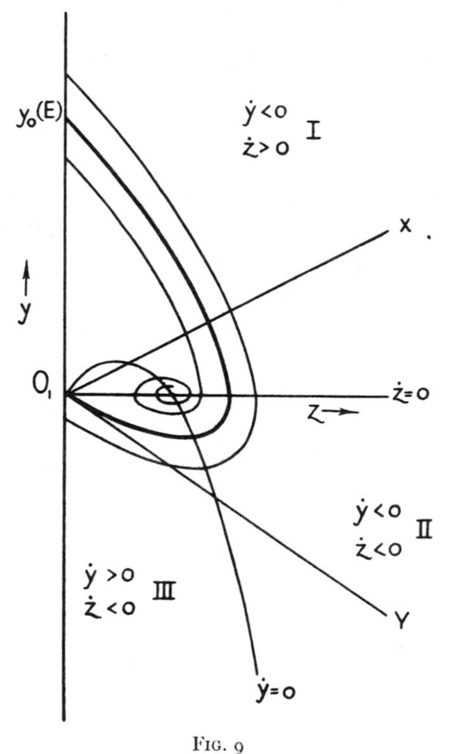

FIG. 9

Let $(y_1, 0)$ be the starting-point, $y_1 < 0$. It is clear that in region *III*, as t decreases, y decreases while z increases. From the equation

$$\frac{dy}{dz} = -(2\bar{\omega} - 1) + \frac{\bar{\omega}(1 - \bar{\omega})z}{y} - \frac{z^n}{y} \tag{236}$$

it follows that, so long as $y < 0$,

$$\frac{dy}{dz} > -(2\bar{\omega} - 1) + \frac{\bar{\omega}(1 - \bar{\omega})z}{y}, \tag{237}$$

which we can also write as (for $y < 0$)

$$\frac{dy}{dz} > -(2\bar{\omega} - 1) - \frac{\bar{\omega}(1 - \bar{\omega})z}{|y|} . \tag{238}$$

Further, in region *III*, $|y|$ increases and hence $|y_1|$ is the minimum value of $|y|$ in this region. Hence,

$$\frac{dy}{dz} > -(2\bar{\omega} - 1) - \frac{\bar{\omega}(1 - \bar{\omega})z}{|y_1|} , \tag{239}$$

or

$$y > y_1 - (2\bar{\omega} - 1)z - \frac{\bar{\omega}(1 - \bar{\omega})}{2|y_1|} z^2 . \tag{240}$$

On the other hand, since we are in region *III*, $\dot{y} > 0$, and we should have

$$y < \frac{\bar{\omega}(1 - \bar{\omega})z - z^n}{(2\bar{\omega} - 1)} . \tag{241}$$

Equations (240) and (241) are obviously contradictory for large values of z. Hence, the solution-curve must leave the region considered for a finite z, and it is clear that it intersects the $\dot{y} = 0$ curve (horizontally) at a point where $z > z_s$. For further decrease in t, z continues to increase, while y now begins to increase, $|y|$ decreasing. If $y_2 < y$, then so long as we are in region *II*, $|y_2| > |y|$; and by (236) we have (for sufficiently large z)

$$\frac{dy}{dz} > -(2\bar{\omega} - 1) + \frac{\bar{\omega}(\bar{\omega} - 1)z + z^n}{|y_2|} > Cz^n \tag{242}$$

for some positive constant, C. Thus,

$$y > \frac{Cz^{n+1}}{n + 1} - D \quad (D \text{ a constant}) , \tag{243}$$

and we conclude that the curve must cross the z-axis. Further, the curve must intersect the z-axis at a point $z > z_s$, since, as we have already seen, the curve already intersects the curve $\dot{y} = 0$ at a point whose z-co-ordinate is greater than z_s, and in the region *II*, z increases with decreasing t. After crossing the z-axis (vertically), y increases, while z now begins to decrease; and, since $|dy/dz|$ is bounded, the solution can be continued to a point on the positive y-axis.

Consider, now, the solution $y(z; y_1)$, which intersects the y-axis at a point $y_1 < 0$. The lemma has described the character of such a solution. Then, since the (y, z) curves must form a one-parametric family (except at the singular points), it is clear that for sufficiently small z we have

$$y(z; y_1) < y(z; \grave{y}_1) \qquad (y_1 < y_1 < 0) . \tag{244}$$

We can therefore construct the limit function

$$y_E(z) = \lim_{y_1 \to -0} y(z; y_1) . \tag{245}$$

We have already shown that for $\bar{\omega} \geqslant 1$ we obtain in this way the E-curve. We shall see presently that this is generally true. The E-curve is, of course, tangential to the direction $y + \bar{\omega} z = 0$ at the origin.

Draw the complete E-curve. From the lemma it follows that the E-curve must cut the y-axis at a point $y_0(E) > 0$. Further, it is also clear from the lemma that the curve surrounds the singular point $(0, z_s)$. Again, it follows from the lemma that a solution starting with a value $y_0 > y_0(E)$ remains outside the E-curve. This solution intersects the y-axis for two values of t, and hence corresponds to a solution in the (ξ, θ) plane which has at least two zeros. We shall refer to these as "F-solutions."

On the other hand, if we consider a solution starting at a point $0 < y_0 < y_0(E)$ on the y-axis, then the solution must be entirely inside the region bounded by the E-curve and the part of the y-axis $0 \leqslant y \leqslant y_0(E)$. Hence, as $t \to \infty$ the curve must tend to one or other of the singular points inside the region described. If $\bar{\omega} \geqslant 1$, there is only one singular point, namely, the origin; and, as we have already seen, the solutions approach the origin. On the other hand, there are two singular points in the region under consideration for $\frac{1}{2} < \bar{\omega} < 1$. In this case, however, the solutions cannot approach the origin. This follows from the analysis of § 12. We have shown that the behavior of the solutions in the immediate neighborhood of the origin is specified by

$$Y = CX^{\bar{\omega}/(\bar{\omega}-1)}\,, \tag{246}$$

where X and Y are the co-ordinates in the oblique frame of reference defined by the lines $y + (\bar{\omega} - 1)z = 0$ and $y + \bar{\omega}z = 0$. If $\bar{\omega} < 1$, we can write

$$YX^{\bar{\omega}/(1-\bar{\omega})} = \text{constant}\,, \tag{247}$$

where the exponent of X is positive. It follows that, in general, $Y = 0$ corresponds to $X = \infty$, and similarly $X = 0$ implies $Y = \infty$. Hence, the solutions approaching the origin must be such that they approach it up to a certain minimum distance and then recede from it. The exception is the E-curve which corresponds to $C = \infty$ in (246); this solution touches the Y-axis at the origin and hence, being tangential to $y + \bar{\omega}z = 0$ at $y, z = 0$ must be the E-curve (§ 11). Hence, the solutions in the region bounded by the E-curve and $0 \leqslant y \leqslant y_0(E)$ must approach the second singular point O_2 $(0, z_s)$ on the z-axis for $1 > \bar{\omega} > \frac{1}{2}$ as $t \to \infty$. We shall refer to these solutions as thc "M-solutions."

The behavior near the singular point O_2 for $1 > \bar{\omega} > \frac{1}{2}$ will be examined below, but it is clear that in the terminology of Fowler the E-solutions form a grid for use in analyzing the nature of the solutions passing through a given point in the (ξ, θ) plane, for $\bar{\omega} > \frac{1}{2}$.

18. *The case* $1 > \bar{\omega} > \frac{1}{2}$, $3 < n < 5$.—The behavior of the solutions as they approach the singular point O_2 will now be investigated. Let us first examine the possible directions of y' at O_2.

Near O_2 we can write

$$y' = \lim_{\substack{y \to 0 \\ \Delta z \to 0}} \left[-(2\bar{\omega} - 1) + \frac{\bar{\omega}(1 - \bar{\omega})(z_s + \Delta z) - (z_s + \Delta z)^n}{y} \right], \tag{248}$$

or, since $z_s^{n-1} = \bar{\omega}(1 - \bar{\omega})$, we have at $(0, z_s)$

$$y' = -(2\bar{\omega} - 1) + \frac{\bar{\omega}(1 - \bar{\omega})(1 - n)}{y'} \qquad (0, z_s)\,, \tag{249}$$

or we can re-write the foregoing as

$$y'^2 + (2\bar{\omega} - 1)y' + 2(1 - \bar{\omega}) = 0 . \tag{250}$$

Solving for y', we have

$$y' = \tfrac{1}{2}[-(2\bar{\omega} - 1) \pm \sqrt{4\bar{\omega}^2 + 4\bar{\omega} - 7}] . \tag{251}$$

The values of y' at $(0, z_s)$ given by (251) are real only if

$$4\bar{\omega}^2 + 4\bar{\omega} - 7 \geqslant 0 . \tag{252}$$

Let $\bar{\omega} = \bar{\omega}^*$ be such that

$$4\bar{\omega}^{*2} + 4\bar{\omega}^* - 7 = 0 . \tag{253}$$

The positive root of the foregoing equation is given by

$$\bar{\omega}^* = \frac{2\sqrt{2} - 1}{2} = 0.91421 \ldots . \tag{254}$$

$\bar{\omega}^*$ corresponds to a value of n^* where

$$n^* = \frac{11 + 8\sqrt{2}}{7} = 3.18767 \ldots . \tag{255}$$

If $\bar{\omega} < \bar{\omega}^*$, the directions specified by (251) are imaginary. It follows that the solutions approaching the singular point O_2 must spiral around $(0, z_s)$. On the other hand, if $1 > \bar{\omega} > \bar{\omega}^*$, the directions specified by (251) are real. Let X_1 and Y_1 denote these two directions. Then,

$$y'_{Y_1} = -\tfrac{1}{2}[(2\bar{\omega} - 1) + \sqrt{4\bar{\omega}^2 + 4\bar{\omega} - 7}] , \tag{256}$$

$$y'_{X_1} = -\tfrac{1}{2}[(2\bar{\omega} - 1) - \sqrt{4\bar{\omega}^2 + 4\bar{\omega} - 7}] . \tag{257}$$

We shall examine in greater detail the behavior of the solutions at O_2. First consider the case $\bar{\omega} < \bar{\omega}^*$:

Case 1: $\frac{1}{2} < \bar{\omega} < \bar{\omega}^* < 1$.—Write

$$z = z_s + z_1 ; \qquad z_s = [\bar{\omega}(1 - \bar{\omega})]^{\bar{\omega}/2} . \tag{258}$$

If z_1 is sufficiently small, we can write the differential equation (130) at $z = z_s + z_1$ (where $z_1 \to 0$), as

$$y \frac{dy}{dz_1} + (2\bar{\omega} - 1)y - \bar{\omega}(1 - \bar{\omega})(z_s + z_1) + z_s^n + n z_s^{n-1} z_1 = 0 \, ; \tag{259}$$

or, remembering that $z_s = [\bar{\omega}(1 - \bar{\omega})]^{1/(n-1)}$, we have

$$y \frac{dy}{dz_1} + (2\bar{\omega} - 1)y + 2(1 - \bar{\omega})z_1 = 0 \, ; \tag{260}$$

or, since $y = dz/dt = dz_1/dt$, we can re-write the foregoing as

$$\frac{d^2 z_1}{dt^2} + (2\bar{\omega} - 1) \frac{dz_1}{dt} + 2(1 - \bar{\omega})z_1 = 0 \, . \tag{261}$$

The roots of the equation

$$q^2 + (2\bar{\omega} - 1)q + 2(1 - \bar{\omega}) = 0 \tag{262}$$

are imaginary (cf. Eq. [250]) when $\bar{\omega} < \bar{\omega}^*$. The roots can be written as

$$-\tfrac{1}{2}(2\bar{\omega} - 1) \pm i\tfrac{1}{2}\sqrt{7 - 4\bar{\omega} - 4\bar{\omega}^2} \, . \tag{263}$$

The solution of (261) can therefore be written as

$$z_1 = A e^{-\frac{1}{2}(2\bar{\omega}-1)t} \cos\left[\tfrac{1}{2}\sqrt{7 - 4\bar{\omega} - 4\bar{\omega}^2}\, t + \delta\right] , \tag{264}$$

where A and δ are integration constants. Remembering that $\bar{\omega} = 2/(n - 1)$, we verify that

$$7 - 4\bar{\omega} - 4\bar{\omega}^2 = \frac{7n^2 - 22n - 1}{(n - 1)^2} \, ; \qquad 2\bar{\omega} - 1 = \frac{5 - n}{n - 1} \, . \tag{265}$$

We can now write (264) as

$$z_1 = A e^{-\frac{5-n}{2(n-1)} t} \cos\left[\frac{\sqrt{7n^2 - 22n - 1}}{2(n - 1)}\, t + \delta\right] \tag{266}$$

Since $y = dz/dt = dz_1/dt$, we have

$$\left. \begin{aligned} y = -A e^{-\frac{5-n}{2(n-1)} t} \Bigg\{ &\frac{5 - n}{2(n - 1)} \cos\left[\frac{\sqrt{7n^2 - 22n - 1}}{2(n - 1)}\, t + \delta\right] \\ &+ \frac{\sqrt{7n^2 - 22n - 1}}{2(n - 1)} \sin\left[\frac{\sqrt{7n^2 - 22n - 1}}{2(n - 1)}\, t + \delta\right] \Bigg\} . \end{aligned} \right\} \tag{267}$$

We thus see that the singular point $(0, z_s)$ is approached spirally as $t \to \infty$, and, since $\xi = e^{-t}$, approaching the singular point corresponds to $\xi \to 0$.

As t increases by $4(n-1)\pi/\sqrt{7n^2-22n-1}$, the representative point in the (y, z) plane makes a complete revolution, while the "amplitude" decreases in the ratio

$$1 : e^{-\frac{2(5-n)\pi}{\sqrt{7n^2-22n-1}}} . \tag{268}$$

Again, if for a certain value of t, $z_1 = 0$, then if t_1 increases by $2(n-1)\pi/\sqrt{7n^2-22n-1}$, z_1 will again be zero; this means that, as $t \to \infty$, $z_1 = 0$ for ξ asymptotically decreasing geometrically in the ratio

$$1 : e^{-\frac{2(n-1)\pi}{\sqrt{7n^2-22n-1}}} . \tag{269}$$

As z_1 becomes successively zero, the y-co-ordinate asymptotically decreases geometrically in the ratio

$$1 : e^{-\frac{(5-n)\pi}{\sqrt{7n^2-22n-1}}} . \tag{270}$$

Again, since

$$\theta = (z_s + z_1)e^{\bar{\omega} t} = (z_s + z_1)\xi^{-\bar{\omega}} , \tag{271}$$

we have, according to (266),

$$\left.\begin{aligned} \theta \sim \left[\frac{2(n-3)}{(n-1)^2}\right]^{\frac{1}{n-1}} \frac{1}{\xi^{\frac{2}{n-1}}} \Big\{ 1 & \\ + C\xi^{\frac{5-n}{2(n-1)}} \cos\left[\frac{\sqrt{7n^2-22n-1}}{2(n-1)} \log \xi - \delta\right]\Big\}, & \end{aligned}\right\} \tag{272}$$

where C is a constant.

From the foregoing it follows that, as $\xi \to 0$, the solution crosses the singular solution at points which asymptotically decrease geometrically in the ratio (269), while (as $\xi \to 0$) the solution becomes asymptotic to the singular solution

$$\theta_s = \left[\frac{2(n-3)}{(n-1)^2}\right]^{1/(n-1)} \frac{1}{\xi^{2/(n-1)}} , \tag{273}$$

$$\theta \sim \theta_s \qquad (\xi \to 0) \tag{274}$$

Case 2: $1 > \bar{\omega} \geqslant \bar{\omega}^*$.—In this case the roots of the quadratic equation (262) are real, and we can write for the solution

$$z_1 = Ae^{-q_1 t} + Be^{-q_2 t}, \tag{275}$$

where $-q_1$ and $-q_2$ are the roots of the equation (262):

$$q_1 = \tfrac{1}{2}[(2\bar{\omega} - 1) + \sqrt{4\bar{\omega}^2 + 4\bar{\omega} - 7}], \tag{276}$$

$$q_2 = \tfrac{1}{2}[(2\bar{\omega} - 1) - \sqrt{4\bar{\omega}^2 + 4\bar{\omega} - 7}]. \tag{277}$$

From (275) we have

$$y = -q_1 Ae^{-q_1 t} - q_2 Be^{-q_2 t}. \tag{278}$$

From (275) and (278) we derive

$$(y + q_2 z_1)^{q_2} = C(y + q_1 z_1)^{q_1}, \tag{279}$$

where C is a constant.

If we choose the directions y'_{X_1}, y'_{Y_1} (cf. Eqs. [256] and [257]) at $(0, z_s)$ to define a new oblique frame of reference and denote by X_1 and Y_1 the co-ordinates of a point with respect to this new frame of reference, we obtain, as in § 12,

$$Y_1^{q_2} = CX_1^{q_1}; \tag{280}$$

or by (276) and (277) we have

$$Y_1 = CX_1^{\frac{(2\bar{\omega}-1)+\sqrt{4\bar{\omega}^2+4\bar{\omega}-7}}{(2\bar{\omega}-1)-\sqrt{4\bar{\omega}^2+4\bar{\omega}-7}}}. \tag{281}$$

Hence,

$$\left(\frac{dY_1}{dX_1}\right)_{O_2} = 0. \tag{282}$$

Hence, all the solutions except one (which corresponds to $C = \infty$ in [281]) touch the X_1 axis at $(0, z_s)$. A closer examination shows that this happens as $t \to \infty$. From (275), (276), and (277) it follows that, since

$$\theta = ze^{\bar{\omega} t} = (z_s + z_1)e^{\bar{\omega} t}, \tag{283}$$

we have

$$\theta \sim z_s e^{\bar{\omega} t} + Ae^{\frac{1}{2}[1-\sqrt{4\bar{\omega}^2+4\bar{\omega}-7}]\, t} + Be^{\frac{1}{2}[1+\sqrt{4\bar{\omega}^2+4\bar{\omega}-7}]\, t} \quad (t \to \infty); \tag{284}$$

or, remembering that $\xi = e^{-t}$, we have

$$\theta \sim \frac{[\bar{\omega}(1 - \bar{\omega})]^{\bar{\omega}/2}}{\xi^{\bar{\omega}}} + \frac{A}{\xi^{\frac{1}{2}[1-\sqrt{4\bar{\omega}^2+4\bar{\omega}-7}]}} + \frac{B}{\xi^{\frac{1}{2}[1+\sqrt{4\bar{\omega}^2+4\bar{\omega}-7}]}}, \qquad (285)$$

as $\xi \to 0$. The "exceptional" case referred to above, which is tangential to the Y_1-axis at O_2, corresponds to $B = 0$ in (285). Equation (285) gives the behavior of the solutions which tend to infinity as $\xi \to 0$; these are the M-solutions.

It is now clear from the lemma of § 17 that the arrangement of the solutions is the same as for the case $\bar{\omega} \geqslant 1$, the difference arising only from the different asymptotic behavior of the M-solutions as $\xi \to 0$. Instead of (221) and (227) for $\bar{\omega} > 1$ and $\bar{\omega} = 1$, respectively, we now have (272) and (285) for $\frac{1}{2} < \bar{\omega} < \bar{\omega}^*$ and $1 > \bar{\omega} > \bar{\omega}^*$, respectively. Fowler's theorem, as stated in § 16, holds good except for (*e*), which describes the asymptotic behavior of the solutions.

19. *Case* $\bar{\omega} = \frac{1}{2}$, n = 5.—In this case the (y, z) differential equation

$$y\frac{dy}{dz} - \tfrac{1}{4}z + z^5 = 0 \qquad (286)$$

can be integrated, and we have

$$y^2 = \tfrac{1}{4}z^2 - \tfrac{1}{3}z^6 + D\,, \qquad (287)$$

where D is an integration constant. As we have already seen in § 4, $D = 0$ leads to the Schuster-Emden integral, and hence in the (y, z) plane the E-curve is given by

$$y_E^2 = \tfrac{1}{4}z^2 - \tfrac{1}{3}z^6\,. \qquad (288)$$

This represents a closed symmetrical curve tangential to the lines $y \pm \frac{1}{2}z = 0$ at the origin. The lines $y \mp \frac{1}{2}z = 0$ for the case under consideration ($\bar{\omega} = \frac{1}{2}$) define the X- and the Y-directions. The origin approached along the line $y + \frac{1}{2}z = 0$ for increasing t corresponds to $\xi \to 0$ in the (ξ, θ) plane. At the same time, the origin approached along the line $y = \frac{1}{2}z$ for decreasing t corresponds to $\xi \to \infty$ with $\theta \to 0$.

The E-curve intersects the z-axis at

$$z_0(E) = (\tfrac{3}{4})^{1/4} . \tag{289}$$

If D in (287) is positive, we get curves (symmetrical about the z-axis) which lie entirely outside the E-curve and which intersect the y-axis at the points $\pm\sqrt{D}$. These are the F-solutions. If D is negative, we obtain closed curves inside the E-curve; these are the M-solutions. As in the previous cases, the region of the F- and of the M-solutions are separated by the E-curve. Thus, the Schuster-Emden solutions in the (ξ, θ) plane also form a grid, and the arrangement of the solutions is of the same nature as in the cases already discussed. The critical curve in the (ξ, θ) plane above which the solutions are all of the F-character is defined by (cf. Eq. [289])

$$\theta = (\tfrac{3}{4})^{1/4}\xi^{-1/2} , \tag{290}$$

which is the envelope of all the Schuster-Emden integrals (Eq. [59]). Again, for $\bar{\omega} = \frac{1}{2}$ the singular point is $[0, (\frac{1}{4})^{1/4}]$, and the corresponding singular solution is

$$\theta_s = (\tfrac{1}{4})^{1/4}\xi^{-1/2} , \tag{291}$$

which lies below the critical curve (290).

If we consider the M-solutions in somewhat greater detail, we see that, as $\xi \to 0$, the solutions lie below the curve (291), while with increasing ξ they cut the singular solution (291) and rise above it, and when ξ still further increases, they cut the singular solution again and tend to zero as $\xi \to \infty$. Since the value of z is the same both when $t \to \infty$ and when $t \to -\infty$, we have

$$\lim_{\substack{\theta \to \infty \\ \xi \to 0}} \xi^{1/2}\theta = \lim_{\substack{\theta \to 0 \\ \xi \to \infty}} \xi^{1/2}\theta . \tag{292}$$

The general run of the curves in the (y, z) plane is illustrated in Figure 11.

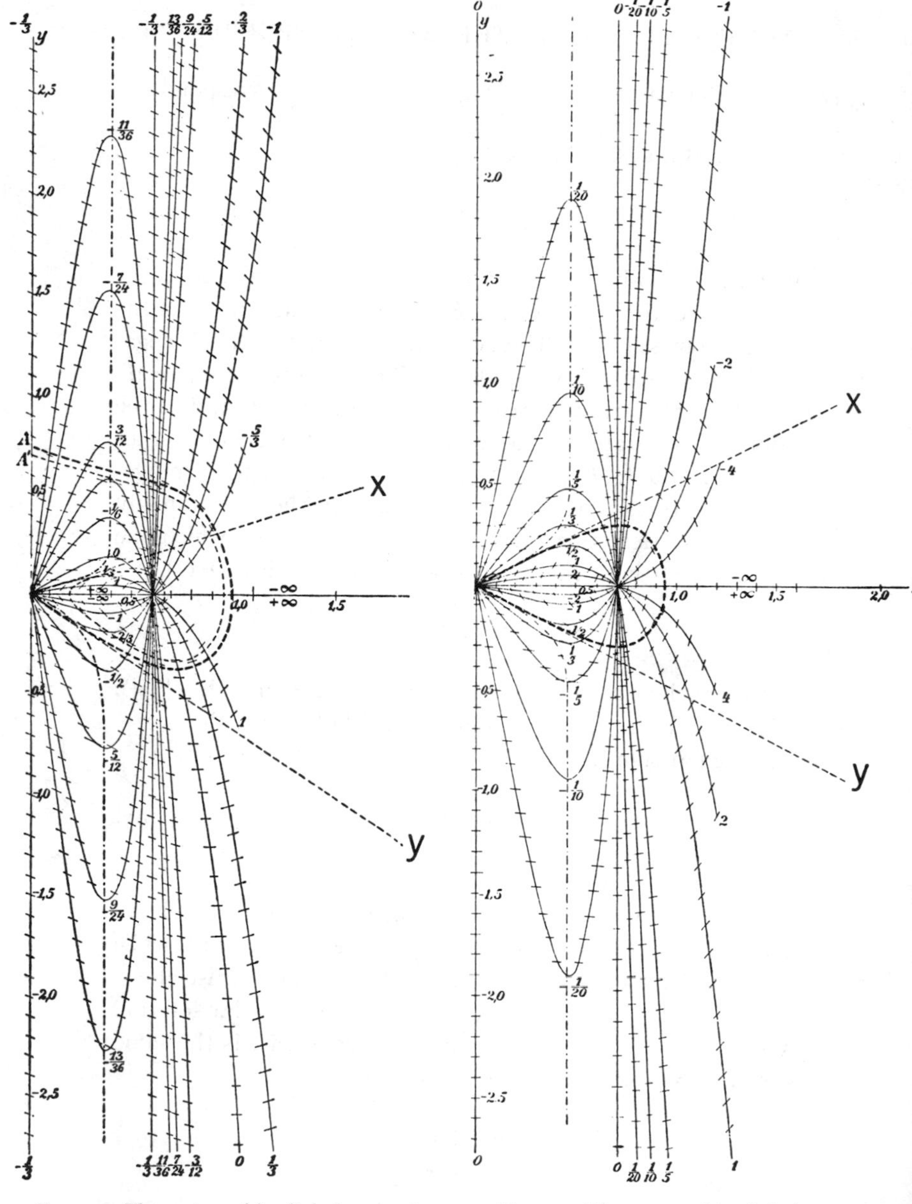

FIG. 10.—The system of isoclinical curves in the (y, z) plane for $n = 4$. (The diagram is reproduced from Emden's *Gaskugeln.*)

FIG. 11.—The system of isoclinical curves in the (y, z) plane for $n = 5$. (The diagram is reproduced from Emden's *Gaskugeln.*)

20. *The case* $\bar{\omega} < \frac{1}{2}$.—The lemma proved in § 17 is valid no longer. This can be seen in the following way:

The equations are

$$\dot{y} = \frac{dy}{dt} = (1 - 2\bar{\omega})y + \bar{\omega}(1 - \bar{\omega})z - z^n , \tag{293}$$

$$\dot{z} = \frac{dz}{dt} = y , \tag{294}$$

$$\frac{dy}{dz} = (1 - 2\bar{\omega}) + \frac{\bar{\omega}(1 - \bar{\omega})z - z^n}{y} . \tag{295}$$

The curve $\dot{y} = 0$ is now given by

$$(1 - 2\bar{\omega})y = -\bar{\omega}(1 - \bar{\omega})z + z^n , \tag{295'}$$

and this curve is initially in the lower negative quadrant; it intersects the z-axis at $(0, z_s)$ and tends to infinity in the upper quadrant. The signs of $\dot{y}$ and $\dot{z}$ are now as shown in Figure 12.

In contrast to the case $\bar{\omega} < \frac{1}{2}$, we can now prove the following lemma:

LEMMA.—Any solution-curve $[y(t), z(t)]$ starting at a point $y_0 > 0$ on the positive y-axis, with increasing t, monotonically rises (both z and y increasing) until it intersects, for a finite t, the curve $\dot{y} = 0$. After crossing this horizontally, the curve monotonically descends with y decreasing but z still continuing to increase, until it intersects the z-axis at a finite point $z > z_s$. After crossing the z-axis (vertically), it continues to descend, with both y and z decreasing, until it finally intersects the negative y-axis at a finite point.

Let the initial starting-point be $(y_0 > 0, 0)$. In the region I both y and z increase with increasing t, so that the minimum value of y in this region is y_0 itself. Hence, from (295) we have

$$\frac{dy}{dz} < (1 - 2\bar{\omega}) + \frac{\bar{\omega}(1 - \bar{\omega})z}{y_0} , \tag{296}$$

or

$$y < y_0 + (1 - 2\bar{\omega})z + \frac{\bar{\omega}(1 - \bar{\omega})}{2y_0} z^2 . \tag{297}$$

But so long as we are in region I,

$$y > \frac{-\bar{\omega}(1-\bar{\omega})z + z^n}{(1-2\bar{\omega})} . \tag{298}$$

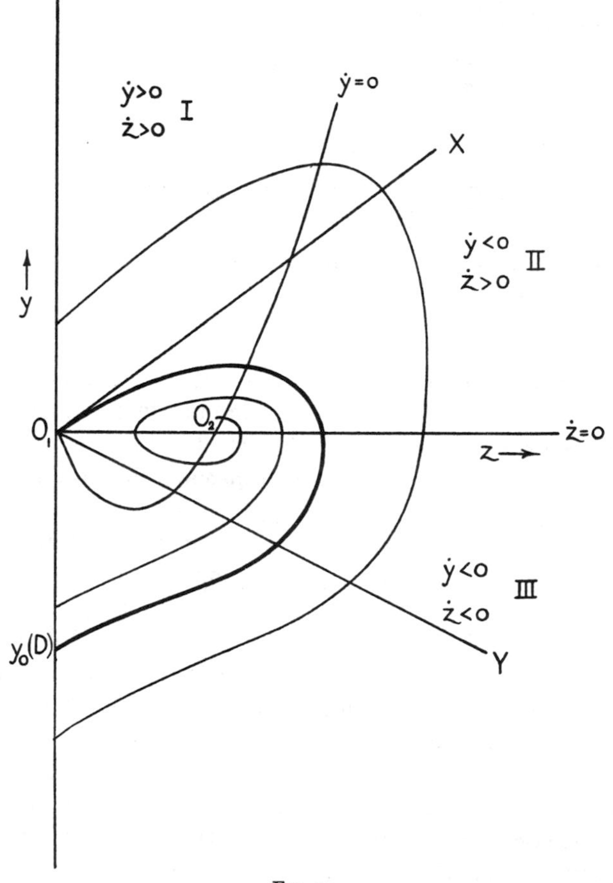

FIG. 12

Equations (297) and (298) are contradictory for large values of z. Hence, the solution must leave the region considered and for a finite t must intersect the locus $\dot{y} = 0$. This point of intersection, (y_1, z_1), must have its z-co-ordinate z_1 greater than z_s.

In the region II, as t continues to increase, y decreases while z continues to increase. From (295) we have

$$\tfrac{1}{2}y^2 = (1 - 2\bar{\omega})\int y dz + \tfrac{1}{2}\bar{\omega}(1 - \bar{\omega})z^2 - \frac{1}{n+1} z^{n+1} + C \,, \quad (299)$$

where C is a constant. Since y decreases in this region, the maximum value of y is y_1, and hence

$$\tfrac{1}{2}y^2 < (1 - 2\bar{\omega})y_1 z + \tfrac{1}{2}\bar{\omega}(1 - \bar{\omega})z^2 - \frac{1}{n+1} z^{n+1} + C \,, \quad (300)$$

which shows that y^2 must become negative for a sufficiently large value of z, i.e., the solution curve cannot avoid the z-axis. Hence, the solution curve must leave the region considered and cross over into region III. In this region y and z both decrease with increasing t; since $|\dot{y}|$ and $|\dot{z}|$ are bounded and finite in the region considered (III), the solution can be continued to intersect the y-axis at a finite nonzero point on the negative y-axis. This proves the lemma.

The foregoing lemma is of importance because it shows that solutions starting on the positive y-axis avoid both singular points, just as the solutions starting on the negative y-axis for $1 > \bar{\omega} > \frac{1}{2}$ avoid the two singular points.

Now since the (y, z) curves form a one-parametric family (except at the singular points) for sufficiently small values of z, we should have

$$y(z; y_0) < y(z; \dot{y}_0) \quad (0 < y_0 < \dot{y}_0) \,. \quad (301)$$

We now construct the limit function

$$y_D(z) = \lim_{y_0 \to +0} y(z; y_0) \,. \quad (302)$$

Our discussion of § 12 has shown that for $\bar{\omega} \neq 1$, we have

$$YX^{\bar{\omega}/(1-\bar{\omega})} = \text{constant} \,, \quad (303)$$

where X and Y are the co-ordinates of the point with respect to the frame of reference defined by $y = (1 - \bar{\omega})z$ and $y = -\bar{\omega}z$ (the

X and the Y directions, respectively). Equation (303) shows that in our case ($\bar{\omega} < \frac{1}{2}$), there can only be (exactly) two solution-curves passing through the origin, one each in the directions X and Y.

It is clear that the solution $y_D(z)$, as defined in (302), which passes through the origin, must be tangential to the X-direction or

$$[y'_D(z)]_{y,\, z=0} = 1 - \bar{\omega} \, . \tag{304}$$

We shall call this the "D-solution," and the corresponding curve $y_D(z)$ the "D-curve."

Draw the complete D-curve and let this cut the negative y-axis at $y_0(D)$. An examination shows that the D-curve which passes through the origin in the direction $y = (1 - \bar{\omega})z$ must do this as $t \to -\infty$ (this arises from the circumstance that, if we trace the solution from $y_0(D)$ "backward" for decreasing t, it can tend toward the singular point O_1 only as $t \to -\infty$ [or $\xi \to +\infty$]). In other words, the origin approached along the D-curve in the $y = (1 - \bar{\omega})z$ direction as y, $z \to 0$ corresponds to approaching the "boundary" in the (ξ, θ) plane. From equation (159) we find that, as $\xi \to \infty$, we should have

$$z = Be^{(1-\bar{\omega})t} \qquad (t \to -\infty) \, . \tag{305}$$

From this it follows that along the solutions in the (θ, ξ) plane which correspond to the D-curve in the (y, z) plane, we should have asymptotically,

$$\theta \sim \frac{B}{\xi} \qquad (\xi \to \infty) \, . \tag{306}$$

Finally, any D-solution has a zero for a finite value of ξ.

From the lemma it follows that the solutions starting on the negative y-axis at $y_0 < y_0(D)$ must lie entirely outside the D-curve; since these solutions intersect the y-axis twice, they correspond in the (θ, ξ) plane to θ having two zeros for finite values of ξ. In other words, outside the D-curve we have F-solutions.

Now consider a solution which starts on the negative y-axis for $0 > y_0 > y_0(D)$. Such a solution must remain entirely in the region bounded by the D-curve and the part of the y-axis

$0 \geqslant y \geqslant y_0(D)$. We shall call this region the "O-region." As we continue the solutions in the O-region from the negative y-axis for decreasing t, it is clear that, as $t \to -\infty$, these solutions must approach one or the other of the two singular points in the O-region. They cannot, however, approach the origin; the D-curve is the only one which does this (this follows from Eq. [303]). Hence, the solutions must approach the second singular point $(0, z_s)$ as $t \to -\infty$. The discussion in § 18 of the possible directions at $(0, z_s)$ now applies; we conclude that, since $\bar{\omega} < \frac{1}{2} < \bar{\omega}^*$, the solutions in the O-region must spiral around the singular point $(0, z_s)$. The discussion of the behavior of the spiraling at $(0, z_s)$ runs parallel to the discussion of the spiraling for the case $\bar{\omega}^* > \bar{\omega} > \frac{1}{2}$; the important difference, however, is that in the previous case $(\bar{\omega}^* > \bar{\omega} > \frac{1}{2})$ the singular point is approached as $t \to +\infty$, while in our present case $(\bar{\omega} < \frac{1}{2})$ the singular point is approached for $t \to -\infty$. In other words, the solutions in the O-region approach the singular solution

$$\theta_s = [\bar{\omega}(1 - \bar{\omega})]^{\bar{\omega}/2}\xi^{-\bar{\omega}} , \tag{307}$$

oscillating as $\xi \to \infty$.

More explicitly, the formulae (266) and (267) are now valid as they stand, but the interpretation is different: while then $t \to +\infty$, now $t \to -\infty$. In particular, we have the solution (cf. Eq. [272])

$$\left.\begin{aligned} \theta \sim \left[\frac{2(n-3)}{(n-1)^2}\right]^{\frac{1}{n-1}} \frac{1}{\xi^{\frac{2}{n-1}}} \Bigg\{ & 1 \\ & + \frac{C}{\xi^{\frac{n-5}{(n-1)}}} \cos\left[\frac{\sqrt{7n^2 - 22n - 1}}{2(n-1)} \log \xi - \delta\right]\Bigg\} , \end{aligned}\right\} \tag{308}$$

where C and δ are constants.

From the foregoing, it follows that, as $\xi \to \infty$, the solution crosses the singular solution θ_s at points which asymptotically increase in the ratio

$$1 : e^{\frac{2(n-1)\pi}{\sqrt{7n^2-22n-1}}} . \tag{309}$$

As $\xi \to \infty$, the difference between θ and θ_s also tends to zero.

Finally, if we make $y_0 \to -0$, we obtain the solution passing through the origin tangential to $y + \bar{\omega} z = 0$ at O_1. This is the E-curve. This E-curve also spirals around the singular point O_2 as $\xi \to \infty$. In the (ξ, θ) plane this means that a solution belonging to the family $\{\theta_n(\xi)\}$ asymptotically approaches the singular solution θ_s oscillating as $\xi \to \infty$. Further, the points of intersection of any E-solution and the singular solution asymptotically increase geometrically in the ratio (309). The E-solution differs from the other solutions in the O-region in that a solution belonging to the family $\{\theta_n(\xi)\}$ has no zero for any finite value of ξ, while in general (i.e., $y_0(D) < y_0 < 0$) the O-solution has a zero. We have, incidentally, shown that configurations with $n > 5$ all extend to infinity. (For $n = 5$ we found that the mass of the configuration was finite; but for $n > 5$, one can easily show from [308] that $-\xi^2\theta' \to \infty$ as $\xi \to \infty$, and that therefore the mass is also infinite.)

It is now clear that the D-solution separates the region of the O-solutions and the region of the F-solutions, and hence the D-solutions in the (ξ, θ) plane form a grid for use in analyzing the other solutions, just as the E-solutions formed a grid with the necessary properties for $\bar{\omega} \geqslant \frac{1}{2}$. The arrangement of the solutions can be stated in the following way, which is very similar to Fowler's theorem:

a) Starting-points (ξ_0, θ_0) are divided into two classes by a critical curve $\theta = z_0(D)\xi^{-\bar{\omega}}$, where $z_0(D)$ is the value of z for which the D-curve in the (y, z) plane intersects the z-axis. The curve

$$\theta = z_0(D)\xi^{-\bar{\omega}} \tag{310}$$

is the envelope of all the D-solutions. A D-solution has a zero for a finite ξ and, after attaining a maximum, tends to zero monotonically as $\xi \to \infty$; and the asymptotic behavior at $\xi = \infty$ is given by

$$\theta \sim \frac{B}{\xi}. \tag{311}$$

All the D-solutions form a homologous family.

b) Any solution $\theta(\xi)$ starting at a point (ξ_0, θ_0) above the critical curve (310) is an F-solution with two zeros at finite points.

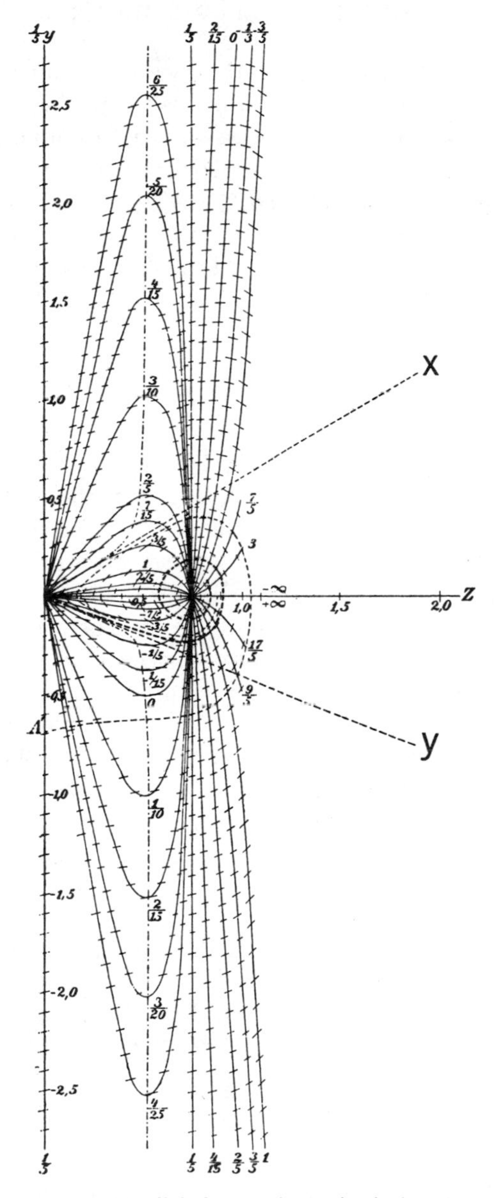

FIG. 13.—The system of isoclinical curves in the (y, z) plane for $n = 6$. (The diagram is reproduced from Emden's *Gaskugeln*.)

c) If the starting-point lies on the critical curve, all the solutions are *F*-solutions, except the one which is tangential at its starting-point to the curve (310); in this case it is a *D*-solution.

d) If the point (ξ_0, θ_0) lies below the critical curve (310), two starting-slopes, $\theta_0' = \theta_0'^{(1)}(D)$ and $\theta_0' = \theta_0'^{(2)}(D)$, exist which lead to *D*-solutions. Let $\theta_0'^{(1)}(D) > \theta_0'^{(2)}(D)$. All solutions corresponding to $\theta_0' > \theta_0'^{(1)}(D)$ and $\theta_0' < \theta_0'^{(2)}(D)$ lead to *F*-solutions. Slopes between $\theta_0'^{(1)}(D)$ and $\theta_0'^{(2)}(D)$ lead to *O*-solutions, which are described below.

e) Any *O*-solution tends asymptotically to the singular solution

$$\theta_s = \left[\frac{2(n-3)}{(n-1)^2}\right]^{1/(n-1)} \frac{1}{\xi^{2/(n-1)}}, \tag{312}$$

as $\xi \to \infty$. The singular solution is, however, approached in an oscillating manner, i.e., an *O*-solution intersects the singular solution again and again, and, as $\xi \to \infty$, the points at which an *O*-solution intersects the singular solution increase asymptotically in a definite ratio.

f) There exists a special class of *O*-solutions—the *E*-solutions—which form a homologous family, and which are finite at $\xi = 0$, and which have, further, $d\theta/d\xi = 0$ at $\xi = 0$. Any *O*-solution which is not an *E*-solution has a zero for some finite ξ.

21. *Discussion in the* (u, v) *plane.*—We shall now briefly discuss the arrangement of the solutions in the (u, v) plane. As has been already defined in § 8, we have

$$u = -\frac{\xi\theta^n}{\theta'}; \qquad v = -\frac{\xi\theta'}{\theta}. \tag{313}$$

Further, we saw that u and v satisfy the differential equation (Eq. [122])

$$\frac{u}{v}\frac{dv}{du} = -\frac{u+v-1}{u+nv-3}. \tag{314}$$

As may be verified easily, the (u, v) variables are related to the (y, z) variables according to the relations

$$z = \xi^{\bar{\omega}}\theta = [uv]^{\bar{\omega}/2} \qquad \left(\bar{\omega} = \frac{2}{n-1}\right), \tag{315}$$

$$y = [uv^n]^{\bar{\omega}/2} - \bar{\omega}[uv]^{\bar{\omega}/2}. \tag{316}$$

The (u, v) variables, first introduced by Milne, have the great advantage that the positive quadrant $(u \geqslant 0, v \geqslant 0)$ contains only such parts of the (ξ, θ) solution which are of astrophysical interest, i.e., points in the positive (u, v) quadrant correspond to $\theta \geqslant 0$, $\theta' \leqslant 0$.

We shall now consider some general properties of the (u, v) plane.

a) The locus of points at which the solution-curves have horizontal tangents is given by

$$u + v = 1 . \tag{317}$$

b) The locus of points at which the solution-curves have vertical tangents is given by

$$u + nv = 3 . \tag{318}$$

c) For $n < 3$ the loci (317) and (318) do not intersect in the positive (u, v) quadrant.

d) For $n = 3$ the loci are

$$u + v = 1 , \qquad u + 3v = 3 . \tag{319}$$

Hence, they both pass through the point $(v = 1, u = 0)$ on the v-axis.

e) For $n > 3$ the two loci intersect at the point (u_s, v_s), which is easily verified to be

$$u_s = \frac{n-3}{n-1} ; \qquad v_s = \frac{2}{n-1} = \bar{\omega} . \tag{320}$$

This intersection of the two loci, (317) and (318), in the positive quadrant corresponds to the existence of the proper singular solution for $n > 3$:

$$\theta_s = \left[\frac{2(n-3)}{(n-1)^2}\right]^{1/(n-1)} \frac{1}{\xi^{2/(n-1)}} , \tag{321}$$

$$\theta_s' = -2\left[\frac{2(n-3)}{(n-1)^{n+1}}\right]^{1/(n-1)} \frac{1}{\xi^{(n+1)/(n-1)}} . \tag{322}$$

If we form the variables u and v as defined in (313), we readily find from the foregoing expressions that u and v reduce to u_s and v_s as defined by (320).

f) Let us consider the E-solutions at $\xi \sim 0$. It is sufficient to consider $\theta_n(\xi)$, since any other member of the family $\{\theta_n(\xi)\}$ will lead to the same (u, v) curve. As $\xi \to 0$, we have (Eq. [62])

$$\theta_n(\xi) = 1 - \tfrac{1}{6}\xi^2 + \frac{n}{120}\xi^4 - \dots \quad (\xi \to 0)\,. \quad (323)$$

From the foregoing, we find

$$\theta_n^n(\xi) \sim 1 - \frac{n}{6}\xi^2\,; \qquad \theta_n'(\xi) \sim -\tfrac{1}{3}\xi + \frac{n}{30}\xi^3\,. \quad (324)$$

Hence, as $\xi \to 0$,

$$u_E = -\frac{\xi\theta^n}{\theta'} \sim 3\left(1 - \frac{n}{15}\xi^2\right) \quad (\xi \to 0)\,, \quad (325)$$

$$v_E = -\frac{\xi\theta'}{\theta} \sim \tfrac{1}{3}\xi^2 \quad (\xi \to 0)\,. \quad (326)$$

We see that the E-curve passes through the point

$$u_E = 3\,, \qquad v_E = 0 \qquad (\xi = 0)\,, \quad 327)$$

for all values of n. At this point the E-curve has a definite slope determined by

$$\frac{du_E}{d\xi} \sim -\frac{2n}{5}\xi\,; \qquad \frac{dv_E}{d\xi} \sim \tfrac{2}{3}\xi\,. \quad (328)$$

Hence,

$$\left(\frac{dv_E}{du_E}\right)_{\xi=0} = -\frac{5}{3n}\,. \quad (329)$$

g) Let us consider a solution for $n < 5$ which starts with a definite slope on the ξ-axis at $\xi = \xi_1$. According to Fowler's theorem, there is exactly one E-solution through the point. All solutions with starting-slopes below the E-curve are M-solutions, and all those with starting-slopes above the E-curve are F-solutions.

Given the slope at ξ_1, we can easily form a Taylor series at this point. From the Lane-Emden equation we have

$$\theta'' = -\theta^n - \frac{2}{\xi}\theta'. \quad (330)$$

Since $\theta = 0$ at $\xi = \xi_1$, we have

$$\theta''_{\xi_1} = -\frac{2}{\xi_1}\,\theta'_{\xi_1}\,. \tag{331}$$

Differentiate (330) and set $\xi = \xi_1$. We find

$$\theta'''_{\xi_1} = \frac{2}{\xi_1^2}\,\theta'_{\xi_1} - \frac{2}{\xi_1}\,\theta''_{\xi_1}\,, \tag{332}$$

or by (331)

$$\theta'''_{\xi_1} = \frac{6}{\xi_1^2}\,\theta'_{\xi_1}\,. \tag{333}$$

The Taylor series in the neighborhood of ξ_1 is given by

$$\theta(\xi) = (\xi - \xi_1)\theta'_{\xi_1} + \frac{(\xi - \xi_1)^2}{2}\,\theta''_{\xi_1} + \frac{(\xi - \xi_1)^3}{6}\,\theta'''_{\xi_1} + \ldots\,; \tag{334}$$

or, using (331) and (333), we obtain

$$\theta(\xi) = (\xi - \xi_1)\theta'_{\xi_1} - \frac{(\xi - \xi_1)^2}{\xi_1}\,\theta'_{\xi_1} + \frac{(\xi - \xi_1)^3}{\xi_1^2}\,\theta'_{\xi_1} - \ldots\,, \tag{335}$$

or

$$\theta(\xi) = -\xi_1\theta'_{\xi_1}\left[\frac{\xi_1 - \xi}{\xi_1} + \left(\frac{\xi_1 - \xi}{\xi_1}\right)^2 + \left(\frac{\xi_1 - \xi}{\xi_1}\right)^3 + \ldots\right]. \tag{336}$$

Let

$$\omega_n = -\xi_1^{\frac{n+1}{n-1}}\left(\frac{d\theta}{d\xi}\right)_{\xi_1}\,. \tag{337}$$

Now the function $-\xi^{\bar{\omega}+1}\theta'$ is a homology-invariant function, and, since the zeros of two members of a homologous family are "corresponding points," it follows that any homology-invariant function (and therefore also $-\xi^{\bar{\omega}+1}\theta'$) will have the same value for all the members of a homologous family at their respective zeros. Since each homologous family yields only one curve in the (u, v) plane, and since, further, every member of the homologous family will have the same value for ω_n, it is clear that we can choose ω_n to label the solution-curves in the (u, v) plane. In particular, the E-curve will be labeled by the quantity ${}_0\omega_n$ already introduced in § 6 (cf. Eq. [73]).

From (336) and (337) we can write

$$\theta(\xi) \sim \frac{\omega_n}{\xi_1^{\bar{\omega}}}\left(\frac{\xi_1 - \xi}{\xi_1}\right) \qquad (\xi \to \xi_1) , \quad (338)$$

$$\theta'(\xi) \sim -\frac{\omega_n}{\xi_1^{\bar{\omega}+1}} \qquad (\xi \to \xi_1) . \quad (339)$$

Hence, we have

$$u(\xi) \sim -\frac{\xi_1 \theta^n}{\theta'} \sim \omega_n^{n-1}\left(\frac{\xi_1 - \xi}{\xi_1}\right)^n , \quad (340)$$

$$v(\xi) \sim -\frac{\xi_1 \theta'}{\theta} \sim \frac{\xi_1}{\xi_1 - \xi} , \quad (341)$$

as $\xi \to \xi_1$. From (340) and (341) we finally have

$$uv^n \sim \omega_n^{n-1} . \quad (342)$$

In other words, as $u \to 0$, $v \to \infty$. In particular, along the E-curve

$$[uv^n]_E \sim {}_0\omega_n^{n-1} \qquad (u \to 0) . \quad (343)$$

Now, from Fowler's theorem we have (cf. Eq. [231], [232], [233], and [234])

$${}_M\omega_n < {}_0\omega_n < {}_F\omega_n . \quad (344)$$

The M-curves, therefore, lie inside the region bounded by the E-curve, the v-axis, and the part of the u-axis $0 \leqslant u \leqslant 3$.

The F-curves, on the other hand, lie entirely outside the E-curve. Also, when an F-solution attains its maximum in the (θ, ξ) plane at, say, ξ_0, θ_0 (where $\xi_0 > 0$), it is clear from the definitions of the variables u and v that along the corresponding F-curve, $u \to \infty$ and $v \to 0$. Further, since

$$uv = -\xi^2 \theta^{n-1} , \quad (345)$$

it follows that the F-curves tend to become like rectangular hyperbolas when they asymptotically approach the u-axis.

h) Finally, let us consider the behavior of the (u, v) curves near the point $(u = 3, v = 0)$. Write

$$u = 3 + u_1 . \quad (346)$$

The (u, v) differential equation (314) now takes the form

$$\frac{3 + u_1}{v} \frac{dv}{du_1} = -\frac{2 + u_1 + v}{nv + u_1} . \tag{347}$$

If u_1 and $v \to 0$, we can write, approximately,

$$\frac{3}{v} \frac{dv}{du_1} = -\frac{2}{nv + u_1} . \tag{348}$$

A separation of the variables can be effected by the substitution

$$v = u_1 w . \tag{349}$$

Equation (348) reduces to

$$\frac{3}{w} \frac{dw}{du_1} = -\frac{5 + 3nw}{u_1(1 + nw)} , \tag{350}$$

or

$$\left[\frac{6n}{5 + 3nw} + \frac{3}{w}\right] dw + 5 \frac{du_1}{u_1} = 0 ; \tag{351}$$

integrating, we find,

$$(5 + 3nw)^2 w^3 u_1^5 = \text{constant} , \tag{352}$$

which we write as

$$(5u_1 + 3nwu_1)^2 (wu_1)^3 = \text{constant} . \tag{353}$$

Returning to our variables u and v, we have

$$v^{3/2}[5(u - 3) + 3nv] = \text{constant} . \tag{354}$$

Hence, near $u = 3$, $v = 0$, the (u, v) curves resemble portions of generalized hyperbolas asymptotic to the u-axis and the line

$$5(u - 3) + 3nv = 0 . \tag{355}$$

We get the F-curves when the constant in (354) is positive, and the M-curves when it is negative. When the constant is zero, we get the E-curve represented near $(u = 3, v = 0)$ by its tangent (355); thus we arrive at our earlier result (cf. Eq. [329]).

We will now consider a little more carefully the different cases: $1 < n < 3$, $n = 3$, $3 \leqslant n \leqslant n^*$, $n^* < n < 5$, $n = 5$, and $n > 5$. The principal question to be considered is the nature of the M-curves.

Case i: $1 < n < 3$.—We have already described the E- and the F-curves (f, g, and h above). Now, along an M-solution as $\xi \to 0$, we have the asymptotic relation (Eq. [221])

$$\theta \sim \frac{C}{\xi}, \tag{356}$$

and correspondingly, as may be verified,

$$u = 0, \qquad v = 1. \tag{357}$$

Hence, the general nature of the (u, v) curves are as shown in Figure 14.

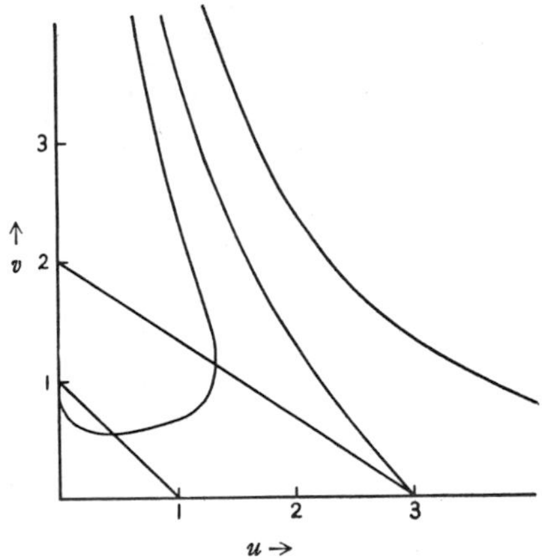

FIG. 14.—The (u, v) curves ($n = 1.5$)

Case ii: $n = 3$.—According to (225), along an M-curve we have

$$y \sim -z^3 \qquad (y \to 0, \quad z \to 0). \tag{358}$$

From (316) it follows that (as $\bar{\omega} = 1$ in our present case)

$$-(uv)^{3/2} = (uv^3)^{1/2} - (uv)^{1/2}, \tag{359}$$

or

$$uv + v - 1 = 0. \tag{360}$$

But we should also have (Eq. [315])

$$z = (uv)^{1/2} \to 0 \qquad \text{as} \qquad \xi \to 0. \tag{361}$$

From (360) and (361) it follows that the M-curve tends to ($u = 0$, $v = 1$) as $\xi \to 0$.

Hence, the arrangement of the curves is qualitatively the same as in case i above (cf. Fig. 15).

Case iii: $3 < n < n^* = 3.18 \ldots$—The two loci $u + v = 1$ and $u + nv = 3$ intersect at the point (u_s, v_s) (given by Eq. [320]) in the positive quadrant. As may be verified from our asymptotic formulae for the M-solutions for this case (Eq. [285]), the M-curves in the (u, v) plane tend to the point (u_s, v_s) in a definite direction.

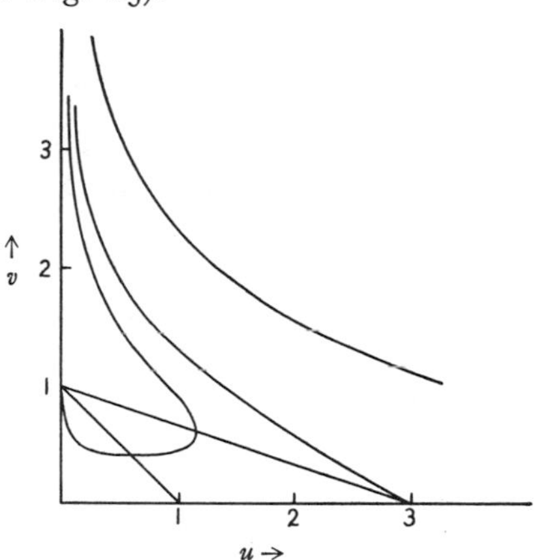

FIG. 15.—The (u, v) curves $(n = 3)$

Case iv: $n^* < n < 5$.—The M-curves now spiral around the point (u_s, v_s), but the nature of the E- and F-curves are as beforc (cf. Fig. 16).

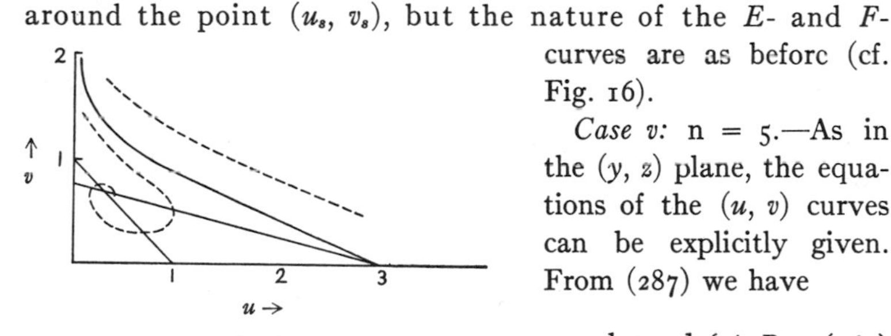

FIG. 16.—The (u, v) curves $(n = 4)$

Case v: $n = 5$.—As in the (y, z) plane, the equations of the (u, v) curves can be explicitly given. From (287) we have

$$y^2 = \tfrac{1}{4}z^2 - \tfrac{1}{3}z^6 + D\,. \qquad (362)$$

Further, for this case we have (Eqs. [315] and [316])

$$z = (uv)^{1/4}\,; \qquad y = (uv^5)^{1/4} - \tfrac{1}{2}z\,. \qquad (363)$$

Substituting (363) in (362), we have

$$(uv^5)^{1/2} + \tfrac{1}{4}z^2 - z(uv^5)^{1/4} = \tfrac{1}{4}z^2 - \tfrac{1}{3}(uv)^{3/2} + D\,, \qquad (364)$$

or

$$(uv^5)^{1/2} - (uv^3)^{1/2} = -\tfrac{1}{3}(uv)^{3/2} + D\,. \qquad (365)$$

Dividing throughout by $(uv^3)^{1/2}$ and rearranging the terms, we find that

$$u + 3v = 3 + \frac{3D}{(uv^3)^{1/2}} \cdot \tag{366}$$

The Schuster-Emden integrals correspond to $D = 0$; hence for $n = 5$ the E-curve is the straight line

$$u + 3v = 3 \cdot \tag{367}$$

The arrangement of the (u, v) curves is as shown in Figure 17.

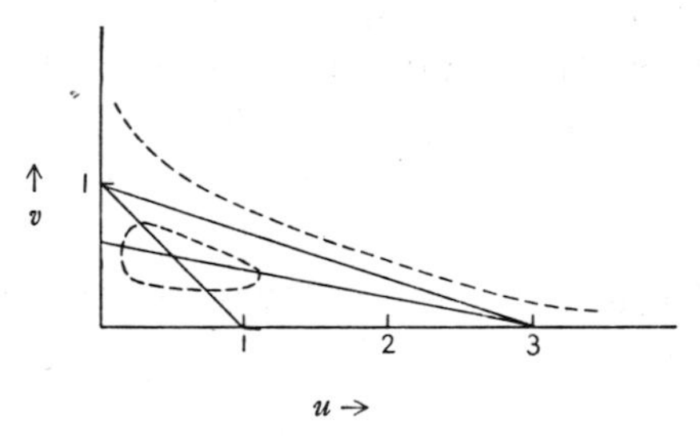

FIG. 17.—The (u, v) curves $(n = 5)$

Case vi: n > 5.—Now the general nature of the (u, v) curves is completely different.

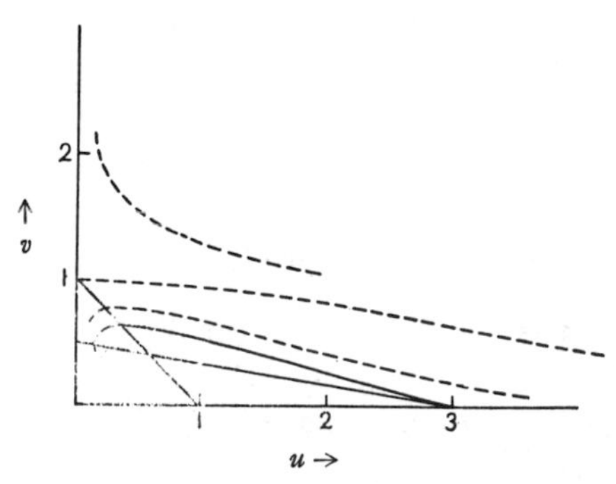

FIG. 18.—(The (u, v) curves $(n = 6)$

The E-curve starts at the point $u = 3, v = 0$ with a slope $-5/3n$ and approaches the singular point (u_s, v_s) by spiraling around it.

The D-curve, as may be verified from our relation of the behavior of these solutions as $\xi \to \infty$, comes from infinity ($u = \infty$, $v = 0$) and joins the point ($v = 1$, $u = 0$). The O-curves all lie inside the region bounded by the D-curve, the u-axis, and the part of the v-axis $0 \leqslant v \leqslant 1$. All these curves approach the point (u_s, v_s) by spiraling around it. The F-curves, which lie outside the D-curve, are of the same general nature as in the previous cases. The general nature of the (u, v) curves are as shown in Figure 18. This concludes our discussion of the Lane-Emden equation.

22. *The isothermal gas sphere.*—We shall now consider an isothermal gas sphere in gravitational equilibrium. We then have

$$P = \left(\frac{k}{\mu H}\right) \rho T + \frac{a}{3} T^4 , \tag{368}$$

where T is assumed to be constant. We can write the foregoing in the standard form

$$P = K\rho + D , \tag{369}$$

where, for the case in hand,

$$K = \frac{k}{\mu H} T ; \qquad D = \frac{a}{3} T^4 . \tag{370}$$

The equation of equilibrium,

$$\frac{1}{r^2} \frac{d}{dr}\left(\frac{r^2}{\rho} \frac{dP}{dr}\right) = -4\pi G \rho , \tag{371}$$

can be written as

$$K \frac{1}{r^2} \frac{d}{dr}\left(r^2 \frac{d \log \rho}{dr}\right) = -4\pi G \rho . \tag{372}$$

Make the substitutions

$$\rho = \lambda e^{-\psi} ; \qquad r = \left[\frac{K}{4\pi G \lambda}\right]^{1/2} \xi = a\xi , \tag{373}$$

where λ is, for the present, an arbitrary constant. Equation (372) now reduces to

$$\frac{1}{\xi^2}\frac{d}{d\xi}\left(\xi^2\frac{d\psi}{d\xi}\right) = e^{-\psi} . \tag{374}$$

Equation (374), which is our present analogue of the Lane-Emden equation, must govern the density distribution in any region in which a relation of the kind (369) is valid. If, however, we consider a complete isothermal gas sphere (or a configuration in which the central regions are isothermal), we can choose λ to be the central density, in which case $\psi = 0$ at $\xi = 0$. Further, it is clear that $d\psi/d\xi$ must vanish at the origin. Thus, with the normalization $\lambda = \rho_c$ we must find a solution of (374) which satisfies the boundary conditions

$$\psi = 0 , \qquad \frac{d\psi}{d\xi} = 0 \qquad \text{at} \qquad \xi = 0 . \tag{375}$$

The structure of the complete isothermal gas sphere can be determined when a solution of (374) satisfying the boundary conditions (375) can be obtained. It does not appear that the equation (374) can be explicitly integrated, and recourse must be made to numerical methods. We start the integration by computing the values of ψ near $\xi = 0$ by means of a power series.

Assuming an expansion of the form

$$\psi = a\xi^2 + b\xi^4 + c\xi^6 + \dots , \tag{376}$$

we substitute it in the isothermal equation and determine the coefficients $a, b, c, \dots$, successively, by equating the coefficients of the like powers of ξ. The first three terms of the series are found to be

$$\psi = \tfrac{1}{6}\xi^2 - \tfrac{1}{120}\xi^4 + \tfrac{1}{1890}\xi^6 + \dots . \tag{377}$$

A few terms of the foregoing series will enable us to compute ψ for $\xi < 1$. For $\xi > 1$ the solution so obtained must be continued by standard methods. We shall denote this function by $\Psi(\xi)$.

We shall show in § 26 that the complete isothermal gas sphere extends to infinity. Here we may notice the following formulae for the mass, $M(\xi)$, interior to ξ and the mean density, $\bar{\rho}(\xi)$, of the matter interior to ξ, which are derived in the same way as the corresponding formulae (Eqs. [67], [77]) for the polytropes

$$M(\xi) = 4\pi a^3 \lambda \xi^2 \frac{d\psi}{d\xi}, \tag{378}$$

$$\bar{\rho}(\xi) = \lambda \frac{3}{\xi} \frac{d\psi}{d\xi}. \tag{379}$$

Before we proceed to discuss the nature of the function $\Psi(\xi)$ as $\xi \to \infty$ and the general solution $\psi(\xi)$, we shall consider some convenient transformations of the isothermal equation.

23. *Transformations of the isothermal equation.*—

a) Put

$$\psi = \frac{\chi}{\xi}. \tag{380}$$

Equation (374) takes the form

$$\frac{d^2\chi}{d\xi^2} = \xi e^{-\chi/\xi}. \tag{381}$$

b) Put

$$x = \frac{1}{\xi}. \tag{382}$$

Equation (374) now takes the form

$$x^4 \frac{d^2\psi}{dx^2} = e^{-\psi}. \tag{383}$$

c) We can verify that equation (383) is satisfied by the following singular solution:

$$e^{-\psi_s} = 2x^2; \qquad \frac{d\psi_s}{dx} = -\frac{2}{x}. \tag{384}$$

d) *Emden's transformation.*—Because of the existence of the singular solution (384) we introduce the new variable z defined by

$$-\psi = 2 \log x + z. \tag{385}$$

We find

$$\frac{d\psi}{dx} = -\frac{2}{x} - \frac{dz}{dx}; \qquad \frac{d^2\psi}{dx^2} = \frac{2}{x^2} - \frac{d^2z}{dx^2}. \tag{386}$$

Equation (383) now takes the form

$$x^2 \frac{d^2 z}{dx^2} + e^z - 2 = 0 . \qquad (387)$$

We can eliminate x from the foregoing equation by the transformation

$$x = \frac{1}{\xi} = e^t . \qquad (388)$$

We have

$$\frac{dz}{dx} = e^{-t} \frac{dz}{dt} ; \qquad \frac{d^2 z}{dx^2} = e^{-2t} \left(\frac{d^2 z}{dt^2} - \frac{dz}{dt} \right) , \qquad (389)$$

and (387) now reduces to

$$\frac{d^2 z}{dt^2} - \frac{dz}{dt} + e^z - 2 = 0 . \qquad (390)$$

24. *The homology theorem for the isothermal equation.*—The isothermal equation admits of a constant of homology quite similar to the homology properties of the Lane-Emden equation. In our present case we have: *If* $\psi(\xi)$ *is a solution of the isothermal equation, then* $\psi(A\xi) - 2 \log A$ *is also a solution of the equation, where* A *is an arbitrary constant.*

To show this, write

$$\eta = A\xi , \qquad \psi = \psi^* - 2 \log A . \qquad (391)$$

These transformations lead to an equation in the (ψ^*, η) variables which is identical in form with the equation in the (ψ, ξ) variables. Hence, if $f(\xi)$ is a solution of the original equation, we can choose as a solution for ψ^* the function $f(\eta)$. Returning to our original variables, we now have

$$\psi(\xi) = \psi^*(\eta) - 2 \log A = f(A\xi) - 2 \log A , \qquad (392)$$

while $f(\xi)$ has already been assumed to be a solution. This proves the theorem.

From (392) it follows that, if we choose for $f(\xi)$ the function $\Psi(\xi)$, then we can derive a whole continuous family of solutions which are finite at the origin and which have, further, $d\psi/d\xi = 0$ at $\xi = 0$. We shall denote this family of solutions by $\{\Psi(\xi)\}$ and refer to them as the "E-solutions."

As in the case of the Lane-Emden equation, we should be able to reduce the isothermal equation to one of the first order. The variables introduced in § 23 (case d) enable us to effect this reduction.

Introduce the variable y defined by

$$y = \frac{dz}{dt} . \tag{393}$$

Then

$$\frac{d^2z}{dt^2} = \frac{dy}{dt} = \frac{dy}{dz}\frac{dz}{dt} = y\frac{dy}{dz} . \tag{394}$$

Equation (390) can now be written as

$$y\frac{dy}{dz} - y + e^z - 2 = 0 , \tag{395}$$

an equation analogous to the (y, z) differential equation we had before. This reduction to a first-order equation is due to the fact that the functions y and z are homology invariant. According to equation (385),

$$z = -\psi + 2\log\xi . \tag{396}$$

Hence,

$$y = \frac{dz}{dt} = -\xi\frac{dz}{d\xi} = \xi\frac{d\psi}{d\xi} - 2 . \tag{397}$$

To show that y and z are homology invariant, we notice that, if ξ and ξ/A are the corresponding points along two solutions ψ and ψ^* (which can be transformed one into the other by means of a homologous transformation), then we have

$$\psi^*(\xi/A) = \psi(\xi) - 2\log A , \tag{398}$$

$$\left(\frac{d\psi^*}{d\xi}\right)_{\xi=\xi/A} = A\left(\frac{d\psi}{d\xi}\right)_{\xi=\xi} . \tag{399}$$

Using (398) and (399), we can easily show that $z^*(\xi/A)$ and $y^*(\xi/A)$ (defined with respect to the function ψ^*) are identical with $z(\xi)$ and $y(\xi)$ (defined with respect to the function ψ).

As another example of the reduction of the isothermal equation to an equation of the first order, let us consider the functions u and v defined as follows:

$$u = \frac{\xi e^{-\psi}}{\psi'} \,; \qquad v = \xi\psi' \,, \tag{400}$$

where we have used ψ' to denote $d\psi/d\xi$. These functions are easily seen to be homology invariant, and the first-order equation between u and v can be obtained as follows: We have

$$\frac{1}{u}\frac{du}{d\xi} = \frac{1}{\xi} - \frac{d\psi}{d\xi} - \frac{\psi''}{\psi'} \,. \tag{401}$$

Since, according to the isothermal equation,

$$\psi'' = e^{-\psi} - \frac{2}{\xi}\psi' \,, \tag{402}$$

we can re-write (401) as

$$\frac{1}{u}\frac{du}{d\xi} = \frac{1}{\xi}\left(3 - \xi\psi' - \frac{\xi e^{-\psi}}{\psi'}\right) , \tag{403}$$

or

$$\frac{1}{u}\frac{du}{d\xi} = \frac{1}{\xi}\,(3 - u - v) \,. \tag{404}$$

Similarly, we find that

$$\frac{1}{v}\frac{dv}{d\xi} = \frac{1}{\xi}\,(u - 1) \,. \tag{405}$$

Hence, combining (404) and (405), we have

$$\frac{u}{v}\frac{dv}{du} = -\frac{u - 1}{u + v - 3} \,. \tag{406}$$

We shall return to this equation in § 27.

25. *The isothermal* E-*solutions.*—We shall prove that *the solutions of the isothermal equation which are finite at the origin have necessarily* $d\psi/d\xi = 0$ *at* $\xi = 0$, and that, consequently, the homologous family $\{\Psi(\xi)\}$ includes all solutions which are finite at the origin.

Consider the isothermal equation in the form (Eq. [381])

$$\frac{d^2\chi}{d\xi^2} = \xi e^{-\chi/\xi} \qquad (\chi = \psi\xi) . \tag{407}$$

Such solutions $\psi(\xi)$ which are finite at $\xi = 0$ correspond to solutions passing through the origin in the (χ, ξ) plane; as in § 9, we now have

$$\left(\frac{d\psi}{d\xi}\right)_{\xi=0} = \frac{1}{2} \lim_{\xi=0} [\xi e^{-\chi/\xi}] = 0 , \tag{408}$$

since $\chi/\xi = \psi$ is finite at the origin, $\xi = 0$; this proves the theorem.

26. *The discussion of the isothermal equation in the* (y, z) *plane.*—The functions y and z as we have defined them are homology-invariant functions, and consequently each solution curve in the (y, z) plane corresponds to a complete homologous family of solutions in the (ξ, θ) plane. In particular, there is just one curve in the (y, z) plane which corresponds to the E-solutions which are included in the homologous family $\{\Psi(\xi)\}$. We shall call the curve which corresponds to the family $\{\Psi(\xi)\}$ the "E-curve" and denote it by $y_E(z)$.

To repeat, our equations are

$$y\frac{dy}{dz} - y + e^z - 2 = 0 , \tag{409}$$

$$z = -\psi + 2 \log \xi , \tag{410}$$

$$y = \xi\psi' - 2 . \tag{411}$$

Further,

$$y = \frac{dz}{dt} ; \qquad \xi = e^{-t} . \tag{412}$$

From (409) we have

$$y' = \frac{2 + y - e^z}{y} , \tag{413}$$

where we have denoted dy/dz by y'. Substituting for z and y according to (410) and (411) in (413), we have

$$y' = \frac{\xi\psi' - \xi^2 e^{-\psi}}{\xi\psi' - 2} . \tag{414}$$

We see that the point $(0, z_s)$, where

$$e^{z_s} = 2 ; \qquad z_s = \log 2 , \tag{415}$$

is a singular point of the differential equation (409). There are no other singular points in the finite part of the (y, z) plane. The existence of the singular point corresponds to the existence of the singular solution (384), for by (410) and (415) we have

$$-\psi_s = z_s - 2 \log \xi = \log \frac{2}{\xi^2} . \tag{416}$$

Now the E-curve is characterized by

$$\psi_E \text{ finite} ; \qquad \psi' = 0 \qquad \text{as} \qquad \xi \to 0 . \tag{417}$$

From (410), (411), and (417) we have

$$z \to -\infty ; \quad y \to -2 , \quad y' \to 0 , \quad t \to \infty . \tag{418}$$

Hence, the E-curve touches the line $y = -2$ asymptotically as $z \to -\infty$. On the other hand, we can show that there cannot be two solution-curves which are both asymptotic to the line $y = -2$ as $z \to -\infty$. For, if there were, let y and y^* be two different solutions such that

$$y \sim -2 , \qquad y^* \sim -2 , \qquad z \to -\infty . \tag{419}$$

We may suppose that $y < y^*$ as $z \to -\infty$. Then we should have

$$\Delta = y^* - y > 0 ; \qquad \lim_{z \to -\infty} \Delta = 0 . \tag{420}$$

From the differential equation (409) we derive

$$\frac{d\Delta}{dz} = -\frac{2 - e^z}{yy^*} \Delta . \tag{421}$$

Hence,

$$\lim_{z \to -\infty} \frac{d \log \Delta}{dz} = - \lim_{z \to -\infty} \left[\frac{2 - e^z}{yy^*} \right] , \tag{422}$$

or by (419)

$$\lim_{z \to -\infty} \frac{d \log \Delta}{dz} = -\tfrac{1}{2} , \tag{423}$$

or

$$\Delta \sim \text{constant } e^{-\frac{1}{2}z} \qquad (z \to -\infty) , \tag{424}$$

which contradicts our assumption that $\Delta \to 0$ as $z \to -\infty$. We have thus proved the uniqueness of the E-curve.

We shall next examine the behavior of the solution-curves near the singular point $(0, z_s)$. Write

$$z = z_1 + z_s = z_1 + \log 2 , \tag{425}$$

where we now regard z_1 as small. Equation (409) now reduces to

$$y \frac{dy}{dz_1} - y + 2e^{z_1} - 2 = 0 . \tag{426}$$

The foregoing equation is exact. Since z_1 is now considered small, we can expand the exponential in a power series and retain only the first two terms. We have, approximately,

$$y \frac{dy}{dz_1} - y + 2z_1 = 0 ; \tag{427}$$

or, since $y = dz/dt = dz_1/dt$, we have, instead of (427),

$$\frac{d^2 z_1}{dt^2} - \frac{dz_1}{dt} + 2z_1 = 0 . \tag{428}$$

The general solution of (428) can be written as

$$z_1 = A e^{q_1 t} + B e^{q_2 t} , \tag{429}$$

where A and B are integration constants and q_1 and q_2 are the roots of the equation

$$q^2 - q + 2 = 0 , \tag{430}$$

or

$$q_1, q_2 = \tfrac{1}{2} \pm i\tfrac{1}{2}\sqrt{7} . \tag{431}$$

The roots are imaginary, and the solution (429) can therefore be written in the form

$$z_1 = Ae^{t/2} \cos \left[\frac{\sqrt{7}}{2} t + \delta \right] , \tag{432}$$

where δ is a constant. We see that (432) is exactly the limiting form of our earlier equation (264) as $\bar{\omega} \to 0$, $n \to \infty$. From (432) we have

$$y = \frac{dz_1}{dt} = \tfrac{1}{2} Ae^{t/2} \left[\cos \left(\frac{\sqrt{7}}{2} t + \delta \right) - \sqrt{7} \sin \left(\frac{\sqrt{7}}{2} t + \delta \right) \right] . \tag{433}$$

We see that the singular point $(0, z_s)$ is approached spirally as $t \to -\infty$, $\xi \to \infty$. The general run of the solution-curves is illustrated in Figure 19.

From (410) we have

$$-\psi = z - 2 \log \xi = z + 2t . \tag{434}$$

From (425), (432), and (434), we have

$$-\psi = 2t + \log 2 + Ae^{t/2} \cos \left[\frac{\sqrt{7}}{2} t + \delta \right] ; \tag{435}$$

or, since $\xi = e^{-t}$, we can also write

$$-\psi = \log \frac{2}{\xi^2} + \frac{A}{\xi^{1/2}} \cos \left[\frac{\sqrt{7}}{2} \log \xi - \delta \right] . \tag{436}$$

Finally, since

$$\rho = \lambda e^{-\psi} ,$$

we have for the Law of Density Distribution:

$$\rho = \lambda \frac{2}{\xi^2} \exp \left\{ \frac{A}{\xi^{1/2}} \cos \left[\frac{\sqrt{7}}{2} \log \xi - \delta \right] \right\} \quad (\xi \to \infty) . \tag{437}$$

Since the exponent tends to zero as $\xi \to \infty$, we can further expand the exponential and retain only the first two terms. We find in this way that

$$\rho = \lambda \frac{2}{\xi^2} \left\{ 1 + \frac{A}{\xi^{1/2}} \cos \left[\frac{\sqrt{7}}{2} \log \xi - \delta \right] \right\} \quad (\xi \to \infty) . \tag{438}$$

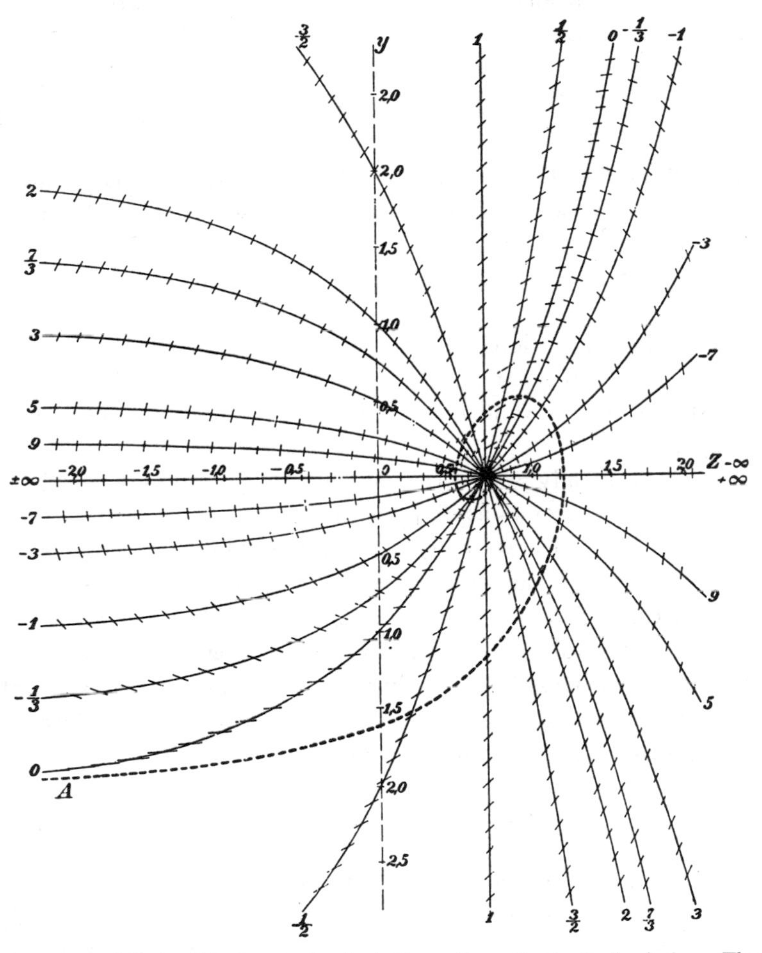

FIG. 19.—The isoclinical curves for the isothermal equation in the (y, z) plane. The diagram is reproduced from Emden's *Gaskugeln*.

From (438) it follows that as $\xi \to \infty$ the density distribution approaches that corresponding to the singular solution, namely,

$$\rho_s = \lambda \frac{2}{\xi^2}, \tag{439}$$

asymptotically. The solution (438) intersects the singular solution (439) at points which asymptotically increase geometrically in the ratio

$$1 : e^{2\pi/\sqrt{7}} = 1 : 10.749 \ldots\ldots \tag{440}$$

Equation (432) describes the general behavior of the solutions near the singular point $(0, z_s)$; hence the law of density distribution (438) as $\xi \to \infty$ is valid quite generally and for the E-solution in particular.

We have already proved the uniqueness of the E-curve, which, as we have seen, becomes asymptotic to the line $y = -2$ as $z \to -\infty$, $t \to +\infty$. As t decreases from $+\infty$, the E-curve monotonically rises, intersecting the y-axis at a definite point; at this point we have, according to (410), $\psi = 2 \log \xi$ or $\rho = \lambda/\xi^2$, which is exactly one-half of the value of ρ on the singular solution (439) (cf. Fig. 19). For further decrease in t the solution approaches the singular point by spiraling around it; after each revolution $z_1 (= z - z_s)$ and y both decrease asymptotically in the ratio

$$1 : e^{2\pi/\sqrt{7}} = 0.09303 \ldots\ldots \tag{441}$$

The density distribution as $\xi \to \infty$ is given by a law of the kind (438).

All solutions other than the E-solution come from $y = -\infty$, and, as $t \to \infty$, they again spiral round the singular point. The behavior of these solutions as $z \to -\infty$ can also be specified by an appeal to Hardy's theorem. For, by an application of Hardy's theorem to (409), it follows that we should ultimately have one of the following three possibilities:

$$\frac{dy}{dz} \to 0\,; \qquad \frac{dy}{dz} \to \infty\,; \qquad \frac{dy}{dz} \to 1 \quad (z \to -\infty)\,. \tag{442}$$

The first possibility yields the E-curve; the second is clearly impossible; hence the only remaining possibility is the third, which gives

$$y \sim z + C_1 \quad (z \to -\infty \,, \xi \to 0) \,, \tag{443}$$

where C_1 is a constant.

From (412) and (442) we also have

$$\frac{1}{y}\frac{dy}{dt} \sim 1 \quad (y \to -\infty \,, t \to +\infty) \,, \tag{444}$$

or, integrating,

$$y = -Ce^t \quad (t \to \infty \,, y \to -\infty) \,, \tag{445}$$

where $C > 0$ is a positive constant. Hence, since $y = dz/dt$, we have

$$\frac{dz}{dt} = -Ce^t \qquad (t \to \infty) \,, \tag{446}$$

or

$$z = -Ce^t + C_1 \,, \tag{447}$$

where C_1 is a constant. Remembering that $\xi = e^{-t}$, we now have

$$z = -\frac{C}{\xi} + C_1 \qquad (\xi \to 0) \,. \tag{448}$$

From (410) and (448), we have, accordingly,

$$\psi = 2 \log \xi + \frac{C}{\xi} - C_1 \qquad (\xi \to 0) \,. \tag{449}$$

The corresponding law for the density distribution is given by

$$\rho = \lambda e^{-\psi} = \lambda \frac{1}{\xi^2} e^{C_1 - C\xi^{-1}} \qquad (\xi \to 0) \,, \tag{450}$$

which can also be written as

$$\rho = \frac{A}{\xi^2} e^{-C/\xi} \quad (C > 0) \ (\xi \to 0) \,, \tag{451}$$

where A and C are constants. From (451) it follows that along all solutions of the isothermal equation except the E-solutions and the singular solution, $\rho \to 0$ as $\xi \to 0$. However, all the solutions have the same behavior at infinity; they asymptotically approach the singular solution, oscillating with respect to it and intersecting it at points which asymptotically increase geometrically in the ratio $e^{2\pi/\sqrt{7}}$.

27. *Discussion of the isothermal equation in the* (u, v) *plane.*—We shall conclude our discussion of the isothermal equation by a brief description of the solution-curves in the (u, v) plane.

Our variables are

$$u = \frac{\xi e^{-\psi}}{\psi'} ; \qquad v = \xi\psi' , \tag{452}$$

where u and v satisfy the first-order equation

$$\frac{u}{v}\frac{dv}{du} = -\frac{u-1}{u+v-3} . \tag{453}$$

a) The locus of points at which the curves have horizontal tangents is given by

$$u = 1 , \tag{454}$$

which is a line parallel to the v-axis.

b) The locus of points at which the curves have vertical tangents is given by

$$u + v = 3 . \tag{455}$$

c) The two loci (454) and (455) intersect at the point

$$u_s = 1 ; \qquad v_s = 2 . \tag{456}$$

This point of intersection corresponds to the existence of the singular solution

$$\psi_s = \log\frac{\xi^2}{2} ; \qquad \frac{d\psi_s}{d\xi} = \frac{2}{\xi} , \tag{457}$$

or

$$u_s = \frac{\xi e^{-\psi_s}}{\psi_s'} = 1 ; \qquad v_s = \xi\psi_s' = 2 . \tag{458}$$

d) Consider the E-solutions at $\xi \sim 0$. It is sufficient to consider $\Psi(\xi)$, since any other member of the family $\{\Psi(\xi)\}$ will lead to the same (u, v) curve. As $\xi \to 0$, we have, according to equation (377),

$$\Psi(\xi) = \tfrac{1}{6}\xi^2 - \tfrac{1}{120}\xi^4 + \tfrac{1}{1890}\xi^6 \ldots \quad (\xi \to 0) . \tag{459}$$

From the foregoing, we find

$$e^{-\Psi(\xi)} \sim 1 - \tfrac{1}{6}\xi^2 + \tfrac{1}{45}\xi^4 - \dots \quad (\xi \to 0), \quad (460)$$

$$\Psi'(\xi) \sim \tfrac{1}{3}\xi(1 - \tfrac{1}{10}\xi^2) \quad (\xi \to 0). \quad (461)$$

Hence, as $\xi \to 0$,

$$u_E = \frac{\xi e^{-\Psi}}{\Psi'} \sim 3(1 - \tfrac{1}{15}\xi^2) \quad (\xi \to 0), \quad (462)$$

$$v_E = \xi\Psi' \sim \tfrac{1}{3}\xi^2 \quad (\xi \to 0). \quad (463)$$

Therefore, as $\xi \to 0$, $u \to 3$ and $v \to 0$; in other words, the E-curve passes through the point

$$u_E = 3, \quad v_E = 0 \quad (\xi = 0). \quad (464)$$

At this point the E-curve has a definite slope determined by

$$\frac{du_E}{d\xi} \sim -\tfrac{2}{5}\xi; \quad \frac{dv_E}{d\xi} \sim \tfrac{2}{3}\xi \quad (\xi \to 0), \quad (465)$$

or

$$\left(\frac{dv_E}{du_E}\right)_{\xi=0} = -\tfrac{5}{3}. \quad (466)$$

It is clear, therefore, that the E-curve starts at the point ($u = 3$, $v = 0$) with a negative slope of 5/3 and approaches the point ($u = 1$, $v = 2$) by spiraling around it (cf. Fig. 20).

e) All the other solutions also spiral around this point, and it is clear that along these curves $v \to 0$ as $u \to \infty$. This arises because, as we have already seen, these solutions correspond to a ρ which vanishes at $\xi = 0$ and at $\xi = \infty$, and hence ψ' must vanish for some finite ξ; for this value of ξ, $v = 0$ and $u = \infty$.

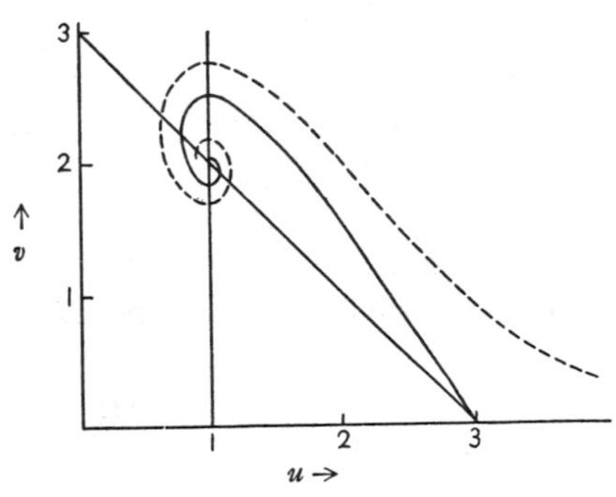

FIG. 20.—The (u, v) curves ($n = \infty$)

f) Finally, we may consider the behavior of the (u, v) curves near the point ($u = 3, v = 0$). Write $u = 3 + u_1$. Equation (453) can be written as

$$\frac{3 + u_1}{v}\frac{dv}{du_1} = -\frac{2 + u_1}{u_1 + v}. \tag{467}$$

If u_1 and $v \to 0$, we can write approximately

$$\frac{3}{v}\frac{dv}{du_1} = -\frac{2}{u_1 + v}, \tag{468}$$

which is of the same form as (348), an equation which arose in a similar connection when discussing the Lane-Emden equation. We therefore have (cf. Eq. [354] and replace n by unity)

$$v^{3/2}[5(u - 3) + 3v] = \text{constant}. \tag{469}$$

Hence, near ($u = 3$, $v = 0$) the (u, v) curves resemble portions of generalized hyperbolas asymptotic to the u-axis and the line

$$5(u - 3) + 3v = 0. \tag{470}$$

When we put the constant in (469) equal to zero, we get the E-curve represented near ($u = 3$, $v = 0$) by its tangent (470), thus arriving at the earlier result (466). This completes our discussion of the polytropic and the isothermal distributions of matter.

28. *Composite configurations.*—So far we have considered only complete polytropes. We shall now proceed to a consideration of composite polytropes, i.e., configurations which consist of different zones each characterized by a different value of the index n. Thus, we can consider configurations consisting of a core of a given index n_2 surrounded by an envelope of another index n_1; in such cases it is clear that the core will be described by the Lane-Emden function of index n_2, while the envelope will be described by a solution (in general not an E-solution) of the Lane-Emden equation of index n_2. In this section we shall consider such configurations.

We suppose that we have the equations of state

$$P = K_1 \rho^{(n_1+1)/n_1}; \qquad P = K_2 \rho^{(n_2+1)/n_2}, \tag{471}$$

for the envelope and the core, respectively. Further, K_1 and K_2 are assumed to be constants. It can happen that one or both of them are "universal" constants. We shall return to this question later, but for the present we shall formulate the problem in the following manner:

To construct an equilibrium configuration of a prescribed mass M and radius R, such that it consists of a polytropic core of index n_2 surrounded by a polytropic envelope of index n_1, it being further specified that the envelope is to extend inward to a fraction $(1 - q)$ of the radius, R.

Let us first consider the envelope of index n_1. The reduction to the Lane-Emden equation of index n_1 is made by the substitutions (Eqs. [8] and [10])

$$\rho = \lambda_1 \theta^{n_1} ; \qquad P = K_1 \lambda_1^{1+\frac{1}{n_1}} \theta^{n_1+1} , \tag{472}$$

$$r = \left[\frac{(n_1 + 1)K_1}{4\pi G}\right]^{\frac{1}{2}} \lambda_1^{\frac{1-n_1}{2n_1}} \xi . \tag{473}$$

Further, we have (Eq. [68])

$$M(r) = -4\pi \left[\frac{(n_1 + 1)K_1}{4\pi G}\right]^{\frac{3}{2}} \lambda_1^{\frac{3-n_1}{2n_1}} \left(\xi^2 \frac{d\theta}{d\xi}\right) . \tag{474}$$

θ need no longer be the Lane-Emden function, since a solution which does not extend to the center can have a singularity at the origin. λ_1 is, for the present, an arbitrary constant and can be adjusted to select any particular solution out of a homologous family i.e., we can regard λ_1 as the constant of homology. Let θ have its zero at $\xi = \xi_1$. Then

$$M = -4\pi \left[\frac{(n_1 + 1)K_1}{4\pi G}\right]^{\frac{3}{2}} \lambda_1^{\frac{3-n_1}{2n_1}} \left(\xi^2 \frac{d\theta}{d\xi}\right)_{\xi=\xi_1} , \tag{475}$$

$$R = \left[\frac{(n_1 + 1)K_1}{4\pi G}\right]^{\frac{1}{2}} \lambda_1^{\frac{1-n_1}{2n_1}} \xi_1 . \tag{476}$$

Eliminating λ_1 between (475) and (476), we obtain (cf. Eq. [74])

$$K_1 = \frac{1}{n_1 + 1} \left[\frac{4\pi}{\omega_{n_1}^{n_1-1}}\right]^{\frac{1}{n_1}} G M^{\frac{n_1-1}{n_1}} R^{\frac{3-n_1}{n_1}} , \tag{477}$$

where, as in equation (337),

$$\omega_{n_1} = -\xi_1^{\frac{n_1+1}{n_1-1}} \left(\frac{d\theta}{d\xi}\right)_{\xi=\xi_1} . \tag{478}$$

As we have already pointed out (§ 21, g), ω_{n_1} can be used to label the different solution-curves in any plane in which the Lane-Emden equation reduces to one of the first order.

The problem now is to determine ω_{n_1} in such a way that the configuration consists of an n_2-core occupying a fraction q of the radius. To do this it is necessary to write down the equations governing the structure of the core. To avoid confusion we shall use the variables ϕ and η to describe the core. With the substitutions

$$\rho = \lambda_2 \phi^{n_2} ; \qquad P = K_2 \lambda_2^{1+\frac{1}{n_2}} \phi^{n_2+1} , \tag{479}$$

$$r = \left[\frac{(n_2+1)K_2}{4\pi G}\right]^{\frac{1}{2}} \lambda_2^{\frac{1-n_2}{2n_2}} \eta , \tag{480}$$

we reduce the equation of equilibrium (7) to the Lane-Emden equation (in ϕ and η) of index n_2. Further, we have

$$M(r) = -4\pi \left[\frac{(n_2+1)K_2}{4\pi G}\right]^{\frac{3}{2}} \lambda_2^{\frac{3-n_2}{2n_2}} \left(\eta^2 \frac{d\phi}{d\eta}\right) . \tag{481}$$

In the foregoing equations λ_2 is the constant of homology. Let the values of the variables at the interface between the core and the envelope be θ, ξ, ϕ, η. At the interface the values of P, ρ, r, and $M(r)$, given by the two sets of formulae ([472], [473], [474] and [479], [480], [481]) should be identical. These "equations of fit" are

$$\lambda_1 \theta^{n_1} = \lambda_2 \phi^{n_2} , \tag{482}$$

$$K_1 \lambda_1^{1+\frac{1}{n_1}} \theta^{n_1+1} = K_2 \lambda_2^{1+\frac{1}{n_2}} \phi^{n_2+1} , \tag{483}$$

$$\left[\frac{(n_1+1)K_1}{4\pi G}\right]^{\frac{1}{2}} \lambda_1^{\frac{1-n_1}{2n_1}} \xi = \left[\frac{(n_2+1)K_2}{4\pi G}\right]^{\frac{1}{2}} \lambda_2^{\frac{1-n_2}{2n_2}} \eta , \tag{484}$$

$$\left[\frac{(n_1+1)K_1}{4\pi G}\right]^{\frac{3}{2}} \lambda_1^{\frac{3-n_1}{2n_1}} \xi^2 \frac{d\theta}{d\xi} = \left[\frac{(n_2+1)K_2}{4\pi G}\right]^{\frac{3}{2}} \lambda_2^{\frac{3-n_2}{2n_2}} \eta^2 \frac{d\phi}{d\eta} . \tag{485}$$

We will now show that we can eliminate the constants of homology λ_1 and λ_2 and reduce the system to one involving only the homology-invariant functions u and v. Raise (484) to the third power, multiply by (482), and divide by (485). We are left with

$$\frac{\xi\theta^{n_1}}{\theta'} = \frac{\eta\phi^{n_2}}{\phi'}, \tag{486}$$

which can be written as

$$u(n_1; \xi) = u(n_2; \eta) . \tag{487}$$

From (482) and (483) we have

$$K_1\lambda_1^{1/n_1}\theta = K_2\lambda_2^{1/n_2}\phi . \tag{488}$$

Divide (485) by the product of (484) and (488). We then have

$$(n_1 + 1)\frac{\xi\theta'}{\theta} = (n_2 + 1)\frac{\eta\phi'}{\phi}, \tag{489}$$

which can be written as

$$(n_1 + 1)v(n_1; \xi) = V(n_1; \xi) = V(n_2; \eta) = (n_2 + 1)v(n_2; \eta) . \tag{490}$$

Our equations of fit then are

$$u(n_1; \xi) = u(n_2; \eta) ; \qquad V(n_1; \xi) = V(n_2; \eta) , \tag{491}$$

ξ and η still referring to the interface.

We will now show how the equations (491) enable us to solve the problem. In § 21 we have already described the nature of the solution-curves in the (u, v) plane. The solution-curves in our present (u, V) plane can be obtained very simply by multiplying the ordinates in the (u, v) plane by $(n + 1)$. It is therefore clear that all the characteristic features of the (u, v) plane are retained in the (u, V) plane; in particular, the E-curve separates the region of the F-curves from the region of the M-curve.

Since the n_2-polytropic equilibrium extends to the center we can choose for ϕ the Lane-Emden function θ_{n_2} and let λ_2 denote the central density. Thus, the $[u(n_2; \eta), V(n_2, \eta)]$ curve to be considered is the E-curve. Now the $[u(n_1; \xi), V(n_1; \xi)] \equiv \Gamma_{n_1}$ curves form a one-parametric family and some (or all) of these will intersect the $E(n_2)$ curve. A point of intersection between the $E(n_2)$ curve and any of the Γ_{n_1} curves corresponds to a particular solution of the equations of fit. On the other hand, each point of a (u, V) curve corresponds to a definite value of ξ/ξ_1, where ξ_1 defines the corresponding zero of the solution in the (θ, ξ) plane; the value of ξ/ξ_1 on a solution-curve in the (u, V) plane is the same for all members of the homologous family in the (θ, ξ) plane which it represents. Hence, the point of intersection between the $E(n_2)$ curve and a Γ_{n_1} curve defines the value ξ/ξ_1 which defines the fraction q. Since this ratio is prescribed, it is clear that only certain of the Γ_{n_1} curves will intersect the $E(n_2)$ curve in such a way that the point of intersection will define for the n_1 envelope an extent equal to that specified. We select, then, each such solution (there can be more than one solution), and the value of ω_{n_1} which labels it is the one appropriate for use in the mass-radius relation (477). This is the procedure to solve the equations of fit.

After solving the equations of fit in the manner described, it is readily seen that the configuration becomes determinate.

Let us assume that we have integrations for all solutions of the Lane-Emden equation of index n_1 which passes through some fixed point $\xi_1 = 1$ (say) on the ξ-axis. A solution of the equations of fit selects one (or more) of the solutions with certain definite value (or values) for ω_{n_1}. This value, substituted in (477), gives K_1, and (476) now determines λ_1, the homology constant. A knowledge of ω_{n_1}, K_1, and λ_1 will determine the structure of the envelope completely, and in particular the interfacial density ρ_i.

Also, as we have already pointed out, we can choose $\rho_c = \lambda_2$. Then $\phi = \theta_{n_2}$ is the Lane-Emden function of index n_2. The solution of the equations of fit provides the value of $\eta = \xi_i^{(n_2)}$ at the interface corresponding to the point on the $E(n_2)$ curve through which the appropriate Γ_{n_1} curve passes. Hence, we have

$$\rho_i = \rho_c \theta_{n_2}^{n_2}(\xi_i^{(n_2)}) . \tag{492}$$

Since ρ_i is already known from the structure of the envelope, ρ_c can now be determined, thus making the structure determinate.

The procedure of solving the equations of fit becomes slightly altered if, instead of the extent $(1 - q)$, the constant K_1 is assigned some definite value. In this case, equation (477) determines ω_{n_1}, and therefore the particular Γ_{n_1} curve. We must find whether this Γ_{n_1} curve intersects the $E(n_2)$ curve. If it does not, then a composite configuration of the character contemplated is impossible. If, however, the Γ_{n_1} curve intersects the $E(n_2)$ curve, then the equations of fit have a solution and the point of intersection will determine the fraction of the radius occupied by the core. The following theorem is of interest in this connection.

If $[u(n_2; \eta), V(n_2; \eta)]$ *corresponds to the* $E(n_2)$ *curve, then the equations of fit have a solution if, and only if, the* n_1*-envelope is described by an* F- *or an* M-*solution according as* n_2 *is less than or greater than* n_1. *Further, it is assumed that* $n_1 < 5$, $n_2 < 5$, *and* $n_1 \neq n_2$.

From our discussion in § 21, it is clear that to prove the foregoing theorem we have only to show that the $E(n_2)$ curve lies entirely below or above the $E(n_1)$ curve according as n_2 is greater or less than n_1.

Since the ordinates in the (u, v) plane are increased by the factor $(n + 1)$ to transform to the (u, V) plane, it is clear that the starting-slope of the $E(n)$ curve is given by (Eq. [329]),

$$\frac{dV_E}{du_E} = -\frac{5}{3}\left(1 + \frac{1}{n}\right). \tag{493}$$

Consequently, the $E(n_2)$ curve lies initially below the $E(n_1)$ curve if

$$1 + \frac{1}{n_2} < 1 + \frac{1}{n_1}, \tag{494}$$

or

$$n_2 > n_1 . \tag{495}$$

Similarly, the $E(n_2)$ curve lies initially above the $E(n_1)$ curve if $n_2 < n_1$. From our discussion in § 21, it follows that, if $n_2, n_1 < 5$, then an $E(n_2)$ curve lies entirely below (or above) an $E(n_1)$ curve if it lies initially below (or above) it.

Finally, since the M-curves lie entirely below the E-curve while the F-curves lie entirely above it, it follows that the $E(n_2)$ curve intersects all the M-curves belonging to n_1 if $n_2 > n_1$, while the

$E(n_2)$ curve intersects all the F-curves belonging to n_1 if $n_2 < n_1$. This proves the theorem.

For $n > 5$ the arrangement and the character of the solutions become different and the enumeration of the different possibilities becomes more complicated. Nothing new in principle, however, arises.

BIBLIOGRAPHICAL NOTES

As has been pointed out, the notion of convective equilibrium is due to LORD KELVIN (*Mathematical and Physical Papers*, **3**, 255–260, first published in 1862), whose investigation may properly be described as the real forerunner of the subsequent studies by Lane, Ritter, and Emden. In view of the fundamental character of the work of these authors we shall describe in some detail the actual contributions of each of them.

I. J. HOMER LANE, *Amer. J. Sci.*, 2d ser., **50**, 57–74, 1869. Lane's paper, "On the Theoretical Temperature of the Sun under the Hypothesis of a Gaseous Mass Maintaining Its Volume by Its Internal Heat and Depending on the Laws of Gases Known to Terrestrial Experiment," considers for the first time the equilibrium of a stellar configuration. It should, however, be pointed out that the problem of the gravitational equilibrium of a gas sphere is considered only incidentally and that the "theoretical temperature" refers to the surface temperature of the sun. Lane's principal object in his investigation was to determine the temperature and the density at the surface of the sun. In order to determine these quantities, he adopts the following procedure. From the value of the solar constant as known at that time (Herschel and Pouillet's determination), he attempts to derive the surface temperature. Stefan's law was still unknown (Stefan published his law in 1879), and Lane therefore uses certain experimental results of Dulong and Petit and of Hopkins on the rate of emission of radiant energy by heated surfaces. Using the empirical law derived by Hopkins, Lane estimates the surface temperature of the sun to be 54,000° F or 30,000° Kelvin. Lane realizes that this estimate depends on a gross extrapolation of the experimental results, but it is interesting to note that, in principle, his method is not different from modern methods of determining the effective temperatures from Stefan's law.

Lane's next problem is to determine the density corresponding to his derived surface temperature. To this end he solves the equilibrium of the sun as a whole. Assuming that the sun is in "convective equilibrium," according to the then recent ideas of Lord Kelvin and referring to the work of Clausius, Lane argues that, because of the "fierce collisions of compound molecules with each other at the temperatures supposed to exist in the sun's body, their component atoms might be torn assunder" it would be safe to assume for the ratio of the specific heats the value for a monatomic gas namely $\gamma = 5/3$. (Lane also considers, formally, the case $\gamma = 1.4$.) The mathematical problem of the equilibrium of a configuration under its own gravitation with an underlying law $P \propto \rho^{1+(1/n)}$ is formally solved and the appropriate "Lane-Emden" function numerically isolated. He gives the graphs of ρ and T as functions of the radius vector. From these graphs he finally reads off the value of ρ corresponding to $T = 30{,}000°$ and obtains $\rho = 0.00036$. When the crudeness of the then available data is considered, Lane's success in estimating ρ and T at the surface of the sun is a remarkable achievement. It may thus be said that his paper has made him the author not only of the first investigation of the physical conditions in the solar atmosphere but also of the first investigation on stellar interiors though the latter was not his primary concern.

Lane's paper contains no explicit reference to what is generally called "Lane's law." Indeed, he does not consider homologous transformations at all—it was Ritter who first considered such transformations explicitly; the reason the law $r\,T(r) =$ constant (for a uniform expansion or contraction of a gaseous configuration) is called "Lane's law" is explained in SIMON NEWCOMB's *The Reminiscences of an Astronomer* (1903, Cambridge, Mass., U.S.A.). The following is an extract from Newcomb's book (pp. 245–249):

> After the paper in question appeared I called Mr. Lane's attention to the fact that I did not find any statement of the theorem which he had mentioned to me to be contained in it. He admitted that it was contained in it only *impliedly* and proceeded to give me a brief and simple demonstration. So the matter stood until the centennial year 1876 when Sir William Thomson paid a visit to this country. . . . . Among other matters I mentioned this law originating with Mr. J. Homer Lane. He did not think it could be well founded and when I attempted to reproduce Mr. Lane's verbal demonstration I found myself unable to do so. . . . . When I again met Mr. Lane I told him of my difficulty and asked him to repeat the demonstration. He did so at once and I sent it off to Sir William. The latter immediately accepted the result and published a paper on the subject in which the theorem was made public for the first time.

Newcomb concludes:

> Altogether I feel it eminently appropriate that his name should be perpetuated by the theorem of which I have spoken.

Three points of historical interest should be noted: (1) Lane's interest was primarily with the solar atmosphere; (2) his interest in the gravitational equilibrium of a gaseous configuration was incidental to the main object of his published investigation; and, finally (3), Lane must have derived the law associated with his name essentially from an argument involving the homology invariance of the equilibrium configurations built on the law $P \propto \rho^{1+(1/n)}$. In KELVIN's paper, *Phil. Mag.*, **22**, 287, 1887, the homology theorem (as we have proved it in § 8) is explicitly proved; and since, further, Kelvin's paper contains a reference to Lane's paper and also to a letter from Newcomb, it is clear that Newcomb's reference to Kelvin's paper as making "public for the first time" Lane's law must refer to Kelvin's proof of the homology theorem.

II. A. RITTER. Ritter's investigations are very remarkable in their range and depth; through his papers he shows himself to be a pioneer of very great originality. Unlike Lane, Ritter was primarily interested in the equilibrium of stellar configurations, and his contribution to the formal mathematical theory is so great that such aspects of the theory of gaseous configurations built on the law $P \propto \rho^{1+(1/n)}$, as are commonly known are almost entirely due to Ritter. It should be noted further that Ritter's work was all done independently and without knowledge of Lane's paper. Ritter's studies extended over a period of six years, and his eighteen communications on "Untersuchungen über die Höhe der Atmosphäre und die Constitution gasförmiger Weltkörper" presented to the *Wiedemann Annalen* during the years 1878–1889 form a classic the value of which has never been adequately recognized, though Emden refers very enthusiastically and at great length to the wealth of material that is contained in Ritter's work. The following is a list of Ritter's papers; the most important of them are starred (*), and the essential results contained in them are briefly reviewed.

1.	**5**, 405, 1878	*8.	**11**, 332, 1880	14.	**17**, 332, 1882
*2.	**5**, 543, 1878	*9.	**11**, 978, 1880	15.	**18**, 488, 1883
*3.	**6**, 135, 1878	10.	**12**, 445, 1881	*16.	**20**, 137, 1883
4.	**7**, 304, 1879	11.	**13**, 360, 1881	*17.	**20**, 897, 1883
*5, *6.	**8**, 157, 1880	12.	**14**, 610, 1881	18.	**20**, 910, 1883
7.	**10**, 130, 1880	*13.	**16**, 166, 1882		

In (2) the uniform expansion and contraction of gaseous configurations are considered, and Lane's law (independently of Lane) explicitly proved. The cosmogenetic equation of state is here defined, and what we have called "Ritter's theorem" (chap. ii) is also proved in this paper.

In (3) the fundamental differential equation for $n = 2.44$ is established, and the appropriate "Lane-Emden" function obtained. This paper also contains the derivation of the Helmholtz-Kelvin time scale (cr. chap. xii).

In (5) and (6) the equation $\Omega = -3\int PdV$ is obtained (his Eq. [186]) and what we have called "Ritter's relation," namely,

$$U = -\frac{1}{3(\gamma - 1)}\,\Omega\,,$$

is also obtained (his Eq. [190]). In this paper the adiabatic pulsation of a gas sphere is considered for the first time, and the fundamental result is proved that ($\gamma = 4/3$) separates the configurations which are stable ($\gamma > 4/3$) from those which are unstable ($\gamma < 4/3$). Ritter also proves the important result that the period of oscillation of a gas sphere is inversely proportional to the square root of its mean density. It should be noted that Ritter develops the theory of pulsating configurations with a definite view toward a theory for the variable stars.

In (8) Ritter establishes (explicitly for the first time) the fundamental differential equation governing the structure of gaseous configurations with an underlying law $P \propto \rho^{1+(1/n)}$. His equation (295) is what we have called the "Lane-

Emden" equation, though it should have been more appropriate to have called it the "Lane-Ritter" equation. The "Lane-Emden" functions for $n = 1$, 1.5, 2, 2.44, 3, and 4 are obtained—for $n = 1$, he uses $\theta_1 = \sin \xi/\xi$. This paper also contains proofs of the important formulae

$$\Omega = -\frac{3}{5-n}\frac{GM^2}{R}\,; \qquad U = \frac{1}{(5\gamma-6)}\frac{GM^2}{R}\,.$$

Ritter also discusses the importance of the case $n = 5$.

In (9) Ritter considers composite configurations consisting of incompressible cores and gaseous envelopes. In this connection Ritter draws attention to the importance of the solutions of the fundamental differential equation other than those which are finite at the origin. In particular he uses the general solution $\theta = A \sin(\xi - \delta)/\xi$ when considering the case $n = 1$. Ritter was thus not only the first to consider composite configurations but also the first to recognize the importance of solutions which have a singularity at the origin.

In (13) Ritter considers the isothermal gas sphere and isolates the singular solution $e^{-\psi} = 2/\xi^2$. In this paper he also proves the integral theorems which are referred to in the bibliographical note for chapter iii.

In (16) and (17) what is generally called the "giant-dwarf" theory of stellar evolution was originated and considered for the first time.

From this very brief and inadequate summary of the important results that are contained in Ritter's papers, it should be clear that almost the entire foundation for the mathematical theory of stellar structure was laid by him. His papers contain, in addition, discussions of a variety of both stellar and meteorological phenomena which are beyond the scope of our present note.

III. Lord Kelvin. It is somewhat surprising that twenty-five years should have elapsed before Lord Kelvin applied his idea of convective equilibrium to the study of gaseous configurations. His paper in the *Philosophical Magazine* in 1887, to which we have already referred several times, is still of interest because of the very short space in which he (independently of his predecessors) derived many of the essential results. It is interesting to recall that Kelvin's interest in the problem "of the equilibrium of a gas under its own gravitation only" originated in a question set by P. G. Tait in an examination paper (Ferguson Scholarship Examination, Glasgow, October 2, 1885). Tait's question reads:

> Assuming Boyle's Law for all pressures form the equation for the equilibrium-density at any distance from the centre of a spherical attracting mass, placed in an infinite space filled originally with air. Find the special integral which depends on a power of the distance from the centre of the sphere alone.

In his 1887 paper Lord Kelvin promises a further paper, but actually he returned to the subject only twenty years later in his posthumous paper on "The Problem of a Spherical Gaseous Nebula" (*Collected Papers*, **5**, 254–283), which appeared in 1908. This last paper contains an extremely attractive summary of the state of the subject prior to the publication of Emden's book. Finally, it may be noted that an appendix to Kelvin's paper by G. Green contains fairly accurate tables of the "Homer Lane" functions; actually, Emden has a slight priority over Green.

IV. Among the early works published prior to the publication of Emden's book, reference should be made to the following papers.

1. E. ZÖLLNER, *Über die stabitität kosmischer Massen*, Leipzig, 1871. Zöllner appears to have been the first to consider the equilibrium of an isothermal gas sphere and isolate the singular solution $e^{-\psi} = 2/\xi^2$. Zöllner, however, draws a number of wrong conclusions on the basis of this singular solution.

2. M. THIESEN, *Über die Verbreitung der Atmosphäre*, Berlin, 1878. This paper is not available to the writer, but from references to it in the literature it appears that Thiesen treated the problem of an isothermal gas sphere very powerfully, and that he was aware of the oscillating nature of the general solution about the singular solution.

3. E. BETTI, *Nuovo cim.*, **7**, 26, 1880. Betti seems to have been the first to discover the expressions for the potential and the internal energies for a polytrope, though his priority over Ritter is only very slight.

4. A. SCHUSTER, *Brit. Assoc. Rept.*, p. 427, 1883. The Lane-Emden function for $n = 5$ is obtained.

5. G. W. HILL, *Collected Papers*, p. 125, 1888. The isothermal gas sphere is considered. Hill does not seem to have been aware of the existence of either Ritter's or Thiesen's work.

6. G. H. DARWIN, *Collected Papers*, **4**, 362, first published in 1889. The cases $n = 3/2$ and $n = \infty$ are considered, and many of Ritter's results are used.

7. P. RUDZKI, *B.A.*, **19**, 134, 1902. Homologous transformations are introduced as a general notion.

8. T. J. J. SEE, *A.N.*, **169**, 322, 1905. In this paper the starting series for the function $\theta_{3/2}$, including the first eleven terms, is given. This series enables See to compute $\theta_{3/2}$ up to the boundary.

V. R. EMDEN. The publication of Emden's *Gaskugeln* marks the end of the first epoch in the study of stellar configurations. Emden's book not only systematizes the earlier work but also contains a fair proportion of new results and a wealth of material, including accurate and extensive tables of the necessary functions. This is not the place to describe the contents of Emden's book, but we may refer specifically to such parts of the analysis contained in our chapter iv which are due to Emden. They are:

1. The use of the (y, z) variables introduced in § 3. Indeed, Emden was the first to reduce the equation to one of the first order.

2. The discovery of the explicit formula for θ_5, independently of Schuster.

3. The discussion given in §§ 9, 10, 11, and 12, and also the discussion in § 13, leading up to the two lemmas which in the form given are due to E. HOPF, *M.N.*, **91**, 653, 1931. These lemmas without rigorous proofs are already implicit in Emden's book (chap. xiii), and Emden himself uses them.

4. The analysis in § 14, and in particular the discovery of the behavior $\theta \sim C/\xi$ for $n < 3$ as $\xi \to 0$.

5. Emden was fully aware of the fact that the E-solutions form a "grid," though the explicit theorem is due to R. H. Fowler.

6. The analysis in § 18. In particular, Emden was the first to isolate the critical role which $n = 3.18767$ (Eq. [255]) plays in the subsequent discussion.

7. The discovery of the behavior near the origin of the general solutions for $3 < n < 5$. In particular, equations (272) and (285), which describe the behavior of θ as $\xi \to 0$.

8. The analysis of § 19.

9. The behavior of the general solutions as $\xi \to \infty$ for $5 < n < \infty$.

10. The use of the (y, z) variables in § 24 for the isothermal gas sphere and the behavior of the general solution as $\xi \to \infty$.

It is thus seen that Emden's own investigations in this field have consisted almost entirely in the discussion of the general solutions; this aspect of his investigations has never been adequately recognized. Though there are a great number of references to *Gaskugeln* in the literature, it is unfortunate that what are generally associated with Emden's name have been derived by the earlier investigators. This is stated, not with a view to minimizing the value of Emden's very great work, but only to draw attention to the fundamental character of his own original contributions.

A fairly good idea as to what is new and original in Emden's book can be obtained by a comparison of Kelvin's posthumous paper (already referred to) with a review of Emden's book by K. Schwarzschild, *V.J.S.*, **43**, 26–55, 1908.

VI. *Recent work.*—Recent work has consisted almost entirely in the rediscovery of a part of Emden's work, and we shall not attempt to give a complete bibliography. We shall, however, refer to such investigations as have been incorporated in our chapter.

1. R. H. Fowler. *M.N.*, **91**, 63, 1930; *Quart. J. Math.* ("Oxford Series"), **2**, 259, 1931; also *Quart. J. Math.*, **45**, 289, 1914. Fowler's work is the most important of the recent investigations. Its importance lies not so much in the discovery of new results as in the mathematical rigor with which even known results are obtained. He obtains the behavior of the general solution for $n = 3$ as $\xi \to 0$. Fowler also explicitly states the grid properties of the E-solutions and gives an enumeration of the arrangement of the general solutions somewhat more complete than Emden's. Finally, his work brings out the importance of Hardy's theorem for the discussion of the differential equations which are of astrophysical importance. Fowler's discussion also includes differential equations which are somewhat more general than Emden's, but these are beyond the scope of the present chapter.

2. E. Hopf, *M.N.*, **91**, 653, 1932. The lemmas given in § 13 are proved here. Hopf considers only $1 < n \leqslant 3$.

3. E. A. Milne, *M.N.*, **91**, 4, 1930; **92**, 610, 1932. In these papers Milne introduces the variables u and v, which are particularly suited for the discussion of composite configurations. His method is largely used in § 28.

4. S. Chandrasekhar. The author has included in this chapter several of his results on the Lane-Emden equation. The discussion in §§ 17 and 20, where Hopf's methods are generalized to cover the cases $n > 3$, is mostly new. The D-solutions with the behavior $\theta \sim C/\xi$ as $\xi \to \infty$ are isolated here for the first time, as are also the grid properties of these D-solutions. The complete discussion in the (u, v) plane (§ 21), the introduction of the (u, v) variables for the isothermal gas sphere (§ 24), most of the discussion of §§ 25 and 26 [with the exception of the part dealing with the derivation of equation (438) describing the behavior $e^{-\psi}$ as $\xi \to \infty$, which is due to Emden], and the results contained in § 27 are all new.

5. H. N. Russell, *M.N.*, **91**, 741, 1931.

6. N. Fairclough, *M.N.*, **91**, 62, 1930; **92**, 644, 1932; **95**, 585, 1935. These papers contain the tabulation of the general solutions for $n = 3$ and $n = 3/2$.

CHAPTER V

THE THEORY OF RADIATION AND THE EQUATIONS OF EQUILIBRIUM

We have already shown in chapter ii, by an application of the laws of thermodynamics, that the energy density u of black-body radiation at temperature T is proportional to the fourth power of the temperature (Stefan's law). In this chapter we shall be concerned with a further discussion of radiation problems and the bearing of these problems upon an understanding of the physical conditions that could be encountered in stellar interiors.

1. *Fundamental notions and definitions.*—We shall begin with a few definitions:

a) The specific intensity of radiation at a given point, P, *and in a given direction.*—Let $d\sigma$ be an arbitrarily chosen small element of surface containing the point P. At a given instant of time there will be rays[1] traversing this element in all the different directions. Let us consider a specific direction—say the s-direction. Through every point of $d\sigma$ construct cones abutting on $d\sigma$ having axes parallel to the s-direction with solid angles at the apex all equal to a definite infinitesimal amount $d\omega$. These cones define a semi-infinite volume in the form of a truncated cone.[2] The energy in the form of radiation traversing the element of area $d\sigma$ and in the semi-infinite volume defined, during an interval of time dt, can be written as

$$I \cos \theta \, d\sigma d\omega dt \, , \tag{1}$$

where θ is the angle which the s-direction makes with the normal to $d\sigma$. The quantity I, thus introduced, depends naturally on the position of the point P, the direction s, and (if the state is nonstationary) on the time t. I is said to define the specific intensity of radiation at the point P and in the prescribed direction.

A radiation field is said to be isotropic if I depends only on the

[1] In the language of geometrical optics.

[2] The construction used here is said to define a "pencil of radiation."

position of the point P and is independent of direction at P; if, further, I is independent of the position of the point P as well, then the radiation field is said to be homogeneous and isotropic.

The s-direction can be completely specified by the angle $\theta (0 \leqslant \theta \leqslant \pi)$, which we have already defined, and the "azimuth" ϕ $(0 \leqslant \phi \leqslant 2\pi)$. The element of solid angle $d\omega$, defined by the ranges $(\theta, \theta + d\theta)$ and $(\phi, \phi + d\phi)$, is

$$d\omega = \sin\theta \, d\theta d\phi \, ; \tag{2}$$

and the expression for the energy traversing the area $d\sigma$ in the directions confined by element of solid angle $d\omega$ $(\theta, \theta + d\theta; \phi, \phi + d\phi)$ during a time dt is, then

$$I \sin\theta \cos\theta \, d\theta d\phi d\sigma dt \, . \tag{3}$$

b) The flux of radiation.—The total amount of radiant energy traversing the surface element $d\sigma$ from one side to another, expressed in terms of unit area and unit time, can be written as

$$F_{+} = \int_0^{2\pi} \int_0^{\pi/2} I \sin\theta \cos\theta \, d\theta d\phi \, . \tag{4}$$

In the same way, the amount of radiant energy traversing $d\sigma$ in the opposite direction, expressed also in terms of unit area and unit time, is given by

$$F_{-} = -\int_0^{2\pi} \int_{\pi/2}^{\pi} I \sin\theta \cos\theta \, d\theta d\phi \, . \tag{5}$$

The net flux of radiation, F, across $d\sigma$ per unit area and unit time is, therefore,

$$F = F_{+} - F_{-} \, , \tag{6}$$

or by (4) and (5)

$$F = \int_0^{2\pi} \int_0^{\pi} I \sin\theta \cos\theta \, d\theta d\phi \, , \tag{7}$$

or, again, by (2)

$$F = \int I \cos\theta \, d\omega \, , \tag{8}$$

where the integral is extended over the complete sphere.

If we consider a Cartesian system of co-ordinates, (X, Y, Z), and denote by F_x, F_y, and F_z the net fluxes at a point across elements of surfaces normal to the directions X, Y, and Z, respectively, then we should have

$$F_x = \int Il d\omega \; ; \qquad F_y = \int Im d\omega \; ; \qquad F_z = \int In d\omega \, , \tag{9}$$

where I is the specific intensity at the point under consideration and in the direction specified by the direction cosines l, m, and n. If we consider the flux across any surface $d\sigma$, the normal to which has the direction cosines l_1, m_1, and n_1, then we should have

$$F_{l_1, m_1, n_1} = \int I(x, y, z \, ; l, m, n) \cos \psi \, d\omega \, , \tag{10}$$

where ψ is the angle between the direction (l, m, n) and the direction (l_1, m_1, n_1); hence,

$$\cos \psi = ll_1 + mm_1 + nn_1 \, . \tag{11}$$

By (9), (10), and (11) we now have

$$F_{l_1, m_1, n_1} = l_1 F_x + m_1 F_y + n_1 F_z \, . \tag{12}$$

In other words, we can regard the flux as a projection of a vector which has the components F_x, F_y, and F_z in the three principal directions.

c) Distribution in the frequency of radiation.—The specific intensity which is related to the total energy radiated in a certain direction can be further divided into the intensities of the radiations in the different frequencies which travel independently of one another. If we consider an infinitesimal interval $(\nu, \nu + d\nu)$, then the specific intensity I_ν is so defined that the total energy in the frequency interval $(\nu, \nu + d\nu)$ which crosses an element of area $d\sigma$ in a direction making an angle θ with the normal to $d\sigma$ and in an element of solid angle $d\omega$, is, during a time dt,

$$I_\nu \cos \theta \, d\sigma d\omega dt d\nu \, . \tag{13}$$

Strictly speaking, we can never consider a rigorously monochromatic pencil of radiation. It is always necessary to consider a nonzero, though infinitesimal, frequency interval.

From our definitions it follows that

$$\int_0^\infty I_\nu d\nu = I . \tag{14}$$

We shall refer to I as the "integrated intensity," in contrast to the monochromatic intensity, I_ν.

In the general theory of radiation we have to distinguish, further, the different states of polarization of the radiation, but in the applications that we shall consider it is not necessary to go into these finer details.

d) The amount of radiant energy flowing from one element of surface to another element of surface.—As the treatment of this problem is the same for the integrated intensity, I, as for the monochromatic intensity, I_ν, we shall explicitly consider only the former case.

Let $d\sigma$ and $d\sigma'$ be the two elements of surface surrounding the points P and P', respectively. Let r be the distance between P and P'. Further, let PP' make angles θ and θ' to the directions of the normals to $d\sigma$ and $d\sigma'$ at P and P', respectively. Finally, let I be the specific intensity at P in the direction PP'.

In free space the energy which traverses the element $d\sigma$ in time dt and which also traverses $d\sigma'$ is, according to our definition of intensity,

$$dE = I \cos\theta \, d\sigma d\omega dt , \tag{15}$$

where $d\omega$ is the solid angle which the element $d\sigma'$ makes at P. This is seen to be

$$d\omega = \frac{d\sigma' \cos\theta'}{r^2} . \tag{16}$$

From (15) and (16) we have

$$dE = I \frac{\cos\theta \cos\theta' \, d\sigma d\sigma'}{r^2} dt . \tag{17}$$

An immediate corollary of the foregoing result (17) is that *the specific intensity is constant along the path of any ray in free space.*

For, if dE' is the energy which traverses the element $d\sigma'$ and which also traverses the element $d\sigma$, then, according to (17),

$$dE' = I' \frac{\cos\theta' \cos\theta \, d\sigma' d\sigma}{r^2} dt , \tag{18}$$

where I' is now the intensity at P' in the direction PP'. But it is clear that

$$dE = dE' \,. \tag{19}$$

Comparing (17) and (18), we see that $I = I'$; we thus have

$$dE = dE' = I \frac{\cos\theta \cos\theta' \, d\sigma d\sigma'}{r^2} \, dt \,. \tag{20}$$

We see that equation (20) is symmetrical between the unprimed and the primed quantities and exhibits in this sense a certain reciprocity; equation (20) is in fact a special case of a more general reciprocity theorem.

e) The energy density of radiation at a given point.—The energy density, u, of the integrated radiation at a given point is the amount of radiant energy per unit volume which is in course of transit, per unit time, in the neighborhood of the point considered.

Consider a point P, and construct around it an infinitesimal element of volume v, the bounding surface of which we shall denote by σ. We shall further restrict the surface σ to be convex everywhere. To allow for all the radiation traversing v, we surround σ by another convex surface Σ such that the linear dimensions of Σ are large compared with the linear dimensions of σ; nevertheless, we can arrange, at the same time that the volume element inclosed by Σ is still so sufficiently small, that we can regard the intensity in a given direction as the same for all the points inside Σ.

Now, all the radiation traversing the element v must have crossed some element of the surface Σ. Let $d\Sigma$ be such an element. The energy flowing across $d\Sigma$ which also flows across an element $d\sigma$ of σ per unit time is, according to (20),

$$I \frac{\cos\theta \cos\Theta \, d\sigma d\Sigma}{r^2} \,, \tag{21}$$

where θ and Θ are the angles which the normals to $d\sigma$ and $d\Sigma$ make with the radius vector r which connects the two elements. Let l be the length traversed by the pencil of radiation considered through the volume element v. The radiation incident on $d\sigma$ will have traversed the element in time l/c, where c is the velocity of light.

Hence, the contribution to the total amount of radiant energy in course of transit through the volume element v by the pencil of radiation considered is

$$I \frac{\cos\theta \cos\Theta \, d\sigma d\Sigma}{r^2} \frac{l}{c} . \tag{22}$$

But the volume dv, intercepted by the pencil of radiation from the element v, is

$$dv = l \cos\theta \, d\sigma . \tag{23}$$

Hence, we can write (22) as

$$I \frac{dv}{c} d\omega , \tag{24}$$

where

$$d\omega = \frac{\cos\Theta \, d\Sigma}{r^2} \tag{25}$$

is the element of solid angle subtended by $d\Sigma$ at P. Therefore, the total energy in course of transit through the volume element v by radiation from all directions is

$$\frac{1}{c}\int\int I dv d\omega = \frac{v}{c}\int I d\omega , \tag{26}$$

where the integration with respect to ω is extended over the whole sphere. Hence,

$$u = \frac{1}{c}\int I d\omega , \tag{27}$$

since the energy density is expressed in terms of unit volume.

We can similarly define the energy density $u_\nu d\nu$ of the radiation in a specified frequency interval $(\nu, \nu + d\nu)$. We have, as before,

$$u_\nu = \frac{1}{c}\int I_\nu d\omega ; \qquad u = \int_0^\infty u_\nu d\nu . \tag{28}$$

If the radiation is isotropic,

$$u_\nu = \frac{4\pi}{c} I_\nu ; \qquad u = \frac{4\pi}{c} I . \tag{29}$$

f) The emission coefficient.—Let us consider a small element of mass m which is radiating. Let us further consider the radiation emitted in the directions specified by an element of solid angle $d\omega$ and in a definite frequency interval $(\nu, \nu + d\nu)$. The amount of radiant energy emitted in the element of solid angle in time dt and in the frequency interval $(\nu, \nu + d\nu)$ can be written as

$$j_\nu m d\omega dt d\nu \ . \tag{30}$$

The quantity j_ν, thus introduced, is called the "emission coefficient for frequency ν." It should be remarked that, even if the element of mass is isotropic, it does not necessarily follow that the emission of radiation takes place uniformly in all directions. As we shall see presently, a further necessary condition for the emission of radiation to be uniform in all directions is that the element of radiating mass should itself be in an isotropic field of radiation.

If we consider the emission in a definite frequency ν_{nm}, corresponding to a quantum transition between two definite states, m and n, of the atoms forming the medium (the states need not be discrete states), then, according to the Bohr frequency condition,

$$h\nu_{nm} = E_n - E_m \ , \tag{31}$$

where E_n and E_m are the corresponding energies of the two stationary states. Emission in the frequency ν_{nm} takes place because in a given instant of time there will be a certain number of atoms in the excited state n, and when these atoms jump to the state m, they will emit quanta of energy $h\nu_{nm}$. Quantitatively, the emission of radiation in the frequency ν_{nm} is determined by the Einstein coefficients A_{nm} and B_{nm} of spontaneous and induced emission, respectively. These coefficients are defined as follows: The probability that in an interval of time dt an atom in the excited state n emits a quantum of energy $h\nu_{nm}$, in the directions confined to an element of solid angle $d\omega$ and in the absence of an external field of radiation, is $A_{nm} d\omega dt$. This spontaneous emission takes place uniformly in all directions. The probability of the emission of a quantum $h\nu_{nm}$ is increased if the atom in the state n is exposed to a field of radiation of frequency ν_{nm}. We take account of this induced emission by introducing the coefficient B_{nm}; it is defined in such a way that the probability that an excited atom in state n is stimulated by an external field of radiation to emit a quantum $h\nu_{nm}$ in the directions specified by an element of solid angle $d\omega$, in time dt, is given by

$$B_{nm}I_{\nu_{nm}}d\omega dt\,, \tag{32}$$

where $I_{\nu_{nm}}$ is the intensity of the radiation of frequency ν_{nm} at the point where the atom is located and in the direction defined by $d\omega$. The expression (32) for the probability of induced emission arises because the emission of radiation induced by a given pencil of radiation takes place in exactly the same direction as the incident pencil.

Hence, the total probability per unit time of induced emission is

$$B_{nm}\int I_{\nu_{nm}}d\omega\,. \tag{33}$$

Thus, the total emission of energy by one single atom in the state n per unit time is given by

$$h\nu_{nm}[4\pi A_{nm} + B_{nm}\int I_{\nu_{nm}}d\omega]\,. \tag{34}$$

Finally, if there are N_n atoms per unit volume in the state n, we have

$$j_{\nu_{nm}}d\omega = \frac{N_n}{\rho}[A_{nm} + B_{nm}I_{\nu_{nm}}]h\nu_{nm}d\omega\,, \tag{35}$$

where ρ is the density. From (35) we see that an element of mass radiates uniformly in all directions only if it is in an isotropic field of radiation.

The total emission in all directions per gram of material is given by

$$\int j_{\nu_{nm}}d\omega = 4\pi\frac{N_n}{\rho}\left[A_{nm} + B_{nm}\int I_{\nu_{nm}}\frac{d\omega}{4\pi}\right]h\nu_{nm}\,. \tag{36}$$

g) The absorption coefficient.—A pencil of radiation traversing a medium will be weakened by absorption. If the specific intensity I_ν of radiation at frequency ν becomes $I_\nu + dI_\nu$ after it has traversed a medium of thickness ds, we can write

$$dI_\nu = -\kappa_\nu\rho I_\nu ds\,. \tag{37}$$

It should be remarked that $I_\nu + dI_\nu$ is the intensity of the emergent radiation which is in phase with the incident radiation. The quantity κ_ν so introduced is defined as the "mass absorption coefficient" for radiation of frequency ν.

From (37) we find, on integration, that

$$I_\nu(s) = I_\nu(0)e^{-\int_0^s \kappa_\nu \rho ds} , \tag{38}$$

where $I_\nu(s)$ is the intensity after the radiation has traversed a length s of the medium. Equation (38) is generally written in the form

$$I_\nu(s) = I_\nu(0)e^{-\tau_\nu} , \tag{39}$$

where

$$\tau_\nu = \int_0^s \kappa_\nu \rho ds . \tag{40}$$

The quantity τ_ν is called the "optical depth" of the material traversed to radiation of frequency ν.

If we consider the case of absorption between two stationary states n and m as in section f above, then the absorption of radiation of frequency ν_{nm} arises from the excitation of the atoms from the lower state m to the higher state n. We express this quantitatively in terms of the Einstein coefficient of absorption, B_{mn}, defined in such a way that the probability of an atom in the state m, exposed to radiation of frequency ν_{nm}, absorbing a quantum $h\nu_{nm}$ in time dt, is given by

$$B_{mn} \int I_{\nu_{nm}} d\omega \, dt , \tag{41}$$

where the integral is extended over the complete sphere. The relation of the coefficient B_{mn} to the mass absorption coefficient $\kappa_{\nu_{nm}}$ is easily seen to be (cf., sec. h, below)

$$\kappa_{\nu_{nm}} = \frac{N_m}{\rho} B_{mn} h\nu_{nm} , \tag{42}$$

where N_m is the number of atoms in unit volume in the state m and ρ is the density.

h) Total absorption.—Consider a small element of mass m which is exposed to a field of radiation. Then in order to calculate the total absorption of radiant energy in the frequency ν per unit time, we inclose the element of mass by a larger surface Σ outside the bounding surface of m which we denote by σ; the linear dimensions of σ are taken to be much smaller than those of Σ. Then, proceeding as in the calculation for the energy density, we have for the amount of energy traversing an element of surface $d\Sigma$ of Σ in unit time, and which is incident on an element $d\sigma$ of the bounding surface of m,

$$I_\nu \frac{\cos\theta \cos\Theta \, d\sigma d\Sigma}{r^2} d\nu \, , \tag{43}$$

where we have used the same notation as in section e above. Of the amount of energy (43), the amount absorbed by the element of mass is obtained by multiplying (43) by $\kappa_\nu \rho l$, where l is the length intercepted in m by the pencil of radiation under consideration. Hence, the amount of energy absorbed per unit time from the pencil of radiation under consideration is

$$I_\nu \frac{\cos\theta \cos\Theta \, d\sigma d\Sigma}{r^2} \kappa_\nu \rho l d\nu = \kappa_\nu I_\nu d\omega dm d\nu \, , \tag{44}$$

where

$$dm = \rho l \cos\theta \, d\sigma \tag{45}$$

and

$$d\omega = \frac{\cos\Theta \, d\Sigma}{r^2} \, . \tag{46}$$

Hence, the total energy absorbed is obtained by integrating (44) over m and ω. We thus find that

$$\kappa_\nu m d\nu \int I_\nu d\omega \tag{47}$$

specifies the amount of energy absorbed by the element of mass considered from the radiation field in the frequency interval $(\nu, \nu + d\nu)$.

i) The pressure of radiation.—The existence of light pressure follows from Maxwell's electromagnetic theory of light. It also follows from the quantum theory, according to which a quantum of energy $h\nu$ is associated with a momentum $h\nu/c$ (where c is the velocity of light) in its direction of propagation. From this it follows that radiant energy of amount E traversing a medium in a specific direction carries with it a momentum E/c, the momentum exerted being in the same direction as the pencil of radiation.

To calculate the pressure of radiation at a given point P, we have to consider the net rate of transfer of momentum normal to an arbitrarily chosen element of surface $d\sigma$ containing P.

If we consider radiation of frequency ν as incident on the surface $d\sigma$ and making an angle θ with the normal to $d\sigma$, the amount of radiant energy in the frequency interval $(\nu, \nu + d\nu)$, traversing $d\sigma$ in directions specified by an element of solid angle $d\omega$ in time dt, is

$$I_\nu \cos\theta \, d\omega d\nu d\sigma dt \, . \tag{48}$$

This amount of radiant energy carries with it the momentum

$$\frac{1}{c} I_\nu \cos\theta \, d\omega d\nu d\sigma dt \tag{49}$$

in the direction of I_ν. Hence, the normal component of the momentum transferred across $d\sigma$ by the pencil of radiation under consideration is

$$\frac{1}{c} d\sigma dt \, I_\nu \cos^2\theta \, d\omega \, d\nu \, . \tag{50}$$

Therefore, the net transfer of momentum across $d\sigma$ by the radiation in the frequency interval $(\nu, \nu + d\nu)$ is

$$d\sigma dt \frac{1}{c} \int I_\nu \cos^2\theta \, d\omega \, d\nu \, , \tag{51}$$

where the integration is to be carried over the complete sphere. Since the pressure at a point P is defined as the net rate of transfer of momentum normal to an arbitrarily chosen infinitesimal element of surface containing P and expressed in terms of unit area, we can write for the pressure $[p_r(\nu)d\nu]$ due to radiation in the frequency interval $(\nu, \nu + d\nu)$:

$$p_r(\nu) = \frac{1}{c} \int_0^{2\pi} \int_0^{\pi} I_\nu \cos^2\theta \sin\theta \, d\theta d\phi \, . \tag{52}$$

If the radiation is isotropic, we have

$$p_r(\nu) = 2\pi\, I_\nu \frac{1}{c}\int_0^\pi \cos^2\theta \sin\theta\, d\theta = \frac{4}{3c}\,\pi I_\nu\,. \tag{53}$$

Comparing this with (29), we have

$$p_r(\nu) = \tfrac{1}{3}u_\nu\,. \tag{54}$$

We can define the integrated radiation pressure, p_r, due to radiation in all the frequencies, by

$$p_r = \int_0^\infty p_r(\nu)d\nu\,, \tag{55}$$

or by (52)

$$p_r = \frac{1}{c}\int I\cos^2\theta\, d\omega\,, \tag{56}$$

where I now defines the integrated intensity. For isotropic radiation we have

$$p_r = \tfrac{1}{3}u\,, \tag{57}$$

a result we have already used in chapter ii, § 11, to derive Stefan's law.

j) The pressure tensor.—Let us consider an element of surface normal to the X-direction. The rate of transfer of the x-component of the momentum across the element of surface (per unit area) by the radiation confined to an element of solid angle $d\omega$, about a direction whose direction cosines are l, m, and n, is

$$\frac{1}{c}\, Ild\omega\, l\,. \tag{58}$$

(We are considering integrated radiation but the treatment is equally valid for monochromatic radiation; we need only to replace I by $I_\nu d\nu$.) The total rate of transfer of the x-momentum across the element per unit area is, then,

$$\frac{1}{c}\int Il^2 d\omega\,. \tag{59}$$

The foregoing quantity defines the x-component of the pressure exerted across the element under consideration. We write it as p_{xx} (strictly speaking, we should have a suffix r for radiation, but this would unnecessarily burden the notation).

In the same way, the y- and the z-components of the pressure across the element of surface considered are

$$p_{xy} = \frac{1}{c}\int Ilm d\omega ; \qquad p_{xz} = \frac{1}{c}\int Iln d\omega . \tag{60}$$

Similarly, by considering elements of surface normal to the Y- and the Z-directions, we can define the further sets of quantities (p_{yx}, p_{yy}, p_{yz}) and (p_{zx}, p_{zy}, p_{zz}). The nine quantities we have thus defined are said to form the "stress tensor":

$$\left.\begin{aligned} p_{xx} &= \frac{1}{c}\int Il^2 d\omega ; & p_{xy} &= \frac{1}{c}\int Ilm d\omega ; & p_{xz} &= \frac{1}{c}\int Iln d\omega , \\ p_{yx} &= \frac{1}{c}\int Iml d\omega ; & p_{yy} &= \frac{1}{c}\int Im^2 d\omega ; & p_{yz} &= \frac{1}{c}\int Imn d\omega , \\ p_{zx} &= \frac{1}{c}\int Inl d\omega ; & p_{zy} &= \frac{1}{c}\int Inm d\omega ; & p_{zz} &= \frac{1}{c}\int In^2 d\omega . \end{aligned}\right\} \tag{61}$$

We see that

$$p_{xy} = p_{yx} ; \qquad p_{xz} = p_{zx} ; \qquad p_{yz} = p_{zy} ; \tag{62}$$

in other words, the tensor is symmetrical. The mean pressure $\bar{p}$ is defined by

$$\bar{p} = \tfrac{1}{3}(p_{xx} + p_{yy} + p_{zz}) . \tag{63}$$

From (61) it follows (since $l^2 + m^2 + n^2 = 1$) that

$$\bar{p} = \frac{1}{3c}\int I d\omega = \tfrac{1}{3}u , \tag{64}$$

a relation which is generally true.

If the radiation is isotropic, we have

$$\bar{p} = p_{xx} = p_{yy} = p_{zz} = \tfrac{1}{3}u \tag{65}$$

and

$$p_{xy} = p_{yx} = 0\,; \qquad p_{xz} = p_{zx} = 0\,; \qquad p_{yz} = p_{zy} = 0\,. \tag{66}$$

Whenever relations (65) and (66) are true, we say that the stress tensor reduces to a simple hydrostatic pressure.

There is another simple case in which the system of stresses (61) reduces to a simple hydrostatic pressure, namely when

$$I = I_0 + \sum_{n=1}^{n=\infty} \underset{(\alpha+\beta+\gamma=2n+1)}{\sum\sum\sum} I_{\alpha\beta\gamma} l^\alpha m^\beta n^\gamma \,, \tag{67}$$

where I_0 and the "coefficients" $I_{\alpha\beta\gamma}$ are all arbitrary functions of position only. The triple summation in (67) is extended over all possible sets (α, β, $\gamma \geqslant 0$) such that $\alpha + \beta + \gamma$ is odd. From (67) and (61) it follows that

$$p_{xx} = p_{yy} = p_{zz} = \frac{4\pi}{3c} I_0;\ p_{xy} = p_{yx} = 0\ ;\ \text{etc.} \tag{68}$$

k) The mechanical force exerted by radiation.—To determine the mechanical force exerted by radiation, consider a thin cylinder of cross-section $d\sigma$ and length ds in the direction normal to $d\sigma$.

The amount of radiant energy in the frequency interval (ν, $\nu + d\nu$) incident on $d\sigma$, in the directions specified by an element of solid angle $d\omega$, about a direction making an angle θ with the s-direction, and in unit time, is

$$I_\nu \cos\theta\, d\sigma d\omega d\nu\,. \tag{69}$$

The amount of this energy which is absorbed in the cylinder is

$$I_\nu \cos\theta\, d\sigma d\omega d\nu\, \kappa_\nu \rho \sec\theta\, ds\,, \tag{70}$$

since $\sec\theta\, ds$ is the length of the path intercepted in the cylinder by the pencil of radiation under consideration.[3] The amount of momentum thus communicated in the direction of I_ν is obtained by dividing (71) by the velocity of light, c. The normal component of the

[3] For the validity of (71) (to the first order) it is necessary to assume that ds is of a higher order of smallness than $d\sigma$. This assumption is also made in § 1 (k) and § 2.

momentum thus communicated to the cylinder by the pencil of radiation under consideration is given by

$$I_\nu \cos\theta \, d\sigma d\omega d\nu \, \kappa_\nu \rho \sec\theta \, ds \frac{1}{c} \cos\theta \, . \tag{71}$$

To obtain the normal force per unit area, we have to divide the foregoing by $d\sigma$. Finally, integrating over all the directions of the incident radiation, we obtain for the mechanical force per unit area of a cylindrical slab of thickness *ds:*

$$\frac{\kappa_\nu \rho ds}{c} \int I_\nu \cos\theta \, d\omega \, d\nu \, . \tag{72}$$

The foregoing force per unit area in the s-direction, on the cylindrical slab considered, arises from absorption.

We shall now examine the possibility of there being some additional mechanical force arising from emission. The spontaneous emission which takes place uniformly in all directions will not give any net resultant force. On the other hand, the induced emission which takes place in exactly the same direction as the incident stimulating radiation will give a net resultant if the incident field of radiation is not isotropic. From equation (32), which gives the atomic probability of induced emission, we obtain for the normal component of momentum communicated to the cylinder by the emission induced by the pencil of radiation defined in (70) the expression

$$-\frac{N_n}{\rho} B_{nm} h\nu_{nm} I_{\nu_{nm}} d\nu d\omega \, \rho d\sigma ds \frac{1}{c} \cos\theta \quad (\nu_{nm} = \nu) \, . \tag{73}$$

We have the negative sign in (73) because the emission takes place in the forward direction and corresponds to a loss of momentum by the infinitesimal cylindrical slab considered.

Using (42) for our definition of the absorption coefficient and integrating (73) over all the directions, we obtain for the normal force per unit area on the slab considered and in the s-direction:

$$-\kappa_\nu \frac{N_n B_{nm}}{N_m B_{mn}} \rho ds \frac{1}{c} \int I_\nu \cos\theta \, d\omega \, d\nu \quad (\nu_{nm} = \nu) \, . \tag{74}$$

Combining (72) and (74), we have for the net normal force per unit area acting in the s-direction on a cylindrical slab of thickness ds:

$$\frac{1}{c}\kappa_\nu\left(1-\frac{N_n B_{nm}}{N_m B_{mn}}\right)\rho ds\, F_s(\nu)d\nu \qquad (\nu_{nm}=\nu)\,, \tag{75}$$

where $F_s(\nu)d\nu$ is the net flux of radiation in the frequency interval $(\nu, \nu + d\nu)$ in the s-direction.

l) *The equation of transfer.*—Consider a small cylinder of cross-section $d\sigma$ and length ds normal to $d\sigma$. Let I_ν be the intensity of the radiation of frequency ν on one face of the cylinder and in the s-direction. Let the intensity emergent through the second face in the same direction be $I_\nu + dI_\nu$. The amount of radiant energy traversing $d\sigma$ in an interval of time dt and in directions confined to an element of solid angle $d\omega$ about the s-direction is $I_\nu d\nu d\omega d\sigma dt$. Of this, the amount of energy $\kappa_\nu \rho ds\, I_\nu d\nu d\omega d\sigma dt$ is absorbed by the cylinder.

Let j_ν be the coefficient of emission. The mass of the material inside the cylinder is $\rho d\sigma ds$; hence, the amount of radiant energy emitted by this element of mass in the frequency interval $(\nu, \nu + d\nu)$ and in directions confined to the element of solid angle $d\omega$ is

$$\rho d\sigma ds\, j_\nu d\omega d\nu dt\,. \tag{76}$$

Therefore, if the state is steady, we should have

$$dI_\nu d\nu d\omega d\sigma dt = \rho j_\nu d\nu d\omega d\sigma dt ds - \rho\kappa_\nu I_\nu d\nu d\omega d\sigma dt ds\,, \tag{77}$$

or

$$\frac{dI_\nu}{\rho ds} = j_\nu - \kappa_\nu I_\nu\,. \tag{78}$$

The foregoing equation is generally referred to as the "equation of transfer." Of course, we have to consider I_ν as a function of position and of direction; if it is necessary to refer to this explicitly, we may write

$$I_\nu \equiv I_\nu(x, y, z; l, m, n)\,, \tag{79}$$

where the direction cosines (l, m, n) refer to the direction we are considering. In a Cartesian system of co-ordinates we can write the equation of transfer in the form

$$\left(l\frac{\partial}{\partial x} + m\frac{\partial}{\partial y} + n\frac{\partial}{\partial z}\right)I_\nu = \rho j_\nu - \rho\kappa_\nu I_\nu\,. \tag{80}$$

In terms of the Einstein coefficients introduced in (*f*) and (*g*), we can write the equation of transfer (78) in the form (cf. Eqs. [35] and [42])

$$\frac{dI_{\nu_{nm}}}{ds} = N_n(A_{nm} + B_{nm}I_{\nu_{nm}})h\nu_{nm} - N_m B_{mn} h\nu_{nm} I_{\nu_{nm}} , \qquad (81)$$

or

$$\frac{dI_{\nu_{nm}}}{ds} = N_n A_{nm} h\nu_{nm} - N_m B_{mn} h\nu_{mn}\left(1 - \frac{N_n B_{nm}}{N_m B_{mn}}\right) I_{\nu_{nm}} . \qquad (82)$$

2. *The thermodynamics of radiation.*—We shall now investigate in some detail the properties of radiation fields in systems adiabatically inclosed. We shall first consider the case of a homogeneous isotropic medium which, since we assume it to be adiabatically inclosed, must be characterized by the same temperature T throughout the medium. If we restrict ourselves to regions sufficiently distant from the walls of the inclosure, it is clear from considerations of symmetry that in such regions the radiation field must be homogeneous and isotropic. In other words, the specific intensity I_ν of ν-radiation must be independent of the position and the direction of the ray. From the equation of transfer (78) it follows immediately that

$$j_\nu = \kappa_\nu I_\nu . \qquad (83)$$

In other words, *the ratio of the emission to the absorption coefficient for the radiation of frequency* ν *in the interior of a homogeneous isotropic medium adiabatically inclosed is equal to the specific intensity of the radiation for frequency* ν. This is one of Kirchhoff's laws of radiation.

If we express the emission and the absorption coefficients in terms of the Einstein coefficients as we have done in equations (35) and (42), we obtain from (83)

$$N_n[A_{nm} + B_{nm}I_{\nu_{nm}}] = N_m B_{mn} I_{\nu_{nm}} ; \qquad (84)$$

or, solving for $I_{\nu_{nm}}$, we have

$$I_{\nu_{nm}} = \frac{N_n A_{nm}}{N_m B_{mn} - N_n B_{nm}} , \qquad (85)$$

which can also be written as

$$I_{\nu_{nm}} = \frac{A_{nm}}{B_{nm}} \frac{1}{\left(\frac{N_m B_{mn}}{N_n B_{nm}}\right) - 1} . \tag{86}$$

Kirchhoff's law in the form we have now derived is stated only for those regions of the medium which are very far from the walls of the inclosure, since it is only in these regions that we can derive the homogeneity and the isotropy of the radiation field. It is, however, relatively simple to remove this restriction and to show that I_ν has the value j_ν/κ_ν for all directions and all points arbitra ily near the walls of the inclosure. For, in an adiabatic inclosure, every pencil of radiation must be characterized by the same value for I_ν as the pencil of radiation traveling in the opposite direction, since otherwise there would be a unidirectional transport of energy. Hence, a pencil of radiation emergent from an element of the surface on the walls of the inclosure must be characterized by the same value for I_ν as the pencil traveling in the opposite direction and coming from the interior of the medium. An immediate consequence of this result is that *the state of the radiation is the same on the surface of the walls of the inclosure as in the interior.* This result is also due to Kirchhoff.

We have thus shown that the specific intensity, I_ν, of radiation of frequency ν in an isotropic homogeneous medium adiabatically inclosed depends only on the temperature and the nature of the medium. We will now show, following Kirchhoff, that I_ν does not also depend on the nature of the medium.

For this purpose consider a small element of mass dm in the form of an infinitesimal cylinder of cross-section $d\sigma$ and height ds normal to $d\sigma$. Let ρ be the density of the material, so that $dm = \rho d\sigma ds$. Let the element dm be at the center of a hollow spherical reflector of unit radius which has, at the opposite ends of a diameter, two small equal infinitesimal openings of area ω; as the notation implies, we assume that $d\sigma$ is very small compared to ω. Let the whole system be adiabatically inclosed by an inclosure the inner surface of which is "perfectly absorbing" while the outer surface is a "perfect reflector." We shall further suppose that the inclosure is completely evacuated, so that a pencil of radiation is not weakened by absorption except when it strikes the element of mass dm, or the spherical reflector (the outer surface of which is also perfectly absorbing), or the inner walls of the inclosure itself. Finally, let the whole system considered be at temperature T (see Fig. 21).

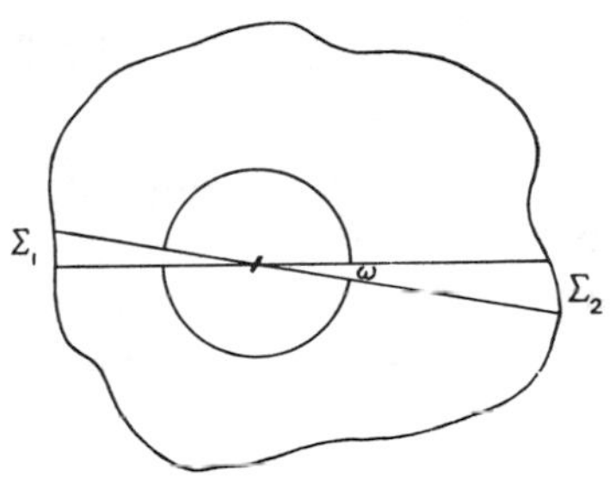

FIG. 21

Since the radiation field inside the inclosure is isotropic, it follows from our remarks in § 1 (*f*) that the element of mass dm will radiate energy uniformly in all directions. Let j_ν be the emission coefficient. The energy radiated by dm in unit time through each of the infinitesimal openings ω in the frequency interval $(\nu, \nu + d\nu)$ is

$$j_\nu \omega dm d\nu = \rho j_\nu \omega d\sigma ds d\nu \,. \tag{87}$$

The energy emitted by dm in all the other directions is reflected at the inner surface of the spherical mirror and, after repeated reflections, will again be incident on dm; it will thus be reabsorbed eventually by dm.

Now the walls of the inclosure radiate toward the interior only since the outer surface is a perfect reflector. Part of the energy emitted by the walls passes through the two openings on the outer surface of the spherical inclosure containing dm, strikes the element dm, and is partially absorbed. The elements of surface of the walls, Σ_1 and Σ_2, which are accessible to the element dm have the areas

$$\Sigma_1 = \frac{r_1^2}{\cos \Theta_1}\,\omega \,; \qquad \Sigma_2 = \frac{r_2^2}{\cos \Theta_2}\,\omega \,, \tag{88}$$

where r_1 and r_2 are the distances of dm from Σ_1 and Σ_2, respectively, and Θ_1 and Θ_2 are the angles which the normals to the elements Σ_1 and Σ_2 respectively make with the direction r which connects the middle point of $d\sigma$ and ω.

Now the total energy of ν-radiation emitted by the element Σ_1, which is incident on $d\sigma$ per unit time, is, according to equation (17),

$$B_\nu^{(1)} \frac{\cos\theta \cos\Theta_1 \, \Sigma_1 d\sigma}{r_1^2} d\nu \, , \tag{89}$$

where $B_\nu^{(1)}$ is the specific intensity of the ν-radiation emergent from Σ_1 in the direction making an angle Θ_1 to its normal, and θ is the angle which the normal to $d\sigma$ makes with the direction r. By (88) we can write, instead of (89),

$$B_\nu^{(1)} \cos\theta \, \omega d\sigma d\nu \, . \tag{90}$$

The amount of this energy absorbed by the element dm is given by

$$B_\nu^{(1)} \cos\theta \, \omega d\sigma d\nu \, \kappa_\nu \rho \sec\theta \, ds = \kappa_\nu \rho B_\nu^{(1)} \omega d\sigma ds d\nu \, , \tag{91}$$

In the same way, the amount of energy absorbed by dm in unit time from the total radiant energy of ν-radiation emitted by Σ_2 and incident on $d\sigma$ is given by

$$\kappa_\nu \rho B_\nu^{(2)} \omega d\sigma ds d\nu \, , \tag{92}$$

where $B_\nu^{(2)}$ is the specific intensity of the ν-radiation emitted by Σ_2 in a direction making an angle Θ_2 with its normal. Hence, the total amount of energy absorbed by dm in unit time from the ν-radiation is given by

$$\kappa_\nu \rho (B_\nu^{(1)} + B_\nu^{(2)}) \omega d\sigma ds d\nu \, . \tag{93}$$

Now, since the system is in a steady state, the energy emitted by the element through the two openings must be equal to (93). From (87) and (93) we have

$$\kappa_\nu (B_\nu^{(1)} + B_\nu^{(2)}) = 2 j_\nu \, . \tag{94}$$

The foregoing equation remains unaltered if the walls of the inclosure are deformed, thus varying the angles Θ_1 and Θ_2. It follows, then, that

$$B_\nu^{(1)} = B_\nu^{(2)} = B_\nu \, , \tag{95}$$

or the *intensity of the radiation emergent from a black surface is independent of the direction of the radiation.* We can now write (94) as

$$\kappa_\nu B_\nu = j_\nu . \tag{96}$$

If different black surfaces are taken for the inner surfaces of the walls of the inclosure while the element dm is kept unchanged, it follows that B_ν remains constant. In other words, *the intensity* I_ν *of the radiation emitted by a black surface is independent of its nature and is a function only of the temperature.* Finally, comparing (96) with (83), we see that $I_\nu = B_\nu$. We have thus proved: *The ratio* j_ν/κ_ν *of the emission to the absorption coefficient of any body in thermodynamical equilibrium depends on the temperature only and is independent of the nature of the body; further,* j_ν/κ_ν *is equal to the specific intensity* B_ν *of the* ν*-radiation emitted by a black surface.* This is Kirchhoff's law in its complete form.

Thus we have shown that $B_\nu = j_\nu/\kappa_\nu$ is a universal function of temperature and frequency. About this function B_ν thermodynamics makes one important prediction. The energy density, u, of radiation in an adiabatic inclosure at temperature T is, according to equation (29),

$$u = \frac{4\pi}{c}\int_0^\infty B_\nu d\nu ; \tag{97}$$

and by Stefan's law (proved in chap. ii) we have

$$u = \frac{4\pi}{c}\int_0^\infty B_\nu d\nu = aT^4 . \tag{98}$$

Hence, if we denote by B the integrated black-body intensity, we have

$$B(T) = \int_0^\infty B_\nu(T) d\nu = \frac{ac}{4\pi} T^4 . \tag{99}$$

For the integrated intensity, B, it is customary to write

$$B = \frac{\sigma}{\pi} T^4 ; \qquad \sigma = \frac{ac}{4} . \tag{99'}$$

σ is called the "radiation constant."

We shall not go into the details of the derivation of the function $B_\nu(T)$ at this place; the derivation is given in chapter x on the basis of the quantum statistics. We note here that the quantum theory predicts for $B_\nu(T)$ the expression

$$B_\nu(T) = \frac{2h\nu^3}{c^2} \frac{1}{e^{h\nu/kT} - 1}, \tag{100}$$

where h is the Planck constant and k is the Boltzmann constant. Equation (100) expresses the well-known Planck law, and the expression for $B_\nu(T)$ is often referred to as the "Planck function."

Comparing (86) and (100), we see that in thermodynamical equilibrium we should have

$$\frac{A_{nm}}{B_{nm}} = \frac{2h\nu_{nm}^3}{c^2}; \qquad \frac{N_m B_{mn}}{N_n B_{nm}} = e^{h\nu_{nm}/kT}. \tag{101}$$

Finally, we observe that Planck's law enables us to evaluate the radiation constant σ and the Stefan-Boltzmann constant a in terms of the fundamental constants h, c, and k. For

$$B(T) = \int_0^\infty B_\nu(T)d\nu = \frac{2h}{c^2}\int_0^\infty \frac{\nu^3 d\nu}{e^{h\nu/kT} - 1}; \tag{102}$$

or, writing $x = h\nu/kT$, we have

$$B(T) = \frac{2h}{c^2}\left(\frac{kT}{h}\right)^4 \int_0^\infty \frac{x^3 dx}{e^x - 1}. \tag{103}$$

Now,

$$\int_0^\infty \frac{x^3 dx}{e^x - 1} = \int_0^\infty x^3 e^{-x}(1 + e^{-x} + e^{-2x} + \dots); \tag{104}$$

or, integrating term by term, we obtain for the integral, the series

$$6\left(1 + \frac{1}{2^4} + \frac{1}{3^4} + \frac{1}{4^4} + \dots\right) = \frac{\pi^4}{15}. \tag{105}$$

Hence, we have

$$B(T) = \frac{2\pi^4 k^4}{15 c^2 h^3} T^4. \tag{106}$$

Comparing this with (99′), we obtain

$$\sigma = \frac{2\pi^5 k^4}{15 c^2 h^3} ; \qquad a = \frac{8\pi^5 k^4}{15 c^3 h^3} . \tag{107}$$

3. *Local thermodynamic equilibrium.*—The thermodynamical theory of radiation described in the previous section is valid only when the system is adiabatically inclosed, and, as a result, when all parts of the system are at the same temperature. Nevertheless, we often encounter physical systems which, though they cannot be described as being in rigorous thermodynamical equilibrium, may yet permit the introduction of a temperature T to describe the local properties of the system to a very high degree of accuracy. The interior of a star, *if* in a steady and static state, is a case in point. For, even if the temperature at the center of the sun, for instance, were 10^8 degrees, the mean temperature gradient would correspond to a change of only 6 degrees in the temperature over a distance of 10^4 cm. This fact, coupled with a probably high value for the stellar absorption coefficient, enables us to ascribe a temperature T at each point P such that the properties of an element of mass in the neighborhood of P are the same as if it were adiabatically inclosed in an inclosure at a temperature T. Under these circumstances we shall say that the material in the neighborhood of the point P is in "local thermodynamical equilibrium." In particular, if κ_ν and j_ν are the coefficients of absorption and emission of an element of mass, we should have

$$j_\nu = \kappa_\nu B_\nu(T) , \tag{108}$$

where $B_\nu(T)$ is the Planck function and T is the local temperature. In using the foregoing equation, we have to remember that j_ν will depend on the incident intensity of the radiation in the frequency interval $(\nu, \nu + d\nu)$ (cf. § 1 [*f*]). It is therefore more convenient to use, instead of (108), the equivalent relations (Eq. [101]) between the Einstein coefficients:

$$\frac{A_{nm}}{B_{nm}} = \frac{2h\nu_{nm}^3}{c^2} ; \qquad \frac{N_m B_{mn}}{N_n B_{nm}} = e^{h\nu_{nm}/kT} , \tag{109}$$

where T is the local temperature.

Let us now examine the steady-state set up in a medium in a static condition and which is in local thermodynamical equilibrium; this type of equilibrium was first studied by Schwarzschild.

Consider the equation of transfer in the form (cf. Eq. [82])

$$\frac{dI_{\nu_{nm}}}{\rho ds} = \frac{N_n A_{nm} h\nu_{nm}}{\rho} - \frac{N_m B_{mn} h\nu_{nm}}{\rho} \left(1 - \frac{N_n B_{nm}}{N_m B_{mn}}\right) I_{\nu_{nm}} . \quad (110)$$

Introduce the absorption coefficient $\kappa_{\nu nm}$ as defined in equation (42):

$$\kappa_{\nu_{nm}} = \frac{N_m B_{mn} h\nu_{nm}}{\rho} . \quad (111)$$

We can write

$$\frac{N_n}{\rho} A_{nm} h\nu_{nm} = \kappa_{\nu_{nm}} \frac{A_{nm}}{B_{nm}} \frac{N_n B_{nm}}{N_m B_{mn}} , \quad (112)$$

or by (109)

$$\frac{N_n}{\rho} A_{nm} h\nu_{nm} = \kappa_{\nu_{nm}} \frac{2h\nu_{nm}^3}{c^2} e^{-h\nu_{nm}/kT} . \quad (113)$$

Hence, we can write the equation of transfer in the form

$$\frac{dI_{\nu_{nm}}}{\rho ds} = \kappa_{\nu_{nm}} \frac{2h\nu_{nm}^3}{c^2} e^{-h\nu_{nm}/kT} - \kappa_{\nu_{nm}}(1 - e^{-h\nu_{nm}/kT}) I_{\nu_{nm}} . \quad (114)$$

Suppressing the suffixes n and m and remembering that

$$B_{\nu_{nm}} = \frac{2h\nu_{nm}^3}{c^2} (e^{h\nu_{nm}/kT} - 1)^{-1} , \quad (115)$$

we can re-write (114) as

$$\frac{dI_\nu}{\rho ds} = \kappa'_\nu B_\nu - \kappa'_\nu I_\nu , \quad (116)$$

where

$$\kappa'_\nu = \kappa_\nu (1 - e^{-h\nu/kT}) . \quad (117)$$

It will now be clear why we did not take the equation of transfer in the form

$$\frac{dI_\nu}{\rho ds} = j_\nu - \kappa_\nu I_\nu \quad (118)$$

and simply insert for j_ν the Kirchhoff expression (108); the reason is that j_ν, which includes the induced emission, is in general not a scalar but depends on the incident intensity I_ν. But in the form (116) we have allowed for the induced emission by reducing κ_ν by the appropriate factor $[1 - \exp(-h\nu/kT)]$. The equation of transfer in the form (116), with κ'_ν defined as in (117), is due to Rosseland.

4. *The equation of radiative equilibrium and the solution of the equation of transfer for the far interior.*—We shall now solve (116) under the circumstances applicable to the interior of a star. For this purpose consider a medium which is in a static state and which extends (for all practical purposes) to infinity in all directions. Let us further assume that the material is in local thermodynamical equilibrium. We shall suppose that a gram of material generates per unit time an amount of energy $\epsilon_\nu d\nu$ by processes of an irreversible character. (We shall refer to $\epsilon_\nu d\nu$ as the "heat liberated," including in this term the net gain of heat per unit mass by an element of mass by "convection,"[4] "conduction," and finally by the internal energy converted into heat. Under the last item we include the [subatomic] energy sources of a star.)

Now the total spontaneous emission per gram per unit time by an element of mass at temperature T is

$$4\pi\kappa'_\nu B_\nu(T) \; . \tag{119}$$

The total absorption, less the total induced emission, is given by (cf. Eqs. [33] and [47])

$$\kappa'_\nu \int I_\nu d\omega \; . \tag{120}$$

Hence, the excess of emission over absorption, given by

$$\kappa'_\nu \int (B_\nu - I_\nu) d\omega \; , \tag{121}$$

must, in a steady state, equal the heat liberated, ϵ_ν. Hence, the condition for a steady state is

$$\kappa'_\nu \int (B_\nu - I_\nu) d\omega = \epsilon_\nu \; . \tag{122}$$

The foregoing equation is generally referred to as the "equation of radiative equilibrium."

[4] This term is used loosely.

We shall now write the equation of transfer (116) in a Cartesian system of co-ordinates:

$$\left(l\frac{\partial}{\partial x}+m\frac{\partial}{\partial y}+n\frac{\partial}{\partial z}\right)I_\nu=\kappa_\nu'\rho(B_\nu-I_\nu)\,. \qquad (123)$$

Multiply (123) by $d\omega$ and integrate over the complete sphere. By (122) the right-hand side reduces to $\epsilon_\nu\rho$. Hence, by equation (9), which defines the flux components F_x, F_y, and F_z, we have

$$\frac{\partial F_x(\nu)}{\partial x}+\frac{\partial F_y(\nu)}{\partial y}+\frac{\partial F_z(\nu)}{\partial z}=\epsilon_\nu\rho\,, \qquad (124)$$

or

$$\operatorname{div}\boldsymbol{F}_\nu=\epsilon_\nu\rho\,. \qquad (125)$$

Equation (125) is simply a statement of the conservation of energy.

Now multiply the equation of transfer successively by $ld\omega$, $md\omega$, and $nd\omega$, and integrate over the complete sphere. By the definitions for the components of the stress tensor (Eq. [61]) we have

$$\frac{\partial p_{xx}(\nu)}{\partial x}+\frac{\partial p_{xy}(\nu)}{\partial y}+\frac{\partial p_{xz}(\nu)}{\partial z}=-\frac{\kappa_\nu'\rho}{c}F_x(\nu)\,, \qquad (126)$$

$$\frac{\partial p_{yx}(\nu)}{\partial x}+\frac{\partial p_{yy}(\nu)}{\partial y}+\frac{\partial p_{yz}(\nu)}{\partial z}=-\frac{\kappa_\nu'\rho}{c}F_y(\nu)\,, \qquad (127)$$

$$\frac{\partial p_{zx}(\nu)}{\partial x}+\frac{\partial p_{zy}(\nu)}{\partial y}+\frac{\partial p_{zz}(\nu)}{\partial z}=-\frac{\kappa_\nu'\rho}{c}F_z(\nu)\,, \qquad (128)$$

or

$$\operatorname{div}\boldsymbol{p}_\nu=-\frac{\kappa_\nu'\rho}{c}\boldsymbol{F}_\nu\,. \qquad (129)$$

We shall now proceed to solve the equation of transfer. Let

$$\tau_\nu=\int_{x_0,\,y_0,\,z_0}^{x,\,y,\,z}\kappa_\nu'\rho ds\,; \qquad ds=ldx+mdy+ndz\,, \qquad (130)$$

the integral being taken from a fixed point (x_0, y_0, z_0) and in the direction (l, m, n). The solution of the equation of transfer can be written as

$$I_\nu(x_0,\,y_0,\,z_0;\,l,\,m,\,n)=\int_0^\infty B_\nu(-\tau_\nu)e^{-\tau_\nu}d\tau_\nu\,. \qquad (131)$$

The physical meaning of (131) is: The specific intensity at a given point and in a given direction is simply the sum of the contributions of the emission due to all the elements of mass behind (i.e., in the direction negative to the one defined) the point under consideration, after allowance has been made for the weakening of the separate pencils of radiation from the different elements of mass, owing to the appropriate amounts of intercepting material.

We assume that we can expand $B_\nu(-\tau_\nu)$ as a Taylor series. We shall retain only the first three terms in the Taylor expansion, and we shall presently verify that this is sufficient for a high degree of accuracy. Hence, we write

$$B_\nu(-\tau_\nu) = B_\nu(0) - \tau_\nu \frac{dB_\nu}{d\tau_\nu} + \tfrac{1}{2}\tau_\nu^2 \frac{d^2 B_\nu}{d\tau_\nu^2} . \tag{132}$$

Inserting the foregoing in (131), we have

$$I_\nu(x_0, y_0, z_0; l, m, n) = B_\nu(0) - \left(\frac{dB_\nu}{d\tau_\nu}\right)_{\tau_\nu=0} + \left(\frac{d^2 B_\nu}{d\tau_\nu^2}\right)_{\tau_\nu=0} . \tag{133}$$

Now

$$\frac{dB_\nu}{d\tau_\nu} = \frac{1}{\kappa_\nu' \rho} \frac{dB_\nu}{ds} = \frac{1}{\kappa_\nu' \rho}\left[l \frac{\partial B_\nu}{\partial x} + m \frac{\partial B_\nu}{\partial y} + n \frac{\partial B_\nu}{\partial z}\right] , \tag{134}$$

or

$$\frac{dB_\nu}{d\tau_\nu} = \frac{1}{\kappa_\nu' \rho} \operatorname{grad}_s B_\nu . \tag{135}$$

In the same way,

$$\left.\begin{aligned}\frac{d^2 B_\nu}{d\tau_\nu^2} = \frac{1}{\kappa_\nu' \rho}\Bigg[& \sum_{l, m, n} l^2 \frac{\partial}{\partial x}\left(\frac{1}{\kappa_\nu' \rho} \frac{\partial B_\nu}{\partial x}\right) \\ & + \sum_{l, m, n} lm \left\{ \frac{\partial}{\partial x}\left(\frac{1}{\kappa_\nu' \rho} \frac{\partial B_\nu}{\partial y}\right) + \frac{\partial}{\partial y}\left(\frac{1}{\kappa_\nu' \rho} \frac{\partial B_\nu}{\partial x}\right)\right\}\Bigg] ,\end{aligned}\right\} \tag{136}$$

or

$$\frac{d^2 B_\nu}{d\tau_\nu^2} = \frac{1}{\kappa_\nu' \rho} \operatorname{grad}_s \left(\frac{1}{\kappa_\nu' \rho} \operatorname{grad}_s B_\nu\right) . \tag{137}$$

Instead of (133) we can now write

$$I_\nu = B_\nu - \frac{1}{\kappa'_\nu \rho} \operatorname{grad}_s B_\nu + \frac{1}{\kappa'_\nu \rho} \operatorname{grad}_s \left(\frac{1}{\kappa'_\nu \rho} \operatorname{grad}_s B_\nu \right) . \tag{138}$$

Inserting the foregoing in the equation of radiative equilibrium, we have

$$\epsilon_\nu \rho = \int \left\{ \operatorname{grad}_s B_\nu - \operatorname{grad}_s \left(\frac{1}{\kappa'_\nu \rho} \operatorname{grad}_s B_\nu \right) \right\} d\omega . \tag{139}$$

Since

$$\int lm d\omega = \int mn d\omega = \int nl d\omega = 0 \tag{140}$$

and

$$\int l^2 d\omega = \int m^2 d\omega = \int n^2 d\omega = \frac{4\pi}{3} , \tag{141}$$

we have by (139), (134), (136), (140), and (141)

$$\epsilon_\nu \rho = -\frac{4\pi}{3} \sum_{x, y, z,} \frac{\partial}{\partial x} \left(\frac{1}{\kappa'_\nu \rho} \frac{\partial B_\nu}{\partial x} \right) . \tag{142}$$

From (138), (139), and (142) it follows that the ratio of the successive terms in (138) are of the order of magnitude

$$\frac{\epsilon_\nu}{\kappa'_\nu B_\nu} \sim \frac{\epsilon}{\bar{\kappa} B} , \tag{143}$$

where ϵ is the total amount of heat liberated in all the frequencies; B, the integrated Planck intensity; and $\bar{\kappa}$, a mean absorption coefficient. Now in the interior of a star $\epsilon \sim 100$ ergs per gram per second, $\kappa \sim 100$ gm^{-1} cm^2 and $T \sim 10^6$ degrees. Hence, the ratio of the successive terms in the expansion is of the order of magnitude

$$\frac{100}{100 \times \frac{\sigma}{\pi} \times 10^{24}} \sim 10^{-19} . \tag{144}$$

Therefore for all practical purposes it would be sufficient to write

$$I_\nu = B_\nu - \frac{1}{\kappa'_\nu \rho} \operatorname{grad}_s B_\nu . \tag{145}$$

From the foregoing it follows that

$$u_\nu = \frac{1}{c}\int I_\nu d\omega = \frac{4\pi}{c} B_\nu , \tag{146}$$

$$u = \int u_\nu d\nu = \frac{4\pi}{c} B = \frac{4\sigma}{c} T^4 = aT^4 , \tag{147}$$

$$p_{xx}(\nu) = p_{yy}(\nu) = p_{zz}(\nu) = p_r(\nu) = \frac{4\pi}{3c} B_\nu , \tag{148}$$

and

$$p_{xy}(\nu) = p_{yz}(\nu) = p_{zx}(\nu) = 0 . \tag{149}$$

In other words, the stress tensor reduces to a simple hydrostatic pressure. From (126), (148), and (149) (or more directly from [145]) we now have

$$\left.\begin{aligned} F_x(\nu) &= -\frac{4\pi}{3\kappa_\nu'\rho}\frac{\partial B_\nu}{\partial x} ; \qquad F_y(\nu) = -\frac{4\pi}{3\kappa_\nu'\rho}\frac{\partial B_\nu}{\partial y} ; \\ F_z(\nu) &= -\frac{4\pi}{3\kappa_\nu'\rho}\frac{\partial B_\nu}{\partial z} , \end{aligned}\right\} \tag{150}$$

or

$$\boldsymbol{F}_\nu = -\frac{4\pi}{3\kappa_\nu'\rho}\,\text{grad}\, B_\nu = -\frac{c}{\kappa_\nu'\rho}\,\text{grad}\, p_r(\nu) . \tag{151}$$

Finally, we have the exact relation (Eq. [125])

$$\text{div}\, \boldsymbol{F}_\nu = \epsilon_\nu\rho . \tag{152}$$

Equations (151) and (152) are the fundamental equations of the problem.

We shall next consider the equation for the integrated flux. From the first of the equations in (150) we have

$$F_x = \int_0^\infty F_x(\nu)d\nu = -\frac{4\pi}{3\rho}\int_0^\infty \frac{1}{\kappa_\nu'}\frac{\partial B_\nu}{\partial x}\,d\nu , \tag{153}$$

or

$$F_x = -\frac{4\pi}{3\rho}\int_0^\infty \frac{1}{\kappa_\nu'}\frac{\partial B_\nu}{\partial T}\,d\nu\,\frac{\partial T}{\partial x} . \tag{154}$$

We now define the coefficient of opacity, κ, by

$$\frac{1}{\kappa}\int_0^\infty \frac{\partial B_\nu}{\partial T}\, d\nu = \int_0^\infty \frac{1}{\kappa'_\nu}\frac{\partial B_\nu}{\partial T}\, d\nu\,. \tag{155}$$

Equation (154) can now be written as

$$F_x = -\frac{4\pi}{3\kappa\rho}\int_0^\infty \frac{\partial B_\nu}{\partial x}\, d\nu = -\frac{4\pi}{3\kappa\rho}\frac{\partial B}{\partial x}\,, \tag{156}$$

where B is now the integrated Planck intensity. Similarly, if F_y and F_z are the integrated fluxes in the Y- and the Z-directions, we have

$$F_y = -\frac{4\pi}{3\kappa\rho}\frac{\partial B}{\partial y}\,; \qquad F_z = -\frac{4\pi}{3\kappa\rho}\frac{\partial B}{\partial z}\,. \tag{157}$$

Thus,

$$\boldsymbol{F} = -\frac{4\pi}{3\kappa\rho}\,\text{grad}\, B = -\frac{c}{\kappa\rho}\,\text{grad}\, p_r\,, \tag{158}$$

where p_r is the radiation pressure $\frac{1}{3}aT^4$. Thus for the integrated flux we have an equation of exactly the same form as (151), provided we average $(\kappa'_\nu)^{-1}$ over the frequencies suitably. According to (155), the coefficient of opacity, κ, is a sort of harmonic mean of κ'_ν. Explicitly,

$$\frac{1}{\kappa} = \frac{\displaystyle\int_0^\infty \frac{\nu^3 e^{2h\nu/kT}}{\kappa_\nu(e^{h\nu/kT}-1)^3}\frac{h\nu}{kT^2}\, d\nu}{\displaystyle\int_0^\infty \frac{\nu^3 e^{h\nu/kT}}{(e^{h\nu/kT}-1)^2}\frac{h\nu}{kT^2}\, d\nu}\,, \tag{159}$$

where we have substituted the Planck function B_ν in (155). The formula (159) for κ is due to Rosseland, and for this reason κ is also called the "Rosseland mean absorption coefficient."

Equation (152) in the integrated form can be written as

$$\text{div}\, \boldsymbol{F} = \epsilon\rho\,, \tag{160}$$

where ϵ is the total amount of heat energy liberated by unit mass in unit time over the whole frequency interval ($0 \leqslant \nu \leqslant \infty$).

For a spherically symmetrical distribution of matter, equations (151), (158), and (160) take the simpler forms

$$F_r(\nu) = -\frac{c}{\kappa'_\nu \rho} \frac{dp_r(\nu)}{dr} , \tag{161}$$

$$F_r = -\frac{c}{\kappa \rho} \frac{d}{dr} (\tfrac{1}{3} a T^4) , \tag{162}$$

and

$$\epsilon \rho = \frac{1}{r^2} \frac{d}{dr} (r^2 F_r) , \tag{163}$$

where $F_r(\nu)$ and F_r are the monochromatic and the integrated fluxes, respectively, across elements of surface normal to the direction of the radius vector r.

5. *The equations of hydrostatic equilibrium.*—Consider a thin cylinder of unit cross-section and height ds in the direction (l, m, n). By equations (75), (109), and (117) the mechanical force exerted by radiation on the cylinder in the frequency interval $(\nu, \nu + d\nu)$ in the s-direction is

$$\frac{\kappa'_\nu \rho F_s(\nu) d\nu}{c} ds . \tag{164}$$

By (151) this can be written as

$$- \operatorname{grad} p_r(\nu) \cdot d\boldsymbol{s} . \tag{165}$$

Hence, the mechanical force exerted by radiation in all the frequencies is obtained by integrating (165) over all the frequencies. By Stefan's law we then have

$$- \operatorname{grad} (\tfrac{1}{3} a T^4) \cdot d\boldsymbol{s} . \tag{166}$$

On the other hand, if p_g is the gas pressure, then this material pressure gradient would exert a further mechanical force on the cylinder considered of amount

$$- \operatorname{grad} p_g \cdot d\boldsymbol{s} . \tag{167}$$

Let V be the gravitational potential. For hydrostatic equilibrium we should have

$$\text{grad}\,(\tfrac{1}{3}aT^4 + p_g) = -\rho\,\text{grad}\,V\,. \tag{168}$$

For a spherically symmetrical distribution of matter (cf. Eq. [7], iii) equation (168) is equivalent to

$$\frac{d}{dr}\,(p_g + \tfrac{1}{3}aT^4) = -\frac{GM(r)}{r^2}\,\rho\,. \tag{169}$$

The other equations of equilibrium are equations (158) and (160); for a spherically symmetrical distribution of matter the appropriate equations are (162) and (163). When these equations are used, the quantity $L(r)$, which is the net amount of energy crossing a spherical surface of radius r, is generally introduced instead of F_r. Then,

$$F_r = \frac{L(r)}{4\pi r^2}\,. \tag{170}$$

In terms of $L(r)$ equations (162) and (163) take the forms

$$\frac{d}{dr}\,(\tfrac{1}{3}aT^4) = -\frac{\kappa L(r)}{4\pi c r^2}\,\rho \tag{171}$$

and

$$\frac{dL(r)}{dr} = 4\pi r^2 \epsilon \rho\,. \tag{172}$$

Equations (169), (171), and (172) are the equations of equilibrium for a star in radiative equilibrium.

BIBLIOGRAPHICAL NOTES

§ 1.—The most complete accounts of the theory of radiation are contained in—

1. M. Planck, *Wärmestrahlung*, 5th ed., Leipzig, 1923; English translation (of the 2d German ed.) by M. Masius, *Planck's Heat Radiation*, Philadelphia: Blackiston, 1914.

2. H. A. Lorentz, *Lectures on Theoretical Physics*, **2**, 209–275, 1927.

From the point of view of astrophysical applications the most valuable accounts are contained in—

3. E. A. Milne, *Handb. d. Astrophys.*, **3**, Part I, 1930.

4. S. Rosseland, *Astrophysik auf atomtheoretischer Grundlage*, Berlin: Springer, 1931.

§ 2.—For the thermodynamics of radiation see Planck (1) and Lorentz (2). Also—

5. P. Drude, *The Theory of Optics*, Part III, chap. ii; English translation by C. R. Mann and R. A. Millikan, New York: Longmans, 1922.

The most rigorous proof of Kirchhoff's law is due to—

6. D. Hilbert, *Physik. Zs.*, **13**, 1056, 1912.

7. D. Hilbert, *Physik. Zs.*, **14**, 592, 1913.

See also—

8. H. Straubel, *Physik. Zs.*, **4**, 114, 1903.

Hilbert's papers are very illuminating, and a careful study of them is well worth while.

§ 3.—The conception of local thermodynamical equilibrium is due to—

9. K. Schwarzschild, *Göttinger Nachrichten*, p. 41, 1906.

The derivation of Eq. (116) with κ'_ν as defined in Eq. (117) is due to S. Rosseland:

10. S. Rosseland, *Handb. der Astrophys.*, **3**, Part I, 443–457, 1930.

§§ 4 and 5.—The conception of radiative equilibrium is also due to Schwarzschild (9), but the analysis in the form given in § 4 is due to Rosseland (4); see also E. A. Milne (3).

Equations (162) and (163), which are fundamental in the theory of radiative equilibrium, are due to

11. A. S. Eddington, *M.N.*, **77**, 16, 1916. See also *ibid.*, **77**, 596, 1917; **79**, 22, 1918.

Eddington did not, however, consider the problem of radiative transfer in the separate frequences.

The physically correct derivation is due to Rosseland (10). Also—

12. S. Rosseland, *M.N.*, **84**, 525, 1924.

CHAPTER VI

GASEOUS STARS

In the last chapter we showed that the equation of hydrostatic equilibrium of a spherically symmetrical distribution of matter in radiative equilibrium is

$$\frac{d}{dr}(p_g + p_r) = -\frac{GM(r)}{r^2}\rho , \tag{1}$$

where p_g is the gas pressure, $p_r(=\frac{1}{3}aT^4)$ is the radiation pressure, and the rest of the symbols have their usual meanings. If the stellar material is a perfect gas, we can write

$$P = p_g + p_r = \frac{k}{\mu H}\rho T + \frac{1}{3}aT^4 . \tag{2}$$

The radiative temperature gradient is determined by (cf. Eqs. [171] and [172], v)

$$\frac{dp_r}{dr} = -\frac{\kappa L(r)}{4\pi c r^2}\rho \tag{3}$$

and

$$dL(r) = 4\pi r^2 \rho\epsilon dr . \tag{4}$$

In discussing the structure of model gaseous stars in radiative equilibrium, we have to exercise considerable care, inasmuch as we do not, as yet, know the exact dependence of ϵ on ρ and T.

In this chapter we shall attempt a first discussion of the equations of equilibrium for a gaseous star in radiative equilibrium. The problem of the "stability" of the radiative temperature gradient will also be considered.

We shall begin our discussion by proving a few integral theorems on the radiative equilibrium of a gaseous star.

1. *Integral theorems on the radiative equilibrium of a star.*—Dividing equation (3) by equation (1), we have

$$\frac{dp_r}{dP} = \frac{\kappa L(r)}{4\pi c G M(r)} . \tag{5}$$

We introduce the quantity η, defined by

$$\eta = \frac{\frac{L(r)}{M(r)}}{\frac{L}{M}} , \tag{6}$$

where L is the luminosity of the star. As defined in this manner, η is the ratio of the average rate of liberation of ("the heat") energy $\bar{\epsilon}(r)$ interior to the point r, to the corresponding average $\bar{\epsilon}$ for the whole star:

$$\eta = \frac{\bar{\epsilon}(r)}{\bar{\epsilon}} . \tag{7}$$

From (7) it follows that

$$\eta(R) = 1 ; \qquad \eta_c = \frac{\epsilon_c}{\bar{\epsilon}} , \tag{8}$$

in an obvious notation. Inserting (6) in (5), we have

$$\frac{dp_r}{dP} = \frac{L}{4\pi cGM} \kappa\eta . \tag{9}$$

Integrating the foregoing equation from $r = r$ to $r = R$ and using the boundary condition $p_r = 0$ at $r = R$, we have

$$p_r = \frac{L}{4\pi cGM} \int_0^P \kappa\eta dP . \tag{10}$$

Following Strömgren, we now define the average value $\overline{\kappa\eta}(r)$ by

$$\overline{\kappa\eta}(r) = \frac{1}{P} \int_0^P \kappa\eta dP . \tag{11}$$

(In writing equation [11] we have used the boundary condition that $P = 0$ at $r = R$.) Hence, we can re-write (10) as

$$p_r = \frac{L}{4\pi cGM} P \, \overline{\kappa\eta}(r) . \tag{12}$$

Let

$$p_r = (1 - \beta)P ; \qquad p_g = \beta P . \tag{13}$$

Then by (12) and (13)

$$\overline{\kappa\eta}(r) = \frac{4\pi cGM(1 - \beta)}{L} . \tag{14}$$

We have thus proved the following theorem due to Strömgren.

THEOREM 1.—*The ratio of the radiation pressure to the total pressure at a point inside a star in radiative equilibrium is proportional to the average value of* $\kappa\eta$ *for the regions exterior to the point* r, *the average being taken with respect to* dP, *where* P *is the total pressure.*

As a particular case of (14) we have

$$L = \frac{4\pi cGM(1 - \beta_c)}{\overline{\kappa\eta}} , \tag{15}$$

where $\overline{\kappa\eta}$ now defines the average value for the whole star. Equation (15), which is an exact equation, is a formula for the luminosity, L, of a star in terms of its mass, M, and an average value of $\kappa\eta$. We shall refer to equations (14) and (15) as the *luminosity formulae.* Sometimes it is useful to have an equation similar to (15) but which involves an average value of $(1 - \beta)$ instead of the value of $(1 - \beta)$ at the center. A formula of this kind can be obtained as follows:

Write equation (10) in the form

$$(1 - \beta)P = \frac{L}{4\pi cGM}\int_0^P \kappa\eta dP . \tag{16}$$

Multiply both sides of the equation by $P^q dP$ and integrate from 0 to P_c. We then have

$$\frac{1}{q + 2}\int_0^{P_c}(1 - \beta)d(P^{q+2}) = \frac{L}{4\pi cGM}\int_0^{P_c}P^q\left[\int_0^P \kappa\eta dP\right]dP . \tag{17}$$

Integrating the right-hand side by parts, we obtain

$$\left.\begin{aligned}&\frac{1}{q + 2}\int_0^{P_c}(1 - \beta)d(P^{q+2})\\ &\qquad = \frac{L}{4\pi cGM}\left\{\frac{P_c^{q+1}}{q + 1}\int_0^{P_c}\kappa\eta dP - \frac{1}{q + 1}\int_0^{P_c}\kappa\eta P^{q+1}\, dP\right\}.\end{aligned}\right\} \tag{18}$$

Define the following averages:

$$\overline{(1-\beta)}_n = \frac{1}{P_c^n}\int_0^{P_c}(1-\beta)d(P^n)\,, \tag{19}$$

$$\overline{\kappa\eta}_n = \frac{1}{P_c^n}\int_0^{P_c}\kappa\eta d(P^n)\,. \tag{20}$$

In terms of these averages equation (18) can now be written as

$$\overline{(1-\beta)}_{q+2} = \frac{L}{4\pi cGM}\left(\frac{q+2}{q+1}\overline{\kappa\eta}_1 - \frac{1}{q+1}\overline{\kappa\eta}_{q+2}\right), \tag{21}$$

which is the required formula. If $q = 0$, we have, as a special case of the foregoing,

$$L = \frac{4\pi cGM\overline{(1-\beta)}_2}{2\overline{\kappa\eta}_1 - \overline{\kappa\eta}_2}\,. \tag{22}$$

THEOREM 2.—*If* $\overline{\kappa\eta}(r)$ *decreases outward from the center in a star in radiative equilibrium, then* $(1-\beta)$ *must also decrease outward from the center.*

This is an immediate consequence of equation (14).

The following theorems, 3 and 4, are due to Chandrasekhar.

THEOREM 3.—*In a wholly gaseous configuration in radiative equilibrium in which the mean density* $\overline{\rho}(r)$ *inside* r *does not increase outward, we have*

$$\overline{\kappa\eta} \leqslant \frac{4\pi cGM(1-\beta^*)}{L}\,, \tag{23}$$

where β^* *satisfies the quartic equation*

$$M = \left(\frac{6}{\pi}\right)^{1/2}\left[\left(\frac{k}{\mu_c H}\right)^4 \frac{3}{a}\,\frac{1-\beta^*}{\beta^{*4}}\right]^{1/2}\frac{1}{G^{3/2}}\,. \tag{24}$$

Proof: Since we have assumed that the mean density does not increase outward, we can apply Theorem 7 of chapter iii, according to which

$$1-\beta_c \leqslant 1-\beta^*\,, \tag{25}$$

where β^* satisfies equation (24) and is determined uniquely by the mass M. Combining (25) with the luminosity formula (15), we obtain

$$L = \frac{4\pi cGM(1 - \beta_c)}{\overline{\kappa\eta}} \leqslant \frac{4\pi cGM(1 - \beta^*)}{\overline{\kappa\eta}}, \tag{26}$$

which is the required result.

THEOREM 4.—*In a wholly gaseous configuration in which the mean density* $\overline{\rho}(r)$ *inside* r *and the rate of liberation of energy* ϵ *do not increase outward, we have*

$$\overline{\kappa} \leqslant \frac{4\pi cGM(1 - \beta^*)}{L}, \tag{27}$$

where $\overline{\kappa}$ *is a mean opacity coefficient defined by*

$$\overline{\kappa} = \frac{1}{P_c}\int_0^{P_c} \kappa dP \tag{28}$$

and where the equality sign in (27) is possible only when ϵ *is a constant throughout the configuration.*

Proof: This is an immediate consequence of Theorem 3 above. For, if ϵ decreases outward, then $\eta (= \overline{\epsilon}(r)/\overline{\epsilon})$ must also decrease outward, and consequently the minimum value of η is unity. Hence,

$$\overline{\kappa\eta} \geqslant \overline{\kappa}, \tag{29}$$

where, according to (11), the opacity $\overline{\kappa}$ is to be defined as in (28) above. The equality sign in (29) is possible only when η = constant = 1, that is, when $\overline{\epsilon}(r) = \overline{\epsilon} = \epsilon$ = constant throughout the configuration. Combining (23) and (29), we have the required result.

We shall apply equation (27) to certain practical cases of interest. Numerically, equation (27) reduces to

$$\overline{\kappa} \leqslant 1.318 \times 10^4 \frac{M}{\odot} \frac{L_\odot}{L} (1 - \beta^*), \tag{30}$$

where $L_\odot$ refers to the luminosity of the sun.

For the sun, on the assumption that $\mu_c = 1$, the solution of the quartic equation (25) (obtained by interpolating among the figures given in Table 2, iii) yields $1 - \beta^* = 0.030$. Hence, by (30)

$$\bar{\kappa}_\odot < 391 \text{ gm}^{-1} \text{ cm}^2 . \tag{31}$$

For Capella, on the other hand, $M = 4.18\odot$ and $L = 120L_\odot$; on the assumption that $\mu_c = 1$, $1 - \beta^* = 0.22$. According to equation (30), we now have

$$\bar{\kappa}_{\text{Capella}} < 100 \text{ gm}^{-1} \text{ cm}^2 . \tag{32}$$

We shall see in chapter vii that the stellar opacity coefficient plays a fundamental role in the further development of the theory; it is therefore satisfactory to have the upper limit (30) proved under extremely general circumstances. It should be noticed that the method of averaging weights the central regions of the configuration very heavily, and hence the upper limit (30) is essentially an upper limit to the opacity at the center of the configuration. The inequality (30) can be interpreted in the following manner:

If, for a star of given mass M and luminosity L, $\bar{\kappa}$ should be greater than the limit set by (30), then, either the density or the rate of generation of energy, ϵ, or both, must increase outward in some finite regions of the interior of a star.

We can prove a somewhat less sharp inequality for κ at all points in the star. The following theorem is due to Eddington.

THEOREM 5.—*For a gaseous star in radiative equilibrium in which the density, temperature, and the rate of liberation of energy do not increase outward, we have*

$$\kappa < \frac{4\pi cGM}{L} \tag{33}$$

at all points inside the configuration.

This theorem is an immediate consequence of equation (9). For, if $\rho(r)$ and $T(r)$ do not increase outward, dp_g will be positive (or zero), for positive increments in ρ and T, and must always be less than dP. Hence dp_r/dP must be less than unity. By (9) we should therefore have

$$\frac{L}{4\pi cGM}\kappa\eta < 1 . \tag{34}$$

If, further, the rate of generation of energy, ϵ, does not increase outward, then, as we have already pointed out, $\eta \geqslant 1$; and hence, by (34)

$$\frac{L}{4\pi cGM}\kappa < 1 , \tag{35}$$

which proves the theorem. It should be noticed that the upper limits for κ and $\bar{\kappa}$ differ only by the additional factor $(1 - \beta^*)$ in the expression for the latter. For stars of small mass this factor can be very small (e.g., for the sun $1 - \beta^* = 0.03$), and thus the upper limit for $\bar{\kappa}$ is physically of greater interest.

Finally, we shall derive a very useful alternative form of equation (9). Combining equations (9) and (14), we have

$$\frac{dp_r}{dP} = (1 - \beta)\frac{\kappa\eta}{\overline{\kappa\eta}(r)} , \tag{36}$$

which can also be written in the form

$$\frac{dp_r}{p_r} = \frac{\kappa\eta}{\overline{\kappa\eta}(r)}\frac{dP}{P} ; \tag{37}$$

again, since $p_r = \frac{1}{3}aT^4$,

$$\frac{dT}{T} = \frac{1}{4}\frac{\kappa\eta}{\overline{\kappa\eta}(r)}\frac{dP}{P} . \tag{38}$$

2. *Stability conditions for radiative equilibrium.*—If a spherically symmetrical distribution of matter is in hydrostatic equilibrium, and if further radiative equilibrium obtains, then the radiative temperature gradient is determined by

$$\frac{dT}{T} = \frac{1}{4}\frac{\kappa\eta}{\overline{\kappa\eta}(r)}\frac{dP}{P} , \tag{39}$$

where

$$P = \frac{k}{\mu H}\rho T + \frac{1}{3}aT^4 . \tag{40}$$

We shall now consider the stability of the radiative gradient: To examine this, suppose an element of mass δm, originally at temperature T, density ρ, and pressure P, suffers a sudden increase of temperature of amount $\delta T > 0$. Then this element exerts a pressure of a definite amount, $\delta P > 0$, on its surroundings and expands and becomes less dense than its immediate neighborhood. The element δm will then experience a force tending to displace it into regions of lower density. During such a movement the element continues to expand, the temperature altering in the meantime.

We shall now make the following assumptions: (*a*) that at each instant of time the element δm expands (or contracts) to such an extent that the pressure exerted on the element *by* the surrounding material is the same as that which the element exerts *on* the surrounding material; (*b*) that the process of expansion (or contraction) takes place adiabatically; (*c*) that the viscous forces restricting the movement of the element δm can be neglected. We shall first examine the consequences of these assumptions.

By our second assumption, since the expansion of δm takes place adiabatically, we should have, according to equations (124) and (134) of chapter ii,

$$\left(\frac{dT}{T}\right)_{\delta m} = \frac{\Gamma_2 - 1}{\Gamma_2}\frac{dP}{P}, \tag{41}$$

where

$$\Gamma_2 = 1 + \frac{(4 - 3\beta)(\gamma - 1)}{\beta^2 + 3(\gamma - 1)(1 - \beta)(4 + \beta)}, \tag{42}$$

for the rate of change of the temperature of the element δm as it moves outward into the regions of lower density, expanding in the meantime.

Comparing (39) and (41), we see that the temperature of δm, as it moves outward, alters at a rate different from that of its immediate surroundings because, according to our first assumption, the pressure P alters in the same way for both δm and the surroundings.

Let us now suppose that

$$\frac{\Gamma_2 - 1}{\Gamma_2} > \frac{1}{4}\frac{\kappa\eta}{\overline{\kappa\eta}(r)}. \tag{43}$$

Then it follows that the element δm, after moving outward for a certain distance, will find itself at the same temperature, pressure, and density as its surroundings at that point; consequently the original disturbance dies out.

In the same way, if the element δm originally suffers a decrease of temperature of amount δT, then it will become denser than its immediate neighborhood and consequently will sink to regions of higher density. If we make the three assumptions as before, then the adiabatic compression it experiences as it sinks to regions of higher density increases the temperature of δm at a rate greater than the local temperature gradient. Again, it will soon find itself at a point where δm and its neighborhood at that point have the same density, pressure, and temperature. Thus in either case, i.e., either for a positive or a negative increment δT of an element δm, the disturbance dies out if equation (43) is true. In this sense *the radiative equilibrium is stable if the radiative gradient is less than the corresponding adiabatic gradient.* This result is due to Schwarzschild.

On the other hand, if

$$\frac{\Gamma_2 - 1}{\Gamma_2} < \frac{1}{4} \frac{\kappa\eta}{\overline{\kappa\eta}(r)} \tag{44}$$

and if the element δm suffers an increase of temperature, then, as before, it will move outward to regions of lower density; now, however, the temperature of δm will decrease less rapidly than that of its surroundings, and hence it is always at a temperature higher than its surroundings. In the same way, if the element δm suffers a decrease of temperature, it will sink to regions of higher density; but the adiabatic compression it experiences (according to our assumption [b]) will not ever be sufficient to raise the temperature of δm to that of its surroundings (if equation [44] remains true); consequently, it always remains cooler than its surroundings. Hence, we have proved on the basis of the assumptions that *the radiative equilibrium is stable or unstable according as*

$$4\,\frac{\Gamma_2 - 1}{\Gamma_2} > \text{ or } < \frac{\kappa\eta}{\overline{\kappa\eta}(r)}\,. \tag{45}$$

Table 5 gives the values of $4(\Gamma_2 - 1)/\Gamma_2$ for different values of $(1 - \beta)$.

TABLE 5

$1-\beta$	0	0.1	0.2	0.3	0.4	0.5	0.6	0.7	0.8	0.9	1.0
$\frac{4(\Gamma_2-1)}{\Gamma_2}$	1.6	1.304	1.177	1.108	1.065	1.039	1.022	1.010	1.004	1.000	1.000

3. *The equations of equilibrium when the radiative gradient is unstable.*—Suppose that we have initially a situation in which the radiative gradient obtains but one which exceeds the adiabatic gradient. By our discussion in the previous section, the radiative gradient is unstable, in the sense that a slight alteration of the local temperature will give rise to a system of ascending and descending currents which will have the effect of reducing the existing temperature gradient. Eventually a steady state must be set up, though it is not a priori clear what the nature of that steady state will be. Under the circumstances, the general picture which is adopted is the following one.

We suppose that masses of gas are continually being detached from the surrounding matter and that they move bodily through a certain distance before they are reabsorbed into the main mass of the material. Alternatively, the situation can be described by saying that "eddies" are continually being formed which travel, on the average, a distance l with a certain mean speed u before being reabsorbed into the main mass of the material. The quantities l and u thus defined are referred to as the "mean free path" of the eddies and the "mean speed of turbulent motion," respectively. We further suppose that we can define a certain mean temperature T at each point to describe the local properties.

Consider the transfer of heat energy across an element of surface, Σ, at r, which is large compared with the cross-sections of individual eddies. The eddies which are absorbed into the main mass of material at r will have been formed, on the average, at points distant l from r.

The eddy which is formed at $(r - l)$ will have a temperature

$$T - l\frac{dT}{dr}.$$

When the eddy appears at r, and before it is reabsorbed, it will have a temperature

$$T - l\frac{dT}{dr} + l\left(\frac{dT}{dr}\right)^*,$$

where $(dT/dr)^*$ is the rate at which the temperature of an eddy alters during its motion. If we assume that during its motion an eddy expands or contracts adiabatically to such an extent that the pressure exerted by it on its surroundings is equal to that exerted on it by its surroundings, then

$$\left(\frac{d \log T}{dr}\right)^* = \frac{\Gamma_2 - 1}{\Gamma_2} \frac{d \log P}{dr} .$$

At r the eddies are reabsorbed into the main mass of the surrounding material at constant pressure. Hence, the total energy, Q, crossing the surface, Σ, and expressed in terms of unit area, is

$$Q = C_P \frac{1}{\Sigma} \int_{(\Sigma)} \rho u l \left\{ \left(\frac{dT}{dr}\right)^* - \left(\frac{dT}{dr}\right) \right\} d\Sigma , \tag{46}$$

where C_P is the specific heat at constant pressure of the matter and radiation (cf. Eq. [146], ii). The foregoing expression can be written as

$$Q = C_P \, \sigma \left\{ \left(\frac{dT}{dr}\right)^* - \left(\frac{dT}{dr}\right) \right\} , \tag{46'}$$

where the eddy conductivity, σ, is defined by

$$\sigma = \overline{\rho u l} = \frac{1}{\Sigma} \int_{(\Sigma)} \rho u l \, d\Sigma .$$

It should be mentioned here that σ will itself depend upon the degree of instability as specified by $\{(dT/dr)^* - (dT/dr)\}$. Indeed, we should expect that with increasing instability the turbulent motions will become more violent; this would, in turn, lead to larger values of u and, hence, of σ. In general, the magnitude of the eddy velocities will be determined by the balance of energy which becomes available to the eddies from the mean internal energy, and the energy lost by the eddies through viscous dissipation.

Further, the mean value of the rate of transfer of momentum across Σ is $\overline{\rho u^2}$ per unit area. Also, the transport of turbulent energy is measured by $\frac{1}{2}\overline{\rho u(u^2 + v^2 + w^2)}$.[1]

[1] u, v, and w are the components of the eddy velocity with respect to a fixed frame of reference. Further, it is assumed that u is in the direction of r.

When writing down the equations of equilibrium, it should be remembered that

$$\frac{dp_r}{dr} = -\frac{\kappa \rho F_r}{c}, \tag{47}$$

where p_r is radiation pressure and F_r is the actual net flux of radiant energy. If ϵ has the same meaning as in equation (4), we then have

$$\left.\begin{aligned} 4\pi r^2 F_r &= 4\pi \int_0^r \epsilon \rho r^2 dr \\ &\quad - 4\pi r^2 \left[C_P\, \sigma \left\{ \left(\frac{dT}{dr}\right)^* - \left(\frac{dT}{dr}\right) \right\} + \tfrac{1}{2}\, \overline{\rho u (u^2 + v^2 + w^2)} \right] \end{aligned}\right\}. \tag{48}$$

The equation of hydrostatic equilibrium can now be written as

$$\frac{d}{dr}\left(p_r + p_g + \overline{\rho u^2}\right) = -\frac{GM(r)}{r^2}\,\rho\,. \tag{49}$$

Equations (47), (48), and (49) are quite general. A more detailed discussion is needed to make these equations more explicit. The case of vanishing radiation pressure has been investigated by Cowling,[2] whose results we shall quote:

(*a*) The transport of turbulent energy, $\frac{1}{2}\overline{\rho u(u^2 + v^2 + w^2)}$, in (48), and the turbulent pressure, $\overline{\rho u^2}$, in (49) can be neglected in comparison with the other terms occurring in the respective equations (*b*). The temperature gradient as defined by (47) and (48) differs from the adiabatic gradient $(dT/dr)^*$ only by a very small amount. The temperature gradient set up will therefore be only very slightly superadiabatic.

The foregoing simplifications seem to arise mainly from the circumstance that the internal energy of a gram of the material is so very large compared to the energy loss, ϵ, due to subatomic processes. Consequently, even very slight mass motions are sufficient to reduce a superadiabatic gradient to a stable one which differs from the adiabatic gradient only by an insignificant amount. Thus, on the basis of the Biermann-Cowling analysis we can conclude that

[2] Essentially equivalent results were given by L. Biermann.

when the radiative gradient becomes unstable, we have an adiabatic gradient set up with the relations

$$p \propto \rho^{\gamma} \propto T^{\gamma/(\gamma-1)} .$$

The case when the radiation pressure is comparable to the gas pressure has not yet been fully investigated.

Finally, it should be pointed out that, if the convection currents become violent, we may have to introduce entirely new considerations. In particular, the three fundamental assumptions of § 2 need not necessarily be valid. For, if the "inertia of motions" (to use Kelvin's phrase) is large, then the elements of ascending and descending masses will experience viscous friction, which may result in the communication of probably quite appreciable amounts of heat to the eddies during their motions.

4. *The standard model.*—In the Introduction we pointed out the fact that an attack on the problem of stellar structure is made possible at the present time only on the basis of certain assumed laws concerning the rate of generation of energy, ϵ, or the energy-source distribution, as defined by η. A model which was first introduced by Eddington and which has played an important role in subsequent developments is the so-called "standard model." This is defined as one in which $\kappa\eta$ is a constant throughout a given configuration.

From the luminosity formula (Eq. [14])

$$1 - \beta = \frac{L}{4\pi cGM} \overline{\kappa\eta}(r) \tag{50}$$

we infer that $(1 - \beta)$ is a constant throughout the configuration. Since we can write (cf. Eq. [87], iii)

$$P = \left[\left(\frac{k}{\mu H} \right)^4 \frac{3}{a} \frac{1 - \beta}{\beta^4} \right]^{1/3} \rho^{4/3} , \tag{51}$$

it follows that for the standard model we have the relation

$$P = K\rho^{4/3} ; \qquad K = \left[\left(\frac{k}{\mu H} \right)^4 \frac{3}{a} \frac{1 - \beta}{\beta^4} \right]^{1/3} , \tag{52}$$

where, if we further assume that μ is a constant, K is a constant. The equilibrium configurations are therefore polytropes of index $n = 3$, and the general theory of chapter iv can be applied. In particular, the Lane-Emden function θ_3 completely determines the structure of the configurations. We have

$$P = P_c\theta_3^4 \; ; \qquad \rho = \rho_c\theta_3^3 \; ; \qquad T = T_c\theta_3 \, , \tag{53}$$

where P_c, ρ_c, and T_c are the values of the variables at the center.

From §§ 6 and 7 of chapter iv we have the following results.

a) The mass relation.—By equation (70) of chapter iv and equation (52) above, we have

$$M = -4\pi \left[\left(\frac{k}{\mu H} \right)^4 \frac{3}{a} \frac{1-\beta}{\beta^4} \right]^{1/2} \frac{1}{(\pi G)^{3/2}} \left(\xi^2 \frac{d\theta_3}{d\xi} \right)_{\xi=\xi_1} ; \tag{54}$$

in other words, β is determined uniquely by M and satisfies a quartic equation. Equation (54) was first derived independently by Bialobjesky and Eddington. It is of interest to compare (54) with the quartic equation for β^* (Eq. [24]), which gives the minimum value of β_c at the center of a star of given M, in which the density does not increase outward. By comparison we find that the ratio of the numerical coefficients in (54) and (24) is given by

$$-\frac{4}{\pi^{1/2}} \left(\xi^2 \frac{d\theta_3}{d\xi} \right)_{\xi=\xi_1} : \frac{6^{1/2}}{\pi^{1/2}} = \left(\tfrac{8}{3}\right)^{1/2} \times 2.0182 = 3.296 \, . \tag{55}$$

Table 6 gives the values of M for different values of $1 - \beta$.

TABLE 6

$(1-\beta)$ FOR THE STANDARD MODEL

$1-\beta$	$\left(\frac{M}{\odot}\right)\mu^2$	$1-\beta$	$\left(\frac{M}{\odot}\right)\mu^2$
0.025	2.993	0.5	50.86
.05	4.456	0.6	87.04
.1	7.020	0.7	167.16
.2	12.56	0.8	402.0
.3	20.10	0.9	1705.9
0.4	31.59	1.0	∞

b) *The ratio of the mean to the central density.*—According to equation (78), chapter iv, and Table 4, we have

$$\rho_c = -\left[\frac{\xi}{3}\frac{1}{\dfrac{d\theta_3}{d\xi}}\right]_{\xi=\xi_1}\bar{\rho} = 54.18\bar{\rho}\,. \tag{56}$$

c) *The central pressure.*—According to equations (80) and (81) of chapter iv and Table 4 we have

$$P_c = \frac{1}{16\pi\left[\left(\dfrac{d\theta_3}{d\xi}\right)_{\xi=\xi_1}\right]^2}\frac{GM^2}{R^4} = 11.05\,\frac{GM^2}{R^4}\,, \tag{57}$$

or, numerically,

$$P_c = 1.242 \times 10^{17}\left(\frac{M}{\odot}\right)^2\left(\frac{R_\odot}{R}\right)^4 \text{ dynes cm}^{-2}\,. \tag{58}$$

d) *The central temperature.*—We have for the present case

$$\frac{k}{\mu H}\rho_c T_c = \beta_c P_c = \beta P_c\,. \tag{59}$$

From (56) and (57)

$$T_c = \frac{\beta\mu H}{k}\frac{1}{4\left[-\xi\dfrac{d\theta_3}{d\xi}\right]_{\xi=\xi_1}}\frac{GM}{R}\,, \tag{60}$$

or

$$T_c = 0.8543\,\frac{\beta\mu H}{k}\frac{GM}{R}\,. \tag{61}$$

Numerically, the foregoing is found to be

$$T_c = 19.72 \times \beta\mu\left(\frac{M}{\odot}\right)\left(\frac{R_\odot}{R}\right) \times 10^6 \text{ degrees.} \tag{62}$$

e) *The potential energy.*—By equation (90) of chapter iv we have

$$-\Omega = \frac{3}{2}\frac{GM^2}{R}\,. \tag{63}$$

f) The mean temperature.—If $\overline{T}$ is the mean temperature defined by

$$M\overline{T} = \int T dM(r) , \tag{64}$$

then

$$M\overline{T} = \frac{\mu H}{k}\int p_{\text{gas}}\, dV = \frac{\mu H}{k}\int \beta P dV , \tag{65}$$

or, since β is a constant,

$$M\overline{T} = -\frac{1}{3}\frac{\beta\mu H}{k}\Omega , \tag{66}$$

or by (63)

$$\overline{T} = \frac{1}{2}\frac{\beta\mu H}{k}\frac{GM}{R} . \tag{67}$$

Numerically, the foregoing is found to be

$$\overline{T} = 11.54\beta\mu\left(\frac{M}{\odot}\right)\left(\frac{R_{\odot}}{R}\right)\times 10^{6}\text{ degrees} . \tag{68}$$

g) The internal and the total energy.—The internal energy consists of two parts: the contribution by the gas and the contribution by the radiation. A slight modification of Ritter's relation (Eq. [55], iii) yields

$$U_{\text{gas}} = -\frac{\beta}{3(\gamma - 1)}\Omega , \tag{69}$$

where γ is the ratio of specific heats of the gas. The internal energy, U_{rad}, due to the radiant energy, is

$$U_{\text{rad}} = \int aT^{4} dV = 3\int (1 - \beta) P dV , \tag{70}$$

or for the standard model

$$U_{\text{rad}} = -(1 - \beta)\Omega . \tag{71}$$

Hence, the total internal energy is

$$U = U_{\text{gas}} + U_{\text{rad}} = -\left[1 + \frac{\beta(4 - 3\gamma)}{3(\gamma - 1)}\right]\Omega , \tag{72}$$

where Ω is given by (63). The total energy, E, is seen to be

$$E = U + \Omega = -\beta \frac{4 - 3\gamma}{3(\gamma - 1)} \Omega . \qquad (73)^{3}$$

h) *Numerial applications.*—As an example of the application of the foregoing formulae, we shall calculate the values of P_c, ρ_c, and T_c for three typical stars—the sun, Sirius A, and Capella A. (More extensive tables are given in Appendix III.)

TABLE 7

	$M/\odot$	$R/R\odot$	$L/L\odot$	$(1-\beta)$*	P_c	ρ_c	T_c
Sun..........	1.00	1.00	1.00	0.003	1.2×10^{17}	76.5	20×10^{6}
Sirius A.......	2.34	1.78	38.9	0.016	6.8×10^{16}	31.7	26×10^{6}
Capella A.....	4.18	15.9	120	0.045	3.4×10^{13}	0.080	5×10^{6}

* μ has been assumed to be equal to unity in all cases (cf. chap. vii).

5. *The luminosity formula for the standard model.*—By the luminosity formula we have, quite generally, that

$$L = \frac{4\pi cGM(1 - \beta_c)}{\overline{\kappa\eta}} . \qquad (74)$$

For the standard model, $(1 - \beta)$ and $\kappa\eta$ are constants; and consequently we can write

$$1 - \beta_c = 1 - \beta ; \qquad \overline{\kappa\eta} = \kappa_c\eta_c , \qquad (75)$$

and the luminosity formula can be written as

$$L = \frac{4\pi cGM(1 - \beta)}{\kappa_c\eta_c} . \qquad (76)^{4}$$

[3] If β is not constant, the general relations are

$$U = \int\left[\frac{\beta}{\gamma - 1} + 3(1 - \beta)\right]PdV \qquad (72')$$

and

$$E = \int\beta\frac{4 - 3\gamma}{(\gamma - 1)} PdV . \qquad (73')$$

[4] The reason for writing the equation in this way will be clear from the discussion in § 7.

For the coefficient of opacity we assume a law of the form

$$\kappa = \kappa_0 \rho T^{-3-s} , \tag{77}$$

where κ_0 is a numerical constant which may depend upon μ. We can quite generally re-write (77) as

$$\kappa = \kappa_0 \frac{a}{3} \frac{\mu H}{k} \frac{\beta}{1 - \beta} T^{-s} . \tag{78}$$

Substituting for κ_c (according to [78]) in (76) and using equation (60) for the central temperature T_c, we get

$$L = \frac{4\pi c G M}{\kappa_0 \eta_c} \frac{1}{\left[-4\xi_1 \left(\frac{d\theta_3}{d\xi} \right)_{\xi=\xi_1} \right]^s} \frac{3}{a} \left(\frac{\beta \mu H}{k} \right)^{s-1} \left(\frac{GM}{R} \right)^s (1 - \beta)^2 . \tag{79}$$

On the other hand, from the quartic equation (54) we have

$$M^4 = \frac{256}{\pi^2} \frac{1}{G^6} \left(\frac{k}{\beta \mu H} \right)^8 \left(\frac{3}{a} \right)^2 (1 - \beta)^2 \left(\xi^2 \frac{d\theta_3}{d\xi} \right)^4_{\xi=\xi_1} . \tag{80}$$

Eliminating $(1 - \beta)^2$ from (79) and (80), we have

$$L = \frac{\pi^3}{4^{3+s}} \left(\frac{GH}{k} \right)^{7+s} \frac{ac}{3} \frac{\xi_1^s}{[{}_0\omega_3]^{4+s}} \frac{1}{\kappa_0 \eta_c} \frac{M^{5+s}}{R^s} (\mu\beta)^{7+s} . \tag{81}$$

For the case $s = \frac{1}{2}$, equation (81) is numerically found to be

$$\frac{L}{L_\odot} = 1.793 \times 10^{25} \frac{1}{\kappa_0 \eta_c} \left(\frac{M}{\odot} \right)^{5.5} \left(\frac{R_\odot}{R} \right)^{0.5} (\mu\beta)^{7.5} . \tag{82}$$

Equation (81), then, is the mass-luminosity-radius relation for the standard model. It is, of course, clear that, since on this model the stars form a homologous family, η_c must be the same for all stars; it is a pure number.

6. *Homologous transformations.*—In the previous section the (L, M, R) relation was derived for the standard model. We have now to examine the question as to how general the results based on the standard model can be expected to be.[5] This problem has a twofold character. (1) How general is the form of the relation (81)? (2) How can we apply the standard-model distributions of density and temperature to configurations in which $\kappa\eta$ is not accurately constant but shows only slight variations from constancy?

Regarding the form of (81), we shall prove, following Strömgren, that *for a star in radiative equilibrium, in which the radiation pressure is negligible throughout the configuration,*

$$L = \text{constant}\,\frac{1}{\kappa_0}\frac{M^{5+s}}{R^s}\,\mu^{7+s} \tag{83}$$

if the rate of generation of energy, ϵ, *and the coefficient of opacity follow the laws*

$$\epsilon = \epsilon_0\rho^{\alpha}T^{\nu}\,; \qquad \kappa = \kappa_0\rho T^{-3-s}\,, \tag{84}$$

where α, ν, *and* s *are arbitrary. The constant in (83) depends only on the exponents* α, ν, *and* s.

Proof: The equations of equilibrium can be written as:

$$\frac{dP}{dr} = -\frac{GM(r)}{r^2}\,\rho\,, \tag{85}$$

$$\frac{dM(r)}{dr} = 4\pi r^2\rho\,, \tag{86}$$

$$P = \frac{k}{\mu H}\,\rho T\,, \tag{87}$$

$$\frac{dT}{dr} = -\frac{3}{4ac}\,\kappa_0\rho^2T^{-6-s}\,\frac{\epsilon_0\int_0^r \rho^{\alpha+1}T^{\nu}r^2dr}{r^2}\,. \tag{88}$$

[5] In an investigation (*Ap. J.*, **87**, 535, 1938) completed since the writing of the monograph, an important minimal characteristic of the standard model has been proved. It can be shown that *for gaseous stars in equilibrium in which* ρ *and* $(1-\beta)$ *do not increase outward the minimum value of* $(1-\beta_c)$ *is the constant value of* $(1-\beta)$ *ascribed to a standard-model configuration of the same mass.* For stars in radiative equilibrium the condition that $(1-\beta)$ should not increase outward is equivalent to $\overline{\kappa\eta}(r)$ not increasing outward.

In writing equation (87) we have neglected radiation pressure, according to our assumption.

The system of equations (85)–(88) has to be solved with the boundary conditions

$$M(r) = M\,, \qquad \rho = 0\,, \qquad T = 0 \qquad \text{at} \qquad r = R \tag{89}$$

and

$$M(r) = 0 \qquad \text{at} \qquad r = 0\,. \tag{90}$$

These provide four boundary conditions; and since the system of equations (85)–(88) is equivalent to a single differential equation of the fourth order, it follows that there is just exactly one solution which will satisfy the foregoing boundary conditions. We shall now show that from such a solution we can construct another solution such that it will describe another configuration with a different M, R, and μ; we shall see, in fact, that the transformation required to go over from one set of values, M, R, and μ, to another set, M_1, R_1, and μ_1, is the successive application of three elementary homologous transformations. To show this, we proceed as follows:

Let the physical variables, after a general homologous transformation has been applied, be denoted by attaching a suffix "1". For a general homologous transformation we should have

$$\left.\begin{aligned} r_1 &= y^{n_1} r\,, & \mu_1 &= y^{n_5}\mu\,, \\ P_1 &= y^{n_2} P\,, & T_1 &= y^{n_6} T\,, \\ M(r_1)_1 &= y^{n_3} M(r)\,, & (\kappa_0\epsilon_0)_1 &= y^{n_7}(\kappa_0\epsilon_0)\,, \\ \rho_1 &= y^{n_4}\rho\,, \end{aligned}\right\} \tag{91}$$

where $n_1, \ldots, n_7$ are, for the present, arbitrary constants and y is the transformation constant. The exponents $(n_1, \ldots, n_7)$ should satisfy certain relations, namely, those which are necessary for the continued validity of equations (85)–(88) in the suffixed variables. Substituting (91) in (85), we find that we should have

$$y^{n_2 - n_1} = y^{n_3 + n_4 - 2n_1}\,, \tag{92}$$

or

$$n_2 - n_1 = n_3 + n_4 - 2n_1\,. \tag{93}$$

In the same way, equations (86), (87), and (88) yield:

$$n_3 - n_1 = 2n_1 + n_4 \,, \tag{94}$$

$$n_2 = n_4 + n_6 - n_5 \,, \tag{95}$$

$$n_6 - n_1 = n_7 + (\alpha + 3)n_4 - (6 + s - \nu)n_6 + n_1 \,. \tag{96}$$

We thus have four equations between the seven unknowns. Hence, we should be able to express any four of n's in terms of the other three. We shall choose n_1, n_3, and n_5 as the independent quantities. Solving for n_2, n_4, n_6, and n_7 in terms of n_1, n_3, and n_5, we find:

$$n_2 = -4n_1 + 2n_3 \,, \tag{97}$$

$$n_4 = -3n_1 + n_3 \,, \tag{98}$$

$$n_6 = -n_1 + n_3 + n_5 \,, \tag{99}$$

$$n_7 = -(s - \nu - 3\alpha)n_1 + (4 + s - \nu - \alpha)n_3 + (7 + s - \nu)n_5 \,. \tag{100}$$

If we choose $n_1 = 1$, $n_3 = 0$, and $n_5 = 0$, we have a homologous transformation in which a star of a given mass M and molecular weight μ is expanded or contracted. In the same way, the choice $n_1 = 0$, $n_3 = 1$, and $n_5 = 0$ corresponds to an alteration of M while R and μ are kept fixed. Finally, the choice $n_1 = 0$, $n_3 = 0$, and $n_5 = 1$ corresponds to an alteration of μ while R and M are kept unchanged. These three elementary homologous transformations are schematically represented by

$$\left.\begin{array}{lll}
r_1 = y_R r & r_1 = r & r_1 = r \,, \\
P_1 = y_R^{-4} P & P_1 = y_M^{2} P & P_1 = P \,, \\
M(r_1)_1 = M(r) & M(r_1)_1 = y_M M(r) & M(r_1)_1 = M(r) \,, \\
\rho_1 = y_R^{-3} \rho & \rho_1 = y_M \rho & \rho_1 = \rho \,, \\
\mu_1 = \mu & \mu_1 = \mu & \mu_1 = y_\mu \mu \,, \\
T_1 = y_R^{-1} T & T_1 = y_M T & T_1 = y_\mu T \,, \\
(\kappa_0\epsilon_0)_1 = y_R^{3\alpha+\nu-s}(\kappa_0\epsilon_0) & (\kappa_0\epsilon_0)_1 = y_M^{4+s-\nu-\alpha}(\kappa_0\epsilon_0) & (\kappa_0\epsilon_0)_1 = y_\mu^{7+s-\nu}(\kappa_0\epsilon_0) \,, \\
R_1 = y_R R & M_1 = y_M M & \mu_1 = y_\mu \mu \,.
\end{array}\right\} \tag{101}$$

We have now to consider how the luminosity is changed by a homologous transformation. Since

$$L = 4\pi \int_0^R r^2 \rho \epsilon \, dr \,, \tag{102}$$

we have, according to our law (84) for ϵ,

$$\kappa_0 L = 4\pi\kappa_0\epsilon_0 \int_0^R r^2 \rho^{\alpha+1} T^\nu dr . \tag{103}$$

Hence, by a general homologous transformation, $\kappa_0 L$ alters to $(\kappa_0 L)_1$, where

$$(\kappa_0 L)_1 = y^{n_7+3n_1+(\alpha+1)n_4+\nu n_6}(\kappa_0 L) , \tag{104}$$

or by (97), (98), (99), and (100)

$$(\kappa_0 L)_1 = y^{-sn_1+(5+s)n_3+(7+s)n_5}(\kappa_0 L) . \tag{105}$$

In other words,

$$L = \text{constant} \, \frac{M^{5+s}}{\kappa_0 R^s} \, \mu^{7+s} . \tag{106}$$

It is clear that the constant in (106) can depend only upon the exponents s, ν, and α.

We have thus proved the invariance of the form of the luminosity formula for stars in radiative equilibrium. If, however, the law of energy generation is such that it leads to a sufficiently strong concentration of the energy sources toward the center, then we should reach a stage when

$$\frac{\kappa\eta}{\overline{\kappa\eta}(r)} > 4 \, \frac{\Gamma_2 - 1}{\Gamma_2} . \tag{107}$$

In other words, going inward[6] toward the interior of a star, the radiative gradient will become unstable at some definite point $r = r_i$, (say). For stars with negligible radiation we have from (107) that

$$\frac{\kappa\eta}{\overline{\kappa\eta}(r)} = 1.6 \qquad (r = r_i) . \tag{107'}$$

For $r < r_i$, $\kappa\eta/\overline{\kappa\eta}(r) > 1.6$. Now the right-hand side of (107′) is a pure number, while the quantity on the left-hand side is homology invariant. Hence, *the fraction*, $q = r_i/R$, *of the radius at which the*

[6] We shall see in chap. viii that the radiative gradient is stable in the outer parts, including the stellar envelope.

instability of the radiative gradient sets in is the same for all stars with vanishing radiation pressure. The fraction q depends only on the exponents s, α, and ν, which occur in the laws for κ and ϵ.

According to the discussion in § 3, the material interior to $r_i = qR$ will be in convective equilibrium. Since the radiation pressure is assumed to be negligible, in the convective core we should have

$$\frac{p}{p_i} = \left(\frac{\rho}{\rho_i}\right)^{\gamma}, \tag{108}$$

where p_i and ρ_i refer to the pressure and the density at the "interface," i.e., at $r_i = qR$. Equation (108) is clearly homology invariant. Hence, the structure of the convective core is also homology invariant. We have thus proved the invariance of the form of the luminosity formula (106), quite generally.

We have stated and proved Strömgren's theorem for strictly vanishing radiation pressure. It is, however, clear that if β is very nearly unity, the variation in β can be properly neglected[7] and a mean value chosen. The result is equivalent to defining a new "molecular weight" $\mu\beta$ instead of μ; in other respects, the method of argument remains as before. Hence, we have, more generally than (106),

$$L = \text{constant}\, \frac{M^{5+s}}{\kappa_0 R^s}\, (\beta\mu)^{7+s} \qquad (\beta \sim 1)\,, \tag{109}$$

which is identical in form with the (L, M, R) relation derived for the standard model. The present restriction that the radiation pressure is negligible means that we should restrict ourselves to stars of small mass (cf. Theorem 7, iii). We shall see in chapter vii that the majority of stars for which we have observational material concerning L, M, and R fall into the class of stars with "negligible radiation pressure." Thus, the use of the (L, M, R) relation derived on the basis of the standard model can be largely justified—especially as we now see that the same form for the relation results for a wide class of stellar models.

Again, since the stars form a homologous family under the restric-

[7] This is not the same thing as neglecting the variation of $1 - \beta$.

tions of Strömgren's theorem, we can apply Theorem 13 of chapter iii, according to which

$$\frac{1-\beta_1}{(\mu_1\beta_1)^4}=\frac{1-\beta_0}{(\mu_0\beta_0)^4}\left(\frac{M_1}{M_0}\right)^2, \tag{110}$$

where β_1 and β_0 refer to the values at the corresponding points in two stars of mass M_1 and M_0. Hence, we should have, as a particular case of (110),

$$1-\beta_c=\text{constant } M^2(\mu\beta_c)^4 . \tag{110'}$$

In other words, $(1-\beta_c)$ should satisfy a certain quartic equation—the constant in (110′) will of course be different from that in the quartic equation for the standard model.

We thus see the complete parallelism between the standard model and these more general models.

7. *Perturbation theory for varying $\kappa\eta$.*—The nature of the problem presented can be described in the following way.

We first assume that $\kappa\eta$ is constant; this leads to a perfectly definite distribution of density and temperature. Now a physical theory, on the other hand, may be expected to specify the precise dependence of the opacity and the rate of generation of energy on the density ρ and the temperature T. From the march of ρ and T, derived on the basis of the constancy of $\kappa\eta$, we can calculate κ and η at each point and form the product $\kappa\eta$; a test of the consistency of the model is that the product $\kappa\eta$, determined in this way, should be reasonably constant. If this is so, the question arises as to how we can apply the results based on the hypothesis of $\kappa\eta$ being constant to cases where $\kappa\eta$ shows slight variations from constancy. The answer to this question can be given only on the basis of a perturbation theory, which we shall proceed to outline. The following analysis is a modified version of the theory which was first developed by Strömgren.

Now, the luminosity formula predicts that

$$1-\beta=\frac{L}{4\pi cGM}\,\overline{\kappa\eta}(r) . \tag{111}$$

We can re-write (111) in the form

$$1 - \beta = \frac{\overline{\kappa\eta}(r)}{\overline{\kappa\eta}} (1 - \beta_c) . \tag{112}$$

If we make definite assumptions concerning the dependences of κ and ϵ on ρ and T, we can evaluate the march of the quantity $\overline{\kappa\eta}(r)$ inside the star on the basis of the standard model. Equation (112) will now specify the variations in $(1 - \beta)$ as determined by the luminosity formula.[8] We suppose that the variation of $(1 - \beta)$ thus predicted is small and that we can write

$$P = \left[\left(\frac{k}{\mu H}\right)^4 \frac{3}{a} \frac{1 - \beta_c}{\beta_c^4}\right]^{1/3} (1 + \psi)\rho^{4/3} ; \tag{113}$$

where we can regard ψ as a small quantity of the first order. It will be noticed that in writing (113) we have assumed that the variations of both $1 - \beta$ and β as determined by (112) are small. This implies that for values of $1 - \beta_c$ near unity, the permissible range of variation for $\kappa\eta$ is much narrower than when $1 - \beta_c$ is small; for example, a variation of $\kappa\eta$ by 10 per cent will be permissible for $1 - \beta_c = 0.1$, while a variation even by this amount should be excluded for $1 - \beta_c = 0.8$ or 0.9, if the standard model is to be regarded as a reasonable first approximation.

We assume that ψ, as introduced, is a known function of r; ψ will be simply related to $\overline{\kappa\eta}(r)/\overline{\kappa\eta}$. We write (113) as

$$P = K(1 + \psi)\rho^{4/3} , \tag{114}$$

where

$$K = \left[\left(\frac{k}{\mu H}\right)^4 \frac{3}{a} \frac{1 - \beta_c}{\beta_c^4}\right]^{1/3} . \tag{115}$$

In the equation of hydrostatic equilibrium,

$$\frac{1}{r^2} \frac{d}{dr} \left(\frac{r^2}{\rho} \frac{dP}{dr}\right) = -4\pi G\rho , \tag{116}$$

[8] *Not* as determined by the local values of density and temperature.

set (cf. Eq. [10], iv)

$$\rho = \lambda\Theta^3 ; \qquad P = K\lambda^{4/3}(1 + \psi)\Theta^4 , \tag{117}$$

$$r = \left[\frac{K}{\pi G}\right]^{1/2} \lambda^{-1/3}\xi = \alpha\xi . \tag{118}$$

Equation (116) reduces to

$$\frac{1}{4}\frac{1}{\xi^2}\frac{d}{d\xi}\left[\frac{\xi^2}{\Theta^3}\frac{d}{d\xi}(1 + \psi)\Theta^4\right] = -\Theta^3 . \tag{119}$$

In the foregoing equation ψ is to be regarded as a known function of ξ. At $\xi = 0$, ψ must satisfy the boundary condition,

$$\psi = 0 ; \qquad \frac{d\psi}{d\xi} = 0 \qquad (\xi = 0) . \tag{120}$$

Also, ψ is a small quantity of the first order and is to be regarded as arbitrary otherwise.

In (119) write (Kelvin's transformation)

$$x = \xi^{-1} . \tag{121}$$

We have

$$\frac{x^4}{4}\frac{d}{dx}\left[\frac{1}{\Theta^3}\frac{d}{dx}(1 + \psi)\Theta^4\right] = -\Theta^3 . \tag{122}$$

To solve the foregoing equation, we shall assume that we can write

$$\Theta = \theta + \chi , \tag{123}$$

where θ satisfies the Lane-Emden equation

$$x^4 \frac{d^2\theta}{dx^2} = -\theta^3 \tag{124}$$

and where χ is a small quantity of the first order. Equation (122) can now be written as (if quantities of the second order are neglected)

$$x^4 \frac{d}{dx}\left[\frac{d\theta}{dx} + \frac{d\chi}{dx} + \psi\frac{d\theta}{dx} + \tfrac{1}{4}\theta\frac{d\psi}{dx}\right] = -\theta^3 - 3\theta^2\chi , \tag{125}$$

or

$$x^4 \frac{d^2\theta}{dx^2} + \psi x^4 \frac{d^2\theta}{dx^2} + x^4\left[\frac{d^2\chi}{dx^2} + \frac{5}{4}\frac{d\theta}{dx}\frac{d\psi}{dx} + \tfrac{1}{4}\theta\frac{d^2\psi}{dx^2}\right] = -\theta^3 - 3\theta^2\chi ; \tag{126}$$

or, using (124), we have

$$x^4 \frac{d^2\chi}{dx^2} = -3\theta^2\chi + \psi\theta^3 - \frac{x^4}{4}\left[\theta \frac{d^2\psi}{dx^2} + 5 \frac{d\theta}{dx}\frac{d\psi}{dx}\right] . \qquad (127)$$

We can also write the foregoing as

$$x^4 \frac{d^2\chi}{dx^2} = -3\theta^2\chi + \psi\theta^3 - \frac{x^4}{4\theta^4}\frac{d}{dx}\left(\theta^5 \frac{d\psi}{dx}\right) . \qquad (128)$$

Reverting to the variable in ξ, we have

$$\frac{d^2\chi}{d\xi^2} + \frac{2}{\xi}\frac{d\chi}{d\xi} = -3\theta^2\chi + \psi\theta^3 - \frac{1}{4\xi^2\theta^4}\frac{d}{d\xi}\left(\xi^2\theta^5 \frac{d\psi}{d\xi}\right) , \qquad (129)$$

which is a linear differential equation for χ. Equation (129) has to be solved subject to the boundary conditions

$$\frac{d\chi}{d\xi} = 0 \quad \text{at} \quad \xi = 0 ; \qquad \chi = \theta = 0 \quad \text{at} \quad \xi = \xi_1 , \qquad (130)$$

where ξ_1 is the boundary of the Lane-Emden function θ_3; we have chosen θ_3 for θ in (123). The boundary conditions (130) are clearly necessary and are, further, sufficient to determine χ uniquely. We have thus solved the formal problem of obtaining a second approximation.

The mass relation is easily found to be

$$M = -4\pi\alpha^3\lambda\left[\frac{\xi^2}{4\Theta^3}\frac{d}{d\xi}(1+\psi)\Theta^4\right]_0^{\xi_1} , \qquad (131)$$

or, using (120), (123), (130), and the boundary conditions that θ_3 satisfies (Eq. [67], iv), we have

$$M = -4\pi\alpha^3\lambda\xi_1^2\left[\left(1+\psi_{\xi=\xi_1}\right)\left(\frac{d\theta_3}{d\xi}\right)_{\xi=\xi_1} + \left(\frac{d\chi}{d\xi}\right)_{\xi=\xi_1}\right] , \qquad (132)$$

which, on substituting for α (Eq. [118]), is seen to be a quartic equation for $1 - \beta_c$. Hence for a given M, equation (132) determines $(1 - \beta_c)$. To use the luminosity formula

$$L = \frac{4\pi cGM(1-\beta_c)}{\overline{\kappa\eta}} , \qquad (133)$$

we have to determine $\overline{\kappa\eta}$ by

$$\overline{\kappa\eta} = \frac{1}{P_c}\int_0^{P_c} \kappa\eta dP \, , \tag{134}$$

or by (117)

$$\overline{\kappa\eta} = \frac{1}{(1+\chi_c)^4}\int_0^{1+\chi_c} \kappa\eta d[(1+\psi)\Theta^4] \, , \tag{135}$$

or by (123)

$$\overline{\kappa\eta} = \frac{\kappa_c}{(1+\chi_c)^4}\int_0^{1+4\chi_c} \frac{\kappa}{\kappa_c}\,\eta d(\theta_3^4 + \theta_3^4\psi + 4\theta_3^3\chi) \, . \tag{136}$$

Hence, we can re-write (133) in the form

$$L = \frac{4\pi c G M(1-\beta_c)}{\kappa_c\tilde{\eta}_c} \, , \tag{137}$$

where

$$\tilde{\eta}_c = \frac{1}{(1+\chi_c)^4}\int_0^{1+4\chi_c} \frac{\kappa}{\kappa_c}\,\eta d(\theta_3^4 + \theta_3^4\psi + 4\theta_3^3\chi) \, . \tag{138}$$

a) First approximation.—It is clear from our analysis that a first approximation can be obtained by using Eddington's quartic equation to determine $(1-\beta_c)$ and by evaluating $\tilde{\eta}_c$ in the luminosity formula (137) by (cf. Eq. [138])

$$\tilde{\eta}_c = \int_0^1 \frac{\kappa}{\kappa_c}\,\eta d\theta_3^4 \, . \tag{139}$$

If we assume for κ a law of the form already used in § 5, equations (77) and (78), then, for the standard model distribution of density and temperature

$$\frac{\kappa}{\kappa_c} = \theta_3^{-s} \, . \tag{140}$$

Hence, by (139)

$$\tilde{\eta}_c = 4\int_0^1 \theta_3^{3-s}\eta d\theta_3 \, . \tag{141}$$

The quantity $\tilde{\eta}_c$, determined by (139), is a homology-invariant constant and is the same for all stars. We thus see that the luminosity-mass-radius relation derived in § 5 (Eq. [81]) can be used as it stands if η_c in equations (81) and (82) is replaced by $\tilde{\eta}_c$:

$$L = \frac{\pi^3}{4^{3+s}}\left(\frac{GH}{k}\right)^{7+s}\frac{ac}{3}\frac{\xi_1^s}{[{}_0\omega_3]^{4+s}}\frac{1}{\kappa_0\tilde{\eta}_c}\frac{M^{5+s}}{R^s}(\mu\beta)^{7+s}\,. \tag{142}$$

As an illustration of the method, we shall consider the case where ϵ varies as some power of temperature and the opacity varies according to (140), with $s = \frac{1}{2}$. Table 8 shows the run of $\eta\theta^{-\frac{1}{2}}$ with θ^4 as argument.

TABLE 8

θ_3^4	$\eta\theta^{-1/2}$			
	ϵ = Constant	$\epsilon \propto T$	$\epsilon \propto T^2$	$\epsilon \propto T^4$
1	1.00	1.70	2.57	4.71
0.9	1.01	1.69	2.53	4.40
0.8	1.02	1.69	2.48	4.08
0.7	1.04	1.69	2.40	3.85
0.6	1.06	1.68	2.34	3.65
0.5	1.09	1.68	2.27	3.40
0.4	1.12	1.67	2.20	3.14
0.3	1.16	1.65	2.13	2.87
0.2	1.22	1.67	2.06	2.55
0.1	1.33	1.71	1.96	2.24

An examination of Table 8 shows that for the cases ϵ = constant and $\epsilon \sim T$ the first approximation can be safely used—at any rate, for stars with negligible radiation pressure. For the case ϵ = constant, the value of $\tilde{\eta}_c$ can be evaluated directly from (141):

$$\tilde{\eta}_c\,(\epsilon = \text{constant}) = 4\int_0^1 \theta_3^{2\frac{1}{2}}d\theta_3 = \frac{8}{7} = 1.14\,. \tag{143}$$

If $\kappa\eta$ were accurately constant, then for this model in the luminosity formula (76) (or [142]) we should strictly have $\eta_c = 1$. Thus our approximation probably introduces an error of 10 per cent in the luminosity formula. (A more detailed investigation of the model [$\kappa \propto \rho T^{-3.5}$, η = constant] given in chapter ix confirms the present conclusion.)

For $\epsilon \propto T$, the standard model is a very good approximation with $\tilde{\eta}_c \simeq 1.68$ in the luminosity formula. The model ceases to be good, in the first approximation, when ϵ varies with a higher power of T. For these cases a second approximation will be necessary. However, in using the formula (142), it is customary to adopt (following Eddington) a value of $\tilde{\eta}_c = 2.5$; this probably corresponds to a rather "high value for the concentration of the energy sources toward the centre."

b) *Second approximation.*—To obtain a second approximation, it is necessary to solve equation (129) for χ. We shall write equation (129) in the form

$$\frac{d^2\chi}{d\xi^2} + \frac{2}{\xi}\frac{d\chi}{d\xi} = -3\theta_3^2\chi + \Pi(\xi) , \tag{144}$$

where

$$\Pi(\xi) = \psi\theta_3^3 - \frac{1}{4\xi^2\theta_3^4}\frac{d}{d\xi}\left(\xi^2\theta_3^5\frac{d\psi}{d\xi}\right) . \tag{145}$$

The boundary conditions are (cf. Eq. [130])

$$\frac{d\chi}{d\xi} = 0 \quad \text{at} \quad \xi = 0 ; \qquad \chi = 0 \quad \text{at} \quad \xi = \xi_1 . \tag{146}$$

These boundary conditions are at different points; hence, if we wish to solve for χ directly, it would be necessary to adopt a method of trial and error. This can, however, be avoided by first solving the corresponding homogeneous equation

$$\frac{d^2\chi}{d\xi^2} + \frac{2}{\xi}\frac{d\chi}{d\xi} = -3\theta^2\chi \tag{147}$$

and by then obtaining χ by quadratures. This is the method of the variation of the parameters.

Let χ_1 and χ_2 be any two linearly independent solutions of (147). Then the solution of (144) can be written as

$$\chi = A(\xi)\chi_1 + B(\xi)\chi_2 , \tag{148}$$

where $A(\xi)$ and $B(\xi)$ are, for the present, two unknown functions, which, however, are restricted to satisfy the relation

$$\chi_1\frac{dA}{d\xi} + \chi_2\frac{dB}{d\xi} = 0 . \tag{149}$$

By (148) and (149) we have

$$\frac{d\chi}{d\xi} = A\frac{d\chi_1}{d\xi} + B\frac{d\chi_2}{d\xi} \tag{150}$$

and

$$\frac{d^2\chi}{d\xi^2} = A\frac{d^2\chi_1}{d\xi^2} + B\frac{d^2\chi_2}{d\xi^2} + \frac{dA}{d\xi}\frac{d\chi_1}{d\xi} + \frac{dB}{d\xi}\frac{d\chi_2}{d\xi}\,. \tag{151}$$

By substituting (150) and (151) in (144), and remembering that χ_1 and χ_2 satisfy the corresponding homogeneous equation, we find

$$\frac{dA}{d\xi}\frac{d\chi_1}{d\xi} + \frac{dB}{d\xi}\frac{d\chi_2}{d\xi} = \Pi(\xi)\,. \tag{152}$$

We can now solve for $dA/d\xi$ and $dB/d\xi$ from (149) and (152). We find

$$\frac{dA}{d\xi} = -\frac{\chi_2}{\chi_1\dfrac{d\chi_2}{d\xi} - \chi_2\dfrac{d\chi_1}{d\xi}}\,\Pi(\xi) \tag{153}$$

and

$$\frac{dB}{d\xi} = +\frac{\chi_1}{\chi_1\dfrac{d\chi_2}{d\xi} - \chi_2\dfrac{d\chi_1}{d\xi}}\,\Pi(\xi)\,. \tag{154}$$

Integrating the foregoing equations, we have

$$A = \int_{\xi}^{\xi_1}\frac{\chi_2}{\chi_1\dfrac{d\chi_2}{d\xi} - \chi_2\dfrac{d\chi_1}{d\xi}}\,\Pi(\xi)d\xi + c_1 \tag{155}$$

and

$$B = -\int_{\xi}^{\xi_1}\frac{\chi_1}{\chi_1\dfrac{d\chi_2}{d\xi} - \chi_2\dfrac{d\chi_1}{d\xi}}\,\Pi(\xi)d\xi + c_2\,, \tag{156}$$

where c_1 and c_2 are two integration constants, which have to be determined from the boundary conditions (146). Since χ has to vanish at ξ_1, we immediately have that

$$c_1\chi_1 + c_2\chi_2 = 0 \qquad (\xi=\xi_1) \tag{157}$$

The other boundary condition yields (cf. Eq. [150])

$$\left.\begin{aligned}\left(\frac{d\chi_1}{d\xi}\right)_{\xi=0}&\left[c_1+\int_0^{\xi_1}\frac{\chi_2}{\chi_1\dfrac{d\chi_2}{d\xi}-\chi_2\dfrac{d\chi_1}{d\xi}}\,\Pi(\xi)d\xi\right]\\&+\left(\frac{d\chi_2}{d\xi}\right)_{\xi=0}\left[c_2-\int_0^{\xi_1}\frac{\chi_1}{\chi_1\dfrac{d\chi_2}{d\xi}-\chi_2\dfrac{d\chi_1}{d\xi}}\,\Pi(\xi)d\xi\right]=0\,.\end{aligned}\right\}\quad(158)$$

In choosing the two linearly independent solutions χ_1 and χ_2 of (147), we can arrange that one of them (say χ_1) is such that

$$\frac{d\chi_1}{d\xi}=0\quad\text{at}\quad\xi=0\,.\qquad(159)$$

If χ_1 has been chosen in this manner, then from (157) and (158) we have

$$c_1\left(\frac{\chi_1}{\chi_2}\right)_{\xi=\xi_1}=-c_2=-\int_0^{\xi_1}\frac{\chi_1}{\chi_1\dfrac{d\chi_2}{d\xi}-\chi_2\dfrac{d\chi_1}{d\xi}}\,\Pi(\xi)d\xi\,.\qquad(160)$$

Hence, we have finally

$$A=\int_\xi^{\xi_1}\frac{\chi_2}{\chi_1\dfrac{d\chi_2}{d\xi}-\chi_2\dfrac{d\chi_1}{d\xi}}\,\Pi(\xi)d\xi-\left(\frac{\chi_2}{\chi_1}\right)_{\xi=\xi_1}\int_0^{\xi_1}\frac{\chi_1}{\chi_1\dfrac{d\chi_2}{d\xi}-\chi_2\dfrac{d\chi_1}{d\xi}}\,\Pi(\xi)d\xi\qquad(161)$$

and

$$B=\int_0^{\xi}\frac{\chi_1}{\chi_1\dfrac{d\chi_2}{d\xi}-\chi_2\dfrac{d\chi_1}{d\xi}}\,\Pi(\xi)d\xi\,.\qquad(162)$$

In this way the problem can be formally solved. For applications to practical cases we shall need χ_1 and χ_2. Once χ_1 and χ_2 are known, then for any given $\Pi(\xi)$ two quadratures are sufficient to determine the appropriate solution χ for (144).

BIBLIOGRAPHICAL NOTES

§ 1.—1. B. STRÖMGREN, *Handb. d. Astrophys.*, **7**, 159, 1936, where Theorem 1 is proved.

2. S. CHANDRASEKHAR, *Ap. J.*, **86**, 78, 1937 (Theorems 2, 3, and 4).

3. A. S. EDDINGTON, *Zs. f. Phys.*, **7**, 351, 1921 (Theorem 5).

§§ 2 and 3.—The stability criterion for the existence of radiative equilibrium was first considered by—

4. K. SCHWARZSCHILD, *Göttinger Nachrichten*, 1906, p. 41. The discussion in the text follows—

5. L. BIERMANN, *Zs. f. Ap.*, **5**, 117, 1932. Also *A.N.*, **257**, 270, 1935.

6. H. SIEDENTOPF, *A.N.*, **244**, 273, 1932, where a condition equivalent to equation (45) is given.

7. S. CHANDRASEKHAR, *Proc. Nat. Acad. Sci.* (Washington), **23**, 572, 1937.

8. T. G. COWLING, *M.N.*, **94**, 768, 1934.

§ 4.—The standard model was first considered by Eddington—(ref. 3, above). Also,

9. A. S. EDDINGTON, *M.N.*, **77**, 596, 1917, and *The Internal Constitution of the Stars*, chap. vi, Cambridge, England, 1926.

10. I. BIALOBJESKY, *Bull. Acad. Sci. Cracovie*, May, 1913 (p. 64).

§ 5.—See ref. 9, above. Also,

11. E. A. MILNE, *Handb. d. Astrophys.*, **3**, Part I, particularly pp. 204–222.

§ 6.—See ref. 1, pp. 168–170. Also,

12. B. STRÖMGREN, *Erg. exakt. Naturwiss.*, **16**, 465, 1937.

§ 7.—The discussion in this section, which is based on an unpublished investigation by B. Strömgren, differs in principle from similar discussions in references (9) and (11). Further reference may be made to—

13. R. EMDEN, *Thermodynamik der Himmelskörper* ("Sonderausgabe aus der Encyklopädie der Mathematischen Wissenschaften").

CHAPTER VII

STRÖMGREN'S INTERPRETATION OF THE HERTZSPRUNG-RUSSELL DIAGRAM

In the last chapter the mass-luminosity-radius relation for gaseous stars in radiative equilibrium was derived. The relation in question (due essentially to Eddington) was first derived on the basis of the standard model; but, as we have seen, the same form for the relation results for a wide class of stellar models if the radiation pressure is not very appreciable. Further, by a perturbation method we have seen how the luminosity formula may be applied to cases where $\kappa\eta$ is variable. For most practical purposes it is sufficient to restrict ourselves to the first approximation considered in the last chapter (§ 7).

In this chapter we shall be concerned mainly with concrete applications of the luminosity formula to the available observational material regarding the masses, luminosities, and radii of the stars.

1. *The statement of the problem.*—On the first approximation considered in chapter vi, § 7, we have

$$L = \frac{4\pi cGM(1 - \beta_c)}{\kappa_c \tilde{\eta}_c} , \tag{1}$$

where

$$\tilde{\eta}_c = \int_0^1 \left(\frac{\kappa}{\kappa_c}\right) \eta d\theta_3^4 . \tag{2}$$

Further, $(1 - \beta_c)$ is determined once M and the mean molecular weight are known, as the solution of Eddington's quartic equation.

Now we shall show in § 5 that the physical theory of the stellar opacity coefficient κ leads to a formula of the type

$$\kappa = \frac{\kappa_0(\mu)}{t} \frac{\rho}{T^{3.5}} , \tag{3}$$

where κ_0 is a constant depending on the molecular weight, and t, called the "guillotine factor" (Eddington), is a slowly varying function of ρ and T. We can write equation (3) as (Eq. [78], v)

$$\kappa = \frac{\kappa_0}{t} \frac{a}{3} \frac{\mu H}{k} \frac{\beta}{1 - \beta} T^{-(1/2)} . \tag{4}$$

Inserting the foregoing in the luminosity formula (1), we have

$$L = \frac{4\pi cGM}{(\kappa_0/t_c)\bar{\eta}_c} \frac{3}{a} \frac{k}{\mu H} \frac{(1 - \beta_c)^2}{\beta_c} T_c^{1/2} , \tag{5}$$

where $\bar{\eta}_c$ can now be expressed in the form

$$\bar{\eta}_c = \int_0^1 \left(\frac{t_c}{t}\right) \theta_3^{-(1/2)} \eta d\theta_3^4 , \tag{6}$$

which can also be written as

$$\bar{\eta}_c = \frac{t_c}{\bar{t}} \int_0^1 \theta_3^{-(1/2)} \eta d\theta_3^4 . \tag{7}$$

In the foregoing equation $\bar{t}$ is a certain harmonic mean value of t defined by

$$\frac{1}{\bar{t}} = \frac{\int_0^1 \frac{1}{t} \theta_3^{-(1/2)} \eta d\theta_3^4}{\int_0^1 \theta_3^{-(1/2)} \eta d\theta_3^4} . \tag{8}$$

Now the integral on the right-hand side of (7) is the value of $\bar{\eta}_c$ if the guillotine factor were unity. We shall accordingly define $\bar{\eta}_c(1)$ by

$$\bar{\eta}_c(1) = \int_0^1 \theta_3^{-(1/2)} \eta d\theta_3^4 . \tag{9}$$

By (7), then,

$$\bar{\eta}_c = \frac{t_c}{\bar{t}} \bar{\eta}_c(1) . \tag{10}$$

From (5) and (10) we have

$$L = \frac{4\pi cGM}{(\kappa_0/\bar{t})\bar{\eta}_c(1)} \frac{(1 - \beta_c)^2}{\beta_c} \frac{3}{a} \frac{k}{\mu H} T_c^{1/2} . \tag{11}$$

Proceeding exactly as in chaper vi, § 5, we find (cf. Eq. [81], vi)

$$L = \frac{\pi^3}{4^{3.5}} \left(\frac{GH}{k} \right)^{7.5} \frac{ac}{3} \frac{\xi_1^{0.5}}{[_0\omega_3]^{4.5}} \frac{1}{(\kappa_0/\bar{t})\tilde{\eta}_c(1)} \frac{M^{5.5}}{R^{0.5}} (\mu\beta_c)^{7.5} , \tag{12}$$

or, numerically (cf. Eq. [82], vi),

$$L = 1.79 \times 10^{25} \frac{\bar{t}}{\kappa_0 \tilde{\eta}_c(1)} \frac{M^{5.5}}{R^{0.5}} (\mu\beta_c)^{7.5} , \tag{13}$$

where L, M, and R are expressed in solar units. According to our remarks in § 7 of chapter vi, we shall adopt $\tilde{\eta}_c(1) = 2.5$. Inserting this value in (13), we finally have

$$L = 7.17 \times 10^{24} \frac{\bar{t}}{\kappa_0} \frac{M^{5.5}}{R^{0.5}} (\mu\beta_c)^{7.5} . \tag{14}$$

Now if we know the mass and the radius of a star and if, further, we assume a value for μ, then the foregoing formula enables us to calculate L. In general, the value of L so calculated may not agree with the observed value. We should, however, be able to adjust μ in such a way that the observed and the predicted values of L agree. In other words, a knowledge of L, M, and R should enable us to determine the mean molecular weight of a star, or, what is equivalent to it, the mean chemical composition of the stellar material. This is precisely our present problem. The solution consists essentially in (*a*) determining the appropriate μ for stellar material of a specified chemical composition and at a prescribed density and temperature, (*b*) determining the dependence of κ_0 on the chemical composition, and (*c*) determining the appropriate value of $\bar{t}$ for individual stars. Once these questions have been settled, the determination of μ is immediate. We can then compute the value of μ for a number of stars for which values of L, M, and R are available. Our final problem is to examine if these computed values of μ enable us to give a general interpretation of the characteristic features summarized in the Hertzsprung-Russell diagram.

We shall consider, following Strömgren, these questions in the order stated. It is necessary, however, as a preliminary to the whole discussion, to consider a fundamental theorem due to Vogt and Russell.

2. *The Vogt-Russell theorem.*—The theorem in question states that *if the pressure*, P, *the opacity*, κ, *and the rate of generation of energy*, ϵ, *are functions of the local values of* ρ, T, *and the chemical composition only, then the structure of a star is uniquely determined by the mass and the chemical composition.*

To prove this theorem we shall first consider the case of a gaseous star in radiative equilibrium. For this case the equations of equilibrium can be written in the form

$$\left.\begin{aligned} g &= \frac{GM(r)}{r^2} \\ F_r &= \frac{L(r)}{4\pi r^2} \end{aligned}\right\} \text{I}\,; \qquad \left.\begin{aligned} dP &= -g\rho dr \\ d(\tfrac{1}{3}aT^4) &= -\frac{\kappa F_r}{c}\rho dr \end{aligned}\right\} \text{II}\,; \qquad \left.\begin{aligned} dM(r) &= 4\pi r^2 \rho dr \\ dL(r) &= 4\pi r^2 \epsilon \rho dr \end{aligned}\right\} \text{III}\,.$$

The foregoing system of equations can, in principle, be solved as follows: We choose a definite value for r and prescribe an arbitrary set of values for the variables P, T, g, and F_r. From P and T we can calculate the "local" values for ρ, κ, and ϵ; to deduce these values, we require a knowledge of the chemical composition or its equivalent, the mean molecular weight. The second set of equations above then enables us to compute dP and dT for an increment dr of r. In the same way the third pair of equations enables us to compute $dM(r)$ and $dL(r)$. Thus we have a set of values for the variables P, T, g, and F_r for $r + dr$. We can therefore continue the solution for a further increment of r. In this way we can integrate the solution both inward, toward the center, and outward, toward the boundary. For a solution to be physically possible the following boundary conditions must be satisfied:

$$M(r) = 0 \quad \text{at} \quad r = 0 \tag{15}$$

and

$$\rho \rightarrow 0\,, \qquad T \rightarrow 0 \text{ simultaneously}\,, \tag{16}$$

or, more exactly, $\rho \rightarrow 0$ and $T \rightarrow T_0$ (a definite limit), but this is a refinement hardly ever necessary. We thus see that there are three relations between the four values initially adopted for P, T, g, and F_r, respectively. Hence, we are left with only one disposable constant. Since at the end of the integrations we should be able to find the total mass M of the configuration, it follows that an assumed fixed chemical composition can lead us only to a one-parametric sequence of configurations; the parameter can clearly be chosen to be M. In other words, given M and μ, we should, in principle, be able to calculate the other two observable characteristics of the star, namely, L and R—of course, on the basis of an assumed chemical composition.

We have thus proved the Vogt-Russell theorem for gaseous stars in radiative equilibrium. From our method of argument, however, it is clear that the theorem is valid quite generally, i.e., also when a part of the stellar interior is in convective equilibrium (cf. chap. vi, §§ 2 and 3); for no new parameters are introduced. The theorem essentially arises from the fact that the equations of equilibrium are equivalent to one differential equation of the fourth order, while a solution, to be physically possible, has to satisfy three boundary conditions. Thus we have proved the general validity of the Vogt-Russell theorem.

It is necessary, however, to point out that there are conceivable physical circumstances under which the Vogt-Russell theorem will not be valid. Thus, ϵ need not, in general, depend upon the local values of ρ and T; this would be the case if the origin of stellar energy were due to physical processes occurring at nearly equilibrium rates, e.g., nuclear transmutations occurring at approximately equilibrium rates but slightly more in one direction than in the other. There are, however, good reasons why such cases can be excluded; we shall return to this question in the last chapter. Meanwhile, we shall accept the validity of the Vogt-Russell theorem. The application of the theorem we have in view is this: Does the use of the luminosity formula to determine the chemical composition of the stars allow us either to confirm or to deny the validity of the Vogt-Russell theorem for stars in nature? We shall see that the answer to this question is largely in the affirmative.

In the next two sections we shall consider the question of the mean molecular weight and the theory of the stellar opacity coefficient. Unfortunately, we cannot start from first principles in our discussion of these two quantities, as we have done so far in the treatment of the other problems. The importance of these quantities in the present connection, however, makes a treatment—even if only a partial one—essential. In the following two sections we shall assume a general familiarity with methods of statistical mechanics and of the quantum theory.

3. *The mean molecular weight of highly ionized stellar material.*— The first problem is to determine the appropriate molecular weight that has to be used in the equation of state adopted:

$$p_g = \frac{k}{\mu H} \rho T \, . \tag{17}$$

This problem, as we shall see, is essentially one of determining the number of particles per unit volume. For we have also

$$p_g = NkT \, . \tag{18}$$

Suppose we have a mixture of elements and that an element of atomic number Z occurs with an abundance factor x_Z—in other words, 1 gram of the material contains x_Z grams of the element. Let us suppose, further, that each atom of the element contributes, on the average, $\bar{n}_Z$ free particles per unit atomic weight, i.e., if A is the atomic weight (that of hydrogen being taken as unity), then each atom contributes $A\bar{n}_Z$ free particles. We then have

$$N = \frac{\rho}{H} \Sigma x_Z \bar{n}_Z \, , \tag{19}$$

where the summation is to be extended over all the elements. By (18) and (19) we have

$$p_g = \frac{k}{H} (\Sigma x_Z \bar{n}_Z) \rho T \, . \tag{20}$$

Comparing the foregoing with (17), we find

$$\mu = \frac{1}{\Sigma x_Z \bar{n}_Z} \, . \tag{21}$$

The determination of μ involves, therefore, the specification of the state of ionization of the stellar material at a prescribed density and temperature.

a) *First approximation.*—We shall first give an elementary first approximation. Let us suppose that the conditions are such that the ionization is complete—that is, an element of atomic number Z and atomic weight A gives rise to $Z + 1$ particles. Then

$$\bar{n}_Z = \frac{Z + 1}{A} . \tag{22}$$

It is well known that, except for the lightest elements (hydrogen and helium), the ratio $\bar{n}_Z$, defined as in (22), is approximately 1 : 2. Hence, if we assume that in 1 gram of the stellar material there are X grams of hydrogen, Y grams of helium, and $(1 - X - Y)$ grams of the "heavy" elements, then we can write

$$\bar{n}_1 = 2 ; \qquad \bar{n}_2 = \tfrac{3}{4} ; \qquad \bar{n}_Z = \tfrac{1}{2} . \tag{23}$$

The expression for μ then becomes

$$\mu = \frac{1}{2X + \frac{3}{4}Y + \frac{1}{2}(1 - X - Y)} , \tag{24}$$

or

$$\mu = \frac{2}{1 + 3X + 0.5Y} . \tag{25}$$

If the helium content can be neglected, we shall denote the abundance of hydrogen by X_0; then $(1 - X_0)$ is the abundance of the heavy elements. In this case

$$\mu = \frac{2}{1 + 3X_0} . \tag{26}$$

The general expression for the number N_e of free electrons is

$$N_e = \frac{\rho}{H} \sum \frac{x_Z}{A} (\bar{n}_Z A - 1) . \tag{27}$$

In the present approximation (Eq. [22]) we have

$$N_e = \frac{\rho}{H} \sum \frac{x_Z Z}{A} . \tag{28}$$

If a mixture of hydrogen, helium, and the heavier elements is considered, we have, according to the abundances leading to (24),

$$N_e = \frac{\rho}{H}\left[X + \tfrac{1}{2}Y + \tfrac{1}{2}(1 - X - Y)\right], \tag{29}$$

or

$$N_e = \frac{1}{2}\frac{\rho}{H}(1 + X). \tag{30}$$

b) *Second approximation.*—In order to determine μ more accurately, it is necessary to calculate $\bar{n}_Z$ more accurately, with allowance for the state of ionization. Strömgren has developed the following elegant method of doing this.

In this method an approximation is made in which the differences between the states of the first and the second K-electrons, or between the states of any of the L-electrons, are ignored. The differences between the states of L- (or M-) electrons belonging to normal or excited configurations are also ignored. Any atomic configuration is then specified sufficiently by the numbers $(n_K, n_L, n_M, \ldots)$, which give, respectively, the number of K-, L-, M-, etc., electrons bound to an atomic nucleus of charge Ze. The energy of such a state may be taken to be

$$-n_K\chi_K^{(Z)} - n_L\chi_L^{(Z)} - n_M\chi_M^{(Z)} - \ldots, \tag{31}$$

where the χ's are constants representing the mean ionization potentials of the various shells. Since, as we shall see presently, at most common temperatures and densities in stellar interiors the ionization is generally very far advanced, it is clear that, consistent with our present scheme of approximation, it is sufficient to use for the χ's the expressions derived for hydrogen-like atoms. If the nuclear charge is Ze, then by Bohr's theory we have

$$\chi_n^{(Z)} = \frac{2\pi^2 e^4 m_e Z^2}{n^2 h^2}, \tag{32}$$

where n stands for the principal quantum number ($n = 1$ for the K-electrons, $n = 2$ for the L-electrons, etc.,), m_e is the mass of the electron, and the other symbols have their usual significance.

The statistical weight for the configuration $(n_K, n_L, n_M, \ldots .)$, for short (n), is given by

$$q(K, n_K)q(L, n_L)q(M, n_M) \ldots . , \tag{33}$$

where, according to Pauli's exclusion principle, it is easy to verify that

$$q(K, n_K) = {}_2C_{n_K} ; \qquad q(L, n_L) = {}_8C_{n_L} ; \qquad q(M, n_M) = {}_{18}C_{n_M} , \tag{34}$$

where the C's are the binomial coefficients.

We now consider the equilibrium between the single configurations (n), free electrons, and the corresponding bare nuclei, denoted by $(0, 0, \ldots .)$ or (0). From statistical mechanics (cf. R. H. Fowler, *Statistical Mechanics*, 2d ed., Cambridge, 1936) we have

$$\left.\begin{aligned} &\frac{N(0)N_e^{n_K+n_L+n_M+\cdots}}{N(n)} \\ &\qquad = \frac{[G(T)]^{n_K+n_L+n_M+\cdots} \times e^{-(n_K\chi_K^{(Z)}+n_L\chi_L^{(Z)}+\cdots)/kT}}{q(K, n_K)q(L, n_L)q(L, n_M) \ldots} , \end{aligned}\right\} \tag{35}$$

where $N(0)$ and $N(n)$ are the number of atomic configurations in the states (0) and (n) and N_e is the number of free electrons per unit volume and where $G(T)$ is defined by

$$G(T) = 2\,\frac{(2\pi m_e kT)^{3/2}}{h^3} . \tag{36}$$

Equation (35) can also be expressed in the form

$$\left.\begin{aligned} \frac{N(n)}{N(0)} = \left[\frac{N_e}{G(T)}\right]^{n_K+n_L+n_M+\cdots} &\times [q(K, n_K)q(L, n_L)q(M, n_M) \ldots] \\ &\times e^{(n_K\chi_K^{(Z)}+n_L\chi_L^{(Z)}+n_M\chi_M^{(Z)}+\cdots)/kT} . \end{aligned}\right\} \tag{37}$$

The evaluation of the total number of bound electrons by this method can be sufficiently illustrated by the calculation of $\bar{n}_L^{(Z)}$, which gives the average number of bound electrons with principal quantum number 2 around a nucleus of atomic number Z.

Let

$$\frac{N_e}{G(T)}\, e^{\chi_L^{(Z)}/kT} = y . \tag{38}$$

Then, by (34) and (37) we have

$$\bar{n}_L^{(Z)} = \frac{\sum_0^8 r\,{}_8C_r y^r}{\sum_0^8 {}_8C_r y^r} = \frac{8y(1+y)^7}{(1+y)^8} = \frac{8y}{1+y}. \tag{39}$$

The other shells give similar contributions. Quite generally, we see that the number of bound electrons, $\bar{n}_n^{(Z)}$, of principal quantum number n, around a nucleus of atomic number Z, is given by

$$\bar{n}_n^{(Z)} = \frac{2n^2}{1 + \frac{G(T)}{N_e} e^{-\chi_n^{(Z)}/kT}}. \tag{40}$$

Since in an unionized atom there are $2n^2$ bound electrons with principal quantum number n, equation (40) corresponds to a number of free electrons per atomic nucleus and arising from the ionization of the n-shell:

$$2n^2 - \bar{n}_n^{(Z)} = \frac{2n^2}{1 + \frac{N_e}{G(T)} e^{\chi_n^{(Z)}/kT}}. \tag{41}$$

Finally, the number of free particles per nucleus of charge Ze is

$$1 + \sum \frac{2n^2}{1 + \frac{N_e}{G(T)} e^{\chi_n^{(Z)}/kT}}, \tag{42}$$

where the summation is extended over all the relevant n's. In practice it would suffice to consider only K-, L-, and M-electrons. (If it should be necessary to consider higher orbits, then factors neglected here, such as "excluded volumes," should be taken into account.)

By definition we have

$$\bar{n}_Z = \frac{1}{A_Z}\left\{1 + \sum \frac{2n^2}{1 + \frac{N_e}{G(T)} e^{\chi_n^{(Z)}/kT}}\right\}. \tag{43}$$

If we have a mixture of elements with definite abundance factors, then the molecular weight is, according to (21),

$$\mu = \frac{1}{\Sigma x_Z \bar{n}_Z} = \frac{1}{\bar{n}} , \tag{44}$$

where

$$\bar{n} = \Sigma x_Z \bar{n}_Z . \tag{45}$$

Table 9, due to Strömgren, illustrates the calculation of μ for the so-called "Russell mixture," in which the elements O, $(Na + Mg)$, Si, $(K + Ca)$, and Fe are assumed to occur by weight in the ratio

TABLE 9

Element	n	$\bar{n}_n$	Number of Bound Electrons	Number of Free Electrons	Total Number of Free Particles per Nucleus	$\bar{n}_Z$	$x_Z \bar{n}_Z$
O	1 2 3	0.04 0.07 0.13	0.24	7.76	8.76	0.548	0.274
Na, Mg	1 2 3	0.09 0.09 0.15	0.3	11.7	12.7	.53	.132
Si	1 2 3	0.24 0.11 0.17	0.5	13.5	14.5	.52	.032
K, Ca	1 2 3	1.46 0.23 0.23	1.9	18.1	19.1	.48	.030
Fe	1 2 3	1.99 0.61 0.36	3.0	23.0	24.0	0.43	0.054
$T = 10^7$ degrees; $\log \left[\frac{G(T)}{N_e}\right] = 5$						$\bar{n} = \Sigma x_Z \bar{n}_Z$ $\mu = \bar{n}^{-1}$	0.52 1.92

8 : 4 : 1 : 1 : 2. We thus see that, given the temperature T and the quantity $G(T)/N_e$, we can calculate $\bar{n}$ and μ. Table 10, also due to Strömgren, gives the values of $\bar{n}_R$ calculated for the Russell

mixture for different values of T and $G(T)/N_e$. The reciprocals of the values of $\bar{n}_R$ tabulated in Table 10 give the mean molecular weight, μ, for the Russell mixture. This is not generally far from 2.

TABLE 10*

Log T \ log $\left[\frac{G(T)}{N_e}\right]$	$\bar{n}_R$ for Russell Mixture							
	3	4	5	6	7	8	9	10
6.4	......	......	......	0.46	0.49	0.50	0.51	0.52
6.6	......	......	0.48	.51	.52	.53	.53	.54
6.8	......	0.48	.51	.53	.53	.53	.54	.54
7.0	0.44	.50	.52	.53	.54	.54	.54	.54
7.2	.46	.51	.53	.54	.54	.54	.54	.54
7.4	.47	.51	.53	.54	.54	.54	.54	.54
7.6	0.47	0.51	0.53	0.54	0.54	0.54	0.54	0.54

For a Russell mixture completely ionized, $\bar{n}_R = 0.54$

* We shall use "Log" to denote logarithms to the base 10 and "log" to denote natural logarithms.

On the other hand, if 1 gram of the stellar material contains X grams of hydrogen, Y grams of helium, and $(1 - X - Y)$ grams of the Russell mixture, then

$$\bar{n} = \Sigma \bar{n}_Z x_Z = 2X + \tfrac{3}{4}Y + \bar{n}_R(1 - X - Y) , \tag{46}$$

or

$$\mu = \frac{1}{2X + \frac{3}{4}Y + \bar{n}_R(1 - X - Y)} . \tag{47}$$

If the helium content can be neglected, then $Y = 0$, $X = X_0$, and we have

$$\mu = \frac{1}{2X_0 + \bar{n}_R(1 - X_0)} ; \tag{48}$$

or, solving for X_0, we obtain

$$X_0 = \frac{\mu^{-1} - \bar{n}_R}{2 - \bar{n}_R} . \tag{49}$$

Equation (49) will give the hydrogen content after the value of μ has been found from the mass, luminosity, and radius of a star.

Finally, to determine μ for a given density ρ and temperature T, it is necessary to know N_e. To determine N_e it is sufficient, in the first instance, to use the result of the first approximation, namely, equation (30):

$$N_e = \frac{1}{2}\frac{\rho}{H}(1 + X) . \tag{50}$$

If necessary, it would be a simple matter to use a method of reiteration. Accurately, i.e., in our present scheme of approximation, we have, according to (41),

$$N_e = \sum_Z \frac{x_Z}{A_Z} \sum_n \frac{2n^2}{1 + \frac{N_e}{G(T)} e^{\chi_n^{(Z)}/kT}} , \tag{51}$$

where the summation is extended over all the elements and all the relevant principal quantum numbers, and where it may be recalled that

$$G(T) = 2\frac{(2\pi m_e kT)^{3/2}}{h^3} . \tag{52}$$

c) The accurate determination of μ.—To determine μ more accurately than by Strömgren's method, we must allow for differences between the different electrons in the same shell, for "excluded volumes," and also for electrostatic corrections. Such calculations have been made, for certain special cases, by Fowler and Guggenheim. We shall not go into these refinements here, but reference may be made to Fowler's *Statistical Mechanics*.

4. *The stellar opacity coefficient.*—The main contribution to the opacity for the radiation in stellar interiors arises from photoelectric ionizations of the electrons bound to the nuclei of the highly ionized atoms. If the ionization potential for a particular state of the atom considered is χ, then radiation of frequency $\nu \geqslant \nu_\chi$, where

$$h\nu_\chi = \chi , \tag{53}$$

can ionize the atom photoelectrically. In addition to these bound-free transitions, there is still another kind, which can be described as "free-free" transitions and which also contribute to the opacity.

These free-free transitions correspond to free electrons which absorb quanta of a definite frequency while in the attractive field of an atomic nucleus.

Now, since the atoms in the interior of a star will be highly ionized, it is sufficient to consider for the probability of these bound-free and free-free transitions those computed for hydrogen-like atoms, i.e., for assumed Coulomb fields around charged nuclei. According to the theory of this phenomenon, as developed by Kramers, Gaunt, and others, we have the following results.

a) *Bound-free transitions.*—If an atomic nucleus of charge Ze is considered with an electron in the state of principal quantum number n, then the atomic absorption coefficient $a_0(\nu; Z; n)$ is given by

$$a_0(\nu; Z; n) = \frac{64\pi^4 Z^4 m_e e^{10}}{3\sqrt{3}\, ch^6} \frac{1}{n^5} \frac{1}{\nu^3} g(\nu; n) \tag{54}$$

if

$$\nu \geqslant \nu_n = \frac{\chi_n}{h}\,. \tag{55}$$

In the expression for $a_0(\nu; Z; n)$, the quantity $g(\nu; n)$, called the "Gaunt factor," is a factor which depends on n and ν, and which, for the values of n and ν of importance in contributing to stellar opacity, is very near unity. Table 11, due to Strömgren, gives the

TABLE 11

ν/ν_n	$n=1$	$n=2$	$n=3$	$n=4$
1.0	0.80	0.88	0.9	1.0
1.5	0.89	0.94	....	....
2.0	0.94	0.97	....	....
3.0	0.98	1.02	....	....
4.0	1.00	1.04	....	....
5.0	0.99	1.05	....	....

values of $g(\nu, n)$ for some values of n, with argument ν/ν_n. As $\nu/\nu_n \rightarrow \infty$, $g \rightarrow 0$; but, as will be shown in the subsequent discussion, only such values of g are of significance as are very near those at the series head. Hence, we can, in practice, regard g as independent of ν and take as its value some constant value $\bar{g}$. We shall return to this question later.

b) Free-free transitions.—If we consider an atomic nucleus of charge Ze, then the rate of absorption of energy $a_0(v; \nu)$, from radiation of frequency ν and of unit intensity by electrons of velocity v. and unit mean density, is

$$a_0(v; \nu; Z) = \frac{4\pi Z^2 e^6}{3\sqrt{3}\, hcm_e^2 \nu^3 v} . \tag{56}$$

There is a Gaunt factor for (56) as well, but we shall take it equal to unity (the free-free transitions do not in any case contribute appreciably to the stellar opacity).

We shall now calculate the Rosseland mean coefficient of opacity as a function of density and temperature:

Let us first consider the case of a single element of atomic number Z and atomic weight A. By (56) the contribution to the absorption coefficient per nucleus, expressed per gram of the material, which arises from free-free transitions and which is due to electrons with velocities in the range v, $v + dv$, is

$$\frac{a_0(v; \nu; Z)}{AH} N_v dv , \tag{57}$$

where $N_v dv$ is the number of electrons per unit volume in the specified velocity range. By Maxwell's law of thc distribution of velocities we have

$$N_v dv = 4\pi N_e \left(\frac{m_e}{2\pi kT}\right)^{3/2} e^{-m_e v^2/2kT}\, v^2 dv . \tag{58}$$

The absorption coefficient $\kappa_\nu^{(Z)}(\infty, \infty)$, due to the atoms under consideration and arising from free-free transitions, is given by

$$\kappa_\nu^{(Z)} \quad (\infty, \infty) = \int_0^\infty \frac{a_0(v; \nu; Z)}{AH} N_v dv , \tag{59}$$

or by (56) and (58)

$$\kappa_\nu^{(Z)}(\infty, \infty) = \frac{1}{AH} \frac{16\pi^2 Z^2}{3\sqrt{3}\, \nu^3} \frac{e^6 m_e N_e}{hc(2\pi m_e kT)^{3/2}} \int_0^\infty e^{-m_e v^2/2kT}\, v dv , \tag{60}$$

or, after integration,

$$\kappa_\nu^{(Z)}(\infty, \infty) = \frac{1}{AH} \frac{16\pi^2 Z^2}{3\sqrt{3}} \frac{e^6}{hc(2\pi m_e)^{3/2}} \frac{N_e}{(kT)^{1/2}} \frac{1}{\nu^3} . \tag{61}$$

If we put

$$\frac{h\nu}{kT} = u \,, \tag{62}$$

we can re-write (61) in the form

$$\kappa_\nu^{(Z)}(\infty, \infty) = \frac{D_{ff}^{(Z)}}{u^3} \,, \tag{63}$$

where

$$D_{ff}^{(Z)} = \frac{1}{AH} \frac{16\pi^2 Z^2}{3\sqrt{3}} \frac{e^6 h^2}{c(2\pi m_e)^{3/2}} \frac{N_e}{(kT)^{3.5}} \,. \tag{64}$$

Let us now consider the bound-free transitions. By equation (40), the average number of electrons per atomic nucleus in the state of principal quantum number n is

$$\bar{n}_n^{(Z)} = \frac{2n^2}{1 + \dfrac{G(T)}{N_e} e^{-\chi_n^{(Z)}/kT}} \,. \tag{65}$$

Hence, the contribution to the absorption coefficient which is due to the electrons in the state n, expressed per gram of the material, is

$$\kappa_\nu^{(Z)}(n, \infty) = \frac{a_0(\nu; Z; n)}{AH} \bar{n}_n^{(Z)} \qquad (\nu \geqslant \nu_n^{(Z)}) \,, \tag{66}$$

where ν_n is defined by

$$h\nu_n^{(Z)} = \chi_n^{(Z)} = \frac{2\pi^2 e^4 m_e Z^2}{n^2 h^2} = \frac{\chi_1^{(Z)}}{n^2} \,. \tag{67}$$

By (54) and (66) we have

$$\kappa_\nu^{(Z)}(n, \infty) = \frac{1}{AH} \frac{64\pi^4 Z^4}{3\sqrt{3}} \frac{m_e e^{10}}{ch^6} \frac{\bar{g}}{n^5} \frac{1}{\nu^3} \bar{n}_n^{(Z)} \,, \tag{68}$$

where, according to the remarks on page 262, the quantity $g(\nu; n)$, which occurs in (54), has been replaced by a constant $\bar{g}$. We now re-write (68) in the form

$$\kappa_\nu^{(Z)}(n, \infty) = \frac{1}{AH} \frac{16\pi^2 Z^2}{3\sqrt{3}} \frac{e^6}{hc(2\pi m_e)^{3/2}} \frac{2(2\pi m_e)^{3/2}}{h^3} \frac{2\pi^2 e^4 m_e Z^2}{n^2 h^2} \frac{\bar{g}}{n^3} \frac{h^3}{(kT)^3} \frac{\bar{n}_n^{(Z)}}{u^3} \,. \tag{69}$$

Substituting for $\chi_n^{(Z)}$, according to (67) and remembering the definition of $G(T)$ (Eq. [52]), we can write, instead of (69),

$$\kappa_\nu^{(Z)}(n, \infty) = \frac{1}{AH} \frac{16\pi^2 Z^2}{3\sqrt{3}} \frac{e^6 h^2}{c(2\pi m_e)^{3/2}} \frac{\bar{g} G(T) \chi_n^{(Z)}}{n^3 (kT)^{4.5}} \frac{\bar{n}_n^{(Z)}}{u^3} . \tag{70}$$

Comparing (70) with (64), we can express $\kappa_\nu^{(Z)}(n, \infty)$ in the form

$$\kappa_\nu^{(Z)}(n, \infty) = \frac{D_n^{(Z)}}{u^3} \qquad (u \geqslant u_n^{(Z)} = h\nu_n^{(Z)}/kT) , \tag{71}$$

where

$$D_n^{(Z)} = D_{ff}^{(Z)} \frac{\chi_n^{(Z)}}{n^3} \frac{\bar{g} G(T)}{N_e kT} \frac{2n^2}{1 + \frac{G(T)}{N_e} e^{-\chi_n^{(Z)}/kT}} . \tag{72}$$

We have the following alternative form for $D_n^{(Z)}$:

$$D_n^{(Z)} = D_{ff}^{(Z)} \frac{2\bar{g}}{n^3} \frac{\chi_1^{(Z)}}{kT} \frac{e^{\chi_n^{(Z)}/kT}}{1 + \frac{N_e}{G(T)} e^{\chi_n^{(Z)}/kT}} . \tag{73}$$

Hence, the absorption coefficient due to electrons in all the electronic shells and also due to the free-free transitions can be expressed as

$$\kappa_\nu^{(Z)} = \frac{D^{(Z)}(u)}{u^3} , \tag{74}$$

where

$$D^{(Z)}(u) = D_1^{(Z)} + D_2^{(Z)} + D_3^{(Z)} + \ldots . + D_{ff}^{(Z)} \qquad (u \geqslant u_1) , \tag{75}$$

$$D^{(Z)}(u) = D_2^{(Z)} + D_3^{(Z)} + \ldots . + D_{ff}^{(Z)} \qquad (u_1 > u \geqslant u_2) , \tag{76}$$

$$D^{(Z)}(u) = D_3^{(Z)} + \ldots . + D_{ff}^{(Z)} \qquad (u_2 > u \geqslant u_3) , \tag{77}$$

. . . .

Finally, if we consider a mixture of elements, we have to consider the functions $D^{(Z)}(u)$ for each element defined as above; we then form the net absorption coefficient by weighting each of the values of $\kappa_\nu^{(Z)}$ with the appropriate abundance factors. Thus we have

$$\kappa_\nu = \Sigma x_Z \kappa_\nu^{(Z)} = \frac{D(u)}{u^3} , \tag{78}$$

where

$$D(u) = \Sigma x_Z D^{(Z)}(u) . \tag{79}$$

The summations in (78) and (79) are to be effected over all the elements.

It is clear that $D(u)$, as defined above, is a discontinuous function; it changes discontinuously at the absorption edges but remains con-

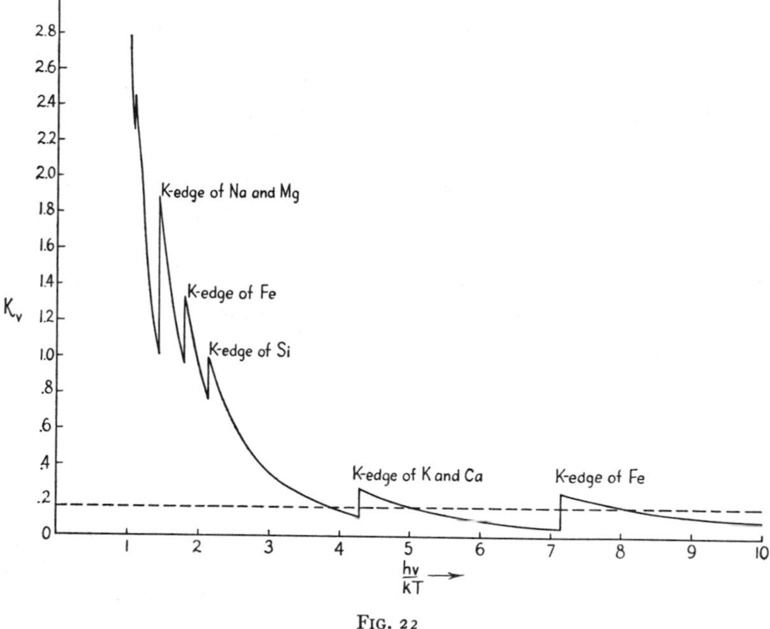

FIG. 22

stant between any two edges. Figure 22 illustrates the variation of κ_ν with frequency for the Russell mixture at $T = 1.4 \times 10^7$ degrees and when $\log [G(T)/N_e] = 3$. [The unit of absorption coefficient in Figure 22 is $3.89 \times 10^{25} \rho T^{-3.5}$.]

We have so far considered only the monochromatic absorption

coefficients. We have now to calculate the Rosseland mean coefficient of opacity, as defined in equation (159) of chapter v. Introducing the variable u (defined as in Eq. [62]) in equation (159) of chapter v, we find that

$$\frac{1}{\kappa} = \frac{\int_0^\infty \frac{1}{\kappa_\nu} \frac{e^{2u} u^4 du}{(e^u - 1)^3}}{\int_0^\infty \frac{e^u u^4 du}{(e^u - 1)^2}} . \tag{80}$$

By (78) we have

$$\frac{1}{\kappa} = \frac{\int_0^\infty \frac{1}{D(u)} \frac{e^{2u} u^7 du}{(e^u - 1)^3}}{\int_0^\infty \frac{e^u u^4 du}{(e^u - 1)^2}} . \tag{81}$$

By what has already been stated, the quantity $D(u)$ is constant between two absorption edges. Let the absorption edges (of all the elements present), arranged in descending order, be $u_1, u_2, \ldots,$ $u_i, \ldots$. Then

$$D(u) = D(u_i, u_{i+1}) \qquad (u_i > u \geqslant u_{i+1}) . \tag{82}$$

We can therefore express κ in the form

$$\frac{1}{\kappa} = \sum_i \frac{S(u_i) - S(u_{i+1})}{D(u_i, u_{i+1})} , \tag{83}$$

where $S(u)$ is the function defined by

$$S(u) = \frac{\int_0^u \frac{e^{2u} u^7 du}{(e^u - 1)^3}}{\int_0^\infty \frac{e^u u^4 du}{(e^u - 1)^2}} . \tag{84}$$

The integral in the denominator is easily evaluated. By a partial integration we have

$$\int_0^\infty \frac{e^u u^4 du}{(e^u - 1)^2} = 4 \int_0^\infty \frac{u^3 du}{e^u - 1} . \tag{85}$$

The integral on the right-hand side has already been evaluated (Eqs. [104] and [105], v); it has the value $\pi^4/15$. Hence, we have

$$S(u) = \frac{15}{4\pi^4}\int_0^u \frac{u^7 e^{2u}}{(e^u - 1)^3}\, du \,. \tag{86}$$

The function $S(u)$ was first introduced by Strömgren, who also tabulated the function sufficiently accurately for purposes of evaluating the stellar opacity coefficient.

Strömgren expresses the opacity in the form

$$\kappa = D_{ff}\left(\frac{Z^2}{A} = 6\right)\frac{\bar{g}}{t}\,, \tag{87}$$

where D_{ff}, defined as in equation (64), is evaluated for $Z^2/A = 6$, and t is a numerical factor (the guillotine factor) depending on $[G(T)/N_e]$ and T. Numerically, it is found that

$$D_{ff}\left(\frac{Z^2}{A} = 6\right) = \frac{144 N_e}{T^{3.5}}\,. \tag{88}$$

So far, we have neglected the g-factor. Now the g-factors, strictly speaking, enter as multipliers of the $D_n^{(Z)}$'s. It is, however, clear that only g-values near the absorption edges are of importance, for g becomes different from (and smaller than) unity only far from the absorption edge; in these regions, however, there will be an absorption edge of another element which will contribute to the stellar opacity for the region considered. In other words, if we go to frequencies $\nu > \nu_n^{(Z)}$, then the particular $D_n^{(Z)}$ (which contributes to the absorption near $\nu \simeq \nu_n^{(Z)}$) becomes small, compared to $D(u)$. An exception occurs for the absorption edges at high frequencies, but for these the weight $\Delta S(u)$ soon becomes negligible. Strömgren estimates that $\bar{g} = 0.90$ will not lead to more than a 2 or 3 per cent error in the final formula for the opacity. Hence, by (87) and (88) we have

$$\kappa = \frac{130 N_e}{T^{3.5}\, t}\,. \tag{89}$$

Table 12, due to Strömgren, gives values of $\text{Log}_{10}\, t$ for different values of T and $G(T)/N_e$, the computation having been carried through for the Russell mixture. Thus, for any given electron concentration the table enables us to calculate the opacity arising from a Russell mixture of elements.

TABLE 12

GUILLOTINE FACTOR $\text{Log}_{10}\, t$

T	log $[G(T)/N_e]$					
	∞	6	5	4	3	2
1×10^6	0.03	....	....	....	....	....
2	.06	....	....	....	....	....
3	.02	....	....	....	....	....
4	.00	0.06	0.13	....	....	....
5	.01	.07	.14	....	....	....
6	.08	.14	.22	0.35	0.56	0.79
8	.25	.30	.37	.50	.68	0.93
10	.33	.38	.45	.58	.76	0.99
12	.37	.41	.48	.60	.78	1.01
14	.39	.43	.49	.61	.79	1.02
16	.40	.43	.49	.60	.78	1.02
18	.40	.44	.49	.58	.78	1.02
20	.41	.44	.49	.60	0.79	1.03
25	0.48	0.51	0.56	0.66	....	....

We must now consider the effect of an admixture of light elements (hydrogen and helium) with the Russell mixture. First of all, it is clear that hydrogen and helium cannot directly contribute to the stellar absorption coefficient, the essential reason being that the absorption edges of these elements lie in a spectral region the absorption in which region does not contribute (for all practical purposes) to the Rosseland mean. The admixture of lighter elements has, however, an indirect effect.

Let us consider a mixture of elements—hydrogen, helium, and the Russell mixture—with the abundance factors X, Y, and $(1 - X - Y)$, respectively. Since, as we have seen, the lighter elements do not contribute to the opacity, the coefficient of opacity is, accordingly,

$$\kappa = \frac{130 N_e}{T^{3.5} t} (1 - X - Y) . \tag{90}$$

On the other hand, we have (cf. Eq. [50])

$$N_e = \frac{1}{2}\frac{\rho}{H}(1 + X) . \tag{91}$$

Combining (90) and (91), we can write

$$\kappa = \frac{65}{H}\frac{1}{t}\frac{\rho}{T^{3.5}}(1 + X)(1 - X - Y) , \tag{92}$$

or, numerically,

$$\kappa = 3.9 \times 10^{25}\frac{1}{t}\frac{\rho}{T^{3.5}}(1 + X)(1 - X - Y) . \tag{93}$$

If the helium content is negligible, then

$$\kappa = 3.9 \times 10^{25}\frac{1}{t}\frac{\rho}{T^{3.5}}(1 - X_0^2) . \tag{94}$$

From (93) (or [94]) and Table 12 we can calculate the stellar opacity as a function of the density, temperature, and chemical composition (here "chemical composition" is essentially equivalent to the abundance of hydrogen and helium). Of course, Strömgren's table of the "guillotine factor" t has been calculated for the case where the heavier elements are assumed to occur in a definite ratio. A closer examination shows, however, that this does not materially affect the formula for the stellar opacity.

We have so far restricted ourselves to photoelectric ionization as contributing to the main source of stellar opacity. At high temperatures, however (higher than in the interiors of the more common stars), there is another physical process which becomes of importance in contributing to stellar opacity. The process in question is the scattering by free electrons.

According to the classical electromagnetic theory, an accelerated electron emits radiation; we are here concerned with the converse phenomenon. J. J. Thomson has given for the scattering coefficient σ_e per electron the formula

$$\sigma_e = \frac{8\pi e^4}{3m_e^2 c^4} . \tag{95}$$

It will be seen that σ is independent of the frequency. Using (91) for the expression of the number of free electrons, we have for the contribution to the mass absorption coefficient by electron scattering the expression

$$\kappa_e(\nu)d\nu = \frac{8\pi e^4}{3m_e^2c^4}\frac{1}{2}\frac{1}{H}(1+X)d\nu\,, \tag{96}$$

or, numerically,

$$\kappa_e(\nu) = 0.20\ (1+X)\,. \tag{97}$$

If electron scattering were the only source of stellar opacity, then the Rosseland mean coefficient κ_e would be given by (cf. Eqs. [80] and [85])

$$\frac{1}{\kappa_e} = \frac{15}{4\pi^4}\int_0^\infty \frac{1}{\kappa_e(u)}\frac{u^4e^{2u}}{(e^u-1)^3}du\,. \tag{98}$$

Since, however, $\kappa_e(u)$ is independent of u, the integral in (98) is easily evaluated. It is found that

$$\frac{15}{4\pi^4}\int_0^\infty \frac{u^4e^{2u}du}{(e^u-1)^3} = \frac{1}{2}\left[1+\frac{1+2^{-3}+3^{-3}+\ldots.}{1+2^{-4}+3^{-4}+\ldots.}\right]. \tag{99}$$

The numerical value of the right side of the foregoing equation is found to be 1.055. Hence, by (97), (98), and (99),

$$\kappa_e = \frac{0.20\rho(1+X)}{1.055} = 0.19\ (1+X)\,. \tag{100}$$[1]

If photoelectric ionization and electron scattering are both about equally important, then we must form the Rosseland mean of the combined absorption coefficient, which by (78) and (97) is

$$\kappa_{e+i}(u) = \frac{D(u)}{u^3}+\kappa_e(u)\,. \tag{101}$$

Thus, the resulting coefficient of opacity is given by

$$\frac{1}{\kappa_{e+i}} = \frac{15}{4\pi^4}\int_0^\infty \frac{1}{D(u)u^{-3}+\kappa_e(u)}\frac{u^4e^{2u}du}{(e^u-1)^3}\,. \tag{102}$$

[1]It is incorrect to allow for "induced emission" for electron scattering when allowance has not been made for the Compton shift in wave length. Equation (100) is therefore incorrect. In this equation, the factor should be 0.20 as in equation (97).

Strömgren has evaluated κ_{e+i} according to (102) for a number of typical temperatures and densities at which electron scattering becomes important, and he has given the following empirical rule:

For a given T and $[G(T)/N_e]$, calculate the coefficient of opacity κ_i due to the ionization process only, using the table of guillotine factors. Then calculate κ_e, which is given by (100). The actual opacity κ_{e+i} is equal to the greater of the two quantities κ_i and κ_e plus 1.5 times the smaller of the two.

This completes the discussion of stellar opacity.

5. *Determination of the mean molecular weights of the stars.*—We have shown in the last two sections how the mean molecular weight and the stellar opacity can be determined in terms of the abundances of hydrogen and helium. In the discussion we shall in the first instance assume that the helium content can be neglected. We shall return to this question in § 9.

Our assumption, therefore, is that the stellar material is a mixture of hydrogen and the heavier elements in the ratio, by weight, $X_0 : 1 - X_0$. We shall further assume that the heavier elements form a Russell mixture. Actually, we make this definite assumption about the heavier elements because that is the ratio in which the elements are approximately present in the sun and in stellar atmospheres (Russell and C. H. Payne), and which further happens to agree roughly with the abundances with which these elements are present in the earth's crust. But this assumption, made for the sake of definiteness, is from the point of view of stellar interiors a very "harmless one," in the sense that any other assumption regarding the abundances of the heavier elements will lead to substantially the same conclusions regarding the abundance of hydrogen. This is seen when we compare our first and second approximations for μ, which are given in § 3. On the first approximation we derived, with no particular assumption regarding the abundances of the heavier elements but only assuming that the material is highly ionized, that

$$\mu = \frac{2}{1 + 3X_0}, \tag{103}$$

while a more rigorous consideration of the state of ionization for the case when the heavier elements form a Russell mixture, led to

$$\mu = \frac{1}{2X_0 + \bar{n}_R(1 - X_0)}, \tag{104}$$

where $\bar{n}_R$ is tabulated in Table 10. An examination of (104) with the values of $\bar{n}_R$ given in Table 10, and a comparison with (103), readily shows that the actual specification of the abundances of the heavier elements is hardly of importance in the present connection. The same thing is true of stellar opacity, for which we can use

$$\kappa = 3.90 \times 10^{25} \frac{1}{t} \frac{\rho}{T^{3.5}} (1 - X_0^2) . \tag{105}$$

It should be mentioned in this connection that if, instead of the Russell mixture, we use one in which there is a higher proportion of the heavier elements, then, though the coefficient $D^{(Z)}$ in the opacity formula (Eqs. [64], [71], [73], and [74]) increases, this effect is largely compensated by a corresponding increase[2] in the mean molecular weight, which works in the opposite direction. We thus see that equation (105) is valid over a wide range of the relative abundances of the heavier elements.

We shall now proceed to outline the method of determining the hydrogen abundance, X_0, for a star of known mass, radius, and luminosity.

We shall write equation (105) in the form

$$\kappa = \kappa_0 \frac{1}{t} \frac{\rho}{T^{3.5}}, \tag{106}$$

where

$$\kappa_0 = 3.9 \times 10^{25}(1 - X_0^2) . \tag{107}$$

[2] That there is an increase of μ is seen as follows: For a Russell mixture we have seen that when it is completely ionized there are 0.54 free particles per unit atomic weight. If there is a higher proportion of the heavier elements than in the Russell mixture, there will be a smaller number of free particles per unit atomic weight; this will increase μ, and therefore decrease the number of free electrons per unit volume.

The formula for κ which was used in § 1 (Eq. [3]) is of the same form as (106), above. We can therefore use the luminosity formula (14), which can be written as

$$\kappa^* = 7.17 \times 10^{24} \frac{M^{5.5}}{LR^{0.5}} (\mu\beta_c)^{7.5} , \tag{108}$$

where L, M, and R are expressed in solar units. Further

$$\kappa^* = \frac{\kappa_0}{\bar{t}} , \tag{109}$$

where $\bar{t}$ is a mean value for the guillotine factor (cf. Eq. [8]).

Thus, for a star of known L, M, and R and an assumed value for μ, equation (108) suffices to determine κ^* and if we can estimate the value of $\bar{t}$, then we have—so to say—an astronomical determination of the physical constant κ_0. Again, an assumed value for μ implies, according to (103) or (104), a definite value for X_0; hence we have, according to (107), a physical determination of κ_0. We now arrange by a proper choice of μ (or X_0) that the astronomical and the physical values of κ_0 agree. The value of X_0 (or μ) which brings about this agreement determines the hydrogen content of the star under consideration. The point which remains to be settled is the determination of the mean value $\bar{t}$ of the guillotine factor.

According to the discussion in § 4, t depends on $[G(T)/N_e]$, which, according to equation (36), is defined by

$$\frac{G(T)}{N_e} = 2 \frac{(2\pi m_e kT)^{3/2}}{h^3} \frac{1}{N_e} . \tag{110}$$

To determine $\bar{t}$ it is clearly sufficient to consider the first approximation of § 4, according to which

$$p_g = \frac{1 + 3X_0}{2} \frac{k}{H} \rho T ; \qquad N_e = \tfrac{1}{2}(1 + X_0) \frac{\rho}{H} . \tag{111}$$

By (110) and (111) we have

$$\frac{G(T)}{N_e} = 2 \frac{(2\pi m_e kT)^{3/2}}{h^3} \frac{1 + 3X_0}{1 + X_0} \frac{kT}{p_g} , \tag{112}$$

which is easily seen to be equivalent to

$$\frac{G(T)}{N_e} = \frac{2(2\pi m_e)^{3/2} k^{5/2}}{h^3} \frac{3}{a} \frac{1 + 3X_0}{1 + X_0} \frac{1 - \beta}{\beta} T^{-3/2} . \tag{113}$$

Numerically, the foregoing is

$$\frac{G(T)}{N_e} = 2.63 \times 10^{14} \times \frac{1 + 3X_0}{1 + X_0} \frac{1 - \beta}{\beta} T^{-3/2} , \tag{114}$$

or

$$\left.\begin{aligned} \log \left[\frac{G(T)}{N_e}\right] = 2.303\Big[14.42 + \text{Log}\, \frac{1 + 3X_0}{1 + X_0} \\ - \text{Log}\, \frac{\beta}{1 - \beta} - \tfrac{3}{2} \text{Log}\, T \Big] . \end{aligned}\right\} \tag{115}$$

Since, according to our first approximation considered in § 7 of chapter vi, the standard-model density distribution is to be regarded as a first approximation, we can use for β in the foregoing equation the value determined by Eddington's quartic equation. We shall then have to study the march of $[G(T)/N_e]$ through the star and by using the table of guillotine factors (Table 12) infer the appropriate mean value, $\bar{t}$. We shall illustrate the estimation of $\bar{t}$ for the sun. If we assume for μ the value 1.05, the solution of the quartic equation yields $1 - \beta = 0.004$, and by equation (62) of chapter vi, it is found that $T_c = 20{,}000{,}000$ degrees. Table 13 (due to Ström-

TABLE 13

T	$\log_e \left[\frac{G(T)}{N_e}\right]$	t	T	$\log_e \left[\frac{G(T)}{N_e}\right]$	t
20×10^6	2.8	7	12×10^6	3.5	5
18	2.9	6	10	3.8	4
16	3.1	6	8	4.1	3
14	3.3	5	6	4.6	2

gren) gives the corresponding variation of t through the star. From the table Strömgren estimates that the appropriate mean value of t is about 5, which is seen to be the value of t at about two-thirds of the central temperature. This appears to be a general rule, so that to determine the order of magnitude of $\bar{t}$ we first determine T_c according to equation (62) of chapter vi and then calculate $[G(T)/N_e]$ for $T = 2/3T_c$ as given by (114) or (115). From Table 12 we then obtain the corresponding value of t. This value of t is used in the luminosity formula as a first approximation to $\bar{t}$. After a first approximation to μ has been obtained, the quantity $\bar{t}$ having been estimated in the foregoing manner, we can then proceed to a second approximation by determining $\bar{t}$ defined appropriately (cf. Eq. [8]).

Using this method, Strömgren has computed the values of μ and has thus inferred the hydrogen contents of those stars for which there is fairly reliable information concerning L, M, and R. Tables 14*a* and 14*b* illustrate the determination of μ for Capella and the sun.

TABLE 14*a*

THE DETERMINATION OF THE HYDROGEN CONTENT OF CAPELLA A

X_0	μ	$1-\beta$	Log κ_0 (Astro.)	Log κ_0 (Physical)
0.34	0.95	0.04	25.37	25.53
.31	1.00	.04	25.51	25.54
.28	1.05	.05	25.64	25.55
.25	1.10	.06	25.76	25.56
0.22	1.15	0.07	25.88	25.56

log $[G(T)/N_e]=7$; $\bar{t}=1$; $X_0=0.30$; $\mu=1.01$

TABLE 14*b*

THE DETERMINATION OF THE HYDROGEN CONTENT OF THE SUN

X_0	μ	$1-\beta$	Log κ_0 (Astro.)	Log κ_0 (Physical)
0.36	1.00	0.003	24.85	24.84
.33	1.05	.004	25.01	24.85
0.29	1.10	0.004	25.16	24.86

log $[G(T)/N_e]=3$; $\bar{t}=5$; $X_0=0.36$; $\mu=1.00$

In connection with the foregoing solutions it is necessary to remark that the values of X_0 given do not represent the only solution to the problem. For a given star there are, in general, two values of

X_0 for which there is agreement between the astronomical and the physical values of κ^*. The second solution, as we shall see presently, however, corresponds to an extremely high abundance of hydrogen. We can best illustrate the existence of this second solution in the case of Capella, for which the guillotine factor is approximately unity.

If there exists a solution corresponding to which X_0 is almost unity, we can put $\mu = 0.5$ in the luminosity formula. Again, for the order of stellar masses we shall normally be interested in (masses less than 10 $\odot$), β_c is very nearly unity. Hence, we can also put $\beta_c = 1$. Equation (108) now reduces to

$$\kappa^* = 4.0 \times 10^{22} \frac{M^{5.5}}{LR^{0.5}} \qquad (\mu = 0.5\,, \quad \beta_c = 1)\,. \quad (108')$$

For Capella, $M = 4.18$, $L = 120$, and $R = 15.8$. Inserting these values in (108′), we find that $\kappa^* = 2.17 \times 10^{23}$. It is also found that l is unity, and hence

$$\kappa_0 \text{ (astronomical)} = 2.17 \times 10^{23}\,,$$

while

$$\kappa_0 \text{ (physical)} = 3.89 \times 10^{25}\,(1 - X_0^2)\,.$$

From the foregoing it follows that $X_0 = 0.997$. In other words, this second solution corresponds to 99.7 per cent abundance of hydrogen, while our first solution corresponds to 30 per cent of hydrogen. Similar results will be obtained for the other stars. Hence, quite generally, the second solution corresponds to an extremely high abundance of hydrogen, and it is improbable that such an extreme abundance of hydrogen can correspond to reality. Actually, there are reasons to believe that the hydrogen abundance in stellar interiors must be less than in stellar atmospheres—the essential ground for this belief being that the heavier elements will "sink" relatively more toward the center of a star than the lighter elements, and it appears that hydrogen is not present in stellar atmospheres to anything approaching 99.7 per cent by weight. We shall therefore restrict ourselves (unless otherwise stated) to the solution which corresponds to a "moderate" abundance of hydrogen.

6. *General remarks.*—Before we proceed to describe Strömgren's results for other stars and the bearing of these calculations toward an interpretation of the Hertzsprung-Russell diagram, it is necessary to make some comments concerning our present attitude, as compared with that generally adopted in an earlier epoch (i.e., prior to Strömgren's systematic work).

In Eddington's earlier work it was assumed that all stars have the same mean molecular weight ($\mu \sim 2$). This implies, for all practical purposes, that the lighter elements (hydrogen in the present connection) are not abundant. The assumption of constant μ is not only characteristic of Eddington's early work but has been implicitly assumed quite generally. This assumption, according to our present point of view, has to be abandoned. The reasons can be briefly summarized as follows:

Let us suppose that the abundance of hydrogen is negligible. Then we immediately come into conflict with the physical theory of the stellar opacity. The nature of the conflict can be illustrated by taking the case of Capella. Observationally, we have

$$L = 120 L_\odot \; ; \qquad M = 4.18 \odot \; ; \qquad R = 15.8 R_\odot \, .$$

Let us assume (as in Eddington's early work) that $\mu = 2.11$. Then we have

$$\beta = 0.717 \; ; \qquad T_c = 7.9 \times 10^6 \, .$$

The guillotine factor $\bar{t}$ is found to be unity, so that, according to (107), ($X_0 = 0$),

$$\kappa^* \text{ (physical)} = 3.9 \times 10^{25} \, .$$

From (108), on the other hand, it is found that

$$\kappa^* \text{ (astronomical)} = 8.8 \times 10^{26} \, . \tag{116}$$

We see that the two values of κ^* differ by a factor of about 23; this is the famous "opacity discrepancy." In spite of this discrepancy, the tendency was to assume that the origin of it is due to the inadequacy of the physical theory, and in Eddington's early work κ^* was assumed to be equal to its "astronomical value." The luminosity formula (108) (with κ^* according to [116]) was therefore used to predict the luminosities of other stars. It was found that the luminosities thus predicted agreed with observation. This was taken to imply that the stars in fact form a one-parametric sequence of configurations.

But the foregoing point of view has to be abandoned, for the more refined theory of the stellar opacity now available leaves no room for doubting the physical theory. Consequently, we must accept an abundance of the lighter elements, in particular of hydrogen, to remove the discordance between the physical and the astronomical value of κ^*. But now it may be argued that we may allow for an abundance of hydrogen but still use a constant μ for all the stars, so that the luminosity formula can still be used to predict the luminosities for stars of known mass and radius. This idea gained some currency when it was found that both Capella and the sun lead to about the same value of X_0. But one important difference has to be noticed. We obtain the same value of X_0 for Capella and the sun because of the guillotine factor. For the sun the guillotine factor is 5, while for Capella it is unity, so that, if we use for κ^* the value derived from Capella, and use (108) to predict the luminosity of the sun, we should be wrong by a factor of 5. Now, this same argument (due originally to Eddington) can be employed to show the necessity for introducing a variable X_0. Consider, for instance, the sun and ζ Herculis A.[3] Both have very nearly the same mass ($M_{\zeta \text{ Her}} = 0.96$), but ζ Herculis A has a radius about twice that of the sun and a luminosity about four times that of the sun. Suppose we assume that the sun and ζ Herculis A have the same value for μ. It is found now that for ζ Herculis l is 2.3. Again, since ζ Herculis has twice the solar radius, the predicted luminosity would be

$$L \text{ (predicted)} = \frac{2.3 \times (0.96)^{5.5}}{2^{0.5} \times 5} = 0.25 ,$$

while L (observed) = 4.0—i.e., a discrepancy of a factor of 16. Thus, ζ Herculis must have a different value of μ from that of the sun. Indeed, calculation shows that for ζ Herculis, $X_0 = 0.11$ and $\mu = 1.45$.

[3] This is only an example. We can give other similar examples.

Thus, for consistency we are forced to accept a variable μ for stars; consequently, the luminosity formula (108) has to be used to determine μ (or X_0) for individual stars, rather than to predict the luminosities for different stars. It is necessary to emphasize this because Eddington, who, independently of Strömgren, introduced the abundance of hydrogen to remove the "opacity discrepancy" for the case of Capella and the sun, seems inclined to the view that the luminosity formula can still be used to predict L for other stars, using a constant $\mu \sim 1.0$ for all stars. The objection to using the luminosity formula in determining μ for the individual stars seems to arise from an uneasiness that μ is used as an "adjustable parameter" to "save" an "inadequate" theory. But the answer to such an objection is that, if we allow μ to be variable, the derived μ's show a systematic variation in the plane of the Hertzsprung-Russell diagram and do not show any randomness. In other words, we derive an interpretation of the characteristic features of the Hertzsprung-Russell diagram in conformity with the Vogt-Russell theorem.

7. *Interpretation of the Hertzsprung-Russell diagram.*—According to the method outlined in § 5, it is relatively simple to determine the hydrogen contents of stars for which the values of the fundamental parameters L, M, and R are known. The computations have been carried out for about forty stars, and the resulting hydrogen contents for some of them are given in Table 34 in the form of an appendix. We shall here be concerned only with the general results, but it may be mentioned that for the B stars it is necessary to take into account the effect of electron scattering, which was considered at the end of § 4.

First of all, the question arises whether we cannot represent the whole observational material—within the limits of the uncertainty of the observations—on the assumption of a constant μ for all the stars. We have already considered this question in § 6, and a closer examination now reveals that, though the most commonly discussed stars—the sun, Capella A, and Sirius A—have about equal hydrogen content ($\sim$35 per cent), it is yet not possible to predict the luminosities of all the stars considered on the basis of a constant μ. If an attempt is made, we encounter discrepancies in the predicted luminosities sometimes amounting to as much as by a factor of 50, and further, these discrepancies show a systematic character. The magnitude of these discrepancies and their systematic nature preclude the possibility of constant μ. It is highly satisfactory that we are led to a variable μ (or X_0) purely from observations, for we are led to precisely the same conclusion by appealing to the Vogt-Russell theorem. For, two stars of equal mass can differ (in radius

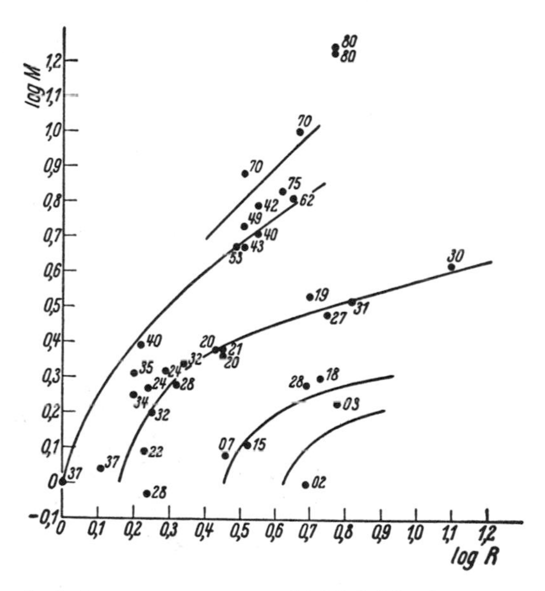

FIG. 23*a*.—Each dot represents a star and is labeled by the computed X_0 (Strömgren, *Zs.f. Ap.*, **7**, 222, 1933).

or/and luminosity) only on account of a difference in chemical composition, and we have now to examine whether the observed masses, radii, and luminosities, *and* the derived hydrogen contents enable us to arrange the stars as a two-parametric family of configurations (the two parameters being M and X_0).

In Figures 23*a* and 23*b* we have plotted Log M against Log R. Each star is represented by a point in this plane, and we label each point by the appropriate X_0. We see that the plot in this diagram enables us—more or less unambiguously—to draw curves of constant

μ (or X_0) in the (Log M, Log R) plane. Schematically, the situation that arises is shown in Figure 24.

It is clear from Figure 23 or Figure 24 that, if we consider a sequence of stars of a given M but of increasing radii, then along this sequence the hydrogen content decreases, while the mean molecular weight increases. This is, indeed, a quite general result: *For a given*

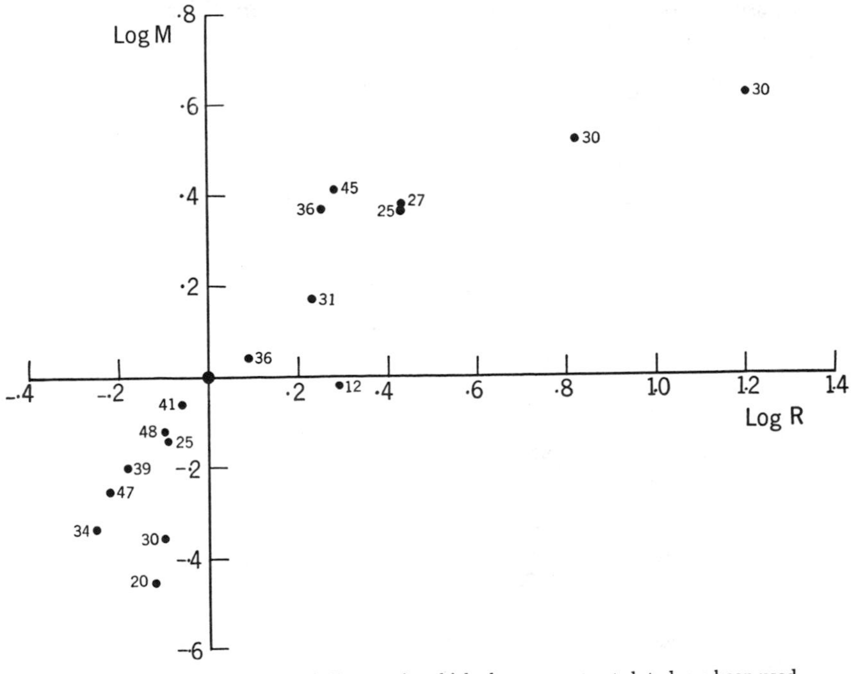

FIG. 23*b*.—This is a revised diagram in which the more recent data have been used. As in Fig. 23*a*, each dot represents a star and is labeled by the computed X_0.

mass, with increasing radius, the hydrogen content decreases. This result is easily understood. Observationally, it is well known that a rough empirical mass-luminosity correlation exists in nature. An increase of radius, then, has two effects: first, the guillotine factor decreases, and, second, the radius factor in the luminosity formula increases. Both these effects act in the same sense—toward lowering the predicted luminosity; to counteract this effect it is necessary to increase μ or, what comes to the same thing, to decrease the hydrogen content. In the case of the massive stars, however, another effect becomes important: For the massive stars of smaller radii (i.e., for the B stars which form the continuation of the main series) the central temperature is sufficiently high to reduce the magnitude of the general (i.e., the Kramers-Gaunt) opacity, thus making the contribution to the absorption by electron scattering important. Thus, while a decreasing radius still corresponds to an increase in the guil-

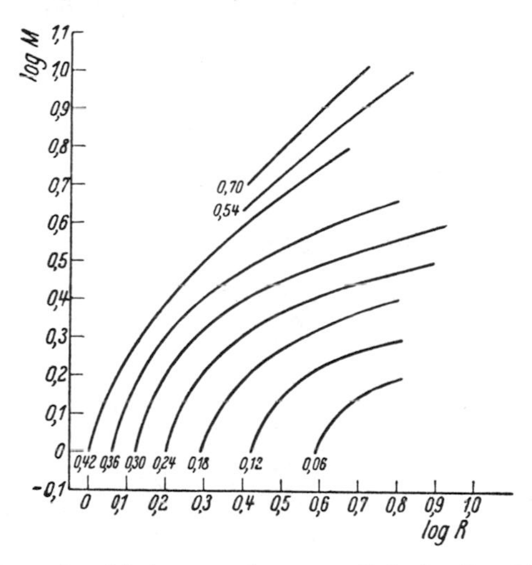

FIG. 24.—The semiempirical curves of constant X_0 in the (Log M, Log R) plane (Strömgren, *Zs. f. Ap.*, **7**, 222, 1933). Each curve is labeled by the corresponding value of X_0.

lotine factor, the increasing importance of electron scattering with decreasing radius acts in the *opposite direction*, in counteracting the decrease of stellar opacity arising from the guillotine factor. Hence, in the case of the massive stars, though decreasing radius still corresponds to increasing X_0, the range of variation in X_0 for given change in R is much less than for stars of "ordinary" masses (i.e., $M < 4 \odot$).

Once we have drawn an empirical set of curves of constant X_0 as in Figure 24 (which, it will be remembered, combines the results derived from a theoretical $[L, M, R, \mu]$ relation and the set of values of L, M, and R that occur in nature), then, if we specify the mass and the radius of a star, the appropriate μ or X_0 can be directly read off from the diagram; from a knowledge of M, R, and μ we can predict L. Further, we can transform the curves of constant X_0 from the (Log M, Log R) plane to a set of curves of constant X_0 in the plane of the Hertzsprung-Russell (for short, "H.R.") diagram. As is well known, the co-ordinates which describe a star in the H.R. plane are the absolute magnitude (essentially -2.5 Log L) and the spectral type (essentially Log T_e), T_e (the effective temperature) increasing

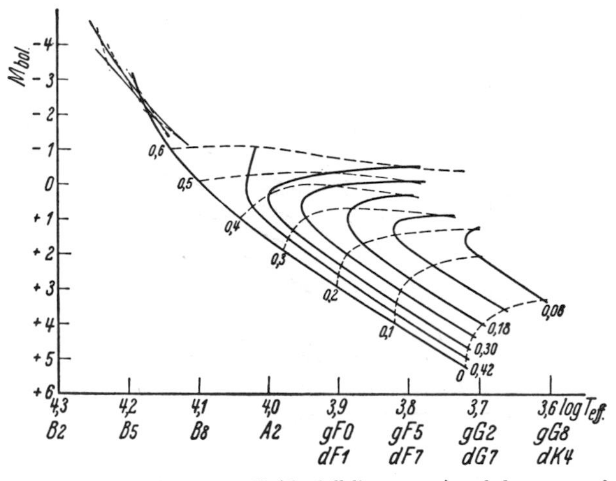

FIG. 25.—The curves of constant X_0 (the full-line curves) and the curves of constant M (the dotted curves) in the plane of the Hertzsprung-Russell diagram (Strömgren, *Zs. f. Ap.*, 7, 222, 1933).

toward the left. To transform the curves of constant X_0 from the (Log M, Log R) plane to curves of constant X_0 in the H.R. plane, we go along each particular curve in the (Log M, Log R) plane, and for each point we calculate L according to the luminosity formula, and hence also Log T_e; for, according to the definition of the effective temperature, we have

$$\sigma T_e^4 = \frac{L}{4\pi R^2} . \tag{117}$$

Similarly, we can draw curves of constant M in the H.R. plane. These two sets of curves will enable us to determine both the mass and the hydrogen content of a star merely from its position in the

Hertzsprung-Russell diagram (see Fig. 25). It may happen that the curves of constant X_0 intersect in the H.R. plane. In such cases reference must be made to the (Log M, Log R) plane. We have thus succeeded in arranging the stars in a two-parametric sequence entirely in conformity with the requirements of the Vogt-Russell theorem.

From the topography of the curves of constant X_0 and constant M, we derive the following interpretation (due to Strömgren) of the characteristic features of the Hertzsprung-Russell diagram:

The main series up to spectral class A is the locus of stars of hydrogen content varying between 25 and 45 per cent—i.e., about a mean of 35 per cent—and masses running up to 2.5 $\odot$. Stars of small mass and low hydrogen content are relatively rare—they occur as subgiants of spectral classes G–K. The gap between the M giants and the corresponding dwarfs (on the main series) arises from the circumstance that not even stars of low hydrogen content "scatter" in this region. The massive stars ($M > 5\ \odot$) occurring in the region of the B stars which are rich in hydrogen (X_0 sometimes going up to 95 per cent) form the continuation of the main series—the continuation arising from the circumstance that massive stars with "medium" hydrogen content ($0.4 < X_0 < 0.8$) which are on the main series occur in a very small region of the H.R. diagram between the B and the A stars. (We shall obtain evidence in chapter viii for the breakdown of the standard model for the very massive stars. Further, along the main series the breakdown probably sets in at about $M = 10\ \odot$. The investigations of the hydrogen content of the B stars is therefore somewhat inconclusive. We shall return to these matters in chapter viii.) The giant branch is characterized by stars having about the same hydrogen content as (or somewhat less than) the main series stars. The giant branch is limited on the side of low luminosity, since stars of low luminosity are relatively rare. On the side of high luminosity it is limited again, because, for X_0 a little greater than 0.3, the characteristic bend of the curves of constant X_0 disappears, and also because the stars of large mass with hydrogen content greater than about 40 per cent scatter over a large area in the H.R. diagram, which must, therefore, be sparsely populated. The gap (the "Hertzsprung gap") in the giant branch in the region of spectral class F is probably due to a real scarcity of stars with masses between 2.5 and 4.5 ⊙. The supergiants, then, are interpreted as massive stars with medium hydrogen content. The "spreading-out" of the curves of constant X_0 in the supergiant region of the H.R. diagram is easily understood from the remarks made on pages 282 and 283.

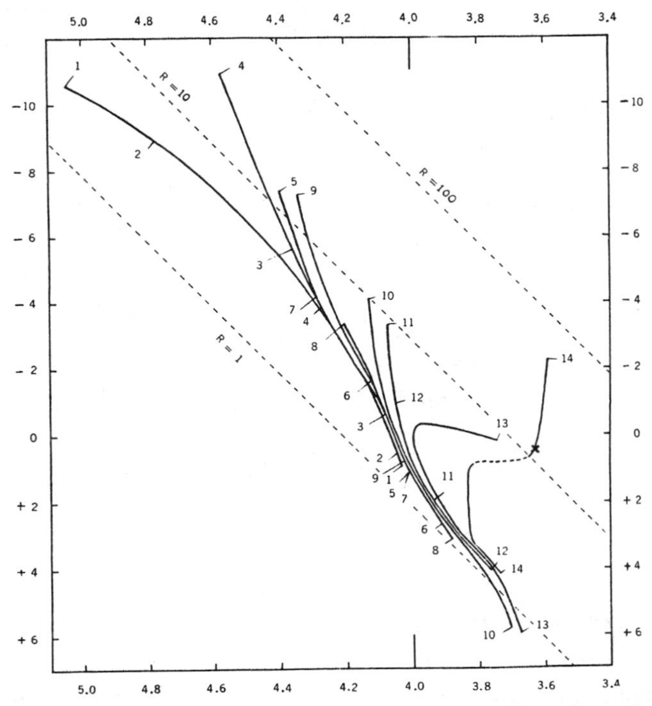

FIG. 26.—The abscissae are Log *Te;* the ordinates are absolute bolometric magnitudes (Kuiper, *Ap. J.*, **86,** 176, 1937). The clusters are identified in Table 2 of Kuiper's paper.

8. *Kuiper's interpretation of the cluster diagrams: the hydrogen content of the Hyades stars.*—So far we have considered only the characteristic features of the general H.R. diagram. It is clear that we can construct the absolute-magnitude–spectral-type diagrams, including in the plot only such stars as are physically associated—like the stars in a cluster. The H.R. diagram for stars in a cluster may be called a "cluster diagram." The pioneering work on this subject was done by Hertzsprung in 1911, and extensive studies of a systematic nature of the cluster diagrams have been carried out by Trumpler; for a general discussion of the subject from the observational side, and for references, see a paper by Kuiper quoted in the Bibliographical Notes at the end of this chapter. Figure 26 is taken from Kuiper's paper. From the similarity of these curves with the Strömgren curves of constant X_0, Kuiper infers that the stars in a cluster are characterized by approximately the same hydrogen content.

A comparison of the diagrams for the Pleiades and the Hyades indicate that Hyades stars should have relatively low hydrogen content. In the case of the Hyades cluster, Kuiper's suggestion is capable of verification. There are six stars in this cluster for which Kuiper has derived the values of L, M, and R. For these stars Strömgren has computed the hydrogen content X_0, and the results are given in Table 15. The uncertainties in the values of L, M, and R used arise essentially from the uncertainties in the parallax of the individual stars (though the parallax of the cluster itself is fairly reliably known). According to Kuiper, the weighted mean of the six determinations, which gives X_0 (mean) = 0.16, may be considered as a reliable estimate of hydrogen content of the Hyades stars.

TABLE 15*

THE HYDROGEN CONTENT OF THE STARS IN HYADES

ADS	Log $L/L_\odot$	Log $R/R_\odot$	Log $(M/\odot)$	X_0	Weight
3264.........	+0.69	+0.11	+0.07	<0.23	$1\frac{1}{2}$
3483.........	+ .40	+ .09	+ .04	.30	1
3135.........	+ .07	− .06	− .195	.14	3
3169.........	+ .06	− .03	− .21	< .13	$\frac{1}{2}$
3475.........	+ .03	− .06	− .19	.15	3
3210.........	−0.37	−0.13	−0.44	0.02	2

* For ADS 3135 Kuiper has revised the data on the basis of additional information. The new value has been used here.

9. *The abundance of helium in stellar interiors.*—We found in § 3 that for purposes of the analysis of stellar interiors it is sufficient to consider the abundances of hydrogen, helium, and all the other heavier elements (Russell mixture) lumped into one group. In the discussions in §§ 5, 6, 7, and 8 we assumed that the helium content

could be neglected. This assumption is justifiable in the first instance in so far as investigations of the stellar atmospheres seem to indicate that hydrogen is very much more abundant than helium. But it should be remembered that the determination of the abundances of hydrogen and helium in stellar atmospheres is a matter of great complexity. This has to be borne in mind, especially in view of the fact that in a recent investigation on the transmutation of elements in stellar interiors (chap. xii) by von Weizsäcker the suggestion is made that helium may be very much more abundant than all the other heavier elements put together. Indeed, von Weizsäcker's theory seems to require helium to be as much as eight to ten times as abundant as the Russell mixture; and Strömgren has examined whether this requirement of von Weizsäcker's theory is compatible with the data concerning the masses, luminosities, and radii of the stars. We shall follow Strömgren's discussion of this matter.

Let us consider the "second solution" for the hydrogen content, the existence of which we pointed out in § 5. In the case of Capella we found that the "second solution" corresponds to an abundance of 99.7 per cent of hydrogen. But this extreme abundance of hydrogen corresponds to the relative abundance of hydrogen to the Russell mixture which is so high as to be quite improbable. However, if we reduce the amount of hydrogen by a small amount and replace it by helium, then we increase μ so that the predicted luminosity, in the first instance, is greater. Hence, in order to predict the correct luminosity, we must increase the absorption and hence increase the abundance of the heavier elements. It is clear that by a suitable increase of the amount of the Russell mixture present we can again obtain agreement between the observed and the predicted luminosities. We now have two unknowns—the hydrogen content, X, and the helium content, Y. There is only one relation—the theoretical (L, M, R, μ) relation—available, so that we can determine the hydrogen content X and the ratio of Russell mixture to helium $(U : Y)$ as functions of the helium content. Tables 16a and 16b (due to Strömgren) illustrate the results of such calculations for the case of the sun and Capella.

The table shows that the (L, M, R, μ) relation can be used to determine the maximum value of the ratio $Y : U$. Also, a definite physical theory of the transmutation of elements in stellar interiors would provide a theoretical value for the ratio $Y : U$. For a specified value of the ratio $Y : U$, we have, in general, two solutions, but it will be more difficult to decide between them now. For example, if the ratio were prescribed to be 10, then for the sun we have for the relative abundances of hydrogen, helium, and the Russell mixture

TABLE 16*a*

THE HYDROGEN AND HELIUM CONTENT OF THE SUN

Helium Content (Y)	Hydrogen Content (X)	Russell-Mixture Content (U)	$Y : U$	$U : X$
0.00	1.00	0.002	0	0.002
.09	0.90	.004	22	.005
.19	0.80	.009	21	.011
.28	0.70	.02	15	.026
.36	0.61	.04	10	.059
.42	0.51	.07	6	.14
0.43	0.43	0.14	3	0.32

TABLE 16*b*

THE HYDROGEN AND HELIUM CONTENT OF CAPELLA A

Helium Content (Y)	Hydrogen Content (X)	Russell-Mixture Content (U)	$Y : U$	$U : X$
0.00	1.00	0.001	0	0.001
.10	0.90	.002	46	.002
.19	0.80	.005	37	.006
.29	0.70	.012	24	.02
.37	0.60	.025	15	.04
.42	0.51	.07	6	.13
0.47	0.42	0.11	4	0.26

the two solutions 60 : 36 : 4 and 90 : 10 : $\sim$ 0.3. Of the two solutions, the first is probably more consistent with the spectroscopic evidence from the study of stellar atmospheres. It should further be noticed that the observational uncertainties in L affect the derived content of the Russell mixture directly, so that the chemical composition derived on the hydrogen-helium–Russell-mixture hypothesis is very much more sensitive to the uncertainties in the observational values of L than is the case on the hydrogen–Russell-mixture hypothesis.

For the other stars we have similar results; for Capella A, Strömgren finds that the calculations lead to results very similar to those for the sun. For the subgiants, it again appears that they are relatively poorer in hydrogen than the main-series stars. Further, the connection between the radius and the content of the heavy elements appears to be the same as in our earlier discussion on the basis of the hydrogen–Russell-mixture hypothesis. Indeed, the run of the curves of constant abundance of the heavy elements in the plane of the H.R. diagram derived in § 7 (Fig. 25) is seen to be very general: If we consider stars of "medium" masses ($M < 2.5\ \odot$), then an increase in R has two effects: it decreases the guillotine factor l, and it increases the $R^{0.5}$ factor in the luminosity formula. Both these effects, acting in the same sense, make the predicted luminosity too low in the first instance (there being an empirical rough mass-luminosity correlation). To obtain agreement between the predicted and the observed values of L, we must increase the abundance of the heavier elements; i.e., U increases with increasing R. For the more massive stars, on the other hand, the electron scattering—now of importance—acts in the opposite direction to the guillotine factor and the $R^{0.5}$ factor in the luminosity formula. The "spreading-out" of the curves of constant U in the H.R. diagram is thus seen to be very general. The general conclusions, then, are essentially the same as before. The important point to note in the present connection is that an abundance of helium comparable to that of hydrogen is "compatible with the observational data concerning the masses, luminosities and radii of the stars" (Strömgren).

BIBLIOGRAPHICAL NOTES

This chapter deals almost entirely with the results contained in—

1. B. Strömgren, *Zs. f. Ap.* **4**, 118, 1932.

2. B. Strömgren, *Zs. f. Ap.*, **7**, 222, 1933.

3. B. Strömgren, *Erg. exakt. Naturwiss.*, **16**, 465, 1937 (§§ 16, 17, and 18 of this paper).

§ 1.—For the whole discussion the luminosity formula $L = 4\pi cGM(1 - \beta_c)/\kappa\eta$ is made the fundamental starting-point. The exposition of the theory starting with the luminosity formula makes the presentation rather neat, and this particular arrangement of the arguments is believed to be new.

§ 2.—4. H. Vogt, *A.N.*, **226**, 301, 1926.

5. H. N. RUSSELL, in RUSSELL, DUGAN, and STEWART, *Astronomy*, **2**, 910, Boston, 1927.

§§ 3 and 4.—References 1 and 2. Also—

6. R. H. FOWLER and E. A. GUGGENHEIM, *M.N.*, **85**, 939, 1925.

7. A. S. EDDINGTON, *M.N.*, **92**, 364, 1932.

§§ 5, 6, and 7.—References 1 and 2. Also—

8. A. S. EDDINGTON, *M.N.*, **92**, 471, 1932. In this paper Eddington tends to a belief that all stars have the same hydrogen content. This differs from the point of view currently adopted.

§ 8.—9. E. HERTZSPRUNG, *Potsdam Pub.*, **63**, 1911. This classical paper, which is very rarely quoted, contains a summary of Hertzsprung's earlier work (1905–1909), in which the giants and dwarfs were discovered. It gives a clear description of the main series (Hauptserie) as being a group of stars of nearly the same radii but widely different surface temperatures, which results in the observed differences in luminosity. Giants and supergiants (c stars) are also discussed. The paper contains the first diagrams relating color equivalent or spectral equivalent with absolute magnitude.

10. R. J. TRUMPLER, *Pub. A.S.P.*, **40**, 265, 1928.

11. G. P. KUIPER, *Harvard Bull.*, No. 903, 1936.

12. G. P. KUIPER, *Ap. J.*, **86**, 176, 1937.

§ 9.—B. STRÖMGREN, *Ap. J.*, **87**, 520, 1938. Also reference 3

The following further references may be noted:

13. R. H. FOWLER, *Statistical Mechanics*, 2d ed., chaps. xiv, xv, xvi, and xvii, Cambridge, 1936.

14. S. ROSSELAND, *Astrophysik auf Atomtheoretischer Grundlage*, Berlin, 1931 (§ 16 of this book).

In references 13 and 14 problems connected with "excluded volumes," "electrical pressure," etc., are considered—topics which have not been treated in the monograph.

CHAPTER VIII

STELLAR ENVELOPES AND THE CENTRAL CONDENSATION OF STARS

In this chapter we shall discuss the equilibrium of stellar envelopes. By a "stellar envelope" we shall mean the outer parts of a star, which, though containing only a small fraction (for definiteness, we shall assume this fraction to be 10 per cent) of the total mass, M, nevertheless occupy a good fraction of the radius, R. A study of stellar envelopes has a twofold importance for astrophysical theories: first, it extends the region of the study of the conventional stellar atmospheres into the far interior, and second, it has also a very definite bearing on the studies of the deep interiors which are our main concern in this monograph. Thus, the central condensation of a star, defined as the fraction ξ^* of the radius, R, which incloses the inner 90 per cent of the mass, M, must give some indication of the concentration of the mass toward the center of the star under consideration. It is clear that $(1 - \xi^*)$ is a measure of the extent of the stellar envelope. The main problem which we shall consider in the theory of stellar envelopes is the evaluation of the central condensations of stars of known L, M, and R and assumed chemical composition. We shall see that this subject is closely related to the problems discussed in chapter vii.

1. *The equilibrium of stellar envelopes.*—The general theory presented here is due to Chandrasekhar.

In writing down the equations of equilibrium, the following two simplifications will be introduced: (*a*) that there are no sources of energy in the stellar envelope, and (*b*) that the mass contained in the envelope can be neglected in comparison with the mass of the star as a whole. Indeed, these two assumptions can be taken to define the stellar envelope.

The equations of equilibrium of the stellar envelope, then, are

$$\frac{d(p_g + p_r)}{dr} = -\frac{GM}{r^2}\rho \qquad (1)$$

and

$$\frac{dp_r}{dr} = -\frac{\kappa L}{4\pi c r^2}\rho , \tag{2}$$

the symbols having their usual meaning. The formula for the stellar opacity appropriate to the present discussion is (cf. Eq. [94], vii)

$$\kappa = 3.89 \times 10^{25} \frac{(1 - X_0^2)}{t} \frac{\rho}{T^{3.5}} . \tag{3}$$

It is found that, under the circumstances of the stellar envelope, the guillotine factor t does not vary appreciably; in many cases of practical importance it is very near unity, and even under the most unfavorable circumstances it varies only by as much as a factor of 3. Further, the guillotine factor occurs in the final formulae (which determine the central condensation ξ^*) only as a square root, so that we can conveniently replace t by a constant $\bar{t}_e$ throughout the envelope. We shall therefore write (3) as

$$\kappa = \kappa_0 \rho T^{-3.5} ; \qquad \kappa_0 = \frac{3.9 \times 10^{25}(1 - X_0^2)}{\bar{t}_e} . \tag{4}$$

Dividing (1) by (2) and using (4), we have

$$\frac{dp_g}{dp_r} = \frac{4\pi c G M}{\kappa_0 L} \frac{T^{3.5}}{\rho} - 1 . \tag{5}$$

The equations of equilibrium, then, are

$$\frac{k}{\mu H} \frac{3}{a} \frac{d(\rho T)}{d(T^4)} = \frac{4\pi c G M}{\kappa_0 L} \frac{T^{3.5}}{\rho} - 1 , \tag{6}$$

$$\frac{a}{3} \frac{d(T^4)}{dr} = -\frac{\kappa_0 L}{4\pi c r^2} \frac{\rho^2}{T^{3.5}} , \tag{7}$$

and

$$\frac{dM(r)}{dr} = 4\pi r^2 \rho . \tag{8}$$

The foregoing equations are reduced by the substitutions

$$r = R\xi ; \qquad T = T_0 \tau ; \qquad \rho = \rho_0 \sigma ; \qquad M(r) = DM\psi , \tag{9}$$

to the form

$$K\frac{d(\sigma\tau)}{d(\tau^4)}=\frac{\tau^{3.5}}{\sigma}-1\,, \tag{10}$$

$$\xi^2\frac{d(\tau^4)}{d\xi}=-\frac{\sigma^2}{\tau^{3.5}}\,, \tag{11}$$

and

$$\frac{d\psi}{d\xi}=\sigma\xi^2\,, \tag{12}$$

provided

$$\frac{T_0^{3.5}}{\rho_0}=\frac{\kappa_0 L}{4\pi cGM}\,;\qquad \frac{T_0^{7.5}}{\rho_0^2}=\frac{\kappa_0 L}{4\pi c}\,\frac{3}{a}\,\frac{1}{R}\,, \tag{13}$$

$$K=\frac{k}{\mu H}\,\frac{3}{a}\,\frac{\rho_0}{T_0^3}\,;\qquad D=\frac{4\pi R^3\rho_0}{M}\,. \tag{14}$$

The solutions of equations (13) and (14) are found to be

$$\rho_0=\left(\frac{4\pi cGM}{\kappa_0 L}\right)^8\left(\frac{3}{a}\,\frac{GM}{R}\right)^7\,, \tag{15}$$

$$T_0=\left(\frac{4\pi cGM}{\kappa_0 L}\right)^2\left(\frac{3}{a}\,\frac{GM}{R}\right)^2\,, \tag{16}$$

$$D=\frac{4\pi R^3}{M}\left(\frac{4\pi cGM}{\kappa_0 L}\right)^8\left(\frac{3}{a}\,\frac{GM}{R}\right)^7\,, \tag{17}$$

and

$$K=\frac{k}{\mu H}\,\frac{3}{a}\left(\frac{4\pi cGM}{\kappa_0 L}\right)^2\left(\frac{3}{a}\,\frac{GM}{R}\right)\,. \tag{18}$$

Numerically, the foregoing are equivalent to

$$T_0=6.41\times 10^{16}\,\frac{M^4}{L^2R^2}\,(1-X_0^2)^{-2}\,\bar{t}_e^{\,2}\,, \tag{19}$$

$$\rho_0=2.26\times 10^{37}\,\frac{M^{15}}{L^8R^7}\,(1-X_0^2)^{-8}\,\bar{t}_e^{\,8}\,, \tag{20}$$

and

$$K=2.78\times 10^{9}\,\frac{M^3}{L^2R}\,\mu^{-1}(1-X_0^2)^{-2}\,\bar{t}_e^{\,2}\,. \tag{21}$$

In equations (19), (20), and (21), above, we have expressed M, L, and R in solar units.

2. *The solution of the equations of equilibrium.*—Introduce the variable y, defined by

$$y\tau^3 = K\sigma \,. \tag{22}$$

From the equation defining K (Eq. [14]), it can be verified that y, as defined above, is precisely the ratio of the gas pressure to the radiation pressure, i.e., $\beta : (1 - \beta)$ in our usual notation. In terms of y, equation (10) takes the form

$$\frac{d(y\tau^4)}{d(\tau^4)} = \frac{K\tau^{1/2}}{y} - 1 \,, \tag{23}$$

which is equivalent to

$$\tfrac{1}{4}y\tau\,\frac{dy}{d\tau} = K\tau^{1/2} - y(y + 1) \,. \tag{24}$$

Introducing the new variable x, defined by

$$K\tau^{1/2} = x \,, \tag{25}$$

we have the following differential equation for y:

$$\tfrac{1}{8}xy\,\frac{dy}{dx} = x - y(y + 1) \,. \tag{26}$$

Instead of (26), consider the more general equation

$$\delta xy\,\frac{dy}{dx} = x - y(y + 1) \,, \tag{27}$$

where δ is a small positive constant. To solve (27) we shall adopt a method in principle due to Jeans.

Assume a solution of the form

$$y = y_0 + \delta y_1 + \delta^2 y_2 + \ldots \,. \tag{28}$$

Inserting the foregoing in (27) and equating the coefficients of the powers of δ, we find

$$x = y_0(y_0 + 1) \,; \qquad xy_0\,\frac{dy_0}{dx} = -y_1(2y_0 + 1) \,; \ldots \,, \tag{29}$$

or

$$y_0 = (x + \tfrac{1}{4})^{1/2} - \tfrac{1}{2} \; ; \qquad y_1 = -\frac{y_0 x}{4(x + \frac{1}{4})} \; ; \; \ldots \ldots \tag{30}$$

Hence,

$$y = y_0 \left[1 - \tfrac{1}{4}\delta \frac{x}{x + \frac{1}{4}} + \ldots . \right] . \tag{31}$$

The foregoing series converges quite rapidly for small δ. Thus, for the case where $\delta = 1/8$, we have

$$y = y_0 \left[1 - \frac{1}{32} \frac{x}{x + 0.25} + \ldots . \right] . \tag{32}$$

The second term in the brackets in (32) contributes, at most, about 3 per cent. Also, since x is very large, except in the immediate neighborhood of the boundary of the star, it is clear that for our present purposes it would be sufficient to replace the second term in the brackets in (32) by its limiting constant value $-1/32$, in which case

$$y = \tfrac{31}{32} y_0 \; . \tag{33}$$

Reverting to the original variable τ, we have, according to equation (29),

$$K\tau^{1/2} = x = y_0(y_0 + 1) \; , \tag{34}$$

or

$$\tau = \frac{1}{K^2} y_0^2 (y_0 + 1)^2 \; . \tag{35}$$

By (22), then,

$$\sigma = \frac{1}{K} y\tau^3 \; , \tag{36}$$

or by (33) and (35)

$$\sigma = \frac{31}{32} \frac{1}{K^7} y_0^7 (y_0 + 1)^6 \; . \tag{37}$$

Equations (35) and (37) determine τ and σ in terms of y_0. We shall now determine the variation of y_0 with ξ.

From (10) and (11) we have

$$K \frac{\xi^2}{\sigma} \frac{d(\sigma\tau)}{d\xi} = -1 + \frac{\sigma}{\tau^{3.5}} \; , \tag{38}$$

or, in terms of y_0 (cf. Eqs. [35] and [37]),

$$\frac{1}{K}\xi^2\frac{1}{y_0^7(y_0+1)^6}\frac{d}{d\xi}[y_0^9(y_0+1)^8]=-1+\frac{31}{32}\frac{1}{y_0+1}. \qquad (39)$$

Equation (39) is found to reduce to

$$\frac{1}{K}\xi^2 y_0(y_0+1)(17y_0+9)\frac{dy_0}{d\xi}=-\frac{y_0+\frac{1}{32}}{y_0+1}, \qquad (40)$$

or

$$\frac{1}{y_0+\frac{1}{32}}y_0(y_0+1)^2(17y_0+9)dy_0=-K\frac{d\xi}{\xi^2}. \qquad (41)$$

We shall presently see that, except in the very immediate neighborhood of the boundary, $y_0 \sim 10$ (often it is very much larger), so that we can properly neglect the term $1/32$ in comparison with y_0 in equation (41). It should be remembered in this connection that in the immediate neighborhood of the boundary $y \to y_0$ (according to the solution [32]), so that we make the best of both "worlds" by neglecting $1/32$ in (41).[1] We therefore have as the (y_0, ξ) differential equation

$$(y_0+1)^2(17y_0+9)dy_0=-K\frac{d\xi}{\xi^2}. \qquad (42)$$

Integrating the foregoing equation and using the boundary condition that at $\xi = 1$, $y_0 = 0$, we have

$$(y_0+1)^3(51y_0+19)-19=12K\left(\frac{1}{\xi}-1\right). \qquad (43)$$

Equation (43), combined with (35) and (37), determines the physical structure of the stellar envelope completely.

To obtain the mass in the envelope, we have to integrate (12). We have

$$\psi(1;\xi)=\int_\xi^1\sigma\xi^2 d\xi=\frac{31}{32}\frac{1}{K^7}\int_\xi^1 y_0^7(y_0+1)^6\xi^2 d\xi. \qquad (44)$$

[1] It should be noticed that the solution (32) is a singular solution of the differential equation (27) (cf. chap. ix), to which all its other solutions very rapidly converge, so that, in any case, we should be careful not to take the behavior of the solutions in the immediate neighborhood of the boundary too literally.

Using (42) to change the variable of integration to y_0, we have

$$\psi(1;\xi) = \frac{31}{32}\frac{1}{K^8}\int_0^{y_0} \xi^4 y_0^7 (y_0+1)^8(17y_0+9)dy_0\ ; \tag{45}$$

or, finally, using (43), we have

$$\psi(y_0) = \frac{31}{32}\frac{1}{K^8}\int_0^{y_0} \frac{y_0^7(y_0+1)^8(17y_0+9)dy_0}{\left[\frac{17}{4K}\left\{(y_0+1)^3(y_0+\frac{19}{51}) - \frac{19}{51}\right\} + 1\right]^4}\,. \tag{46}$$

Put

$$w = \alpha y_0 = \tfrac{32}{31}\alpha y\ , \tag{47}$$

where

$$\alpha^4 = \frac{17}{4K}\,. \tag{48}$$

Equation (46) now reduces to

$$\psi(w) = \frac{31}{32}\left(\frac{4}{17}\right)^{3.25}\frac{1}{K^{3.75}}f(\alpha;w)\ , \tag{49}$$

where

$$f(\alpha;w) = 4\int_0^w \frac{w^7(w+\alpha)^8(w+\frac{9}{17}\alpha)dw}{[(w+\alpha)^3(w+\frac{19}{51}\alpha) - \frac{19}{51}\alpha^4 + 1]^4}\,. \tag{50}$$

As stated on page 292, we shall define the extent of the stellar envelope by the fraction $(1-\xi^*)$ which contains the outer 10 per cent of the total mass of the star. By (9) this means that

$$\psi(1;\xi^*) = \frac{1}{10D}\,. \tag{51}$$

Let $w = w^*$ where $\xi = \xi^*$. Then by (49)

$$\frac{1}{10} = \frac{31}{32}\left(\frac{4}{17}\right)^{3.25}\frac{D}{K^{3.75}}f(\alpha;w^*)\ ; \tag{52}$$

or, using the explicit expressions for K and D given in equations (17) and (18), we have

$$\frac{1}{10} = \frac{31}{32}\frac{4\pi R^3}{M}\left(\frac{4}{17}\frac{3}{a}\frac{GM}{R}\right)^{3.25}\left(\frac{4\pi cGM}{\kappa_0 L}\right)^{0.5}\left(\frac{\mu H}{k}\frac{a}{3}\right)^{3.75} f(\alpha;w^*)\,. \tag{53}$$

Inserting the numerical values for the various quantities occurring in (48) and (53) and expressing M, L, and R in solar units (which convention we shall adopt hereafter), we have finally the following equations which determine the central condensation of any star for which L, M, and R are known:

$$\alpha = 6.25 \times 10^{-3} \left[\frac{L^2 R\mu(1 - X_0^2)^2}{l_e^2 M^3}\right]^{1/4}, \tag{54}$$

$$f(\alpha; w^*) = 0.0618 \frac{(1 - X_0^2)^{0.5}}{l_e^{0.5}\mu^{3.75}} \left(\frac{LR^{0.5}}{M^{5.5}}\right)^{1/2}, \tag{55}$$

and

$$\xi^* = \frac{1}{(w^* + \alpha)^3(w^* + \frac{19}{51}\alpha) + 1 - \frac{19}{51}\alpha^4}. \tag{56}$$

Equation (56) is obtained from (43), which, in terms of w $(= \alpha y_0)$, has the following form:

$$(w + \alpha)^3(w + \tfrac{19}{51}\alpha) - \tfrac{19}{51}\alpha^4 = \left(\frac{1}{\xi} - 1\right). \tag{57}$$

3. *Stellar envelopes with negligible radiation pressure.*—From (47) it is clear that α is a measure of the importance of the radiation pressure (for some typical stars [cf. § 5] $\alpha \sim 0.05$). We shall consider now the case of negligible radiation pressure, i.e., the case where α is small.

Let us first consider the case of vanishing radiation pressure. Then y_0 is very large, and α can be neglected in comparison with unity. According to (35) and (37), we have for the case under consideration

$$\tau = \frac{1}{K^2} y_0^4 ; \qquad \sigma = \frac{31}{32} \frac{1}{K^7} y_0^{13} \qquad (y_0 \to \infty), \tag{58}$$

so that

$$\sigma = \tfrac{31}{32} K^{-0.5}\tau^{3.25} \qquad (y_0 \to \infty). \tag{59}$$

Again from (43)

$$51 y_0^4 = 12K\left(\frac{1}{\xi} - 1\right) \qquad (y_0 \to \infty), \tag{60}$$

or from the definition of w (Eqs. [47] and [48])

$$1 + w^4 = \frac{1}{\xi} \qquad (y_0 \to \infty) , \quad (61)$$

a relation which could have been obtained directly from (57) by making α tend to zero.

Using (60) to eliminate y_0 from the relations (58) and (59), we have

$$\tau = \frac{4}{17} \frac{1}{K} \left(\frac{1}{\xi} - 1 \right) \qquad (y_0 \to \infty) , \quad (62)$$

and

$$\sigma = \frac{31}{32} \left(\frac{4}{17} \right)^{3.25} \frac{1}{K^{3.75}} \left(\frac{1}{\xi} - 1 \right)^{3.25} . \quad (63)$$

Using (52), equation (63) can be reduced to the form

$$\sigma = \frac{1}{10Df(0; w^*)} \left(\frac{1}{\xi} - 1 \right)^{3.25} \quad (64)$$

Since

$$\rho = \rho_0 \sigma ; \qquad T = T_0 \tau , \quad (65)$$

we have, according to equations (15), (16), (17), (18), (62), and (64),

$$T = \frac{4}{17} \frac{\mu H}{k} \frac{GM}{R} \left(\frac{1}{\xi} - 1 \right) \quad (66)$$

and

$$\rho = \frac{1}{30f(0; w^*)} \bar{\rho} \left(\frac{1}{\xi} - 1 \right)^{3.25} , \quad (67)$$

where $\bar{\rho}$ is the mean density of the whole star. If we put $\xi = \xi^*$ in the foregoing equation and use for f the value given by (55), we shall obtain the density and the temperature at the "base" of the stellar envelope.

Equations (62) and (63) show that *stellar envelopes with negligible radiation pressure form a homologous family.*

For $\alpha = 0$, the equation determining the extent of the stellar envelope simplifies considerably. From (50) we now have

$$f(0; w) = 4 \int_0^w \frac{w^{16} dw}{(w^4 + 1)^4} . \quad (68)$$

Equation (68) is an elementary integral and can be evaluated. The result is

$$
\left.\begin{aligned}
f(0; w) = \frac{585}{96}\left\{w - \frac{1}{4\sqrt{2}}\left[\log\frac{w^2 + w\sqrt{2} + 1}{w^2 - w\sqrt{2} + 1} + 2\tan^{-1}\frac{w\sqrt{2}}{1 - w^2}\right]\right\} \\
- \frac{w^{13}}{3(w^4 + 1)^3} - \frac{13w^9}{24(w^4 + 1)^2} - \frac{117w^5}{96(w^4 + 1)}\,.
\end{aligned}\right\} \qquad (69)
$$

For the case under consideration, namely, that of vanishing radiation pressure, ξ is related to w according to equation (61). In Table 17 the function $f(0; \xi)$ is tabulated.

If α is small (but not vanishingly small), we can obtain for $f(\alpha; w)$ a three-term Taylor expansion in α:

$$
f(\alpha; w) = f(0; w) + \alpha\left(\frac{\partial f}{\partial \alpha}\right)_{\alpha=0} + \frac{\alpha^2}{2!}\left(\frac{\partial^2 f}{\partial \alpha^2}\right), \qquad (70)
$$

when terms of higher order in α are neglected. From the definition of $f(\alpha; w)$ (Eq. [50]), we verify that

$$
\left(\frac{\partial f}{\partial \alpha}\right)_{\alpha=0} = 4\left(8 + \tfrac{9}{17}\right)\int_0^w \frac{w^{15}dw}{(w^4 + 1)^4} - 16\left(3 + \tfrac{19}{51}\right)\int_0^w \frac{w^{19}dw}{(w^4 + 1)^5} \qquad (71)
$$

and

$$
\left.\begin{aligned}
\left(\frac{\partial^2 f}{\partial \alpha^2}\right)_{\alpha=0} = {} & 80\left(3 + \tfrac{19}{51}\right)^2\int_0^w \frac{w^{22}dw}{(w^4 + 1)^6} \\
& - 32\left[\left(8 + \tfrac{9}{17}\right)\left(3 + \tfrac{19}{51}\right) + 3\left(1 + \tfrac{19}{51}\right)\right]\int_0^w \frac{w^{18}dw}{(w^4 + 1)^5} \\
& + 32\left(8 + \tfrac{1}{17}\right)\int_0^w \frac{w^{14}dw}{(w^4 + 1)^4}\,.
\end{aligned}\right\} \qquad (72)
$$

The integrals occurring in (71) and (72) are all elementary and can be evaluated.

We write (70) in the form:

$$
f(\alpha; w) = f(0; w) + \alpha\Delta_1 f_0 + \alpha^2\Delta_2 f_0\,, \qquad (73)
$$

where the explicit expressions for $\Delta_1 f_0$ and $\Delta_2 f_0$ are easily found from equations (70), (71), and (72).

To determine the relation between ξ and w to the order of accuracy we are working, we write

$$\xi = \xi_0 + a\left(\frac{\partial\xi}{\partial a}\right)_{a=0} + \frac{a^2}{2!}\left(\frac{\partial^2\xi}{\partial a^2}\right)_{a=0}, \tag{74}$$

where (cf. Eq. [57])

$$\xi = \frac{1}{(w+a)^3(w+\frac{19}{51}a) - \frac{19}{51}a^4 + 1} \tag{75}$$

and

$$\xi_0 = \frac{1}{1+w^4}. \tag{76}$$

We re-write (74) in the form

$$\xi = \xi_0 + a\Delta_1\xi_0 + a^2\Delta_2\xi_0 , \tag{77}$$

where it is easily found that

$$\Delta_1\xi_0 = -(3+\tfrac{19}{51})\frac{w^3}{(w^4+1)^2} \tag{78}$$

and

$$\Delta_2\xi_0 = (3+\tfrac{19}{51})^2\frac{w^6}{(w^4+1)^3} - 3(1+\tfrac{19}{51})\frac{w^2}{(w^4+1)^2}. \tag{79}$$

In Table 17 the values of $f(0; w)$, $\Delta_1 f_0$, $\Delta_2 f_0$, $\Delta_1\xi_0$, and $\Delta_2\xi_0$ are tabulated with argument ξ_0. This table, combined with equations (73)

TABLE 17

ξ_0	$f(0; \xi)$	$\Delta_1 f_0$	$\Delta_2 f_0$	$\Delta_1\xi_0$	$\Delta_2\xi_0$
0.90	+0.000014	+0.00020	+0.00114	−0.5257	−0.8047
.85	.000085	+ .00100	+ .00436	.6634	− .7319
.80	.00032	+ .00303	+ .01166	.7631	− .5897
.75	.00087	+ .00711	+ .01875	.8322	− .4138
.70	.00201	+ .01405	+ .02709	.8753	− .2263
.65	.00415	+ .02461	+ .03143	.8957	− .0424
.60	.00789	+ .03925	+ .02586	.8958	+ .1270
.55	.01414	+ .05788	+ .00304	.8776	+ .2738
.50	.02417	+ .07942	− .04523	.8431	+ .3924
.45	.03991	+ .10125	− .12674	.7939	+ .4787
.40	.06421	+ .11815	− .24684	.7314	+ .5304
.35	.10148	+ .12068	− .40521	.6572	+ .5468
.30	.15875	+ .09222	− .59024	.5730	+ .5285
.25	.24771	+ .00342	− .76909	.4805	+ .4777
0.20	+0.39051	−0.19996	−0.86837	−0.3816	+0.3985

and (77), will enable us to determine $f(\alpha; \xi)$ for small values of α. Actual comparisons with the values of $f(\alpha; \xi)$, computed accurately from the integral which defines it, show that the approximate solution obtained by using the table is correct to within 1 per cent for $\alpha \leqslant 0.1$; for $\alpha = 0.15$, a maximum error of about 3 per cent is made.

If α cannot be neglected in comparison with unity, recourse must be had to numerical methods to evaluate the function $f(\alpha; w)$. For practical purposes it is convenient to tabulate $f(\alpha; w)$ for different

TABLE 18*a*

x	$\alpha=0.05$		$\alpha=0.10$		$\alpha=0.15$		$\alpha=0.20$		$\alpha=0.25$	
	ξ	f	ξ	f	ξ	f	ξ	f	ξ	f
0.35	.....	.....	.....	.....	.....	.....	0.935	0.0000002	.....	.....
0.40	.....	.....	0.948	.....	0.930	0.0000006	.908	.0000012	0.882	0.000002
0.45	.....	.....	.925	0.000001	.902	.000003	.875	.000005	.844	.000009
0.50	.....	.....	.896	.000007	.868	.000012	.836	.000021	.801	.000034
0.55	.....	.....	.861	.000025	.828	.000043	.792	.000069	.753	.000105
0.60	.....	.....	.821	.000084	.783	.000134	.744	.000202	.702	.000287
0.65	0.813	0.0002	.775	.000246	.735	.000368	.692	.000520	.649	.000700
0.70	.767	.0005	.726	.000641	.683	.000902	.639	.001206	.596	.00154
0.75	.718	.0011	.674	.001502	.630	.002000	.586	.002544	.543	.00311
0.80	.666	.0025	.621	.003205	.577	.00406	.534	.004936	.492	.00578
0.85	.612	.0050	.568	.006290	.525	.00762	.483	.008894	.444	.01005
0.90	.559	.0095	.516	.01146	.475	.01333	.436	.01502	.399	.01643
0.95	.508	.0169	.467	.01953	.428	.02194	.391	.02396	.357	.02548
1.00	.459	.0281	.420	.03139	.384	.03419	0.350	0.03633	0.319	0.03772
1.05	.413	.0441	.377	.04792	0.344	0.05082	.....	.....	.....	.....
1.10	.370	.0661	.337	.06994	.....	.....	.....	.....	.....	.....
1.15	.331	.0948	.301	.09811	.....	.....	.....	.....	.....	.....
1.20	.296	.1311	0.269	0.1330	.....	.....	.....	.....	.....	.....
1.25	0.264	0.1752	.....	.....	.....	.....	.....	.....	.....	.....

specified values of α. The numerical integration has been effected for the cases $\alpha = 0.05$, 0.10, 0.15, 0.20, 0.25, 0.50, and 1.00; the results are tabulated in Tables 18*a* and 18*b*.

4. *General remarks.*—An important quantity which has been isolated is α; this determines the relative importance of the radiation

TABLE 18*b*

x	$\alpha=0.5$		$\alpha=1.0$	
	ξ	f	ξ	f
0	....	....	1.0000	0............
0.05	....	....	0.896	....
0.10	....	....	0.796	0.000000002
0.15	0.935	....	0.703	.00000007
0.20	.902	0.00000002	0.619	.0000006
0.25	.862	.00000017	0.542	.0000033
0.30	.816	.0000011	0.475	.0000126
0.35	.766	.0000052	0.416	.0000375
0.40	.712	.0000195	0.364	.0000936
0.45	.657	.0000617	0.319	.0002044
0.50	.601	.0001675	0.280	.0004017
0.55	.547	.0004011	0.246	.0007253
0.60	.494	.000863	0.217	.001222
0.65	.445	.001694	0.192	.001945
0.70	.399	.003075	0.170	.002952
0.75	.356	.005222	0.151	.004302
0.80	.318	.00837	0.134	.006057
0.85	.284	.01276	0.120	.008278
0.90	.253	.01865	0.107	.011026
0.95	.225	.02624	0.096	.014359
1.00	0.201	0.03576	0.086	0.01834

pressure in the stellar envelope. Let ξ_1 be a point where $w = 1$, or, according to (57),

$$\xi_1 = \frac{1}{(1+\alpha)^3(1+\frac{19}{51}\alpha) + 1 - \frac{19}{51}\alpha^4} . \qquad (80)^2$$

If α is small, we can write, according to equation (77) and Table 17,

$$\xi_1 = 0.5 - 0.8431\alpha + 0.3924\alpha^2 . \qquad (81)$$

By (47) at $\xi = \xi_1$; we have

$$y(\xi_1) = \frac{31}{32} y_0(\xi_1) = \frac{31}{32\alpha} , \qquad (82)$$

or, according to (54),

$$\left(\frac{1-\beta}{\beta}\right)_{\xi=\xi_1} = \frac{32}{31}\alpha = 6.45 \times 10^{-3} \left[\frac{L^2 R\mu(1-X_0^2)^2}{M^3 \bar{l}_e^2}\right]^{1/4} . \qquad (83)$$

[2] This is a purely formal definition. It can happen that $\xi_1 < \xi^*$.

Hence, the particular combination of L, M, and R which occurs on the right-hand side of the foregoing equation determines whether, for a particular star, the radiation pressure is important or not. It is important to notice that a knowledge of all three parameters L, M, and R is required to determine the relative importance of the radiation pressure. It is therefore satisfactory that for normal stars—i.e., ordinary giants and dwarfs—a correspondence is found to exist between $(1 - \beta)$ at ξ_1 (determined according to [83]) and $(1 - \beta_c)$ (determined according to Eddington's quartic equation). Thus, on the assumption that $\mu = 1$, we find that for the sun and Capella A the quantities $(1 - \beta)$ at ξ_1 are 0.004 and 0.041, respectively, while the quartic equation yields for $(1 - \beta_c)$ the values 0.003 and 0.046. The fact that the observed sets of values for L, M, and R for the normal stars predict values for $(1 - \beta)$ at $\xi = \xi_1$ in such close correspondence with the values of $(1 - \beta_c)$ according to the quartic equation is a confirmation of the adequacy of the standard model (in its first approximation cf. § 7, chap. vi) for these stars. On the other hand, we shall see that this correspondence fails when the very massive Trumpler stars are considered. For these stars we should normally expect $(1 - \beta)$ to be quite near unity, while observationally the radiation pressure is, in fact, quite negligible in the envelopes of these stars; we have here, therefore, a breakdown of the theory which has been found to be applicable to stars of ordinary mass. We shall return to these questions in § 6 (cf. Table 21).

Let us now consider stellar envelopes with negligible radiation pressure, i.e., stars for which $\alpha \sim 0$. The equation determining the central condensation of the star can be written as (Eq. [55])

$$f(0; \xi^*) = 0.0618 \frac{(1 - X_0^2)^{0.5}}{\bar{t}_e^{0.5}\mu^{3.75}} \left(\frac{LR^{0.5}}{M^{5.5}}\right)^{1/2} . \tag{84}$$

The occurrence of $(\kappa_0 LR^{0.5}/M^{5.5})$ on the right-hand side of (84) is easily understood. For, according to the discussion in chapter vi, § 6, stars with negligible radiation pressure form a homologous family, and, further,

$$\frac{\kappa_0 LR^{0.5}}{\mu^{7.5}M^{5.5}} \tag{85}$$

is a homology invariant. Also, we have already shown in § 3 that stellar envelopes with negligible radiation pressure form a homologous family. Hence, ξ^* is also a homology invariant; and, as it depends on L, M, and R, we should have

$$\xi^* = \text{function}\left[\left(\frac{\kappa_0 L R^{0.5}}{\mu^{7.5} M^{5.5}}\right)\right] . \tag{86}$$

Equation (84) is simply the explicit expression of this form of dependence.

Another feature of (84) which should be noticed is its remarkable similarity to the luminosity formula (Eq. [14], vii) for the case $\beta \sim 1$. By equations (14) and (107) of the preceding chapter we have

$$L = 0.184 \frac{\bar{t}}{(1 - X_0^2)} \frac{M^{5.5}}{R^{0.5}} \mu^{7.5} . \tag{87}$$

It will be remembered that $\bar{t}$, which occurs in (87), is a certain harmonic mean value of t taken through the star (cf. Eq. [8], vii); it is accordingly different from t_e, which occurs in (84). Equation (87) can be re-written in the form

$$\left(\frac{LR^{0.5}}{M^{5.5}}\right)^{1/2} \frac{(1 - X_0^2)^{1/2}}{\bar{t}^{0.5} \mu^{3.75}} = 0.429 . \tag{88}$$

Comparing (84) and (88), we have

$$f(0; \xi^*) = 0.0618 \times 0.429 = 0.0265 ; \tag{89}$$

or, interpolating among the values of $f(0; \xi)$ in Table 17, we find that

$$\xi^* = 0.496 . \tag{90}$$

Now the model specifically underlying the luminosity formula (87) is the standard model with a density distribution corresponding to the polytrope $n = 3$. An examination of the Lane-Emden function θ_3 shows that the polytrope $n = 3$ has a central condensation of approximately 0.504. The agreement of $\xi^* = 0.496$ with the "theoretical" central condensation of 0.504 proves the consistency of the model for the stars of negligible radiation pressure; the consistency here proved may be compared with the discussion of the assumption "$\kappa\eta$ = constant" in chapter vi, § 7.

5. *Central condensations of some typical normal stars: dependence on chemical composition.*—We shall now proceed to apply the theory we have developed to derive the central condensations of some typical stars. The data on the masses, luminosities, and radii of the stars has been supplied to the writer by Dr. Kuiper, who has undertaken a critical re-examination and rediscussion of the relevant observational material. It is beyond the scope of the present monograph to include Kuiper's discussion; such discussions should, however, be regarded as an integral part of the study of stellar structure. For the derivation of the data used here, reference is made to Kuiper's investigation in the *Astrophysical Journal*, **88,** 472, 1938. (It should be pointed out that the absolute bolometric magnitude of the sun which Kuiper adopts is +4.63.)

a) *Capella A.*—To illustrate the method of calculating the central condensations of stars we shall first consider the case of Capella A. This star presents an exceptionally "pure" case, in so far as a preliminary examination shows that the guillotine factor $\bar{t}_e$ equals unity.[3] (Cf. Table 14*a*, chap. vii).

For the case of Capella A we have

$$\text{Log}\, L = 2.08\;; \qquad \text{Log}\, M = 0.62\;; \qquad \text{Log}\, R = 1.20\,. \tag{91}$$

Substituting the foregoing values in equations (54) and (55) and putting $t_e = 1$, we find that

$$a = 0.0469[\mu(1 - X_0^2)^2]^{1/4} \tag{92}$$

and

$$f = 0.0268(1 - X_0^2)^{0.5}\mu^{-3.75}\,. \tag{93}$$

To evaluate ξ^*, we shall have to make some assumption concerning μ and X_0. To examine first the nature of the dependence of ξ^* on μ and X_0, it is sufficient to use the "first approximation" considered

[3] This is, in fact, a general characteristic of the normal giants, subgiants, and M supergiants.

in § 3 of chapter vii, according to which μ and X_0 are related by (Eq. [26], vii)

$$\mu = \frac{2}{1 + 3X_0} \cdot \tag{94}$$

Equations (92), (93), and (94) and the tables of the function $f(\alpha; \xi)$ are sufficient to determine ξ^* as a function of X_0. Table 19 shows the result of the computations.

TABLE 19

THE CENTRAL CONDENSATION OF CAPELLA A

X_0	α	f	ξ^*
0	0.056	0.00199	0.678
.2	.049	.0114	.542
.4	.042	.0351	.441
.6	.035	.0757	.363
.8	.025	.118	.321
.9	.018	.117	.325
0.95	0.012	0.098	0.347

In Figure 29 the corresponding (ξ^*, X_0) curve is drawn. The following two important features of the (ξ^*, X_0) curve should be noted; they are, as we shall see, quite general for normal giants and dwarfs: (*a*) The quantity ξ^* as a function of X_0 has a minimum; (*b*) ξ^* (X_0) intersects the line $\xi^* = 0.5$ at two points, one of which corresponds to the extreme abundance of hydrogen.

An immediate consequence, then, of the theory of stellar envelopes is the prediction for normal stars of a minimum possible value for ξ^*. Thus, Capella A cannot be centrally condensed to a degree greater than that corresponding to $\xi^* = 0.32$. The existence of the minimum is easily understood:

If the radiation pressure is negligible, the minimum value of ξ^* corresponds to the maximum value of f, or, according to (55) and (94), to the maximum of

$$(1 - X_0^2)^{0.5}(1 + 3X_0)^{3.75} . \tag{95}$$

It is easily found that the maximum of (95) is attained for

$$X_0 = \frac{\sqrt{2569} - 2}{57} = 0.84 ; \qquad \mu = 0.561 . \tag{96}$$

For Capella A, α is certainly not "vanishing"; yet the maximum is not appreciably shifted from the value given by (96). However, if α is sufficiently large, it can happen that the increase of α with μ is sufficient to compensate for the decrease in f so that the (ξ^*, X_0) curve shows only a very shallow minimum, or even no minimum at all (cf. the case of HD 1337, the infrared component of ϵ Aurigae, and the M-component of VV Cephei considered in § 7).

Finally, if the values of X_0 and μ derived by Strömgren ($\mu = 1.04$, $X_0 = 0.29$) are adopted, it is found that $\xi^* = 0.486$; this confirms the model underlying Strömgren's theory. We now see that the two intersections of the (ξ^*, X_0) curve with the line $\xi^* \sim 0.5$ precisely correspond to the two solutions for the hydrogen content discussed in § 5 of chapter vii.

b) *The sun.*—According to (54) and (55), we have

$$\alpha = 0.00625 \left[\frac{\mu(1 - X_0^2)^2}{\bar{t}_e^2} \right]^{1/4} \tag{97}$$

and

$$f = 0.0618 \frac{(1 - X_0^2)^{0.5}}{\mu^{3.75} \bar{t}_e^{0.5}} . \tag{98}$$

We see that in this case α is quite negligible for the possible range of μ and X_0. To calculate ξ^* we shall adopt Strömgren's values of μ and X_0, namely, $\mu = 0.98$ and $X_0 = 0.37$ (cf. Table 14*b*, chap. vii). In the case of the sun, the guillotine factor is *not* entirely negligible, and to estimate $\bar{t}_e$ we proceed as follows: On the standard model at the base of the envelope $T^* = 0.3T_c$, and for the sun we find $T^* = 6 \times 10^6$ degrees. From Table 13 we see that here $t = 2$. Since this represents the maximum value of t, we may choose the mean value of t_e to be about 1.5. The maximum value of $\bar{t}_e$ can be taken to be 2. Using Strömgren's value of μ and X_0, we find that for

$$\left. \begin{array}{l} \bar{t}_e = 1 , \quad 1.5 , \quad 2.0 , \\ \xi^* = 0.40 , \ 0.42 , \ 0.44 . \end{array} \right\} \tag{99}$$

We see that the uncertainty in the guillotine factor l_e does not introduce any substantial uncertainty in the derived values of ξ^*; this arises from the circumstance that l_e occurs in the square root. Taking the case $l_e = 1.5$ as typical, we have, according to equations (66) and (67), the following values for the density and temperature at the base of the envelope:

$$T^* = 7.3 \times 10^6 \text{ degrees} ; \qquad \rho = 1.8\bar{\rho}_\odot = 2.54 \text{ grams cm}^{-3} . \qquad (100)$$

c) *ζ Herculis A*.—As we have seen in § 6 of chapter vii, ζ Herculis is considered in Strömgren's theory to be poorer in hydrogen than the sun. We shall see that we can confirm this conclusion independently. According to Kuiper's discussion of the star,

$$\text{Log } M = -0.02 ; \qquad \text{Log } L = 0.596 ; \qquad \text{Log } R = 0.28 . \qquad (101)$$

Since the radius is about twice that of the sun, the guillotine factor can be put equal to unity. Using the foregoing values for L, M, and R, we find that

$$a = 0.0152[\mu(1 - X_0^2)^2]^{1/4} \qquad (102)$$

and

$$f = 0.164(1 - X_0^2)^{0.5}\mu^{-3.75} . \qquad (103)$$

Comparing (102) and (103) with (97) and (98), we infer (*a*) that for equal values of μ and X_0 the envelope of ζ Herculis has a larger radiation pressure than has the solar envelope, although the two stars have about the same mass; this arises from the circumstance that ζ Herculis has a much larger radius; (*b*) if the sun and ζ Herculis were characterized by the same values of μ and X_0, then ζ Herculis would have a value for ξ^* much less than that for the sun. But we have already seen that stars with negligible radiation pressure should be homologous. Hence, ζ Herculis must be characterized by a larger value of μ, in order that its ξ^* may be (approximately) equal to that for the sun; this confirms the conclusion based on Strömgren's theory. Actually, Strömgren finds for ζ Herculis, $\mu = 1.45$ and $X_0 = 0.11$. Using these values in (102) and (103), we find that $\xi^* = 0.44$, which makes it, in fact, approximately homologous with the sun.[4]

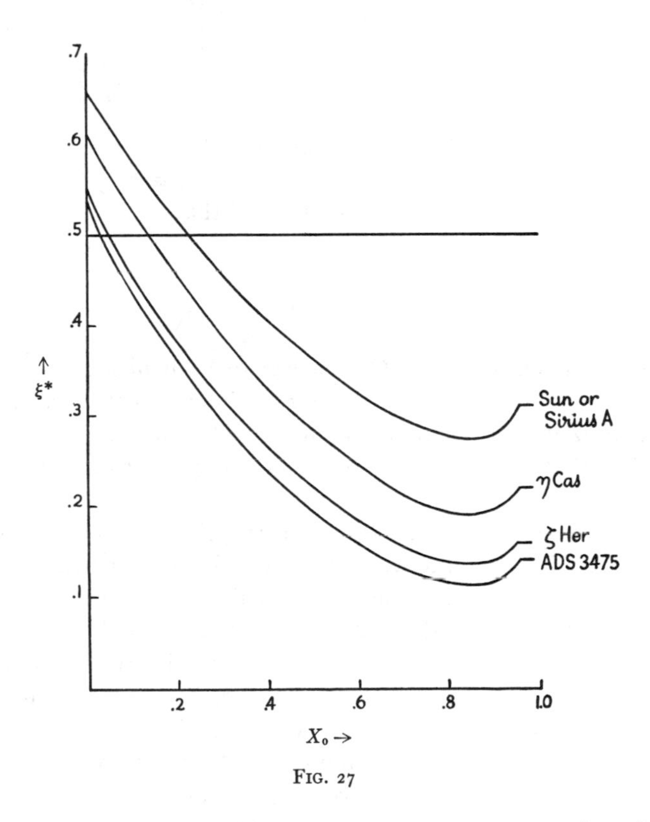

FIG. 27

d) *η Cassiopeiae A*.—This star presents a case somewhat similar to ζ Herculis A. For η Cassiopeiae, we have, according to Kuiper,

$$\text{Log } M = -0.14 \; ; \qquad \text{Log } L = -0.08 \; ; \qquad \text{Log } R = -0.08 \; .$$

[4] It should perhaps be mentioned in this connection that at this stage in the development of the theory it would be unwise to stress the slight "discrepancies"—the difference, for instance, between $\xi^* = 0.44$ and $\xi^* = 0.42$, just noticed. It would first be necessary to examine carefully the state of ionization in the stellar envelope. It is clear, however, that with further refinements the theory of stellar envelopes is capable of including finer details than are considered in this chapter.

Using the foregoing values, we find that

$$a = 0.007\left[\frac{\mu(1 - X_0^2)^2}{\bar{t}_e^2}\right]^{1/4}$$

and

$$f = 0.131\,\frac{(1 - X_0^2)^{1/2}}{\mu^{3.75}\bar{t}_e^{0.5}}\,.$$

With $\mu = 1.25$ and $X_0 = 0.20$ it is found that

$$\bar{t}_e = 1\,; \qquad \bar{t}_e = 2\,,$$
$$\xi^* = 0.42\,; \qquad \xi^* = 0.45\,.$$

Since we should expect η Cassiopeiae to be homologous with the other stars we have considered so far, we infer that this star must be characterized by a value of the mean molecular weight somewhat less than that of ζ Herculis A but definitely greater than that of the sun.

e) A star in the Hyades cluster (ADS 3475).—In chapter vii we referred to Kuiper's discovery of the relatively poor hydrogen content of the stars in the Hyades. We shall consider one of the stars for which Kuiper has derived values for L, M, and R. ADS 3475 presents a case very similar to that of ζ Herculis. With Strömgren's values for μ and X_0 for this star ($\mu = 1.42$, $X_0 = 0.15$) we find that $\xi^* = 0.42$ or 0.44, according as $\bar{t}_e$ is taken to be 1 or 1.5.

f) Sirius A.—As a final example of a "typical" normal star we shall consider Sirius A. We have for this star

$$\text{Log } L = 1.59\,; \qquad \text{Log } M = 0.37\,; \qquad \text{Log } R = 0.25\,. \tag{104}$$

With Strömgren's values of μ and X_0 ($\mu = 0.95$, $X_0 = 0.36$) we find that

$$\left.\begin{array}{lll} \bar{t}_e = 1\,; & \bar{t}_e = 1.5\,; & \bar{t}_e = 2\,, \\ \xi^* = 0.42\,; & \xi^* = 0.44\,; & \xi^* = 0.46\,. \end{array}\right\} \tag{105}$$

The (ξ^*, X_0) curves are shown in Figure 27 for the stars considered in *b* to *f* above.

6. *The structure of the Trumpler stars.*—So far we have considered only normal stars, and the theory of stellar envelopes essentially confirms the theory described in chapters vi and vii. As an extreme on the other side, we shall consider the very massive Trumpler stars, for which the usual theory seems to break down completely. Table 20 contains the data for the Trumpler stars as revised by Kuiper on the basis of his temperature scale, which should be more reliable than that originally adopted by Trumpler. These stars occur in the region of the M–R diagram (see Fig. 2), where one would expect

TABLE 20*

THE TRUMPLER STARS

Star	Log M	Log L	Log R
T_1	1.74	5.88	0.64
T_2	1.99	4.72	0.86
T_3	2.14	5.60	0.78
T_4	2.45	5.76	1.26
T_5	2.35	5.36	1.18
T_6	2.60	5.04	1.22
T_7	1.89	4.92	0.96

* The stars are numbered $T_1, \ldots, T_7$ in the order in which they are contained in Trumpler's Table III (*Pub. A.S.P.*, **47**, 254, 1935).

the hydrogen content to be fairly high, from an extrapolation of the Strömgren curves of constant X_0 (established in the region of the normal stars). The calculation for the central condensations of these stars has been carried out for two values of X_0 ($X_0 = 0.95$ and 0.60). The results are summarized in Table 21. It is at once clear that the theory applicable to the normal stars breaks down for these objects.

In the calculations, electron scattering has been neglected; but it is clear, from the empirical rule stated on page 272, that we can take this into account by allowing the guillotine factor to be less than unity. This would cause the ξ^*'s (for $X_0 = 0.95$) given in Table 21 to lie between the tabulated values of ξ^* for $X_0 = 0.95$ and $X_0 = 0.60$; for the case $X_0 = 0.6$, electron scattering is seen to be negligible.

What is, perhaps, most striking is the systematic increase of ξ^* with increasing mass. The conclusion, then, is that the Trumpler stars are more or less homogeneous gaseous configurations. This conclusion, it should be pointed out, is an almost immediate inference from the observations. We encounter the "breakdown" nature of the Trumpler stars also when we attempt to calculate their hydrogen content by the Strömgren method. As Beer and Chandrasekhar showed, the problem has no solution.

7. *Further applications.*—In the last two sections we considered, on the one hand, the normal giants, subgiants, and dwarfs (for which the standard model was seen to be a sufficiently good approximation), and, on the other, the Trumpler stars, for which the model certainly breaks down. We shall now consider some intermediate cases.

TABLE 21

CENTRAL CONDENSATIONS FOR THE TRUMPLER STARS

No.	Mass/$\odot$	$X_0=0.95$			$X_0=0.60$		
		α	f	ξ^*	α	f	ξ^*
T_6..........	400	0.01	0.000011	0.90	0.04	0.000008	0.90
T_4..........	280	.04	.000066	.85	.10	.000051	.84
T_5..........	220	.03	.000073	.85	.07	.000057	.84
T_3..........	140	.04	.00029	.79	.11	.00023	.78
T_2..........	100	.02	.00028	.80	.06	.00022	.79
T_7..........	80	.03	.00071	.75	.10	.00055	.73
T_1..........	55	0.10	0.0047	0.59	0.29	0.0036	0.51

a) *VV Cephei: the B-component.*—The system of VV Cephei is a spectroscopic binary which Gaposchkin discovered to be an eclipsing system with a period of about twenty years. The brighter component is an M supergiant, while the fainter is a B star. The observational data for this system are of a provisional character, the chief uncertainty being in the mass ratio. A value of about 1.6 appears to be the best estimate. It may, however, be as high as 2.2. We shall first consider the B star and return to the M star later.

For the B star we have, according to Kuiper,

$$M = 31; \qquad R = 28; \qquad \text{Log } L = 4.22 . \tag{106}$$

The foregoing values correspond to an assumed value of 1.6 for the mass ratio. If the higher value of 2.2 is adopted, we have

$$M = 46; \qquad R = 35; \qquad \text{Log } L = 4.40 . \tag{107}$$

The guillotine factor is found to be unity, and we find that for

$$\left.\begin{aligned} X &= 0. \quad 0.2,\ 0.4,\ 0.6,\ 0.8,\ 0.9,\ 0.95, \\ \xi^* &= \begin{cases} 0.78, 0.71, 0.64, 0.60, 0.58, 0.59, 0.61, \\ 0.82, 0.76, 0.70, 0.67, 0.65, 0.66, 0.68. \end{cases} \end{aligned}\right\} \tag{108}$$

The first set of values for ξ^* are derived on the basis of (106), while the second set corresponds to (107). The B star is thus seen to be similar to the Trumpler stars. This is what we should have expected from the mass and the radius of this star as compared to the Trumpler stars.

b) V Puppis.—The two components of this system are nearly identical, so that for the present purpose it is sufficient to consider the average of the two components. We have $M = 18.6$, $R = 6.8$, and Log $L = 3.87$. The (ξ^*, X_0) curve is shown in Figure 28. We infer from this curve that V Puppis is probably somewhat more homogeneous than the less massive stars.

c) μ_1 Scorpii.—This star has been recently investigated by Elvey and Rudnick; a rereduction of their data by Kuiper leads to $M = 12.0$, $R = 5.50$, and Log $L = 3.37$. The (ξ^*, X_0) curve is shown in Figure 28.

d) The B-component of ζ Aurigae.—The system ζ Aurigae is of the first class. The observations by Guthnick and his collaborators and by Christie and Wilson have been rereduced by Kuiper. We shall first consider the B-component. For this star we have, according to Kuiper,

$$M = 8.1 ; \qquad R = 5.1 ; \qquad \text{Log } L = 3.01 . \tag{109}$$

This star has a mass less than that of the other massive stars we have considered so far. The (ξ^*, X_0) curve is shown in Figure 28.

It will be seen that the Trumpler stars, VV Cephei (B star), V Puppis, μ_1 Scorpii, ζ Aurigae (B star), Sirius A, and the sun represent a sequence along which the (ξ^*, X_0) curves change continuously; this strongly suggests that the breakdown of the standard model for

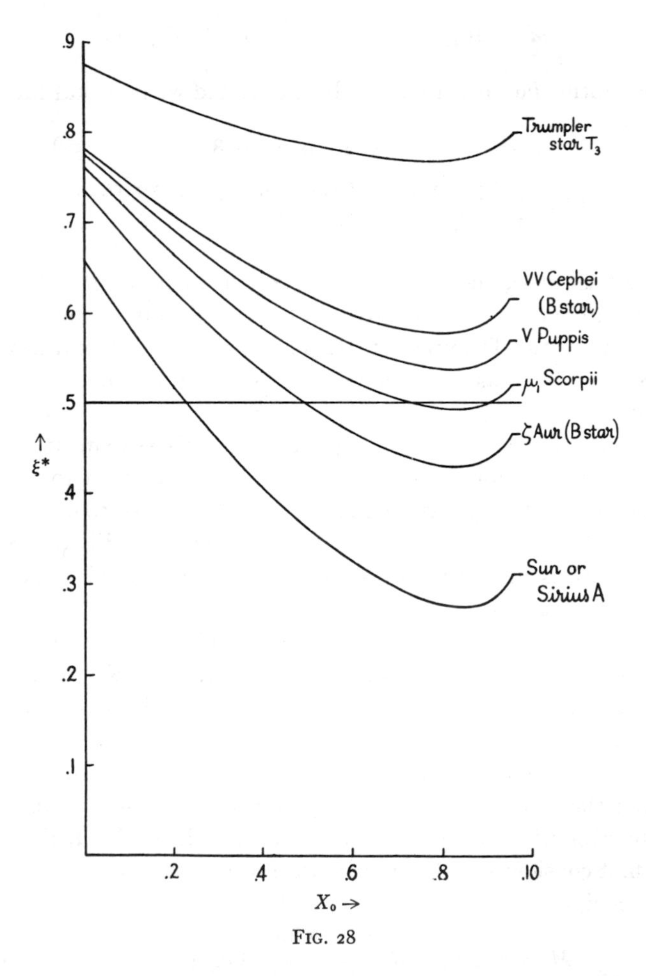

FIG. 28

stars on the main series sets in at about $M = 10\odot$, becoming more and more pronounced on passing toward the larger masses.

e) The K-component of ζ Aurigae.—For this star we have

$$M = 14.8; \qquad R = 200; \qquad \text{Log } L = 3.80 . \tag{110}$$

The (ξ^*, X_0) curve is shown in Figure 29. We see that, though this star has a mass about equal to that of μ_1 Scorpii, it seems to have a normal density distribution. Indeed, with Strömgren's value of μ and X_0 ($\mu = 1.07$, $X_0 = 0.34$) we find that

$$\alpha = 0.24 \; ; \qquad f = 0.0082 \; ; \qquad \xi^* = 0.466 \; ; \tag{111}$$

this confirms the standard model for the interior of this star.[5] The K star here considered is, in fact, just as "pure" a case as Capella A: it is a first-class determination, and the guillotine factor is unity. We have here a suggestion that the massive stars in the supergiant region of the Hertzsprung-Russell diagram are probably different from the equally massive stars forming the extension of the main series. Though the two cases considered below are somewhat uncertain from the observational side, they seem to lend support to the suggestion.

f) The infrared component of ϵ Aurigae.—The data for ϵ Aurigae have been derived partly by an indirect method and are less reliable than for most of the other stars we have considered. The mass and the radius seem to be fairly well determined according to the investigation of Kuiper, Struve, and Strömgren.[6] The luminosity is only approximately known through the recent measures by Hall, which have been discussed by Kuiper.[7] We have

$$M = 24.6 \; ; \qquad R = 2140 \; ; \qquad \text{Log } L = 4.46 \; . \tag{112}$$

We find that for

$$\left.\begin{array}{l} X_0 = \quad 0,\ 0.2,\ 0.4,\ 0.6,\ 0.8,\ 0.9, \\ \xi^* = 0.34,\ 0.28,\ 0.23,\ 0.22,\ 0.24,\ 0.29, \end{array}\right\} \tag{113}$$

respectively. It is thus seen that the (ξ^*, X_0) curve for this star (cf. Fig. 29) shows a very shallow minimum; further, for a wide range in X_0, ξ^* does not differ appreciably from the value 0.23. We can therefore conclude that the infrared component of ϵ Aurigae is probably

[5] Since $\alpha = 0.24$, the radiation pressure is quite appreciable; as such, the envelope of ζ Aurigae (K star) is not strictly homologous with that of a star having negligible radiation pressure.

[6] *Ap. J.*, **86**, 570, 1937.

[7] *Ibid.*, **87**, 209, 1938.

much more centrally condensed than the normal stars. This result is easily understood. According to our definition of α (Eq. [54]), a large R (and/or L) implies that the radiation pressure is important; this has the effect of forming an extended stellar envelope for the star.

A case similar to ϵ Aurigae is presented by the M-component of VV Cephei.

g) The M-component of VV Cephei.—As already indicated in (*a*), above, the data for this system are of a provisional character. For the brighter component of the system (which is an M supergiant) we have

$$M = 49 \; ; \qquad R = 2130 \; ; \qquad \text{Log}\, L = 5.62 \,, \tag{114}$$

or

$$M = 102 \; ; \qquad R = 2630 \; ; \qquad \text{Log}\, L = 5.80 \,, \tag{115}$$

according as the adopted mass ratio is 1.6 or 2.2. Computing ξ^* for different values of X_0, we find that for

$$\left.\begin{array}{l} X_0 = 0.4, \quad 0.6, \quad 0.8, \quad 0.9, \quad 0.95, \\ \xi^* = \begin{cases} 0.065, 0.072, 0.12, 0.20, 0.29, \\ 0.21, \quad 0.26, \quad 0.32, 0.41, 0.50. \end{cases} \end{array}\right\} \tag{116}$$

The first set of values for ξ^* are derived on the basis of (114), while the second set correspond to (115).

We thus see that in spite of the uncertainty in the observational material we can conclude that the M-component of VV Cephei must be highly centrally condensed. Indeed, the possibility that 90 per cent of its mass is concentrated within 5 per cent of its radius cannot be overlooked. The system of VV Cephei is therefore of quite unusual interest from the theoretical viewpoint, inasmuch as we have indications that the standard model breaks down in the opposite directions for the two components.

Considering, now, the stars in the sequence the sun, ζ Herculis, Capella, ζ Aurigae (K star), ϵ Aurigae (I star), and VV Cephei (M star), we infer the possibility of a breakdown of the standard model also in the region of the massive supergiants (stars of high luminosity and large radius). The breakdown is now, however, in the sense of becoming more centrally condensed; this differs from the case of the massive stars which form an extension of the main series; the latter are certainly more homogeneous than the normal stars.

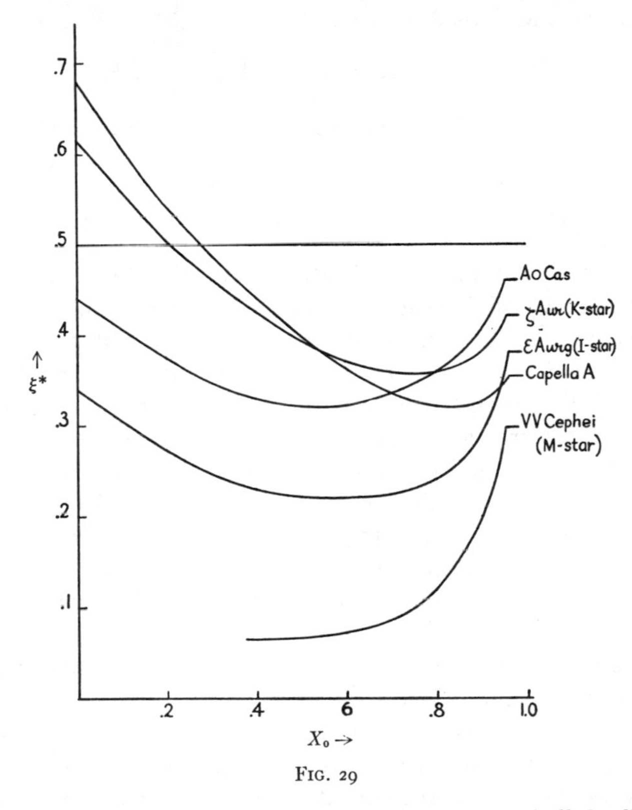

FIG. 29

h) AO Cassiopeiae and 29 Canis Majoris.—We shall finally consider "overluminous" stars, of which AO Cassiopeiae is an example. In this case we have

$$M = 40 \; ; \qquad R = 19 \; ; \qquad \text{Log } L = 5.77 \, . \tag{117}$$

This star is therefore almost as luminous as the most massive of the Trumpler stars. Because of this high luminosity with respect to its mass, we should expect AO Cassiopeiae to be more centrally condensed than the massive stars in the sequence μ_1 Scorpii, V Puppis,

VV Cephei (B star), and the Trumpler stars. The (ξ^*, X_0) curve for this star (see Fig. 29) confirms this expectation.

A star quite similar to AO Cassiopeiae is 29 Canis Majoris A. For 29 Canis Majoris A we have

$$M = 46\ ; \qquad R = 20\ ; \qquad \text{Log}\ L = 5.84\ . \tag{118}$$

Computing ξ^* for different values of X_0 we find that for

$$\left.\begin{array}{l} X_0 = 0, \quad 0.2,\ 0.4,\ 0.6,\ 0.8,\ 0.9,\ 0.95, \\ \xi^* = 0.44, 0.38, 0.34, 0.33, 0.38, 0.43, 0.49. \end{array}\right\} \tag{119}$$

The (ξ^*, X_0) curve for 29 Canis Majoris A is therefore quite similar to that for AO Cassiopeiae. Both of these stars are probably somewhat more centrally condensed than even the normal stars.

i) *Y Cygni*.—A case intermediate to those considered in (*h*) above is provided by the system, Y Cygni. The two components of this system are nearly identical, and, taking their average, we have $M = 17$, $R = 5.9$, and Log $L = 4.51$. Thus, Y Cygni, though it has a mass less than V Puppis, is yet very much more luminous. We can therefore expect Y Cygni to be more centrally condensed than V Puppis. The calculation of ξ^* for different values of X_0 confirms this conclusion. Y Cygni has probably a "normal" density distribution.

8. *Concluding remarks*.—The main results of the foregoing discussion can be summarized as follows:

a) The general way in which the theory of stellar envelopes supports the essential conclusions reached in chapter vii concerning the structures and the hydrogen contents of the normal stars like the sun, Sirius, ζ Herculis, and Capella;

b) The increasing homogeneity of the massive stars on the main series, the breakdown of the standard model setting in probably at values of the mass of about 10 $\odot$;

c) The centrally condensed nature of the massive supergiants

We may also infer from the examples discussed that a certain systematic variation of the stellar model in the (M, R) plane exists. Only more extended observations will show whether this is a legitimate generalization.

BIBLIOGRAPHICAL NOTES

Stellar envelopes with negligible radiation pressure have been considered by—

1. A. S. EDDINGTON, *M.N.*, **91**, 109, 1930.

2. B. STRÖMGREN, *Zs. f. Ap.*, **2**, 345, 1931.

A treatment which takes into account the radiation pressure exactly was first given by—

3. S. CHANDRASEKHAR, *M.N.*, **96**, 647, 1936.

This chapter represents a hitherto unpublished investigation by the author, except in so far as it overlaps his earlier paper (ref. 3).

§§ 1 and 2.—Reference 3. The analysis is carried through somewhat more rigorously than was necessary for the purposes of reference 3. The treatment of the differential equation (27) is similar to a method used by Jeans.

4. J. H. JEANS, *M.N.*, **85**, 201, 1925. Also, *Astronomy and Cosmogony*, §§ 78–86, Cambridge, England. 1929.

The discussion contained in the rest of the chapter is published here for the first time. The writer is indebted to Dr. Kuiper for providing the observational material. Some further references may be noted:

5. G. P. KUIPER, *Ap. J.*, **86**, 166, 1937, where L, M, and R values for certain members of the Hyades are given.

6. R. J. TRUMPLER, *Pub. A.S.P.*, **47**, 254, 1935, where Trumpler's massive stars are described.

7. H. VOGT, *A.N.*, **261**, 73, 1936.

8. E. A. MILNE, *M.N.*, **90**, 17, 1929.

9. A. BEER and S. CHANDRASEKHAR, *Observatory*, **59**, 168, 1936.

10. G. P. KUIPER, O. STRUVE, and B. STRÖMGREN, *Ap. J.*, **86**, 570, 1937. This paper contains the interpretation of ϵ Aurigae.

CHAPTER IX

STELLAR MODELS

In this chapter we shall consider some classes of stellar models. The interest in some of them may be of a rather formal character, but for the future development of the theory of gaseous stars the properties of the models considered may afford some guidance. A variety of stellar models has been investigated by several writers, of which only a limited number will be included in this chapter.

1. *The model ϵ = constant.*—This model corresponds to a uniform distribution of the sources of energy; as such, it represents one limiting case of the possible stellar models, the other limiting case being the "point-source model" (cf. §§ 3 and 4).

For the model ϵ = constant we have

$$\epsilon = \frac{L(r)}{M(r)} = \frac{L}{M} \, . \tag{1}$$

We shall assume for the law of opacity

$$\kappa = \kappa_0 \rho T^{-3.5} \, . \tag{2}$$

The equations of equilibrium are, as usual,

$$\frac{d}{dr}\left(\frac{k}{\mu H}\,\rho T + \tfrac{1}{3} a T^4\right) = -\frac{GM(r)}{r^2}\,\rho \, , \tag{3}$$

$$\frac{d}{dr}\,(\tfrac{1}{3} a T^4) = -\frac{\kappa_0 L(r)}{4\pi c r^2}\,\rho^2 T^{-3.5} \, , \tag{4}$$

and

$$\frac{d}{dr}\,M(r) = 4\pi r^2 \rho \, . \tag{5}$$

From (1), (3), and (4) we have an alternative form for one of the equations of equilibrium:

$$\frac{k}{\mu H}\,\frac{3}{a}\,\frac{d(\rho T)}{d(T^4)} = \frac{4\pi c G M}{\kappa_0 L}\,\frac{T^{3.5}}{\rho} - 1 \, . \tag{6}$$

Let

$$y = \frac{\beta}{1 - \beta} = \frac{k}{\mu H} \frac{3}{a} \frac{\rho}{T^3}. \tag{7}$$

Equation (6) can now be written as

$$\frac{d(yT^4)}{d(T^4)} = \frac{4\pi cGM}{\kappa_0 L} \frac{k}{\mu H} \frac{3}{a} \frac{T^{1/2}}{y} - 1, \tag{8}$$

or

$$\tfrac{1}{4} yT \frac{dy}{dT} = \frac{4\pi cGM}{\kappa_0 L} \frac{k}{\mu H} \frac{3}{a} T^{1/2} - y(y + 1). \tag{9}$$

Introduce the new variable x, defined by

$$x = \frac{4\pi cGM}{\kappa_0 L} \frac{k}{\mu H} \frac{3}{a} T^{1/2}. \tag{10}$$

We then have, instead of (9),

$$\tfrac{1}{8} xy \frac{dy}{dx} = x - y(y + 1), \tag{11}$$

an equation identical with the one we have discussed in the last chapter (Eq. [26], viii). The solution is accordingly given by (Eqs. [29] and [33], viii)

$$x = y_0(y_0 + 1); \qquad y_0 = \tfrac{32}{31} y. \tag{12}$$

By (7), (10), and (11) we have

$$\frac{4\pi cGM}{\kappa_0 L} \frac{k}{\mu H} \frac{3}{a} T^{1/2} = \frac{32}{31} \frac{\beta}{(1 - \beta)^2} \left(1 + \frac{\beta}{31}\right). \tag{13}$$

Eliminating T between (7) and (13), we find

$$\frac{4\pi cGM}{\kappa_0 L} \left(\frac{k}{\mu H} \frac{3}{a}\right)^{7/6} \rho^{1/6} = \frac{32}{31} \frac{\beta^{7/6}}{(1 - \beta)^{13/6}} \left(1 + \frac{\beta}{31}\right). \tag{14}$$

Differentiating the foregoing expression, we find that

$$\frac{1}{6} \frac{d\rho}{\rho} = \left[(1 + \beta) + \tfrac{1}{6} + \frac{\beta(1 - \beta)}{31 + \beta}\right] \frac{d\beta}{\beta(1 - \beta)}. \tag{15}$$

On the other hand, we have

$$P = \left[\left(\frac{k}{\mu H}\right)^4 \frac{3}{a}\,\frac{1-\beta}{\beta^4}\right]^{1/3} \rho^{4/3}\,, \tag{16}$$

from which we derive

$$\frac{dP}{P} = \frac{1}{3}\left[4\,\frac{d\rho}{\rho} - \frac{4-3\beta}{\beta(1-\beta)}\,d\beta\right]. \tag{17}$$

Eliminating $d\beta$ between (15) and (17), we obtain

$$\frac{dP}{P} = \frac{1}{3}\left[4 - \frac{4-3\beta}{6(1+\beta)+1+6\beta(1-\beta)(31+\beta)^{-1}}\right]\frac{d\rho}{\rho}\,, \tag{18}$$

which can be written in the form

$$\frac{dP}{P} = \left(1 + \frac{1}{n_{\text{eff}}}\right)\frac{d\rho}{\rho}\,, \tag{19}$$

where the effective polytropic index n_{eff} is given by

$$n_{\text{eff}} = \frac{(7+6\beta)(31+\beta)+6\beta(1-\beta)}{(1+3\beta)(31+\beta)+2\beta(1-\beta)}\,. \tag{20}$$

From (20) it follows that for the cases $\beta = 1$ and $\beta = 0$ we have

$$n_{\text{eff}} = 3.25 \qquad (\beta = 1)\,; \qquad\qquad n_{\text{eff}} = 7 \qquad (\beta = 0)\,. \tag{21}$$

Since, in general, we are interested in values of $(1-\beta) \sim 0.05$ (or less), it would be sufficient to consider the case of $(1-\beta) \sim 0$. Then it is a sufficiently close approximation to regard the configurations as polytropes of index $n = 3.25$ (which is constant throughout the configuration), and therefore is described completely by the Lane-Emden function $\theta_{3.25}$:

$$\rho = \rho_c \theta_{3.25}^{3.25}\,; \qquad T = T_c \theta_{3.25}\,; \qquad P = P_c \theta_{3.25}^{4.25}\,. \tag{22}$$

We may note the form which the mass relation takes. Since there is a relation of the form

$$P = K\rho^{1+1/3.25}\,, \tag{23}$$

where K is a constant, we have, on comparing (16) and (23) at $\rho = \rho_c$,

$$K = \left[\left(\frac{k}{\mu H}\right)^4 \frac{3}{a} \frac{1 - \beta_c}{\beta_c^4}\right]^{1/3} \rho_c^{1/39} . \tag{24}$$

Inserting the foregoing value of K in the mass relation (Eq. [69], iv) and putting $n = 3.25$, we have

$$M = -4\pi \left(\frac{4.25}{4}\right)^{3/2} \left[\left(\frac{k}{\mu H}\right)^4 \frac{3}{a} \frac{1 - \beta_c}{\beta_c^4}\right]^{1/2} \frac{1}{(\pi G)^{3/2}} \left(\xi^2 \frac{d\theta_{3.25}}{d\xi}\right)_{\xi=\xi_1(\theta_{3.25})} . \tag{25}$$

This is a quartic equation for $(1 - \beta_c)$. Comparing (25) with the corresponding equation for the standard model, we have, in an obvious notation,

$$M(3.25; 1 - \beta_c) = M(3; 1 - \beta_c) J_M , \tag{26}$$

where

$$J_M = \left(\frac{4.25}{4}\right)^{3/2} \frac{(\xi^2 \theta'_{3.25})_{\xi=\xi_1(\theta_{3.25})}}{(\xi^2 \theta'_3)_{\xi=\xi_1(\theta_3)}} . \tag{27}$$

From the constants of the Lane-Emden functions given in Table 4, chapter iv (p. 96) we find that

$$J_M = 1.0581 . \tag{28}$$

In a manner quite analogous to the standard model (chap. vi, § 4), similar formulae for the physical characteristics may be found. We shall consider here explicitly only the luminosity formula

$$L = \frac{4\pi c G M (1 - \beta_c)}{\overline{\kappa \eta}} , \tag{29}$$

which can be written in the form (cf. chap. vi, § 7, first approximation)

$$L = \frac{4\pi c G M (1 - \beta_c)}{\kappa_c \tilde{\eta}_c} , \tag{30}$$

where

$$\tilde{\eta}_c = \frac{1}{P_c} \int_0^{P_c} \left(\frac{\kappa}{\kappa_c}\right) \eta \, dP . \tag{31}$$

Equation (30) is, of course, an exact equation. According to (2) and (22), we have, for the model under consideration,

$$\tilde{\eta}_c = \int_0^1 \theta_{3.25}^{-0.25} d(\theta^{4.25}) \,, \tag{32}$$

or, as is easily found,

$$\tilde{\eta}_c = \frac{4.25}{4} \,. \tag{33}$$

Since (cf. Eqs. [60] and [78], vi)

$$T_c = \frac{\beta_c \mu H}{k} \frac{1}{[-4.25\, \xi \theta'_{3.25}]_{\xi=\xi_1(\theta_{3.25})}} \frac{GM}{R} \tag{34}$$

and

$$\kappa_c = \kappa_0 \frac{a}{3} \frac{\mu H}{k} \frac{\beta_c}{1 - \beta_c} T_c^{-1/2} \,, \tag{35}$$

we can re-write the luminosity formula as

$$L = \frac{4\pi c GM}{\kappa_0 \tilde{\eta}_c} \frac{1}{[-4.25 \xi \theta'_{3.25}]^{1/2}_{\xi=\xi_1(\theta_{3.25})}} \frac{3}{a} \left(\frac{k}{\beta_c \mu H}\right)^{1/2} \left(\frac{GM}{R}\right)^{1/2} (1 - \beta_c)^2 \,. \tag{36}$$

On the other hand, the quartic equation (25) can be written in the form

$$M^4 = \frac{256}{\pi^2} \left(\frac{4.25}{4}\right)^6 \frac{1}{G^6} \left(\frac{k}{\beta_c \mu H}\right)^8 \left(\frac{3}{a}\right)^2 (1 - \beta_c)^2 (\xi^2 \theta'_{3.25})^4_{\xi=\xi_1(\theta_{3.25})} \,. \tag{37}$$

Eliminating $(1 - \beta_c)$ between (36) and (37) and using (33), we find that

$$L = \frac{\pi^3}{4^{3.5}} \left(\frac{4}{4.25}\right)^{7.5} \left(\frac{GH}{k}\right)^{7.5} \frac{ac}{3\kappa_0} \frac{[\xi_1(\theta_{3.25})]^{0.5}}{[-(\xi^2 \theta'_{3.25})_{\xi=\xi_1(\theta_{3.25})}]^{4.5}} \frac{M^{5.5}}{R^{0.5}} (\mu \beta_c)^{7.5} \,. \tag{38}$$

Comparing (38) with the luminosity formula on the standard model (Eq. [142], vi), we can re-write the former as

$$L = \frac{\pi^3}{4^{3.5}} \left(\frac{GH}{k}\right)^{7.5} \frac{ac}{3} \frac{[\xi_1(\theta_3)]^{0.5}}{[{}_0\omega_3]^{4.5}} \frac{1}{\kappa_0 \tilde{\eta}_c(3.25)} \frac{M^{5.5}}{R^{0.5}} (\mu \beta_c)^{7.5} \,, \tag{39}$$

where

$$\tilde{\eta}_c(3.25) = \left(\frac{4.25}{4}\right)^{7.5} \left[\frac{(\xi^2 \theta'_{3.25})_{\xi=\xi_1(\theta_{3.25})}}{(\xi^2 \theta'_3)_{\xi=\xi_1(\theta_3)}}\right]^{4.5} \left[\frac{\xi_1(\theta_3)}{\xi_1(\theta_{3.25})}\right]^{0.5} \,. \tag{40}$$

Numerically,

$$\eta_c(3.25) = 1.251 . \quad (41)$$

But, according to the first approximation considered in § 7 of chapter vi, for the model ϵ = constant, we found $\tilde{\eta}_c = 1.14$ (Eq. [143]). It is thus seen that the first approximation in the perturbation theory considered in chapter vi leads to a luminosity formula practically identical with the one just obtained by a rigorous method. It should be remembered in this connection that, since the quartic equations which determine β_c differ by the factor J_M (Eq. [27]), the β_c in (39) is somewhat *larger* than the β_c derived on the standard model, and as such the difference between $\tilde{\eta}_c(3) = 1.14$ and $\tilde{\eta}_c\,(3.25) = 1.25$ is almost completely compensated.

2. *The models $\eta \propto \rho^a T^\nu$.*—By definition we have

$$\epsilon = \frac{dL(r)}{dM(r)}\,; \qquad \eta = \frac{\dfrac{L(r)}{M(r)}}{\dfrac{L}{M}}\,. \quad (42)$$

It is therefore clear that in the immediate neighborhood of the center, ϵ and η become identical, apart from a constant factor. Morc precisely,

$$\epsilon = \frac{L}{M}\,\eta\left[1 + \frac{d\log\eta}{d\log M(r)}\right]; \quad (43)$$

or, if $\bar{\epsilon}$ is the average rate of liberation of energy for the whole star, (43) can be expressed in the form

$$\epsilon = \bar{\epsilon}\eta\left[1 + \tfrac{1}{3}r\,\frac{\bar{\rho}(r)}{\rho(r)}\,\frac{d\log\eta}{dr}\right]. \quad (44)$$

Thus, as $r \to 0$, we have

$$\epsilon = \bar{\epsilon}\eta + O(r^2)\,; \qquad \frac{d\epsilon}{dr} = \bar{\epsilon}\,\frac{d\eta}{dr} + O(r)\,. \quad (45)$$

We thus see that in the immediate neighborhood of the center the assumption

$$\eta \propto \rho^a T^\nu \qquad (a \geqslant 0, \nu \geqslant 0) \quad (46)$$

becomes equivalent to

$$\epsilon \propto \rho^a T^\nu\,. \quad (47)$$

But the analysis according to (46) is more elementary than that according to (47). We shall therefore restrict ourselves to a consideration of the models (46), remembering, however, that the analysis may also be regarded as a start on the more difficult (and physically, the more interesting) models (47).[1] For a law of the form (46), equation (44) takes the form

$$\epsilon = \bar{\epsilon}\eta \left[1 + \frac{1}{3} r \frac{\bar{\rho}(r)}{\rho(r)} \left\{ \frac{\alpha}{\rho} \frac{d\rho}{dr} + \frac{\nu}{\rho} \frac{dT}{dr} \right\} \right] . \tag{48}$$

For positive density gradients (i.e., $d\rho/dT \geqslant 0$), the terms in the curly brackets in the foregoing expression are negative; it is, therefore, clear that the right-hand side of (48) will vanish for some value of r (say r^*). For r greater than r^*, equation (48) will give negative values of ϵ. Consequently, for the models of the type considered, we should "break off" the solution at r^* and consider a "point-source" envelope for $r > r^*$.

We shall now proceed to a discussion of the models (46), for which we can write

$$\eta = \eta_0 \rho^\alpha T^\nu \; ; \qquad \eta_0 = \frac{\epsilon_0}{\bar{\epsilon}} . \tag{49}$$

Further, for the law of opacity we shall assume

$$\kappa = \kappa_0 \rho^n T^{-3-s} \qquad (n > 0) . \tag{50}$$

From the equations of equilibrium (3) and (4), and (49) and (50), we derive

$$\frac{k}{\mu H} \frac{3}{a} \frac{d(\rho T)}{d(T^4)} = \frac{4\pi c G M}{\kappa_0 \eta_0} \frac{T^{3+s-\nu}}{\rho^{\alpha+n}} - 1 . \tag{51}$$

The foregoing equation is reduced by the substitutions

$$\rho = \rho_0 \sigma \; ; \qquad T = T_0 \tau , \tag{52}$$

to the form

$$\frac{d(\sigma\tau)}{d(\tau^4)} = \frac{\tau^{3+s-\nu}}{\sigma^{\alpha+n}} - 1 , \tag{53}$$

[1] At the present time no systematic investigations of the models (47) exist. A fairly complete analysis of the models (46) has been given by Chandrasekhar.

provided

$$\frac{\rho_0}{T_0^3} = \frac{\mu H}{k}\frac{a}{3}; \qquad \frac{T_0^{3+s-\nu}}{\rho_0^{\alpha+n}} = \frac{\kappa_0 \eta_0}{4\pi cGM}. \tag{54}$$ [2]

Equation (54) can be expressed alternatively as

$$\rho_0 = \left[\left(\frac{\mu H}{k}\frac{a}{3}\right)^{3+s-\nu}\left(\frac{\kappa_0 \eta_0}{4\pi cGM}\right)^3\right]^{\frac{1}{3(1-\alpha-n)+s-\nu}}, \tag{55}$$

$$T_0 = \left[\left(\frac{\mu H}{k}\frac{a}{3}\right)^{\alpha+n}\frac{\kappa_0 \eta_0}{4\pi cGM}\right]^{\frac{1}{3(1-\alpha-n)+s-\nu}}. \tag{56}$$

Equation (53) now becomes

$$\frac{d(\sigma\tau)}{d\tau} = 4\tau^3\left[\frac{\tau^{3+s-\nu}}{\sigma^{\alpha+n}} - 1\right], \tag{57}$$

or

$$\sigma^{\alpha+n}\tau\frac{d\sigma}{d\tau} + \sigma^{\alpha+n+1} + 4\sigma^{\alpha+n}\tau^3 - 4\tau^{6+s-\nu} = 0. \tag{58}$$

From this equation it follows that two critical cases arise, namely, when $\nu = 3 + s$ and when $\nu = 6 + s$. In the former case, we have

$$\sigma^{\alpha+n}\tau\frac{d\sigma}{d\tau} + \sigma^{\alpha+n+1} + 4\tau^3(\sigma^{\alpha+n} - 1) = 0, \tag{59}$$

from which it follows that along all solutions of the equation (59)

$$\sigma \to 1 \qquad \text{as} \qquad \tau \to \infty \qquad (\nu = 3 + s). \tag{60}$$

Again, if $\nu = 6 + s$, the (σ, τ) differential equation is

$$\sigma^{\alpha+n}\tau\frac{d\sigma}{d\tau} + \sigma^{\alpha+n+1} + 4\sigma^{\alpha+n}\tau^3 - 4 = 0. \tag{61}$$

From equation (61) it follows that for a solution which has no singularity at the origin, σ must tend to a finite limit as $\tau \to 0$:

$$\sigma \to 4^{1/(\alpha+n+1)} \qquad (\tau \to 0). \tag{62}$$

[2] These transformations are not possible for the case $3 + s - \nu = 3(\alpha+n)$.

More generally, we find that all solutions of (58) must tend asymptotically to a certain singular solution whose behavior at infinity is determined by an asymptotic series, the dominant term of which is seen to be

$$\sigma \sim \tau^{\frac{3+s-\nu}{a+n}} \qquad (\tau \to \infty) . \quad (63)$$

In the same way, the solution which has no singularity at the origin[3] has the following behavior at that point:

$$\sigma \sim \left[\frac{4(a+n+1)}{a+s+7+n-\nu}\right]^{\frac{1}{a+n+1}} \tau^{\frac{6+s-\nu}{a+n+1}} . \qquad (64)$$

We shall now consider another form of the differential equation (58) which isolates certain cases for which the integration can be carried out explicitly. Let

$$z = \sigma\tau ; \qquad t = \tau^4 . \qquad (65)$$

Then, instead of (53), we have

$$\frac{dz}{dt} = \frac{t^{\frac{1}{4}(3+a+n+s-\nu)}}{z^{a+n}} - 1 . \qquad (66)$$

From the foregoing equation it follows that if

$$\nu = 3 + a + n + s , \qquad (67)$$

the equation can be integrated. (The case $a + n = 0$ can also be integrated; but this case is not of much physical interest, as we should expect $a \geqslant 0$, $n \geqslant 0$.) For the case (67) the integrated form of (66) can be expressed in the form

$$t = C - z - \int \frac{dz}{z^{a+n} - 1} , \qquad (68)$$

[3] Such solutions exist only for $\nu \leqslant 6 + s$.

where C is a constant of integration. For specified values of $\alpha + n$, the integral occurring in (68) can be evaluated. The following explicit forms of the solutions may be noted:

$$\left. \begin{aligned} t &= C - \{z + 2z^{1/2} + 2 \log |(1 - z^{1/2})|\}, \qquad \alpha + n = \tfrac{1}{2}, \\ \nu &= 3.5 + s, \end{aligned} \right\} \quad (69)$$

$$t = C - \{z + \log |(1 - z)|\}, \qquad \alpha + n = 1, \qquad \nu = 4 + s, \quad (70)$$

$$t = C - \left\{ z + \tfrac{1}{2} \log \left(\left| \frac{1 - z}{1 + z} \right| \right) \right\}, \qquad \alpha + n = 2, \qquad \nu = 5 + s, \quad (71)$$

$$\left. \begin{aligned} t = C - \left\{ z + \tfrac{1}{6} \log \frac{(1 - z)^2}{z^2 + z + 1} - \frac{1}{\sqrt{3}} \tan^{-1} \frac{2z + 1}{\sqrt{3}} + \frac{1}{\sqrt{3}} \tan^{-1} \frac{1}{\sqrt{3}} \right\}, \\ \alpha + n = 3, \qquad \nu = 6 + s, \end{aligned} \right\} \quad (72)$$

and

$$\left. \begin{aligned} t = C - \left\{ z + \tfrac{1}{4} \log \left(\left| \frac{1 - z}{1 + z} \right| \right) - \tfrac{1}{2} \tan^{-1} z \right\}, \\ \alpha + n = 4, \qquad \nu = 7 + s. \end{aligned} \right\} \quad (73)$$

In the foregoing equations, $t(=\tau^4)$ is proportional to the radiation pressure and $z(=\sigma\tau)$ is proportional to the gas pressure (the constant of proportionality in each case is the same [cf. Eq. (54)]; σ/τ^3 is $\beta : (1 - \beta)$ in the usual notation).

As would be expected from the earlier discussion, for the case $\alpha + n = 3$ the solution which has no singularity at the origin tends to a finite limit as $\tau \to 0$; for the case under consideration, $\sigma \to \sqrt{2}$ as $\tau \to 0$. The general nature of the (σ, τ) relations for these models is shown in Figure 30. From an examination of this figure and from the earlier discussion of the (α, ν) models we have the following theorem.

If $\epsilon \propto \rho^\alpha T^\nu$ *and if, further,* $\nu \geqslant 6 + s$, *then in the immediate neighborhood of the center either the density falls off extremely rapidly with temperature*—almost a vertical drop of the density with decreasing temperature[4]—*or the density gradient is always negative*, i.e.,

[4] Actually, it is easy to verify that $d\sigma/d\tau \to \infty$.

$d\rho/dT < 0$. *If* $\nu \leqslant 3 + s$, and $a + n \geqslant 0$, *then we always have (eventually) positive density gradients for increasing temperature.*

3. *The point-source model with* κ = *constant.*—In the point-source model it is assumed that the entire source of energy is liberated at the center of the star; analytically, the assumption is that $L(r)$ = constant = L. The point-source model, then, is another limiting

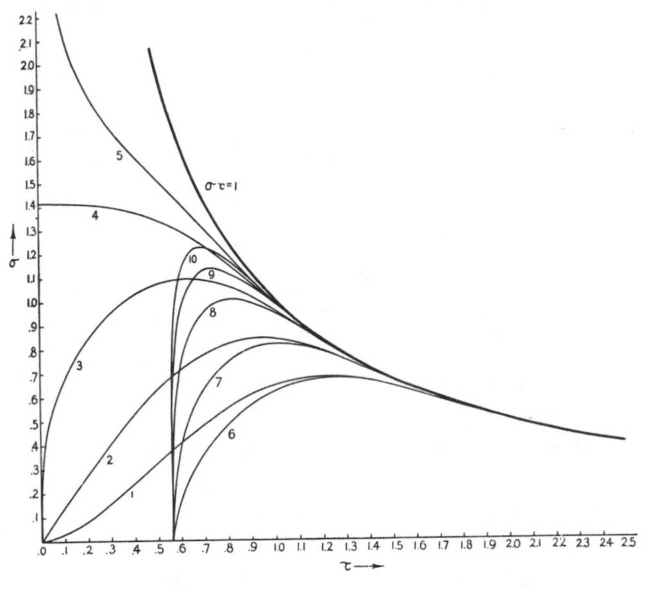

FIG. 30.—(σ, τ) variations for the models ($\kappa = \kappa_0 \rho^n T^{-3-s}$ and $\eta = \eta_0 \rho^a T^\nu$) with $\nu = 3 + a + n + s$. The curves 1 and 6, 2 and 7, 3 and 8, 4 and 9, and 5 and 10 refer to the cases $a + n = 0.5$, 1, 2, 3, and 4, respectively.

case for the possible stellar models; the uniform distribution of the sources $L(r) \propto M(r)$ is another.

The point-source model with κ = constant presents certain simplifying features, and has been studied by Cowling and von Neumann. Von Neumann's treatment of this problem is very powerful and, as such, is instructive as an example of the application of methods and principles which should be of quite general value in the discussion of other stellar models.

The equations of equilibrium for this model are

$$\frac{d(p + p_r)}{dr} = -\frac{GM(r)}{r^2}\,\rho \tag{74}$$

and

$$\frac{dp_r}{dr} = -\frac{\kappa L}{4\pi c r^2}\,\rho\,. \tag{75}$$

From (74) and (75) we have

$$\frac{dp}{dp_r} = \frac{4\pi c G}{\kappa L}\,M(r) - 1\,, \tag{76}$$

or

$$M(r) = \frac{\kappa L}{4\pi c G}\left(\frac{dp}{dp_r} + 1\right). \tag{77}$$

Differentiating (77), we find

$$\frac{\kappa L}{4\pi c G}\,\frac{d^2 p}{dp_r^2} = 4\pi r^2 \rho\,\frac{dr}{dp_r} = -\frac{16\pi^2 c}{\kappa L}\,r^4\,, \tag{78}$$

or

$$\frac{d^2 p}{dp_r^2} = -\frac{64\pi^3 c^2 G}{\kappa^2 L^2}\,r^4\,. \tag{79}$$

Again,

$$\frac{d}{dp_r}\left(\frac{1}{r}\right) = -\frac{1}{r^2}\,\frac{dr}{dp_r} = \frac{4\pi c}{\kappa L}\,\rho^{-1}\,, \tag{80}$$

or

$$\frac{d}{dp_r}\left(\frac{1}{r}\right) = \frac{4\pi c}{\kappa L}\,\frac{k}{\mu H}\left(\frac{3}{a}\right)^{1/4} p^{-1} p_r^{1/4}\,. \tag{81}$$

Equations (79) and (80) are reduced by the substitutions

$$r = \alpha\xi\,; \qquad p = \Pi z\,; \qquad p_r = \Pi_r t \tag{82}$$

to the forms

$$\frac{d^2 z}{dt^2} = -\,\xi^4 \tag{83}$$

and

$$\frac{d}{dt}\left(\frac{1}{\xi}\right) = z^{-1} t^{1/4}\,, \tag{84}$$

provided

$$\alpha^{-4}\Pi\Pi_r^{-2} = \frac{64\pi^3 c^2 G}{\kappa^2 L^2} \tag{85}$$

and

$$a^{-1}\Pi\Pi_r^{-5/4} = \frac{4\pi c}{\kappa L}\frac{k}{\mu H}\left(\frac{3}{a}\right)^{1/4}. \tag{86}$$

Equations (85) and (86) can be expressed in the alternative forms

$$a = \left[\frac{\kappa L}{16\pi^2 cG}\frac{k}{\mu H}\left(\frac{3}{a}\right)^{1/4}\right]^{1/3}\Pi_r^{-1/4} \tag{87}$$

and

$$\Pi = \left[\frac{4\pi c^2}{\kappa^2 L^2 G}\left(\frac{k}{\mu H}\right)^4\frac{3}{a}\right]^{1/3}\Pi_r. \tag{88}$$

The mass relation (77) now becomes

$$M(r) = \frac{\kappa L}{4\pi cG}\left[\Pi\Pi_r^{-1}\frac{dz}{dt} + 1\right], \tag{89}$$

or, according to (88),

$$M(r) = \left[\frac{\kappa L}{16\pi^2 cG^4}\left(\frac{k}{\mu H}\right)^4\frac{3}{a}\right]^{1/3}\left\{\frac{dz}{dt} + \left[\frac{\kappa^2 L^2 G}{4\pi c^2}\left(\frac{\mu H}{k}\right)^4\frac{a}{3}\right]^{1/3}\right\}. \tag{90}$$

Finally, if we introduce the new variable x, defined by ("Kelvin's transformation")

$$x = \xi^{-1}, \tag{91}$$

the fundamental differential equations are

$$\frac{dx}{dt} = z^{-1}t^{1/4}, \quad (\mathrm{I}); \qquad \frac{d^2z}{dt^2} = -x^{-4}, \quad (\mathrm{II}). \tag{92}$$

We shall refer to the foregoing equations as (I) and (II) respectively.

We first notice that equations (I) and (II) admit of a constant of homology; if

$$x, z, t \to C^{1/4}x, Cz, Ct, \tag{93}$$

the differential equations are unaltered, and consequently we can use the foregoing transformation for normalizing purposes—for instance, to make the boundary of the star correspond to $x = 1$.

The problem now is to solve equations (I) and (II) with appropriate boundary conditions. Before we formulate these conditions, however, we shall first consider, following von Neumann, the *general behavior* of the solutions of the foregoing system. For this purpose we shall consider only such solutions as have a physical meaning—i.e., as long as

$$0 < x, z, t < \infty \; ; \; z' \neq \infty \; , \qquad (94)$$

where the prime denotes differentiation with respect to the independent variable t. Let $S : x = x(t)$, $z = z(t)$ be a solution of (I) and (II), and let the maximum t-interval in which (94) holds be denoted by $I(S)$. We shall refer to $I(S)$ as the regularity interval of S. Let the interval be specified by

$$0 \leqslant a < t < b \leqslant +\infty \; . \qquad (95)$$

By definition, at each of the ends a or b, x, z, or t must become 0 or ∞ or $z' = \infty$. We shall first examine the conditions at a.

a) Behavior of the solutions at a.—At a, x, z, or t becomes 0 or ∞ or $z' = \infty$. By definition, t has here the finite limit $a \geqslant 0$. We shall first prove the following lemma.

Lemma 1.—*At* a *the only possibilities are* $x_a = 0$, $z = 0$, *or* $a = 0$; *further,* z'_a *is finite if* $x_a \neq 0$.

Proof: It is clear that as $t \to a$, x tends to a finite limit $x_a \geqslant 0$, for, according to (I), x increases with t and hence decreases as $t \to a$. On the other hand, according to (II), z' decreases with t and hence increases as $t \to a$; hence, as $t \to a$, z' must tend to a limit which may be finite or infinite. But if $x_a \neq 0$, then, according to (II), z'' is bounded, and hence z' must tend to a finite limit z'_a as $t \to a$; this, in turn, implies a finite limit, z_a, for z as $t \to a$. If, however, $z'_a = \infty$, then, since z is increasing, it must decrease as $t \to a$ and z must again tend to a finite limit, z_a. This proves the lemma.

We have now to consider the two cases $a \neq 0$ and $a = 0$, separately.

Let $a \neq 0$. Then by lemma 1, either $x_a = 0$ and/or $z_a = 0$. If $z_a \neq 0$, then $x_a = 0$, and hence,

$$x \sim \text{constant } (t - a) \qquad (t \to a) \; , \quad (96)$$[5]

[5] We shall adopt the convention that the term "constant" refers to an absolute positive constant the actual value of which may change from one equation to another.

or, according to (II),

$$z'' \sim -\text{constant}\,(t-a)^{-4} \qquad (t \to a)\,; \qquad (97)$$

or, integrating twice,

$$z \sim -\text{constant}\,(t-a)^{-2} \qquad (t \to a)\,. \qquad (98)$$

Thus, $z \to -\infty$, which is impossible.

If, on the other hand, $x_a \neq 0$, then (according to the lemma) $z_a = 0$. But if $x_a \neq 0$, then z_a' is finite. As $z_a = 0$, $z'' < 0$, and, since $z > 0$ (for $t > a$), we should have $z_a' > 0$. Hence, we can write

$$z \sim \text{constant}\,(t-a) \qquad (t \to a)\,, \qquad (99)$$

or, according to (I),

$$x' \sim \text{constant}\,(t-a)^{-1} \qquad (t \to a)\,, \qquad (100)$$

or

$$x \sim \text{constant}\,\log\,(t-a) \qquad (t \to a)\,, \qquad (101)$$

which is again impossible, as x would then tend to $-\infty$.

Hence, at $a \neq 0$ the only possibility is that $x_a = 0$, $z_a = 0$, and $z_a' = \infty$. We shall now examine the behavior of such solutions at $a \neq 0$. Put

$$x \sim A(t-a)^m\,; \qquad z = B(t-a)^n\,. \qquad (102)$$

Since $z_a' = \infty$, it is clear that $0 < n < 1$. Also, $m > 0$. Substituting (102) in (I) and (II), we obtain

$$mA(t-a)^{m-1} = a^{1/4}B^{-1}(t-a)^{-n} \qquad (103)$$

and

$$n(n-1)B(t-a)^{n-2} = -A^{-4}(t-a)^{-4m}\,. \qquad (104)$$

Equating the coefficients and the exponents of $(t-a)$, we have

$$m-1 = -n\,; \qquad n-2 = -4m\,, \qquad (105)$$

$$mA = a^{1/4}B^{-1}\,; \qquad n(n-1)B = -A^{-4}\,. \qquad (106)$$

Solving (105) and (106), we find that

$$m = \tfrac{1}{3}\,; \qquad n = \tfrac{2}{3}\,; \qquad A = \left(\frac{3}{2a^{1/4}}\right)^{1/3}\,; \qquad B = (18a)^{1/3}\,. \qquad (107)$$

Hence,

$$x \sim \left(\frac{3}{2a^{1/4}}\right)^{1/3}(t-a)^{1/3} \tag{108}$$

and

$$z \sim (18a)^{1/3}(t-a)^{2/3} . \tag{109}$$

From (108) and (109) it follows that

$$\rho \propto (t-a)^{2/3} \propto x^2 \propto \frac{1}{r^2} . \tag{110}$$

Hence, the models for which T tends to a finite limit as the boundary is approached extend to infinity, and the law of variation of density as $r \to \infty$ is the same as for an isothermal gas sphere (cf. Eq. [439], iv).

Let $a = 0$. We shall show that $x_a \neq 0$. To prove this, consider the cases $z_a \neq 0$ and $z_a = 0$.

First, assume that $z_a \neq 0$. If $x_a = 0$, then (I) implies that

$$x \sim \text{constant } t^{5/4} \qquad (t \to a = 0) , \tag{111}$$

or, according to (II),

$$\frac{d^2z}{dt^2} \sim -\text{ constant } t^{-5} \qquad (t \to a = 0) , \tag{112}$$

or

$$z \sim -\text{ constant } t^{-3} \qquad (t \to a = 0) . \tag{113}$$

Hence, $z \to -\infty$, which is impossible. Therefore, at $t = a = 0$, x_a is finite and $z_a \neq 0$; these solutions correspond to the density tending to infinity as the boundary is approached—the temperature, however, tends to zero.

Let us next assume that $z_a = 0$. By an argument which we have already employed (Lemma 1) under these circumstances $z_a' > 0$. According to (II), z' decreases, and hence, if $t \leqslant t_1$,

$$z \geq tz' \geq tz_1' . \qquad (t \leqslant t_1) \tag{114}$$

Hence, by (I),

$$x' \leqslant (z_1')^{-1} t^{-3/4} , \qquad (t \leqslant t_1) \tag{115}$$

or

$$x \leqslant 4\, (z_1')^{-1} t^{1/4} , \qquad (t \leqslant t_1) \tag{116}$$

if we now assume that $x_a = 0$ and $a = 0$. If in (116) we put $t = t_1$ and take t for t_1, we have

$$x \leqslant 4\,(z')^{-1}t^{1/4}\,. \tag{117}$$

Using the foregoing inequality in (II), we obtain

$$\left(\frac{dz}{dt}\right)^{-4}\frac{d^2z}{dt^2}+\frac{1}{4^4}\frac{1}{t} \geqslant 0\,; \tag{118}$$

or, integrating, we have

$$-\frac{1}{3}\left(\frac{dz}{dt}\right)^{-3}+\frac{1}{4^4}\log t \geqslant 0\,, \tag{119}$$

which is impossible, since z' tends to a finite limit. Hence, x_a cannot vanish for $t = 0$. Along such a solution $\rho \to 0$, $T \to 0$, as $r \to R$.

We can collect the results so far obtained in the following theorem. *If* $0 \leqslant \mathrm{a} < \mathrm{b} \leqslant \infty$ *is the regularity interval of a solution of the differential equations (I) and (II), then at the end* a *there are three possibilities:*

1_a) $0 < \mathrm{a} < \infty$. *Then* $\mathrm{x_a} = 0$, $\mathrm{z_a} = 0$, $\mathrm{z'_a} = \infty$ *and the asymptotic forms of the solution are*

$$x \sim \left(\frac{3}{2a^{1/4}}\right)^{1/3}(t-a)^{1/3}\,; \qquad z \sim (18a)^{1/3}(t-a)^{2/3}\,.$$

2_a) $\mathrm{a} = 0$. *If* $\mathrm{z_a}$ *is finite, then* $\mathrm{x_a}$ *is also finite and* $\mathrm{z'_a} < \infty$:

$$0 < x_a\,, \qquad z_a < \infty\,, \qquad 0 < z'_a < \infty\,.$$

3_a) $\mathrm{a} = 0$. *If* $\mathrm{z_a} = 0$, *then* $\mathrm{x_a}$ *is finite and* $\mathrm{z'_a} < \infty$:

$$0 < x_a < \infty\,, \qquad z_a = 0\,, \qquad 0 < z'_a < \infty\,.$$

b) Behavior of the solutions at b.—At b, by definition, x or z or t becomes 0 or ∞, or z' becomes infinite. Here t has a positive limit $0 < b \leqslant +\infty$. Since x increases with t, it also has a positive limit, $0 < x_b \leqslant +\infty$. z' decreases with t, and we therefore distinguish the two cases: A, $z' \leqslant 0$ does occur and B, $z' > 0$.

In case A, z decreases if t is sufficiently large; thus z' is negative when $t \to b$. In case B, z always increases. Hence, as $t \to b$, z has a limit z_b, which is necessarily finite in case A and necessarily positive in case B.

Case A.—At $t = b$, $z \leqslant 0$, and hence z' is negative for some value of t. If $z' = -c < 0$ for $t = t_1 < b$, then, since z' always decreases, $z' \leqslant -c$ for $t \geqslant t_1$. Hence, z can at best tend only to $-\infty$ if $t \to \infty$. Therefore, for this case $b \neq \infty$; hence, b is finite (and positive) and, as we have already seen, z_b has to be finite as well. Thus at $t = b$, $z_b = 0$ or $x_b = \infty$.

Now, since $x_b > 0$, z'' is bounded; and as $t \to b$, z' tends to a finite limit, z'_b. It is clear that $z'_b \leqslant -c < 0$. If $z_b \neq 0$, then according to (I), x' is bounded and x would tend to a finite limit as $t \to b$; this is impossible since, if z_b is finite, x_b must be infinite. Hence, as $t \to b$, $z \to z_b = 0$ and $x \to x_b = \infty$.

Let us now examine more closely the behavior of the solutions as $z \to 0$ and $x \to \infty$. Let $z'_b = -c$, $(0 < c < \infty)$. Then,

$$z \sim c(b-t) \qquad (t \to b) , \qquad (120)$$

or, according to (I),

$$x' \sim \frac{b^{1/4}}{c}(b-t)^{-1} , \qquad (121)$$

or, again,

$$x \sim -\frac{b^{1/4}}{c}\log(b-t) , \qquad (122)$$

which shows that $x \to \infty$ as $t \to b$. Equations (120) and (122) correspond to the following behavior of ρ and T as the center is approached:

$$\rho T \propto ce^{-(c/b^{1/4})\xi^{-1}} \qquad (\xi \to 0) \qquad (123)$$

and

$$T_b^4 - T^4 \propto e^{-(c/b^{1/4})\xi^{-1}} \qquad (\xi \to 0) . \qquad (124)$$

From (123) and (124) it is clear that this asymptotic behavior corresponds to the case of the density falling off exponentially as $\xi \to 0$, while the temperature very slowly attains its maximum; the central regions will be practically isothermal.

A second approximation to z can be obtained by a process of iteration. According to (II) and (122), we have

$$\frac{d^2z}{dt^2} \sim -\frac{c^4}{b}\,|\log\,(b-t)|^{-4}\,, \tag{125}$$

or

$$\frac{dz}{dt} \sim -c + \frac{c^4}{b}\,(b-t)\,|\log\,(b-t)|^{-4} + \ldots . \tag{126}$$

if terms of higher order are omitted; in (126) the constant of integration has been chosen in a manner such that $z' \to -c$ as $t \to b$. From (126) we have

$$z \sim c(b-t) - \frac{1}{2}\frac{c^4}{b}\,(b-t)^2\,|\log\,(b-t)|^{-4} + \ldots , \tag{127}$$

the integration constant again having been chosen in a manner such that $z_b = 0$. Finally, from (I),

$$\frac{dx}{dt} - \frac{b^{1/4}}{c(b-t)} = \frac{t^{1/4}}{z} - \frac{b^{1/4}}{c(b-t)}\,. \tag{128}$$

The right-hand side of (128) is seen to approach zero as $t \to b$ (cf. Eq. [127]). Hence,

$$\lim_{t\to b}\left[x + \frac{b^{1/4}}{c}\log\,(b-t)\right] = c_0 \tag{129}$$

exists. We can therefore write

$$x = -\frac{b^{1/4}}{c}\log\,(b-t) + c_0 + \ldots . \tag{130}$$

Case B.—As $z' > 0$ and decreases, a finite b implies a finite z_b; this, according to (I), would in turn imply a finite x_b, and this is impossible. Hence, b has necessarily to be infinite.

It is clear that as $t \to b = \infty$, z tends to a limit z_∞, which may be finite or infinite. Suppose z_∞ were finite. Then, as $t \to \infty$, we should have

$$x' \sim z_\infty^{-1} t^{1/4} \qquad \text{or} \qquad x \sim \frac{4}{5z_\infty}\,t^{5/4}\,. \tag{131}$$

By (II), then,

$$z'' \sim -\left(\frac{5z_\infty}{4}\right)^4 t^{-5} ; \qquad (132)$$

or, integrating twice (remembering that $z' \to 0$ as $t \to \infty$), we have

$$z = z_\infty - \frac{1}{12}\left(\frac{5z_\infty}{4}\right)^4 t^{-3} + \ldots . \qquad (133)$$

By (I) and (133) we have

$$\frac{dx}{dt} - \frac{t^{1/4}}{z_\infty} = \left(\frac{1}{z} - \frac{1}{z_\infty}\right) t^{1/4} = O(t^{-11/4}) . \qquad (134)$$

Hence,

$$\lim_{t\to\infty}\left(x - \frac{4}{5z_\infty} t^{5/4}\right) = \gamma_0 \qquad (135)$$

exists and is finite. We can therefore write

$$x = \frac{4}{5z_\infty} t^{5/4} + \gamma_0 + \ldots . \qquad (136)$$

Equations (133) and (136) correspond to the following behavior of ρ and T as the center of the configuration is approached.

$$\rho T \to \text{constant} ; \qquad T \sim \left(\frac{1}{\xi}\right)^{1/5} ; \qquad (137)$$

in other words, along such solutions $T \to \infty$ and $\rho \propto T^{-1} \to 0$.[6]

We have finally to examine the case $z_\infty = \infty$. If x_∞ were finite, then, according to (II),

$$z'' = -x^{-4} \leqslant -(x_\infty)^{-4} < 0 , \qquad (138)$$

or $z' \to -\infty$, which contradicts the hypothesis (case B). Hence, $x \to \infty$ as $z \to \infty$ and $t \to b = \infty$. Let us next examine the way in which x and z tend to infinity. Various trials indicate that z

[6] It may be recalled that in the discussion of the (α, ν) models we have already encountered the behavior $\rho \propto T^{-1}$, $T \to \infty$ in the models $\nu = 3 + \alpha + n + s$ and $\alpha + n \neq 0$. Along any solution (68), $z = \sigma\tau \to 1$ as $t = \tau^4 \to \infty$.

increases at a rate very near to the order of increase of t. We shall therefore try the following behavior:

$$z \sim Ct\,(\log t)^n \qquad (C = \text{constant})\,. \tag{139}$$

By (139)

$$\frac{d^2 z}{dt^2} \sim \frac{nC}{t}\,(\log t)^{n-1}\,, \tag{140}$$

or, according to (II),

$$x^{-4} \sim -\frac{nC}{t}\,(\log t)^{n-1}\,. \tag{141}$$

Hence, $n < 0$, and we have

$$x \sim (-nC)^{-1/4} t^{1/4}\,(\log t)^{(1-n)/4}\,. \tag{141'}$$

Substituting in (I), we have

$$\frac{dx}{dt} = z^{-1} t^{1/4} \sim \tfrac{1}{4}(-nC)^{-1/4} t^{-3/4}\,(\log t)^{(1-n)/4}\,, \tag{142}$$

or

$$z \sim 4(-nC)^{1/4} t\,(\log t)^{(n-1)/4}\,. \tag{143}$$

Comparing this with (139), we have

$$4n = n - 1\,; \qquad 4(-nC)^{1/4} = C\,. \tag{144}$$

Solving (144), we find

$$n = -\tfrac{1}{3}\,; \qquad C = \frac{4^{4/3}}{3^{1/3}}\,. \tag{145}$$

Hence, finally,

$$x \sim \frac{3^{1/3}}{4^{1/3}}\, t^{1/4}\,(\log t)^{1/3} \tag{146}$$

and

$$z \sim \frac{4^{4/3}}{3^{1/3}}\, t\,(\log t)^{-1/3}\,. \tag{147}$$

Equations (146) and (147) correspond to the following behavior of ρ and T as the center of the configuration is approached:

$$T\,(\log T)^{1/3} \propto \frac{1}{\xi} \qquad (\xi \to 0)\,, \tag{148}$$

and

$$\rho \propto T^3\,(\log T)^{-1/3} \qquad (\xi \to 0)\,. \tag{149}$$

Hence, along these solutions both ρ and $T \to \infty$.

This completes the discussion of the behavior of the solutions at the end b of the regularity. We can collect together the results in the following theorem.

If $0 \leqslant \mathrm{a} < \mathrm{b} \leqslant \infty$ *is the regularity interval of a solution of the differential equations (I) and (II), then at the end* b *there are three possibilities:*

1_b) $0 < \mathrm{b} < \infty$. *Then* $x_b \to \infty$, $z_b = 0$, *and* z_b' *finite and negative. More precisely,*

$$x \sim \frac{b^{1/4}}{c} |\log (b - t)| + c_0 + \ldots .$$

and

$$z \sim c(b - t) - \frac{1}{2} \frac{c^4}{b} (b - t)^2 |\log (b - t)|^{-4} + \ldots . .$$

2_b) $\mathrm{b} = +\infty$ *and* $\mathrm{z} \to \mathrm{z}_\infty$ *which is finite. Then* $\mathrm{z}'_\infty = 0$. *More precisely,*

$$x \sim \frac{4}{5z_\infty} t^{5/4} + \gamma_0 + \ldots .$$

and

$$z \sim z_\infty - \frac{1}{12} \left(\frac{5z_\infty}{4}\right)^4 t^{-3} + \ldots . .$$

3_b) $\mathrm{b} = +\infty$ and $\mathrm{z} \to \infty$. *Then* $\mathrm{x} \to \infty$. *More precisely,*

$$x \sim \frac{3^{1/3}}{4^{1/3}} t^{1/4} (\log t)^{1/3}$$

and

$$z \sim \frac{4^{4/3}}{3^{1/3}} t (\log t)^{-1/3} .$$

c) The number of arbitrary parameters.—We have now to determine the number of arbitrary parameters corresponding to each of the different types 1_a, 2_a, 3_a and 1_b, 2_b, 3_b.

Solutions 1_a, 2_a, *and* 3_a.—Solutions of the type 2_a and 3_a satisfy regular initial conditions and are characterized by three (namely, x_0, z_0, and z_0') and two (namely, x_0 and z_0') parameters, respectively. However, in the case 1_a the dominant terms have been uniquely determined without any arbitrary constants, and the number of arbitrary parameters must be determined by a perturbation method. Let

$$x = \bar{x} + \varphi \qquad (\varphi \ll \bar{x}) \qquad (150)$$

and

$$z = \bar{z} + \psi \qquad (\psi \ll \bar{z}) \, , \qquad (151)$$

where $\bar{x}$ and $\bar{z}$ are solutions of (I) and (II) such that, as $t \to a$, their behavior is governed by equations (108) and (109). Substituting (150) and (151) in (I) and (II) and retaining only the terms of the first order of smallness, we obtain

$$\frac{d\varphi}{dt} = -\frac{\psi}{\bar{z}^2} t^{1/4} \; ; \qquad \frac{d^2\psi}{dt^2} = 4\bar{x}^{-5}\varphi \, . \qquad (152)$$

As $t \to a$, we have, according to (108) and (109),

$$\frac{d\varphi}{dt} \sim -\frac{1}{2^{2/3}3^{4/3}a^{5/12}} (t-a)^{-4/3}\psi \qquad (153)$$

and

$$\frac{d^2\psi}{dt^2} \sim \frac{2^{11/3}a^{5/12}}{3^{5/3}} (t-a)^{-5/3}\varphi \, . \qquad (154)$$

Put

$$\varphi \sim A(t-a)^m \; ; \qquad \psi \sim B(t-a)^n \, . \qquad (155)$$

Substituting the foregoing in (153) and (154), we find

$$Am(t-a)^{m-1} = -\frac{1}{2^{2/3}3^{4/3}a^{5/12}} B(t-a)^{n-4/3} \qquad (156)$$

and

$$n(n-1)B(t-a)^{n-2} = \frac{2^{11/3}a^{5/12}}{3^{5/3}} A(t-a)^{m-5/3} \, . \qquad (157)$$

Equating the coefficients and the exponents of $(t - a)$ in the foregoing equations, we find

$$m - n = -\tfrac{1}{3} \qquad (158)$$

and

$$\frac{A}{B} = -\frac{1}{2^{2/3}3^{4/3}a^{5/12}m} = \frac{n(n-1)3^{5/3}}{2^{11/3}a^{5/12}} \, . \qquad (159)$$

From equations (158) and (159) we derive that

$$3n(3n-1)(3n-3) = -8 , \tag{160}$$

or (as can be verified)

$$(3n+1)(9n^2-15n+8) = 0 . \tag{161}$$

Hence,

$$n = -\tfrac{1}{3} \quad \text{or} \quad \frac{5 \pm \sqrt{-7}}{6} . \tag{162}$$

On the other hand, $\varphi \ll \bar{x}$ and $\psi \ll \bar{z}$ implies, according to (108), (109), and (155), that $m > \frac{1}{3}$ and $n > \frac{2}{3}$. Since $m - n = -\frac{1}{3}$, $n > \frac{2}{3}$ would imply that $m > \frac{1}{3}$. This excludes the case $n = -\frac{1}{3}$ in (162). Thus, the only possibilities are

$$n = \frac{5 \pm \sqrt{-7}}{6} ; \qquad m = \frac{3 \pm \sqrt{-7}}{6} . \tag{163}$$

Thus, there are two linearly independent solutions of the type 1_a for any specified $a \neq 0$; these solutions, therefore, are characterized by three parameters. We have thus proved: *Solutions of the type* 1_a, 2_a, *and* 3_a *are characterized by three, three, and two parameters, respectively.*

Now the differential equations (I) and (II) are equivalent to a single differential equation of the third order, and hence the solutions must form a three-parametric family. (In the language of the theory of sets of points, solutions of the type 1_a and 2_a form open domains in the manifold of all solutions. Solutions of the type 3_a form, however, only a two-parametric manifold and hence can contain no "interior points." The boundary of 1_a and 2_a must, therefore, be 3_a.)

Solutions 1_b, 2_b, and 3_b.—Solutions of the types 1_b and 2_b satisfy regular boundary conditions and are therefore characterized by three (namely, b, c, and c_0) and two (namely, z_∞ and γ_0) parameters, respectively. The dominant terms of the solutions of the type 3_b have been uniquely determined, and the number of linearly independent solutions belonging to this class must be determined by the perturbation method, as in case I_a, above. Write

$$x = \bar{x} + \varphi \qquad (\varphi \ll \bar{x}) \tag{164}$$

and

$$z = \bar{z} + \psi \qquad (\psi \ll \bar{z}) \,, \tag{165}$$

where $\bar{x}$ and $\bar{z}$ are now solutions whose behavior at infinity is governed by the equations (146) and (147). The differential equations for φ and ψ are the same as before (Eq. [152]); substituting for $\bar{x}$ and $\bar{z}$ the expressions (146) and (147), we obtain

$$\frac{d\varphi}{dt} \sim -\frac{3^{2/3}}{4^{8/3}} t^{-7/4} (\log t)^{2/3} \psi \tag{166}$$

and

$$\frac{d^2\psi}{dt^2} \sim \frac{4^{8/3}}{3^{5/3}} t^{-5/4} (\log t)^{-5/3} \varphi \,. \tag{167}$$

Put

$$\varphi = At^m (\log t)^r \,; \qquad \psi = Bt^n (\log t)^s \,. \tag{168}$$

If m, n, r, and s are not all zero, then the leading terms in equations (166) and (167) are proportional respectively to

$$mt^{m-1} (\log t)^r \propto t^{n-7/4} (\log t)^{s+2/3} \tag{169}$$

and

$$n(n-1)t^{n-2} (\log t)^s \propto t^{m-5/4} (\log t)^{r-5/3} \,. \tag{170}$$

Equations (169) and (170) imply

$$m - 1 = n - \tfrac{7}{4} \,; \qquad n - 2 = m - \tfrac{5}{4} \,, \tag{171}$$

$$r = s + \tfrac{2}{3} \,; \qquad s = r - \tfrac{5}{3} \,. \tag{172}$$

Equations (172) are inconsistent; hence, according to (169) and (170), either $m = 0$ or $n(n-1) = 0$. Thus, we have the possibilities:

$$\varphi = A (\log t)^r \,; \qquad \psi = Bt^n (\log t)^s \,, \tag{173}$$

$$\varphi = At^m (\log t)^r \,; \qquad \psi = B (\log t)^s \,, \tag{174}$$

$$\varphi = At^m (\log t)^r \,; \qquad \psi = Bt (\log t)^s \,. \tag{175}$$

Substituting (173), (174), and (175) successively in (166) and (167) and equating the coefficients and the exponents of the leading terms, we find:

$$m = 0\,; \quad n = \tfrac{3}{4}\,; \quad r = \tfrac{16}{9}\,; \quad s = \tfrac{1}{9}\,; \quad A : B = -\frac{3^{8/3}}{4^{14/3}}\,, \qquad (173')$$

$$m = -\tfrac{3}{4}\,; \quad n = 0\,; \quad r = \tfrac{2}{9}\,; \quad s = -\tfrac{4}{9}\,; \quad A : B = \frac{1}{3^{1/3}4^{5/3}}\,, \qquad (174')$$

$$m = \tfrac{1}{4}\,; \quad n = 1\,; \quad r = -\tfrac{2}{3}\,; \quad s = -\tfrac{4}{3}\,; \quad A : B = -\frac{3^{2/3}}{4^{5/3}}\,. \qquad (175')$$

With m, n, r, and s defined as in the foregoing equations, φ and ψ (Eq. [168]) satisfy the requirements $\varphi \ll \bar{x}$ and $\psi \ll \bar{z}$, where $\bar{x}$ and $\bar{z}$ are defined according to (146) and (147); thus there are three linearly independent solutions of type 3_b. We have thus proved: *Solutions of the type* 1_b, 2_b, *and* 3_b *are characterized by three, two, and three parameters, respectively.* (In the language of the theory of sets of points, the solutions of the type 1_b and 3_b correspond to open domains in the manifold of all solutions, and the solutions of the type 2_b form a closed set containing no interior points. The "border line" of 1_b and 3_b must therefore correspond to 2_b.)

d) Conditions at the boundary of the configuration.—So far we have considered only the behavior of the general solutions of the differential equations (I) and (II) at the ends "*a*" and "*b*" of the regularity interval. It now remains to select such solutions as can describe a stellar configuration. At the boundary R of the configuration we require both ρ and T to vanish.[7] In other words, the requirement is that, when $t = 0$, $z = 0$. From the earlier discussion (case 3_a), $t = 0$, $z = 0$ necessarily imply the existence of the limit $x(t \to 0) > 0$.

[7] If we require that at the boundary of the configuration, T tends to a finite limit T_0 while at the same time $\rho \to 0$, then the solution must be such that for a finite $t = t_a$, $z_a = 0$. Hence, the solution must be of type 1_a. This solution, as we have shown, corresponds to the case where the configuration extends to infinity with ρ falling off as r^{-2}—i.e., in the same way as in the isothermal gas sphere. Further, according to (89), $M \to \infty$. These are, really, only formal difficulties (cf. Cowling's paper referred to in the Bibliographical Notes at the end of the chapter), and it is safe to use the initial conditions (176), since the ratio of T_0^4 to the values of T^4 occurring in the far interior is of the order of $(10^4/10^6)^4 \sim 10^{-8}$; we can, therefore, certainly put $t_a = 0$.

Further, from the homology argument (Eq. [93]), we can normalize the units in such a way that $x = 1$ corresponds to the boundary

$$t = 0 \,, \qquad x = 1 \,, \qquad z = 0 \,. \tag{176}$$

To make the solution definite, another boundary condition is needed. Assume that

$$z_0' = \delta_1 > 0 \qquad (t = 0) \,. \tag{177}$$

A solution satisfying the initial conditions (176) and (177) must belong to a solution of type 3_a. The problem presented is twofold: (1) How does the regularity interval $(0, b)$ depend on δ_1? (2) For a specified δ_1, what is the type of the solution which we are led to at the end b of the regularity interval? To answer these questions we proceed as follows:

Begin a solution of type 2_b at $t = \infty$ and continue it backward for decreasing t. It is easily verified that as $\gamma_0 \rightarrow +\infty$ (cf. Eq. [136]), the solution is of type 2_a as $t \rightarrow 0$. On the other hand, if $\gamma_0 \rightarrow -\infty$, the solution we are led to is of type 1_a. Hence, an intermediate value of γ_0 which leads to a solution of class 3_a must exist. On the other hand, it is easy to see that as $\delta_1 \rightarrow \infty$ we should eventually have solutions of type 1_b. We can therefore conclude that *there exists a value of* $\delta_1 = \delta_0$ *such that a solution satisfying the boundary conditions* (*176*) *and* (*177*) *with* $\delta_1 = \delta_0$ *is of type* 2_b *as* $t \rightarrow \infty$. *If* $\delta_1 > \delta_0$, *then the solutions are of type* 1_b; *and if* $\delta_1 < \delta_0$, *they are of type* 3_b.

e) *Boundary conditions at the center: discussion of the point-source model.*—We must next consider the boundary conditions at the center. This is a more difficult problem, since we cannot expect that in the point-source model the equations of radiative equilibrium will be valid right up to the center of the configuration; the condition for the stability of the radiative gradient (Eq. [45], vi] will certainly become invalidated as we approach the center. As has already been explained (see p. 228), it is a rather delicate matter to continue the solution beyond the point where the instability of the radiative gradient sets in. For the present, however, we shall continue to discuss the point-source model as though the equations of radiative equilibrium were universally valid, in an attempt toward the enumeration of the possible configurations.

i) *The complete point-source model.*—We require that at the center of the configuration

$$M(r) = 0 \qquad (r = 0) . \quad (178)$$

By (90) this means that

$$\left(\frac{dz}{dt}\right)_{t=t_0} = -\left[\frac{\kappa^2 L^2 G}{4\pi c^2}\left(\frac{\mu H}{k}\right)^4 \frac{a}{3}\right]^{1/3} = -\delta_2 . \quad (179)$$

From (179) it is clear that δ_2 is finite. Thus, a configuration which satisfies the boundary conditions (176), (177), and (178) must be described by a solution which is of type 3_a as $t \to 0$ and is of type 1_b as $t \to t_0$. By the theorem proved in section *d* it follows that $\delta_1 > \delta_0$.

We can re-write (179) as

$$L = \frac{c}{\kappa}\left(\frac{12\pi}{aG}\right)^{1/2}\left(\frac{k}{\mu H}\right)^2 \delta_2^{3/2} , \quad (180)$$

or, numerically,

$$\frac{L}{L_\odot} = \kappa^{-1}\mu^{-2}\delta_2^{3/2} \times 1.471 \times 10^4 . \quad (181)$$

The mass relation (90) can now be written as

$$M = \left[\frac{\kappa L}{16\pi^2 c G^4}\left(\frac{k}{\mu H}\right)^4 \frac{3}{a}\right]^{1/3} (\delta_1 + \delta_2) ; \quad (182)$$

or, eliminating L from (182), we have

$$M = \left(\frac{3}{4\pi a}\right)^{1/2} \frac{1}{G^{3/2}}\left(\frac{k}{\mu H}\right)^2 \delta_2^{1/2}(\delta_1 + \delta_2) , \quad (183)$$

or, numerically,

$$M = 1.117\odot\, \delta_2^{1/2}(\delta_1 + \delta_2)\mu^{-2} . \quad (184)$$

The luminosity formula takes the form

$$L = \frac{4\pi c G M}{\kappa} \frac{\delta_2}{\delta_1 + \delta_2} . \quad (185)$$

It is, of course, clear that a specified $\delta_1 (> \delta_0)$ will lead to a unique value for δ_2. Thus, for the complete point-source model with $\kappa =$ constant, we have an (L, M, κ, μ) relation quite analogous to those for the other models which have been considered.

An important point to note concerning this complete point-source model is that, as $r \to 0$, $\rho \to 0$ and $T \to T_0$ (cf. Eqs. [123] and [124]). This clearly shows that the radiative gradient becomes unstable before the center is reached.

P. C. Keenan has integrated the differential equations I and II and has obtained by numerical methods several solutions corresponding to the model just considered. Table 22 summarizes the results of Keenan's computations.

TABLE 22

SOLUTIONS FOR THE COMPLETE POINT-SOURCE MODEL WITH κ=CONSTANT

δ_1	δ_2	$\frac{M}{\odot}\mu^2$	$\frac{L}{L\odot}\mu^2\kappa$	$\frac{\delta_2}{\delta_1+\delta_2}$
16.9....	5.15	56.0	1.72×10^5	0.234
29.0....	14.4	184	8.00×10^5	0.331
42.6....	25.4	383	1.88×10^6	0.374
54.3....	35.6	600	3.12×10^6	0.396
73.1....	52.7	1020	5.63×10^6	0.419
107.3....	84.9	1954	1.11×10^7	0.442
625.0....	566.0	31500	1.97×10^8	0.475

ii) *The point-source model with point mass at the center.*—If we use the other types of solutions (i.e., solutions which begin as those of type 3_a but are types 2_b or 3_b as $t \to \infty$), then it is clear that $T \to \infty$ as $r \to 0$. Further, for solutions of type 2_b or 3_b, $dz/dt \to 0$ as $t \to \infty$. Hence, by the mass relation (89) it follows that

$$\lim_{r\to 0} M(r) = \frac{\kappa L}{4\pi c G}\,. \tag{186}$$

This point mass at the center does not necessarily imply that $\rho \to \infty$ as $r \to 0$. Indeed, along solutions of type 2_b, $\rho \to 0$ as $r \to 0$, though along solutions of type 3_b, $\rho \to \infty$ as well. It is, however, difficult to interpret these solutions without an adequate examination of the way in which the physical situation alters when new equations of equilibrium are introduced as the instability of the radiative gradient sets in.

4. *The point-source model with negligible radiation pressure and with* $\kappa = \kappa_0 \rho T^{-3.5}$.—The structure of stellar envelopes with negligible radiation pressure has already been investigated in chapter viii. In particular, the temperature and the density distributions in the outer parts are governed by (Eqs. [66], [67] and [53], viii).

$$T = \frac{4}{17} \frac{\mu H}{k} \frac{GM}{R} \left(\frac{R}{r} - 1 \right) , \tag{187}$$

and

$$\rho = \frac{31}{32} \left(\frac{4}{17} \frac{3}{a} \frac{GM}{R} \right)^{3.25} \left(\frac{\mu H}{k} \frac{a}{3} \right)^{3.75} \left(\frac{4\pi c GM}{\kappa_0 L} \right)^{0.5} \left(\frac{R}{r} - 1 \right)^{3.25} . \tag{188}$$

For a star of prescribed M, R, μ, and $\kappa_0 L$, equations (187) and (188) give the initial variations of density and temperature. These equations, however, cease to be good approximations after we have traversed the outer 10 per cent of the mass of the star. On the other hand, we can continue these solutions inward, allowing for the variation of $M(r)$. Since we are considering a point-source model with negligible radiation pressure, the equations of equilibrium which should be used to continue the solutions (187) and (188) are:

$$\frac{k}{\mu H} \frac{d}{dr} (\rho T) = -\frac{GM(r)}{r^2} \rho , \tag{189}$$

$$\frac{dM(r)}{dr} = 4\pi r^2 \rho , \tag{190}$$

and

$$\frac{a}{3} \frac{d}{dr} (T^4) = -\frac{\kappa_0 L}{4\pi c r^2} \frac{\rho^2}{T^{3.5}} . \tag{191}$$

As we have seen, when the variations in $M(r)$ are neglected, we have

$$\rho T^{-3.25} = \text{constant} . \tag{192}$$

When the solutions which describe the stellar envelope are continued inward by means of (189), (190), and (191), the effective polytropic index, n_{eff}, will begin to decrease from its boundary value

3.25. For some definite value of $r = r_i$ (say), n_{eff} will become 1.5. For $r < r_i$, n_{eff} would be less than 1.5, and according to the discussion in § 3, chapter vi, these regions will be in convective equilibrium. The density distribution for $r \leqslant r_i$ should, therefore, be governed by the Lane-Emden function $\theta_{3/2}$. For prescribed values of M, R, and μ and an arbitrarily assigned value of $\kappa_0 L$, we cannot, in general, fit the outer envelope on to a polytropic core of index $n = \frac{3}{2}$. For, an assigned value of $\kappa_0 L$ will lead to definite values for ρ, P, and $M(r)$ at the interface $r = r_i$. Let these quantities have the values ρ_i, P_i, and $M(r_i)$ at $r = r_i$. Now, if the convective core is to be described by the Lane-Emden function $\theta_{3/2}$, then the quantities $-\xi\theta^{3/2}/\theta'$ and $-\xi\theta'/\theta$ should have the following values at the interface:

$$u_i = -\left(\frac{\xi\theta^{3/2}}{\theta'}\right)_{r=r_i} = \frac{4\pi\rho_i r_i^3}{M(r_i)}\,, \tag{193}$$

$$v_i = -\left(\frac{\xi\theta'}{\theta}\right)_{r=r_i} = \frac{2}{5}\,\frac{GM(r_i)\rho_i}{r_i P_i}\,. \tag{194}$$

(The foregoing equations follow from equations [8], [10], and [68] of chapter iv.) In order that a solution be possible, the values of u_i and v_i thus computed should lie on the E-curve in the (u, v) plane (cf. the discussion in § 28, iv). This will not in general be the case. If the values of M, R, and μ are prescribed, we must, therefore, adjust the value of $\kappa_0 L$ until the quantities u_i and v_i computed according to the right-hand sides of the equations (193) and (194) lead to a point on the E-curve in the (u, v) plane. This condition will determine a $(\kappa_0 L, M, R, \mu)$ relation of the same general form as the corresponding relation for the standard model (cf. § 6, vi).

The situation described in the last paragraph can be considered in the following alternative way.

Equations (189) and (190) can be reduced to the forms

$$\frac{d}{d\xi}(\theta\sigma) = -\frac{5\sigma\psi}{2\xi^2}\,; \qquad \frac{d\psi}{d\xi} = \sigma\xi^2\,, \tag{195}$$

by the substitutions

$$\rho = \rho_c\sigma\,, \qquad T = T_c\theta\,, \qquad r = a\xi\,, \qquad M = M_0\psi\,, \tag{196}$$

if ρ_c, T_c, α, and M_0 satisfy the relations

$$\alpha T_c = \frac{2}{5} \frac{\mu H}{k} GM_0 \, ; \qquad \alpha^3 \rho_c = \frac{M_0}{4\pi} \, . \tag{197}$$

In the convective core we should have

$$\sigma = \theta^{3/2} \, . \tag{198}$$

Equation (198) will reduce the equations (195) to the Lane-Emden equation of index $n = \frac{3}{2}$. If ρ_c and T_c correspond to the central values, then the appropriate solution for the convective core is $\theta = \theta_{3/2}$. Outside the convective core the temperature gradient will be governed by equation (191). In terms of the ξ, θ, and ψ variables, equation (191) takes the form

$$\frac{d\theta}{d\xi} = -Q \frac{\sigma^2}{\xi^2 \theta^{6.5}} \, , \tag{199}$$

where

$$Q = \frac{\kappa_0 L}{16\pi} \frac{3}{ac} \frac{\rho_c^2}{\alpha T_c^{7.5}} \, . \tag{200}$$

Q is thus a numerical constant.

Suppose we assume that the convective core extends to $\xi = \xi_i$. At this point the Lane-Emden function will be characterized by definite values for θ, σ, and θ'. Equation (199) will then determine Q. We have

$$Q = -\left(\theta^{3.5} \xi^2 \frac{d\theta}{d\xi}\right)_{\xi = \xi_i} \, . \tag{201}$$

With this value of Q we can numerically integrate the equations (195) and (199) for $\xi \geqslant \xi_i$. Now for a solution to be physically significant σ and θ should tend to zero simultaneously. For an arbitrarily assigned initial value of ξ_i, this will not, in general, be the case. We can, however, adjust ξ_i until σ and θ tend to zero simultaneously. This is the method which Cowling has adopted in his treatment of the equations (195) and (199). The value of ξ_i which leads to the physically significant solution is

$$\xi_i = 1.188 \, . \tag{202}$$

Further, at $\xi = \xi_i$ we have

$$\theta_i = 0.7878; \qquad \theta_i' = -0.3212; \qquad \psi_i = 0.4534 . \tag{203}$$

Equation (201) now gives

$$Q = 0.1968 . \tag{204}$$

Cowling's integration shows that the boundary of the configuration is reached at

$$\xi = \xi_1 = 7.027 . \tag{205}$$

At $\xi = \xi_1$ we have

$$\psi = \psi_1 = 3.1237 . \tag{206}$$

Hence,

$$\frac{\xi_i}{\xi_1} = 0.169 ; \qquad \frac{\psi_i}{\psi_1} = 0.145 . \tag{207}$$

In other words, the convective core occupies 16.9 per cent of radius of the star and incloses 14.5 per cent of the mass.

Equation (197) can now be re-written as

$$T_c = \frac{2}{5} \frac{\xi_1}{\psi_1} \frac{\mu H}{k} \frac{GM}{R} , \tag{208}$$

$$\rho_c = \frac{\xi_1^3}{4\pi\psi_1} \frac{M}{R^3} . \tag{209}$$

From (208) and (209) we derive that

$$P_c = \frac{k}{\mu H} \rho_c T_c = \frac{\xi_1^4}{10\pi\psi_1^2} \frac{GM^2}{R^4} , \tag{210}$$

$$\rho_c = \frac{\xi_1^3}{3\psi_1} \bar{\rho} . \tag{211}$$

Numerically, equations (208), (210), and (211) are found to be

$$\left. \begin{aligned} T_c &= 0.900 \frac{\mu H}{k} \frac{GM}{R} , \\ P_c &= 7.954 \frac{GM^2}{R^4} , \\ \rho_c &= 37.0 \bar{\rho} . \end{aligned} \right\} \tag{212}$$

Comparing the foregoing equations with the corresponding equations for the standard model (Eqs. [61], [57], and [56], vi), we notice that, though this model is less centrally condensed than a polytrope of index 3, it is yet characterized by a higher value for the central temperature.

Equations (200), (204), (208), and (209) lead to the following mass-luminosity-radius relation:

$$L = \frac{4^{11.5}\pi^3 Q}{10^{7.5}} \frac{\xi_1^{0.5}}{\psi_1^{5.5}} \frac{1}{\kappa_0} \frac{ac}{3}\left(\frac{GH}{k}\right)^{7.5} \frac{M^{5.5}}{R^{0.5}} \mu^{7.5} . \tag{213}$$

If we write (213) in the form (cf. Eq. [39])

$$L = \frac{\pi^3}{4^{3.5}} \frac{[\xi_1(\theta_3)]^{0.5}}{[{}_0\omega_3]^{4.5}} \frac{1}{\kappa_0 \tilde{\eta}_c} \frac{ac}{3}\left(\frac{GH}{k}\right)^{7.5} \frac{M^{5.5}}{R^{0.5}} \mu^{7.5} , \tag{214}$$

then we should have

$$\tilde{\eta}_c = \frac{10^{7.5}}{4^{15} Q} \left[\frac{\xi_1(\theta_3)}{\xi_1}\right]^{0.5} \frac{\psi_1^{5.5}}{[{}_0\omega_3]^{4.5}} ; \tag{215}$$

or, introducing the numerical values, it is found that

$$\tilde{\eta}_c = 3.30 . \tag{216}$$

The value of $\tilde{\eta}_c$ for this model is thus seen to be somewhat larger than the value 2.5 adopted in chapter vi. It should, however, be remarked that if we use (213) to determine the hydrogen contents of stars, the appropriate guillotine factors will be less than for the standard model (on account of the higher temperatures and lower densities in the central regions of the model considered as compared to the polytrope of index $n = 3$).

One important characteristic of the model considered in this section must be noticed. The luminosity formula (213) derived for this model will be valid for any stellar configuration (with negligible radiation pressure) in which the energy-generating regions do not occupy more than a fraction 0.17 of the radius of the star. The same is true for the distributions of density and temperature derived for this model. The analysis of this model confirms, therefore, the generality of the conclusions drawn on the basis of the luminosity formula used in the discussion of chapter vii.

BIBLIOGRAPHICAL NOTES

Stellar models have been considered by several writers, and the following references do not exhaust the list.

1. J. H. JEANS, *M.N.*, **85**, 196, 394, 1925.
2. H. N. RUSSELL, *M.N.*, **85**, 935, 1925.
3. T. G. COWLING, *M.N.*, **91**, 92, 1931.
4. B. STRÖMGREN, *Zs. f. Ap.*, **2**, 345, 1931.
5. L. BIERMANN, *Zs. f. Ap.*, **3**, 116, 1931.
6. E. A. MILNE, *Zs. f. Ap.*, **4**, 75, 1932.
7. S. ROSSELAND, *Zs. f. Ap.*, **4**, 255, 1932.
8. E. A. MILNE, *Zs. f. Ap.*, **5**, 337, 1932.
9. S. CHANDRASEKHAR, *M.N.*, **97**, 132, 1936.
10. S. CHANDRASEKHAR, *Zs. f. Ap.*, **14**, 164, 1937.
11. A. B. SEVERNY, *M.N.*, **97**, 699, 1937.
12. J. Tuominen, *Annales Academiae Scientiarum Fenniace*, **48**, No. 16, 1938.

§ 1.—References 1 and 2. The actual form of the analysis and the derivation of the luminosity formula (39) is new.

§ 2.—References 9 and 10. The investigations of Severny and Tuominen appeared after this chapter had been written.

§ 3.—The discussion of the point-source model given in the text is in part based on an unpublished investigation of J. von Neumann. See also reference 3.

§ 4.—The discussion in this section is based on a numerical integration carried out by T. G. Cowling in *M.N.*, **96**, 42, 1936; see the appendix to this paper.

CHAPTER X

THE QUANTUM STATISTICS

In this chapter we shall consider the quantum theory of an ideal gas, with a view toward the applications contained in the next chapter. It was originally intended to make the presentation of statistical mechanics as logically satisfactory as that given (following Carathéodory) of the foundations of thermodynamics in chapter i. This intention, however, had, in part, to be abandoned, owing to the space which such an exposition would require; such a discussion would, also, lead us too far from the main thesis of the present monograph. The most important formula to be established is the relation between the electron pressure, P, and the electron concentration, n, for a completely degenerate electron gas. This formula can be derived in an entirely elementary way, but to appreciate fully the physical meaning of the (P, n) relation and the physical circumstances under which it is applicable a more elaborate treatment is required, which follows the elementary derivation contained in § 1, below. Applications of the physical theory presented here are contained in chapter xi.

1. *A completely degenerate electron gas: elementary treatment.*—A given number N of electrons can be confined in a given volume V by one of two methods: either by means of "potential walls" such that electrons inside the "potential hole" cannot escape, or by means of imposing a certain periodicity condition. We shall consider these restrictions in greater detail in § 2, but the essential result is that we can label the possible energy states for an electron inside a given volume V by means of quantum numbers in somewhat the same manner as the quantum states of an electron in an atom. If we assume that the volume V is large, then it follows from the general

theory that the number of quantum states with momenta between p and $p + dp$ is given by

$$V \frac{8\pi p^2 dp}{h^3} . \tag{1}$$

The meaning of (1) is that the accessible six-dimensional phase-space can be divided into "cells" of volume h^3 and that in each cell there are two possible states. Now the Pauli principle states that *no two electrons can occupy the same quantum state.* This implies that, if $N(p)dp$ denotes the number of electrons in the assembly with momenta between p and $p + dp$, then

$$N(p)dp \leqslant V \frac{8\pi p^2 dp}{h^3} . \tag{2}$$

Now *a completely degenerate electron gas is one in which all the lowest quantum states are occupied.* In other words, we should have

$$N(p) = V \frac{8\pi p^2}{h^3} . \tag{3}$$

It is clear that if there is only a finite number, N, of electrons in the specified volume, then all the electrons must have momenta less than a certain maximum value, p_0, such that

$$N = V \frac{8\pi}{h^3} \int_0^{p_0} p^2 dp , \tag{4}$$

or

$$N = V \frac{8\pi}{3h^3} p_0^3 . \tag{5}$$

The "threshold value," p_0, of p is related to the electron concentration, n, by

$$n = \frac{N}{V} = \frac{8\pi}{3h^3} p_0^3 . \tag{6}$$

To calculate the pressure, we recall that by definition the pressure, P, exerted by a gas is simply the mean rate of transfer of momentum across an ideal surface of unit area in the gas. From this definition it follows quite generally that

$$PV = \frac{1}{3}\int_0^\infty N(p)pv_p dp \,, \tag{7}$$

where v_p is the velocity associated with the momentum p. According to (3), we have for the case under consideration:

$$P = \frac{8\pi}{3h^3}\int_0^{p_0} p^3 \frac{\partial E}{\partial p}\, dp \,, \tag{8}$$

where E is the kinetic energy of the electron which has a momentum p. Finally, if U_{kin} is the internal energy of the gas which is due to the translational energy of the motions of the individual electrons, we have (quite generally)

$$U_{\text{kin}} = \int_0^\infty N(p)E dp \,, \tag{9}$$

or for the completely degenerate case,

$$U_{\text{kin}} = V\,\frac{8\pi}{h^3}\int_0^{p_0} Ep^2 dp \,. \tag{10}$$

From (8) and (10) we find

$$P = \frac{8\pi}{3h^3}\,E(p_0)p_0^3 - \frac{U_{\text{kin}}}{V} \,. \tag{11}$$

So far the results are quite general, in the sense that we have not introduced any relation between E and p. According to the special theory of relativity, we have

$$E = mc^2\left\{\left(1 + \frac{p^2}{m^2c^2}\right)^{1/2} - 1\right\}, \tag{12}$$

which gives

$$\frac{\partial E}{\partial p} = \frac{1}{m}\left(1 + \frac{p^2}{m^2c^2}\right)^{-1/2} p \,. \tag{13}$$

Substituting (13) in (8), we have

$$P = \frac{8\pi}{3mh^3} \int_0^{p_c} \frac{p^4 dp}{\left(1 + \frac{p^2}{m^2c^2}\right)^{1/2}} . \qquad (14)$$

Introduce the variable, θ, defined by

$$\sinh \theta = \frac{p}{mc} ; \qquad \sinh \theta_0 = \frac{p_0}{mc} . \qquad (15)$$

Equation (14) now reduces to

$$P = \frac{8\pi m^4 c^5}{3h^3} \int_0^{\theta_0} \sinh^4 \theta \, d\theta . \qquad (16)$$

On integration, we have

$$P = \frac{8\pi m^4 c^5}{3h^3} \left[\frac{\sinh^3 \theta \cosh \theta}{4} - \frac{3 \sinh 2\theta}{16} + \frac{3\theta}{8} \right]_{\theta=\theta_0} . \qquad (17)$$

Finally, writing

$$x = \frac{p_0}{mc} , \qquad (18)$$

we have

$$P = \frac{\pi m^4 c^5}{3h^3} f(x) = 6.01 \times 10^{22} f(x) , \qquad (19)$$

where

$$f(x) = x(2x^2 - 3)(x^2 + 1)^{1/2} + 3 \sinh^{-1} x . \qquad (20)$$

Again from (6), we have

$$n = \frac{8\pi m^3 c^3}{3h^3} x^3 = 5.87 \times 10^{29} x^3 . \qquad (21)$$

Equations (19), (20), and (21) represent parametrically the equation of state of a completely degenerate electron gas. From (11) it now follows that

$$U_{\text{kin}} = \frac{\pi m^4 c^5}{3h^3} V g(x) , \qquad (22)$$

where

$$g(x) = 8x^3[(x^2 + 1)^{1/2} - 1] - f(x) . \qquad (3)$$

Equation (22) for the internal energy of an electron gas was first derived by E. C. Stoner. In Table 23 the functions $f(x)$ and $g(x)$ are tabulated. The table is more accurate and more extensive than any that has been published so far.

TABLE 23

THE PRESSURE AND THE INTERNAL ENERGY OF A COMPLETELY DEGENERATE GAS

x	$f(x)$	$g(x)$	$\frac{g(x)}{f(x)}$	x	$f(x)$	$g(x)$	$\frac{g(x)}{f(x)}$
0.	0.	0.	1.5	2.7. .	95.17935	200.7327	2.1090
0.1. . .	0.000016	0.000024	1.5	2.8. .	110.8207	235.7072	2.1269
0.2. . .	0.000505	0.000762	1.509	2.9. .	128.3012	275.1070	2.1442
0.3. . .	0.003769	0.005742	1.5233	3.0. .	147.7578	319.2942	2.1609
0.4. . .	0.015527	0.023914	1.5402	3.5. .	279.8113	625.728	2.2363
0.5. . .	0.046093	0.071941	1.5608	4.0. .	484.5644	1114.466	2.2999
0.6. . .	0.111126	0.17604	1.5841	4.5. .	784.5271	1846.997	2.3543
0.7. . .	0.231992	0.27348	1.6099	5.0. .	1205.2069	2893.813	2.4011
0.8. . .	0.435865	0.71358	1.6372	5.5. .	1775.1094	4334.407	2.4418
0.9. . .	0.755661	1.25849	1.6654	6.0. .	2525.7390	6257.275	2.4774
1.0. . .	1.229907	2.0838	1.6943	6.5. .	3491.599	8759.913	2.5089
1.1. . .	1.902586	3.2788	1.7233	7.0. .	4710.192	11948.818	2.5368
1.2. . .	2.82298	4.9468	1.7523	7.5. .	6222.021	15939.488	2.5618
1.3. . .	4.04557	7.2052	1.7810	8.0. .	8070.587	20856.421	2.5842
1.4. . .	5.62991	10.1857	1.8092	8.5. .	10302.39	26833.12	2.6045
1.5. . .	7.64053	14.0344	1.8368	9.0. .	1.296694×10^4	3.401207×10^4	2.6230
1.6. . .	10.14696	18.9115	1.8638	9.5. .	1.611672×10^4	4.254479×10^4	2.6398
1.7. . .	13.22359	24.9920	1.8900	10.0. .	1.980725×10^4	5.2591×10^4	2.6552
1.8. . .	16.94969	32.4649	1.9154	20.0. .	3.192093×10^5	8.9839×10^5	2.8144
1.9. . .	21.40937	41.5338	1.9400	30.0. .	1.618212×10^6	4.6494×10^6	2.8732
2.0. . .	26.69159	52.4168	1.9638	40.0. .	5.116812×10^6	1.48596×10^7	2.9041
2.1. . .	32.89010	65.3462	1.9868	50.0. .	1.249501×10^7	3.6515×10^7	2.9224
2.2. . .	40.10347	80.5689	2.0090	60.0. .	2.591280×10^7	7.6053×10^7	2.9349
2.3. . .	48.43509	98.3463	2.0305	70.0. .	4.801018×10^7	1.41346×10^8	2.9441
2.4. . .	57.99311	118.9541	2.0512	80.0. .	8.190727×10^7	2.41703×10^8	2.9509
2.5. . .	68.89053	142.6823	2.0711	90.0. .	13.12039×10^7	3.87876×10^8	2.9563
2.6. . .	81.24509	169.8355	2.0904	100.0. .	19.9980×10^7	5.9206×10^8	2.9606

The function $f(x)$ has the following asymptotic forms:

$$f(x) \sim \tfrac{8}{5}x^5 - \tfrac{4}{7}x^7 + \tfrac{1}{3}x^9 - \tfrac{5}{22}x^{11} + \ldots \qquad (x \to 0) \qquad (24)$$

and

$$f(x) \sim 2x^4 - 3x^2 + \ldots \qquad (x \to \infty). \qquad (25)$$

Finally, we see that

$$\frac{f(x)}{2x^4} < 1 \qquad \text{for all finite } x\,. \tag{26}$$

The inequality (26) is a strict one. If only the first terms in the expansions (24) and (25) are retained, we can eliminate x between (19) and (21) and obtain the following explicit forms of the equations of state for the two limiting cases:

$$P = \frac{1}{20}\left(\frac{3}{\pi}\right)^{2/3}\frac{h^2}{m}\,n^{5/3} \qquad (x \to 0) \tag{27}$$

and

$$P = \frac{1}{8}\left(\frac{3}{\pi}\right)^{1/3} hc\, n^{4/3} \qquad (x \to \infty)\,. \tag{28}$$

We may note that $g(x)$ has the following asymptotic forms:

$$g(x) \sim \tfrac{12}{5}x^5 - \tfrac{3}{7}x^7 + \tfrac{1}{6}x^9 - \tfrac{15}{176}x^{11} + \ldots \qquad (x \to 0) \tag{29}$$

and

$$g(x) \sim 6x^4 - 8x^3 + 7x^2 - \ldots \qquad (x \to \infty)\,. \tag{30}$$

From (24), (25), (29), and (30) we infer that

$$U_{\text{kin}} = \tfrac{3}{2}\, PV \qquad (x \to 0) \tag{31}$$

and

$$U_{\text{kin}} = 3\;\, PV \qquad (x \to \infty)\,. \tag{32}$$

The elementary derivation of the equation of state of a completely degenerate electron gas should be supplemented in two ways: first, by the enumeration of the states which leads to (1); and second, by the investigation of the physical circumstances under which the equation of state given by (19) and (21) can be considered to be valid. These require a rather elaborate treatment of statistical mechanics, which will now be given. For a more general discussion than that undertaken here, reference may be made to Jordan's book which is referred to in the bibliographical note at the end of the chapter.

2. *The enumeration of the quantum states.*—The wave equation of the electron in free space is, according to Dirac,

$$\left(\alpha_x p_x + \alpha_y p_y + \alpha_z p_z + \alpha_m mc + \frac{W}{c}\right)\psi = 0\,, \tag{33}$$

where α_x, α_y, α_z, and α_m are anticommuting variables whose squares are unity, i.e.,

$$\alpha_\nu \alpha_\mu + \alpha_\mu \alpha_\nu = 2\delta_{\mu\nu} \qquad (\mu, \nu = x, y, z, m)\,, \tag{34}$$

where $\delta_{\mu\nu}$ is the Kronecker symbol:

$$\left.\begin{aligned} \delta_{\mu\nu} &= 0 \qquad (\mu \neq \nu)\,, \\ &= 1 \qquad (\mu = \nu)\,. \end{aligned}\right\} \tag{35}$$

Further,

$$p_x = -i\hbar\frac{\partial}{\partial x}\,; \qquad p_y = -i\hbar\frac{\partial}{\partial y}\,; \qquad p_z = -i\hbar\frac{\partial}{\partial z} \tag{36}$$

and

$$W = i\hbar\frac{\partial}{\partial t}\,, \tag{37}$$

where $\hbar$ is the Planck constant divided by 2π. The wave equation can therefore be written as

$$i\hbar\left\{\frac{1}{c}\frac{\partial}{\partial t} - \alpha_x\frac{\partial}{\partial x} - \alpha_y\frac{\partial}{\partial y} - \alpha_z\frac{\partial}{\partial z} + \alpha_m mc\right\}\psi = 0\,. \tag{38}$$

As is well known, the α's can be represented as matrices with four rows and columns and ψ is to be regarded as a (complex) vector with four components. Choosing a particular representation for the α's we may write equation (33) as

$$\left.\begin{aligned} &\left[\frac{W}{c} + p_x\begin{vmatrix} 0 & 0 & 0 & 1 \\ 0 & 0 & 1 & 0 \\ 0 & 1 & 0 & 0 \\ 1 & 0 & 0 & 0 \end{vmatrix} + p_y\begin{vmatrix} 0 & 0 & 0 & -i \\ 0 & 0 & i & 0 \\ 0 & -i & 0 & 0 \\ i & 0 & 0 & 0 \end{vmatrix}\right. \\ &\left. + p_z\begin{vmatrix} 0 & 0 & 1 & 0 \\ 0 & 0 & 0 & -1 \\ 1 & 0 & 0 & 0 \\ 0 & -1 & 0 & 0 \end{vmatrix} + mc\begin{vmatrix} 1 & 0 & 0 & 0 \\ 0 & 1 & 0 & 0 \\ 0 & 0 & -1 & 0 \\ 0 & 0 & 0 & -1 \end{vmatrix}\right]\begin{vmatrix} \psi_1 \\ \psi_2 \\ \psi_3 \\ \psi_4 \end{vmatrix} = 0\,, \end{aligned}\right\} \tag{39}$$

where ψ_1, ψ_2, ψ_3, and ψ_4 are the four components of the wave function. In (39) the matrix representing the α's should be multiplied by the matrix $|\psi_\lambda|$—of just one column—according to the law of matrix multiplication. Explicitly, (39) takes the form

$$\left.\begin{aligned}\frac{W}{c}\begin{vmatrix}\psi_1\\ \psi_2\\ \psi_3\\ \psi_4\end{vmatrix} + p_x\begin{vmatrix}\psi_4\\ \psi_3\\ \psi_2\\ \psi_1\end{vmatrix} + p_y\begin{vmatrix}-i\psi_4\\ i\psi_3\\ -i\psi_2\\ i\psi_1\end{vmatrix}\qquad\qquad\qquad\qquad\\ + p_z\begin{vmatrix}\psi_3\\ -\psi_4\\ \psi_1\\ -\psi_2\end{vmatrix} + mc\begin{vmatrix}\psi_1\\ \psi_2\\ -\psi_3\\ -\psi_4\end{vmatrix} = 0 .\end{aligned}\right\} \qquad (40)$$

According to (36) and (37), the foregoing equation is equivalent to the following four ordinary partial differential equations:

$$\left(\frac{i\hbar}{c}\frac{\partial}{\partial t} + mc\right)\psi_1 - i\hbar\left(\frac{\partial}{\partial x} - i\frac{\partial}{\partial y}\right)\psi_4 - i\hbar\frac{\partial\psi_3}{\partial z} = 0 , \qquad (41)$$

$$\left(\frac{i\hbar}{c}\frac{\partial}{\partial t} + mc\right)\psi_2 - i\hbar\left(\frac{\partial}{\partial x} + i\frac{\partial}{\partial y}\right)\psi_3 + i\hbar\frac{\partial\psi_4}{\partial z} = 0 , \qquad (42)$$

$$\left(\frac{i\hbar}{c}\frac{\partial}{\partial t} - mc\right)\psi_3 - i\hbar\left(\frac{\partial}{\partial x} - i\frac{\partial}{\partial y}\right)\psi_2 - i\hbar\frac{\partial\psi_1}{\partial z} = 0 , \qquad (43)$$

$$\left(\frac{i\hbar}{c}\frac{\partial}{\partial t} - mc\right)\psi_4 - i\hbar\left(\frac{\partial}{\partial x} + i\frac{\partial}{\partial y}\right)\psi_1 + i\hbar\frac{\partial\psi_2}{\partial z} = 0 . \qquad (44)$$

To solve the foregoing equations, put

$$\psi_\lambda = a_\lambda e^{\frac{i}{\hbar}(p_x x + p_y y + p_z z - Et)} \qquad (\lambda = 1, 2, 3, 4) , \quad (45)$$

where p_x, p_y, and p_z are now ordinary real numbers and the a_λ's are, for the present, arbitrary numbers. On substituting (45) in equations (41), (42), (43), and (44) we find that the a_λ's must satisfy the following set of homogeneous linear equations:

$$\left.\begin{aligned}
\left(\frac{E}{c}+mc\right)a_1 \qquad\qquad + p_z a_3 + (p_x - ip_y)a_4 &= 0\,,\\
\left(\frac{E}{c}+mc\right)a_2 + (p_x+ip_y)\,a_3 \qquad - p_z a_4 &= 0\,,\\
p_z a_1 + (p_x - ip_y)\,a_2 + \left(\frac{E}{c}-mc\right)a_3 \qquad &= 0\,,\\
(p_x+ip_y)a_1 \qquad - p_z a_2 \qquad + \left(\frac{E}{c}-mc\right)a_4 &= 0\,.
\end{aligned}\right\} \quad (46)$$

In order that the a_λ's shall not be identically zero it is necessary that the determinant formed by the coefficients of the a_λ's in (46) be zero. The determinant is found to reduce to

$$\left(\frac{E}{c}+mc\right)^2\left(\frac{E}{c}-mc\right)^2-(p_x^2+p_y^2+p_z^2)^2\,. \qquad (47)$$

Hence, the condition that the a_λ's do not vanish identically is

$$\left(\frac{E}{c}\right)^2 = p_x^2+p_y^2+p_z^2+m^2c^2\,. \qquad (48)$$

In other words, the relativistic expression which connects the total energy, E, and the components of the momentum must be valid in order that (45) may be a solution of Dirac's equation.

Further, we find from the first two equations in (46) that

$$a_1 = -\frac{a_3 p_z + a_4(p_x - ip_y)}{mc+\dfrac{E}{c}} \qquad (49)$$

and

$$a_2 = -\frac{a_3(p_x+ip_y) - a_4 p_z}{mc+\dfrac{E}{c}}\,. \qquad (50)$$

The foregoing values of a_1 and a_2, in terms of a_3 and a_4, also satisfy the last two equations in (46). Hence, of the four a_λ's only two can be arbitrarily specified for a given set of values of (p_x, p_y, p_z). Hence, there are two linearly independent solutions for a given set of values for the components of the momentum which satisfy (48).

We must now obtain some restriction on the possible eigenfunctions due to the presence of the boundary walls. To obtain these restrictions quite generally, we shall follow Dirac in his approach to the problem.

According to the general principles of quantum mechanics, there must be just exactly as many eigenfunctions as should enable one to represent by a matrix any function of the co-ordinates which has a physical meaning. Let us suppose, for definiteness, that each electron is confined between two boundaries at $x = 0$ and $x = l_x$. Then only those functions of x which are defined for $0 < x < l_x$ have a physical meaning and must be capable of being expanded in terms of a complete set of eigenfunctions. It is, of course, obvious that this will require fewer eigenfunctions than would be required for the representation of an arbitrary function. A function $F_\lambda(x)$, defined in the range $0 < x < l_x$, can always be expanded in a Fourier series of the form

$$F_\lambda(x) = \sum_{k_x=-\infty}^{\infty} a_{k_x}^{(\lambda)} e^{2\pi i k_x x/l_x} , \tag{51}$$

where the $a_{k_x}^{(\lambda)}$'s are constants and the k_x's are integers. It is clear, then, that if we choose from the eigenfunctions,

$$\psi_\lambda = a_\lambda e^{\frac{i}{\hbar}(p_x x + p_y y + p_z z - Et)} , \tag{52}$$

those for which

$$\frac{p_x}{\hbar} = \frac{2\pi k_x}{l_x} \qquad (k_x = \pm 1, \pm 2, \ldots \pm \infty) , \tag{53}$$

then, $F_\lambda(x)$ times any of the eigenfunctions so selected can be expanded in a series in terms of the selected eigenfunctions. Thus, the selected eigenfunctions are sufficient and are easily seen to be only just sufficient for the expansion of functions of the form (51).

Similarly, if the y and the z co-ordinates are also bounded, so that

$$0 < y < l_y \, ; \qquad 0 < z < l_z \, , \tag{54}$$

then we should have

$$\frac{p_y}{\hbar} = \frac{2\pi k_y}{l_y} \, ; \qquad \frac{p_z}{\hbar} = \frac{2\pi k_z}{l_z} \, , \tag{55}$$

where k_y and k_z are positive or negative integers. The conditions (53) and (55) can also be written as

$$p_x = \frac{k_x h}{l_x} \, ; \qquad p_y = \frac{k_y h}{l_y} \, ; \qquad p_z = \frac{k_z h}{l_z} \, ; \tag{56}$$

where h is now the usual Planck constant.

We have derived (56) from very general considerations. The following special method of imposing the boundary conditions is illustrative.

We impose the periodicity condition

$$\psi_\lambda(x + l_x; \, y + l_y; \, z + l_z) = \psi_\lambda(x; \, y; \, z) \, . \tag{57}$$

From (52) and (57) it immediately follows that the conditions (56) should be satisfied. We thus see that the state of an electron confined in the volume $l_x \, l_y \, l_z$ can be specified by the quantum numbers k_x, k_y, and k_z, and that to the quantum state (k_x, k_y, k_z) there corresponds the following value for the energy, $\boldsymbol{E}$:

$$\left(\frac{\boldsymbol{E}}{c}\right)^2 = \left(\frac{k_x h}{l_x}\right)^2 + \left(\frac{k_y h}{l_y}\right)^2 + \left(\frac{k_z h}{l_z}\right)^2 + m^2 c^2 \, . \tag{58}$$

From (56) it follows that by making l_x, l_y, and l_z sufficiently large we can make the discrete eigenvalues of the momenta p_x, p_y, and p_z lie as closely together as we may choose. We can therefore ask as to the number of quantum states for the electron corresponding to a specified energy interval, $\boldsymbol{E}, \boldsymbol{E} + \Delta\boldsymbol{E}$, where $\Delta\boldsymbol{E}$ is large compared to the separation between the consecutive eigenvalues for $\boldsymbol{E}$.

Let $Z(\boldsymbol{E})\Delta\boldsymbol{E}$ be the number of quantum states in the specified energy interval, $\boldsymbol{E}, \boldsymbol{E} + \Delta\boldsymbol{E}$. To find $Z(\boldsymbol{E})\Delta\boldsymbol{E}$, we first consider the total number $J(\mathbf{E})$ of the quantum states for $\mathbf{E}$ less than the specified amount:

$$J(\mathbf{E}) = \int_0^{\mathbf{E}} Z(\mathbf{E}) d\mathbf{E} . \tag{59}$$

If we remember that for a given set of values of p_x, p_y, and p_z (which satisfy the relation [48]) there are two linearly independent solutions of the Dirac equation, it is clear that $J(\mathbf{E})$ is simply twice the number of points with integral co-ordinates inside the ellipsoid (58). The equation of the ellipsoid (58) can be re-written in the form

$$\frac{k_x^2}{a_x^2} + \frac{k_y^2}{a_y^2} + \frac{k_z^2}{a_z^2} = 1 , \tag{60}$$

where

$$\left.\begin{aligned} a_x = \frac{l_x}{h}\left(\frac{\mathbf{E}^2}{c^2} - m^2c^2\right)^{1/2} ; \qquad a_y = \frac{l_y}{h}\left(\frac{\mathbf{E}^2}{c^2} - m^2c^2\right)^{1/2} ; \\ a_z = \frac{l_z}{h}\left(\frac{\mathbf{E}^2}{c^2} - m^2c^2\right)^{1/2} . \end{aligned}\right\} \tag{61}$$

If a_x, a_y, and a_z are large compared to unity, the number of points with integral co-ordinates inside the ellipsoid (60) is simply the volume of the ellipsoid, which is

$$\frac{4\pi}{3} a_x a_y a_z . \tag{62}$$

Hence,

$$J(\mathbf{E}) = 2\,\frac{4\pi}{3}\,\frac{l_x l_y l_z}{h^3}\left(\frac{\mathbf{E}^2}{c^2} - m^2c^2\right)^{3/2} . \tag{63}$$

By (59) it now follows that the number of independent eigenfunctions (which is equal to the number of quantum states) belonging to the eigenvalues of $\mathbf{E}$ in the range $\mathbf{E}$, $\mathbf{E} + \Delta\mathbf{E}$ is obtained by differentiating (63) with respect to $\mathbf{E}$:

$$Z(\mathbf{E})\Delta\mathbf{E} = 2\,\frac{4\pi V}{h^3}\left(\frac{\mathbf{E}^2}{c^2} - m^2c^2\right)^{1/2}\frac{\mathbf{E}}{c^2}\,\Delta\mathbf{E} , \tag{64}$$

where $V = l_x l_y l_z$. If we denote the kinetic energy by E, we have

$$\mathbf{E} = E + mc^2 . \tag{65}$$

Equation (64) can now be written alternatively in the form

$$Z(E) = 2\,\frac{4\pi V}{h^3}\left(\frac{E}{c^2} + m\right)\left(2mE + \frac{E^2}{c^2}\right)^{1/2}. \tag{66}$$

On the other hand, if p denotes the absolute magnitude of the momentum, defined by

$$p^2 = p_x^2 + p_y^2 + p_z^2\,, \tag{67}$$

then, according to (48) and (65),

$$\frac{E^2}{c^2} + 2Em = p^2\,; \qquad pdp = \left(\frac{E}{c^2} + m\right) dE\,. \tag{68}$$

Equation (66) can therefore be written in the form

$$Z(p)dp = \frac{8\pi V}{h^3}\,p^2 dp\,, \tag{69}$$

which is the result quoted in § 1 (Eq. [1]).

We have derived the result (69) on the assumption that the electrons are confined in a rectangular box, but it is clear that the result should be quite generally true independent of the shape of the vessel. The most general proof of (66) and (69) is due to Peierls,[1] to whose derivation reference may be made.

3. *The Gibbs canonical ensemble and its properties.*—In the last section we saw that the number of quantum states with energy between E and $E + dE$ is given by

$$Z(E)dE = 2\,\frac{4\pi V}{h^3}\left(\frac{E}{c^2} + m\right)\left(2mE + \frac{E^2}{c^2}\right)^{1/2} dE\,. \tag{70}$$

The foregoing density of the quantum states in the scale of the kinetic energy is, in fact, a very general characteristic for a gas of material "particles."[2] Let the discrete eigenvalues of the energy E be denoted by $\epsilon_0, \epsilon_1, \epsilon_2, \ldots, \epsilon_s, \ldots$.

[1] R. Peierls, *M.N.*, **96**, 780, 1936. The same result is also obtained by E. K. Broch (*Phys. Rev.*, **51**, 586, 1937), who has explicitly solved the Dirac equation in a spherical potential hole and enumerated the states.

[2] We shall use the word "particle" to denote an electron, molecule, or atom. The theory presented in this section deals with a general assembly of similar particles.

Let us consider an assembly of N similar particles in a given volume V and with an internal energy U due to the kinetic energies of the individual particles. Now, since the particles are assumed to be similar, they cannot be distinguished from one another, and a microscopic state of the gas will be completely described by the specification of the number of particles, n_s, belonging to the eigenvalue ϵ_s for the energy E. We should then have

$$N = \sum_s n_s \tag{71}$$

and

$$U = \sum_s n_s \epsilon_s . \tag{72}$$

A possible sequence of numbers $n_1, n_2, \ldots, n_s, \ldots$, must satisfy the restrictions (71) and (72). We shall write the different sequences of values for the n_s's which satisfy (71) and (72) in the form

$$\left.\begin{array}{l} n_0^{(1)},\ n_1^{(1)},\ \ldots,\ n_s^{(1)},\ \ldots, \\ n_0^{(2)},\ n_1^{(2)},\ \ldots,\ n_s^{(2)},\ \ldots, \\ \ldots\ldots\ldots\ldots\ldots\ldots, \\ n_0^{(i)},\ n_1^{(i)},\ \ldots,\ n_s^{(i)},\ \ldots, \\ \ldots\ldots\ldots\ldots\ldots\ldots, \\ n_0^{(W)},\ n_1^{(W)},\ \ldots,\ n_s^{(W)},\ \ldots, \end{array}\right\} \tag{73}$$

where W is the number of different solutions in integers for the equations (71) and (72).

The entropy, S, is now defined by

$$S = k \log W , \tag{74}$$

where k is the Boltzmann constant. Instead of (74), we can write

$$e^{S/k} = W . \tag{75}$$

The actual justification of (74) and (75) will take us too far into the foundations of statistical mechanics in its relation to thermodynamics and for this reason we shall simply assume the validity of (74) and (75)—reference may be made, however, to the literature quoted in the bibliographical note.

Now the restrictions (71) and (72) can be dropped by the passage from the microcanonical state specified by (71) and (72) (according to which both N and U are defined exactly), to a canonical state in which both the energy U and the number of particles N are distributed canonically, i.e., in such a way that U and N have sharp maxima at certain prescribed values—say $\overline{U}$ and $\overline{N}$. This process, due to Gibbs, will become clear from the following discussion where the method of actually carrying out this passage to the canonical distribution is described. Our presentation closely follows a treatment originally due to Pauli.

First, let us try to replace condition (72) by one whereby U must have a sharp maximum at a certain specified value of $U = \overline{U}$ (say), while retaining the condition (71). This means that we make the passage from the microcanonical state in which both N and U are exactly specified to one in which N has the specified value (exactly), while U has an extremely sharp maximum at $\overline{U}$ in such a way that, as we shall see presently, U is appreciably different from zero for

$$U = \overline{U} \pm \Delta U \,, \tag{76}$$

where

$$\frac{\Delta U}{\overline{U}} \sim \frac{1}{\sqrt{N}} \,. \tag{77}$$

According to (72) and (75), we can write, for the microcanonical state,

$$e^{\frac{S}{k} - \vartheta U} = W e^{-\vartheta \sum_s n_s^{(i)} \epsilon_s} \,, \tag{78}$$

where ϑ is, for the present, an arbitrary constant. It is clear that in the summation occurring in the exponent in the right-hand side of (78) we can choose i to be any number from 1, 2, ... , W (cf. the scheme [73]); and, since for each value of i and j

$$\sum_s n_s^{(i)} \epsilon_s = \sum_s n_s^{(j)} \epsilon_s \qquad (i, j = 1, 2, \ldots . W) \,, \tag{79}$$

we can write (78) more "symmetrically" as

$$e^{\frac{S}{k} - \vartheta U} = \sum_{i=1}^{W} e^{-\vartheta \sum_s n_s^{(i)} \epsilon_s} \,. \tag{80}$$

We shall now drop the restriction (72) and write, instead of (80),

$$e^{\frac{S}{k}-\vartheta U} = \sum_q e^{-\vartheta \sum_s n_s^{(q)} \epsilon_s} , \tag{81}$$

where the index "q" means that we should now have

$$\sum_s n_s^{(q)} = N \qquad \text{for all } q , \tag{82}$$

without, however, the restriction (72). Further, the summation with respect to q in (81) is to be carried over all the different solutions in integers of equation (82). The quantity ϑ is now so chosen, that the expression (81), when differentiated with respect to U for fixed ϑ and fixed ϵ_s's (i.e., for a fixed V), vanishes. Hence,

$$\frac{1}{k}\left(\frac{\partial S}{\partial U}\right)_V - \vartheta = 0 . \tag{83}$$

Now according to the first and the second laws of thermodynamics,

$$dQ = TdS = dU + PdV , \tag{84}$$

so that

$$\left(\frac{\partial S}{\partial U}\right)_V = \frac{1}{T} . \tag{85}$$

Hence, according to (83) and (85),

$$\vartheta = \frac{1}{kT} . \tag{86}$$

Since the free energy, F, is defined by (cf. § 11, i)

$$F = U - TS , \tag{87}$$

we can now write (81) in the form

$$e^{-\frac{F}{kT}} = \sum_q e^{-\frac{1}{kT}\sum_s n_s^{(q)} \epsilon_s} . \tag{88}$$

We shall now show that according to (83) and (88), U has, in fact, an extremely sharp maximum at a certain $U = \overline{U}$ (say) and that U is appreciably different from zero only in a range of ΔU such that $\Delta U/U \sim 1/\sqrt{N}$. Further, we shall show that the entropy, defined according to (83) and (88), differs from that defined by (75) only by a quantity of the order of $\log N/N$. To prove these, we first remark that (81) is now interpreted by the statement that the probability of a microscopic state defined by a sequence of numbers $(n_1^{(q)}, \ldots, n_s^{(q)}, \ldots)$ and an energy $U = \sum_s n_s^{(q)} \epsilon_s$, is proportional to

$$e^{-\vartheta \sum_s n_s^{(q)} \epsilon_s} . \tag{89}$$

In order to obtain the probability of a microscopic state with a definite total energy U corresponding to the canonical distribution (89), we have to sum over all sequences $\{n_s^{(q)}\}$ which lead to the energy U.

Let $S_I(U, V)$ be the entropy defined according to (75), and $S_{II}(U, V)$ that defined according to (83) and (88). By our definition of $S_I(U, V)$, according to (89), the rule stated above for determining the probability of a microscopic state with a definite energy U, and (80) we find that

$$W(U) = \text{constant}\, e^{\frac{S_I}{k} - \vartheta U} . \tag{90}$$

If we now regard the right-hand side of (90) as a function of U, we see that $W(U)$ has a maximum at $U = \overline{U}$ (say), where

$$\frac{1}{k}\left(\frac{\partial S_I}{\partial U}\right)_V = \vartheta \qquad (U = \overline{U}) . \tag{91}$$

We therefore expand the exponent occurring in (90) in the neighborhood of $U = \overline{U}$ by a Taylor series in $\Delta U = U - \overline{U}$ and retain terms up to the second order in ΔU:

$$\left.\begin{aligned} \frac{S_I}{k} - \vartheta U = \frac{S_I(\overline{U}, V)}{k} - \vartheta \overline{U} + \left[\frac{1}{k}\left(\frac{\partial S_I}{\partial U}\right)_{V,\, U=\overline{U}} - \vartheta\right]\Delta U \\ + \frac{1}{2k}\left(\frac{\partial^2 S_I}{\partial U^2}\right)_{V,\, U=\overline{U}} (\Delta U)^2 + \ldots . \end{aligned}\right\} \tag{92}$$

By (90), (91), and (92) we now have

$$W(\Delta U) = \text{constant } e^{-\frac{1}{2k}\left|\left(\frac{\partial^2 S_I}{\partial U^2}\right)_{V,\, U=\bar{U}}\right|(\Delta U)^2}. \tag{93}$$

In writing (93) we have used the circumstance that $(\partial^2 S_I/\partial U^2)$ is negative, for, according to (85),

$$\left(\frac{\partial^2 S_I}{\partial U^2}\right)_V = -\frac{1}{T^2}\left(\frac{\partial T}{\partial U}\right)_V. \tag{94}$$

Hence, if we denote by $\overline{(\Delta U)^2}$ the "mean square error," we have

$$\overline{(\Delta U)^2} = \frac{k}{\left|\left(\frac{\partial^2 S_I}{\partial U^2}\right)_{V,\, U=\bar{U}}\right|}, \tag{95}$$

or

$$\overline{\left(\frac{\Delta U}{U}\right)^2} = \frac{k}{\left|\left(\frac{\partial^2 S_I}{\partial U^2}\right)_{V,\, U=\bar{U}}\right|\bar{U}^2}. \tag{96}$$

The right-hand side is easily seen to be of order N^{-1}; and hence,

$$\frac{\Delta U}{U} = O\left(\frac{1}{\sqrt{N}}\right). \tag{97}$$

In order to prove our second statement concerning the entropy, we write (81) in the form

$$e^{\frac{S_{II}}{k} - \vartheta \bar{U}} = \sum_q e^{-\vartheta \sum_s n_s^{(q)} \epsilon_s}. \tag{98}$$

To carry out the summation in (98) we first fix a certain value for U and select from the sequences $\{n_s^{(q)}\}$ those which correspond to a specified U. We then sum over all the possible U's. Equation (98) can then be written as

$$e^{\frac{S_{II}}{k} - \vartheta \bar{U}} = \sum_U W(U) e^{-\vartheta U}, \tag{99}$$

where $W(U)$ has the same meaning as in equations (73) and (75). Hence,

$$e^{\frac{S_{II}}{k}-\vartheta\bar{U}} = \sum_{U} e^{\frac{S_I(U)}{k}-\vartheta U} . \tag{100}$$

Expanding the exponent occurring on the right-hand side of (100) in a Taylor series about $U = \bar{U}$, we have, according to (91) and (92),

$$e^{\frac{S_{II}}{k}-\vartheta\bar{U}} = \sum_{\Delta U} e^{\frac{S_I(\bar{U})}{k}-\vartheta\bar{U}+\frac{1}{2k}\left(\frac{\partial^2 S_I}{\partial U^2}\right)_{V,\,U=\bar{U}}(\Delta U)^2} , \tag{101}$$

or

$$e^{\frac{S_{II}-S_I}{k}} = \sum_{\Delta U} e^{-\frac{1}{2k}\left|\left(\frac{\partial^2 S_I}{\partial U^2}\right)_{V,\,U=\bar{U}}\right|(\Delta U)^2} . \tag{102}$$

Replacing the sum by an integral, we have

$$e^{\frac{S_{II}-S_I}{k}} \simeq \frac{N}{\bar{U}}\int_{-\infty}^{+\infty} e^{-\frac{1}{2k}\left|\left(\frac{\partial^2 S_I}{\partial U^2}\right)_{V,\,U=\bar{U}}\right|(\Delta U)^2} d(\Delta U) , \tag{103}$$

or, finally,

$$e^{\frac{S_{II}-S_I}{k}} = \frac{N}{\bar{U}}\sqrt{\frac{2\pi k}{\left|\left(\frac{\partial^2 S_I}{\partial U^2}\right)_{V,\,U=\bar{U}}\right|}} . \tag{104}$$

Equation (104) is equivalent to

$$\frac{S_{II}-S_I}{S_I} = \frac{k}{S_I}\log\frac{N}{\bar{U}}\sqrt{\frac{2\pi k}{\left|\left(\frac{\partial^2 S_I}{\partial U^2}\right)_{V,\,U=\bar{U}}\right|}} = O\left(\frac{\log N}{N}\right) . \tag{105}$$

Further the maximum error in determining ϑ according to (83) (instead of according to [91] with S_I instead of S_{II} in [83]), will also be of the order (105).

Second, we now try to replace the condition (71) or (82) by the one that N is to have a sharp maximum at a certain specified

$N = \bar{N}$ (say), in such a way that, as we shall see, N is appreciably different from zero for

$$N = \bar{N} \pm \Delta N\ , \tag{106}$$

where

$$\frac{\Delta N}{N} \sim \frac{1}{\sqrt{N}}\ . \tag{107}$$

This corresponds to the passage to a canonical distribution, not only for the energy U but also for the number N of the particles concerned.[3] To do this we proceed as follows:

According to (82) and (88), we have

$$e^{-\frac{F}{kT}-\alpha N} = \sum_q e^{-\frac{1}{kT}\sum_s n_s^{(q)}\epsilon_s - \alpha \sum_s n_s^{(q)}}\ , \tag{108}$$

where α is, for the present, an arbitrary constant. To make the passage to a canonical distribution, we write, instead of (108),

$$e^{-\frac{F}{kT}-\alpha N} = \sum_{n_s} e^{-\sum_s \left(\alpha n_s + \frac{\epsilon_s n_s}{kT}\right)}\ , \tag{109}$$

where we no longer have the restriction (82) but α is now so chosen that the expression (109), when differentiated with respect to N for fixed temperature T and volume V vanishes. Hence,

$$\frac{1}{kT}\left(\frac{\partial F}{\partial N}\right)_{T,\,V} + \alpha = 0\ . \tag{110}$$

We shall now show (following Pauli) that N defined according to (109) and (110) has, in fact, an extremely sharp maximum at a certain $N = \bar{N}$ (say), and that N is appreciably different from zero only in the range $\Delta N/N \sim 1/\sqrt{N}$. Again, we shall show that the free energy defined according to (109) and (110) differs from that defined by (88) only by a quantity of the order $\log N/N$.

[3] This is the essential difference between Gibbs's classical treatment and the quantum mechanical version of Gibbs due to Pauli.

To prove these statements we remark that (109) is now interpreted by the statement that the probability of a microscopic state defined by a definite sequence of numbers $(n_1, \ldots, n_s, \ldots)$ is proportional to

$$e^{-\sum_s \left(\alpha n_s + \frac{\epsilon_s n_s}{kT}\right)}. \tag{111}$$

In order, then, to obtain the probability of a definite total number $N = \sum_s n_s$ of particles corresponding to the canonical distribution (111), we should sum over all such n_s-sequences which belong to the number N.

Let $F_I(N, T, V)$ be the free energy defined according to (88) and $F_{II}(N, T, V)$ that defined according to (109) and (110). By the definition of $F_I(N, T, V)$, and according to (111) and the rule stated above, the probability $W(N)$ for a definite total number N, is seen to be

$$W(N) = \text{constant } e^{-\left[\alpha N + \frac{F_I(N, T, V)}{kT}\right]}. \tag{112}$$

For a fixed α, $W(N)$ has a maximum where

$$\alpha + \frac{1}{kT}\left(\frac{\partial F_I}{\partial N}\right)_{T, V} = 0. \tag{113}$$

Let (113) be satisfied at $N = \overline{N}$ (say). We now expand the exponent occurring in (112) by a Taylor series in the neighborhood $N = \overline{N}$ in terms of $\Delta N = N - \overline{N}$ and obtain

$$\left.\begin{aligned} \alpha N + \frac{F_I(N, T, V)}{kT} &= \alpha \overline{N} + \frac{F_I(\overline{N}, T, V)}{kT} \\ &+ \left[\alpha + \frac{1}{kT}\left(\frac{\partial F_I}{\partial N}\right)_{N=\overline{N}}\right] \Delta N + \frac{1}{2kT}\left(\frac{\partial^2 F_I}{\partial N^2}\right)_{N=\overline{N}} (\Delta N)^2 + \ldots \end{aligned}\right\} \tag{114}$$

By (112), (113), and (114) we now have

$$W(\Delta N) = \text{constant } e^{-\frac{1}{2kT}\left(\frac{\partial^2 F_I}{\partial N^2}\right)_{N=\overline{N}} (\Delta N)^2} \tag{115}$$

Hence, the "mean square error," $\overline{(\Delta N)^2}$, is given by

$$\overline{(\Delta N)^2} = \frac{kT}{\left(\frac{\partial^2 F_I}{\partial N^2}\right)_{T,\,V,\,N=\overline{N}}}\,, \tag{116}$$

or

$$\overline{\left(\frac{\Delta N}{N}\right)^2} = \frac{kT}{\overline{N}^2\left(\frac{\partial^2 F_I}{\partial N^2}\right)_{T,\,V,\,N=\overline{N}}} = O\left(\frac{1}{N}\right), \tag{117}$$

which proves (107).

In order to prove the second statement concerning the free energy, we write (109) in the form

$$e^{-\frac{F_{II}}{kT}-\alpha\overline{N}} = \sum_{n_s} e^{-\sum_s\left(\alpha n_s+\frac{\epsilon_s n_s}{kT}\right)}. \tag{118}$$

To carry out the summation in (118), we first fix a certain value for N and select from all the sequences $\{n_s\}$ those which correspond to a specified N. We then sum over all the possible N's. Equation (118) can then be written as (cf. Eq. [88])

$$e^{-\frac{F_{II}}{kT}-\alpha\overline{N}} = \sum_N e^{-\left[\alpha N+\frac{F_I(N,\,T,\,V)}{kT}\right]}. \tag{119}$$

Expanding the exponent occurring on the right-hand side of (119) as a Taylor series about $N = \overline{N}$, we have, according to (113) and (114),

$$e^{-\frac{F_{II}}{kT}-\alpha\overline{N}} = \sum_{\Delta N} e^{-\alpha\overline{N}-\frac{F_I(\overline{N},\,T,\,V)}{kT}-\frac{1}{2kT}\left(\frac{\partial^2 F_I}{\partial N^2}\right)_{N=\overline{N}}(\Delta N)^2}, \tag{120}$$

or

$$e^{-\frac{F_{II}-F_I}{kT}} = \sum_{\Delta N} e^{-\frac{1}{2kT}\left(\frac{\partial^2 F_I}{\partial N^2}\right)_{N=\overline{N}}(\Delta N)^2}. \tag{121}$$

Replacing the sum by an integral, we have

$$e^{-\frac{F_{II}-F_I}{kT}} \simeq \int_{-\infty}^{+\infty} e^{-\frac{1}{2kT}\left(\frac{\partial^2 F_I}{\partial N^2}\right)_{N=\overline{N}}(\Delta N)^2} d(\Delta N)\,, \tag{122}$$

or

$$e^{-\frac{F_{II}-F_I}{kT}} = \sqrt{\frac{2\pi kT}{\left(\frac{\partial^2 F_I}{\partial N^2}\right)_{N=\bar{N}}}}. \tag{123}$$

Equation (123) is equivalent to

$$\frac{F_{II} - F_I}{F_I} = -\frac{kT}{F_I} \log \sqrt{\frac{2\pi kT}{\left(\frac{\partial^2 F_I}{\partial N^2}\right)_{V,\, T,\, N=\bar{N}}}} = O\left(\frac{\log N}{N}\right). \tag{124}$$

Further, the maximum error in determining α according to (110) instead of (113) with F_I (instead of F_{II} in [110]) will also be of the order (124).

We shall now return to (109). Since there is now no restriction with regard either to $\sum n_s$ or to $\sum n_s \epsilon_s$, we can re-write (109) in the form

$$e^{-\left(\frac{F}{kT}+\alpha N\right)} = \prod_s \sum_{n_s} e^{-n_s\left(\alpha+\frac{\epsilon_s}{kT}\right)}. \tag{125}$$

Now equation (110) is to serve the purpose of determining α. We can transform this into a more convenient form as follows: Differentiate (125) logarithmically with respect to N keeping V and T constant. Then,

$$\frac{1}{kT}\left(\frac{\partial F}{\partial N}\right)_{T,\, V} + \alpha + N\left(\frac{\partial \alpha}{\partial N}\right)_{T,\, V} = \left(\frac{\partial \alpha}{\partial N}\right)_{T,\, V} \sum_s \frac{\sum_{n_s} n_s e^{-n_s\left(\alpha+\frac{\epsilon_s}{kT}\right)}}{\sum_{n_s} e^{-n_s\left(\alpha+\frac{\epsilon_s}{kT}\right)}}. \tag{126}$$

Hence, according to (110), we have, since $(\partial\alpha/\partial N)$ is not in general zero,

$$N = \sum_s \frac{\sum_{n_s} n_s e^{-n_s\left(\alpha+\frac{\epsilon_s}{kT}\right)}}{\sum_{n_s} e^{-n_s\left(\alpha+\frac{\epsilon_s}{kT}\right)}}. \tag{127}$$

The thermodynamical significance of α can be found as follows: As we shall see presently, (F/N) depends, apart from temperature, only on the concentration of the particles N/V. Consequently, we can write

$$F = Nf\left(\frac{N}{V}, T\right). \tag{128}$$

Hence,

$$\left(\frac{\partial F}{\partial N}\right)_{T,V} = f\left(\frac{N}{V}, T\right) + \frac{N}{V}\frac{\partial f}{\partial (N/V)}. \tag{129}$$

But by (128)

$$\left(\frac{\partial F}{\partial V}\right)_{N,T} = -\frac{N^2}{V^2}\frac{\partial f}{\partial (N/V)}. \tag{130}$$

From (129) and (130) we derive

$$N\left(\frac{\partial F}{\partial N}\right)_{T,V} = F - V\left(\frac{\partial F}{\partial V}\right)_{N,T}. \tag{131}$$

Since, however, we have the thermodynamical relation (chap. i, Eq. 110),

$$P = -\left(\frac{\partial F}{\partial V}\right)_T, \tag{132}$$

we have

$$N\left(\frac{\partial F}{\partial N}\right)_{T,V} = F + PV = G, \tag{133}$$

where G is the thermodynamic potential at constant pressure (cf. § 12, i). Hence, by (110) and (133),

$$\alpha = -\frac{G}{NkT} = -\frac{F + PV}{NkT}, \tag{134}$$

which then gives the thermodynamical meaning of the parameter α.

For the calculation of the statistical mean value of any physical quantity it is important to note that the quantity

$$e^{-\sum_s n_s\left(\alpha + \frac{\epsilon_s}{kT}\right)} = \prod_s e^{-n_s\left(\alpha + \frac{\epsilon_s}{kT}\right)}, \tag{135}$$

which occurs in (125), is, apart from a constant, the probability of a definite microscopic state:

$$W(n_1, n_2, \ldots, n_s, \ldots) = \text{constant} \prod_s e^{-n_s\left(\alpha+\frac{\epsilon_s}{kT}\right)}. \quad (136)$$

If we compare two microscopic states $(n_1, \ldots, n_s, \ldots)$ and $(n_1', \ldots, n_s', \ldots)$ for which the total number of particles (71) and the total energy (72) are equal (or nearly equal), then, according to (136), the two states are equally probable; this is, in fact, an assumption implicit in equation (74).

For the internal energy U we have, immediately,

$$U = \sum_s \frac{\sum_{n_s} n_s \epsilon_s e^{-n_s\left(\alpha+\frac{\epsilon_s}{kT}\right)}}{\sum_{n_s} e^{-n_s\left(\alpha+\frac{\epsilon_s}{kT}\right)}}. \quad (137)$$

Equation (137) follows also from the thermodynamical relation (chap. i, Eq. [110])

$$U = F - T\left(\frac{\partial F}{\partial T}\right)_{V,N} = -T^2 \frac{\partial}{\partial T}\left(\frac{F}{T}\right)_{V,N}. \quad (138)$$

On the other hand, differentiating (125) logarithmically with respect to T and keeping N fixed, we have

$$\left.\begin{aligned} &\frac{1}{k}\frac{\partial}{\partial T}\left(\frac{F}{T}\right)_{V,N} + N\left(\frac{\partial \alpha}{\partial T}\right)_{V,N} \\ &= \left(\frac{\partial \alpha}{\partial T}\right)_{V,N} \sum_s \frac{\sum_{n_s} n_s e^{-n_s\left(\alpha+\frac{\epsilon_s}{kT}\right)}}{\sum_{n_s} e^{-n_s\left(\alpha+\frac{\epsilon_s}{kT}\right)}} - \frac{1}{kT^2}\sum_s \frac{\sum_{n_s} n_s \epsilon_s e^{-n_s\left(\alpha+\frac{\epsilon_s}{kT}\right)}}{\sum_{n_s} e^{-n_s\left(\alpha+\frac{\epsilon_s}{kT}\right)}}. \end{aligned}\right\} \quad (139)$$

By (127) the terms proportional to $(\partial \alpha/\partial T)$ cancel, and (138) and (139) together imply precisely the expression (137) for U.

Finally we shall obtain some formulae which are of practical importance in the application of the theory.

By (134)

$$PV = -kT\left(\frac{F}{kT} + \alpha N\right). \tag{140}$$

Hence, according to (125), we can write

$$PV = kT\sum_s \log \sigma_s , \tag{141}$$

where

$$\sigma_s = \sum_{n_s} e^{-n_s(\alpha+\vartheta\epsilon_s)} . \tag{142}$$

Equations (127) and (137) can now be written in the form

$$N = -\sum_s \frac{\partial}{\partial\alpha} \log \sigma_s \tag{143}$$

and

$$U = -\sum_s \frac{\partial}{\partial\vartheta} \log \sigma_s \qquad \left(\vartheta = \frac{1}{kT}\right). \tag{144}$$

Equations (141), (143), and (144) are extremely general and give the physical variables for a system in statistical equilibrium which is also a thermodynamical system.

4. *The symmetrical and the antisymmetrical states; the Einstein-Bose and the Fermi-Dirac distributions.*—If we consider a system containing a number of similar particles, then no observable change is made when two of them are interchanged. A satisfactory theory, then, should consider two such observationally indistinguishable states as really the same state.

Suppose we have a system of N similar particles. Let $q_1, q_2, \ldots\ldots,$ q_N, be the variables describing the first, the second, ... , the Nth particle in the system. Then the Hamiltonian, $\boldsymbol{H}$, of the system will be a function of the variables $q_1, q_2, \ldots\ldots, q_N$:

$$\boldsymbol{H} = \boldsymbol{H}(q_1; q_2; \ldots\ldots; q_N) . \tag{145}$$

Since the particles are indistinguishable from one another it is clear that $\boldsymbol{H}$ should be symmetrical in all the particles, i.e., symmetrical in the variables $q_1, \ldots, q_N$. If Ψ is a wave function describing the system, then we should have

$$\left(\boldsymbol{H} - i\hbar \frac{\partial}{\partial t}\right)\Psi = 0 . \qquad (146)$$

From the foregoing it follows that if $\Psi(q_1, \ldots, q_N)$ is a solution of (146), then so is $\boldsymbol{P}\Psi(q_1, q_2, \ldots, q_N)$, where $\boldsymbol{P}\Psi$ stands for the function obtained by applying the permutation $\boldsymbol{P}$ to the variables $q_1, \ldots, q_N$.

Suppose that at any given time,

$$\boldsymbol{P}\Psi \equiv 0 \qquad (t = t_0) ; \quad (147)$$

then, since $\boldsymbol{H}$ is an operation in the space variables only, we have

$$\boldsymbol{HP}\Psi = [\boldsymbol{H}]0 = 0 , \qquad (148)$$

so that by (146) $\partial(\boldsymbol{P}\Psi)/\partial t \equiv 0$ at $t = t_0$. If, now, $\boldsymbol{H}$ and $\boldsymbol{P}\Psi$ are analytic functions of t for all real values of t, it follows that we can prove by repeated applications of the argument that

$$\frac{\partial^n}{\partial t^n}(\boldsymbol{P}\Psi) \equiv 0 \qquad (t = t_0) , \quad (149)$$

for all n, and that therefore $\boldsymbol{P}\Psi \equiv 0$ for all time. From this it follows that if Ψ is of a given "symmetry character" at a given instant of time, it retains its "symmetry character" for all time. In particular, if the wave function is initially symmetrical (i.e., is unaltered by any permutation of the variables $q_1, \ldots, q_N$), then it is symmetrical for all time. In the same way, if the wave function is initially antisymmetrical (i.e., is unaltered or changes sign according as an even or an odd permutation[4] is applied to the variables), then it is antisymmetrical for all time.

[4] A simple interchange is an odd permutation, while two interchanges will be an even permutation.

The permanency of the symmetry properties of the state means that for some kind of particles only the symmetrical or the antisymmetrical states occur. It is found that light quanta should be described by symmetrical wave functions (as we shall see, it is only then that we have Planck's law for radiation). On the other hand, the electrons should be described by antisymmetrical wave functions, only then can we obtain Pauli's exclusion principle, which states that no two electrons can be described by the same set of quantum numbers. For if two electrons were described by the same set of quantum numbers, then an interchange of the variables corresponding to these two electrons must leave the wave function unaltered; the wave function under these circumstances can vanish identically only if it is antisymmetrical in the variables of the two electrons. Since the "two electrons" can be any two, the wave function must be antisymmetrical in all the variables describing the different electrons.

For our purposes it is only necessary to remark that in the symmetrical case there can be 0, 1, 2, ... , ∞, particles in the same quantum state, while in the antisymmetrical case there can only be 0 or 1 particle in a specified quantum state. The former case leads to the Einstein-Bose statistics while the latter case leads to the Fermi-Dirac statistics.[5] Hence, according to equation (142) of the last section, we have for these two cases,

$$\sigma_s = \sum_{n_s=0}^{\infty} e^{-n_s(\alpha+\vartheta\epsilon_s)} = \frac{1}{1 - e^{-(\alpha+\vartheta\epsilon_s)}} \qquad \text{(symmetrical case)} \qquad (150)$$

and

$$\sigma_s = 1 + e^{-(\alpha+\vartheta\epsilon_s)} \qquad \text{(antisymmetrical case)} . \qquad (151)$$

[5] It is somewhat misleading to use the word "statistics" in "Einstein-Bose statistics" and "Fermi-Dirac statistics." There is only one statistics, namely, the Gibbs statistics described in § 3. The symmetrical and the antisymmetrical cases simply correspond to two different assumptions for the evaluation of σ_s (Eq. [142]); the explicit forms for N, U, and PV naturally differ, but nevertheless we have the same statistical theory (Gibbs) underlying both the cases. It would be more logical to refer to "Einstein-Bose formulae" and "Fermi-Dirac formulae."

From (150) and (151) we have, respectively,

$$-\frac{\partial}{\partial a}\log \sigma_s = \left\{\begin{array}{c}\dfrac{e^{-(a+\vartheta\epsilon_s)}}{1 - e^{-(a+\vartheta\epsilon_s)}} \\ \\ \dfrac{e^{-(a+\vartheta\epsilon_s)}}{1 + e^{-(a+\vartheta\epsilon_s)}}\end{array}\right\} = \frac{1}{e^{a+\vartheta\epsilon_s} \mp 1} \tag{152}$$

and

$$-\frac{\partial}{\partial \vartheta}\log \sigma_s = \frac{\epsilon_s}{\epsilon^{a+\vartheta\epsilon_s} \mp 1} \qquad \left(\vartheta = \frac{1}{kT}\right). \tag{153}$$

In (152) and (153) the minus sign corresponds to the symmetrical (Einstein-Bose) case and the plus sign to the antisymmetrical (Fermi-Dirac) case. Finally, according to (141), (143), and (144), we have

$$\frac{PV}{kT} = \mp\sum_s \log\left(1 \mp e^{-(a+\vartheta\epsilon_s)}\right), \tag{154}$$

$$N = \sum_s \frac{1}{e^{a+\vartheta\epsilon_s} \mp 1}, \tag{155}$$

and

$$U = \sum_s \frac{\epsilon_s}{e^{a+\vartheta\epsilon_s} \mp 1}. \tag{156}$$

5. *The electron gas: general formulae.*—For an electron assembly we should use the results for the antisymmetrical case considered in § 4. The summation over "s" occurring in equations (154), (155), and (156) can be transformed into integrals if we remember that the density of the quantum states is given by $Z(E)$ (the explicit expression for which is derived in § 2):

$$N = \int_0^\infty \frac{Z(E)dE}{e^{a+\vartheta E} + 1}; \qquad U = \int_0^\infty \frac{Z(E)EdE}{e^{a+\vartheta E} + 1}, \tag{157}$$

$$PV = \frac{1}{\vartheta}\int_0^\infty \log\left[1 + e^{-(a+\vartheta E)}\right]Z(E)dE. \tag{158}$$

The expressions take their simplest forms when, instead of the kinetic energy E we choose the momentum p as the variable for the integration. Then, according to (69) and (68),

$$Z(p)dp = \frac{8\pi V}{h^3} p^2 dp \,, \tag{159}$$

where

$$\frac{E^2}{c^2} + 2Em = p^2 \,. \tag{160}$$

Equations (157) and (158) can now be written in the forms

$$N = \frac{8\pi V}{h^3}\int_0^\infty \frac{p^2 dp}{e^{\alpha+\vartheta E}+1} \,; \qquad U = \frac{8\pi V}{h^3}\int_0^\infty \frac{Ep^2 dp}{e^{\alpha+\vartheta E}+1} \,, \tag{161}$$

$$PV = \frac{8\pi V}{h^3 \vartheta}\int_0^\infty \log\left[1 + e^{-(\alpha+\vartheta E)}\right] p^2 dp \,. \tag{162}$$

Equation (162) can be transformed by an integration by parts so that

$$PV = \frac{8\pi V}{3h^3}\int_0^\infty \frac{p^3}{e^{\alpha+\vartheta E}+1}\frac{\partial E}{\partial p}\, dp \,. \tag{163}$$

The equations for U and P can be derived in an elementary way on the basis of the distribution function,

$$N(p)dp = \frac{8\pi V}{h^3}\frac{p^2 dp}{e^{\alpha+\vartheta E}+1} \,, \tag{164}$$

which gives the number of particles in the assembly which have momenta between p and $p + dp$. In particular, equation (163) is consistent with our definition of pressure used in § 1, above.

We can obtain (164), or more generally, an expression for the number of electrons in the assembly with the components of the momentum in the range (p_x, p_y, p_z; $p_x + dp_x$, $p_y + dp_y$, $p_z + dp_z$), as follows:

From equations (56) it follows that the number of quantum states in the specified range is given by

$$2\,\frac{l_x l_y l_z}{h^3}\, dp_x dp_y dp_z = 2\,\frac{V}{h^3}\, dp_x dp_y dp_z \,. \tag{165}$$

The factor 2 in (165) arises from the circumstance that for a given set of values for p_x, p_y, and p_z the Dirac equation has two (or no) linearly independent solutions according as (48) is satisfied (or not). The number of electrons in the range specified is obtained by summing (155), not over all the quantum states but only over those in the range specified. We thus have

$$dN = 2\frac{V}{h^3}\frac{dp_x dp_y dp_z}{e^{\alpha+\vartheta E}+1}, \tag{166}$$

which expresses the Fermi distribution for the momentum components. If α is very large, then we can neglect the term unity occurring in the denominator in (166) and obtain

$$dN = 2\frac{V}{h^3}e^{-\alpha-\vartheta E}dp_x dp_y dp_z, \tag{167}$$

which expresses Maxwell's law of the distribution of momenta. The case $\alpha \gg 1$ is called the nondegenerate case. On the other hand, if α is large and negative the Fermi distribution becomes markedly different from the Maxwell distribution and the gas is then said to be degenerate. We shall consider these questions in greater detail in the following sections, but we shall now obtain a very convenient form of the equations (161) and (163). The transformations to be introduced are due to Juttner. Let

$$\frac{p}{mc} = \sinh\theta, \tag{168}$$

$$E = mc^2(\cosh\theta - 1). \tag{169}$$

Then we easily derive

$$N = \frac{8\pi V m^3c^3}{h^3}\int_0^\infty \frac{\sinh^2\theta\cosh\theta\, d\theta}{\frac{1}{\Lambda}e^{\vartheta mc^2\cosh\theta}+1}, \tag{170}$$

$$U = \frac{8\pi V m^4c^5}{h^3}\int_0^\infty \frac{\sinh^2\theta\cosh\theta(\cosh\theta-1)d\theta}{\frac{1}{\Lambda}e^{\vartheta mc^2\cosh\theta}+1}, \tag{171}$$

and

$$P = \frac{8\pi m^4 c^5}{3h^3} \int_0^\infty \frac{\sinh^4 \theta \, d\theta}{\frac{1}{\Lambda} e^{\vartheta mc^2 \cosh \theta} + 1} , \tag{172}$$

where we have used

$$\frac{1}{\Lambda} = e^{\alpha - \vartheta mc^2} . \tag{173}$$

6. *The degenerate case.*—As we have already pointed out, the degenerate case corresponds to the case where α is large and negative. A condition equivalent to this is that Λ (as defined in [173]) is very large compared to unity.

It is clear that as $\Lambda \to \infty$ the term

$$\frac{1}{\Lambda} e^{\vartheta mc^2 \cosh \theta} \tag{174}$$

occurring in the denominator in (170), (171), and (172) is negligible compared to unity for all $\theta \leqslant \theta_0$ where θ_0 is defined by

$$\log \Lambda = \vartheta mc^2 \cosh \theta_0 . \tag{175}$$

We can therefore write as a first approximation (a rigorous justification is given later in this section)

$$N = \frac{8\pi V m^3 c^3}{h^3} \int_0^{\theta_0} \sinh^2 \theta \cosh \theta \, d\theta , \tag{176}$$

$$U = \frac{8\pi V m^4 c^5}{h^3} \int_0^{\theta_0} \sinh^2 \theta \cosh \theta \, (\cosh \theta - 1) d\theta , \tag{177}$$

and

$$P = \frac{8\pi m^4 c^5}{3h^3} \int_0^{\theta_0} \sinh^4 \theta \, d\theta . \tag{178}$$

The foregoing expressions are precisely those considered in § 1 (eqs. [6], [10], and especially [16]). In order, however, to consider more explicitly the circumstances under which the foregoing approximation becomes valid, we shall have to evaluate the integrals (170),

(171), and (172) to a higher degree of approximation than above. To do this we shall first prove the following lemma (due to Sommerfeld).

SOMMERFELD'S LEMMA.—*If* $\varphi(u)$ *is a sufficiently regular function which vanishes for* $u = 0$, *then we have the asymptotic formula*

$$\int_0^\infty \frac{du}{\frac{1}{\Lambda} e^u + 1} \frac{d\varphi(u)}{du} = \varphi(u_0) + 2[c_2\varphi''(u_0) + c_4\varphi^{(iv)}(u_0) + \ldots.], \quad (179)$$

where $u_0 = \log \Lambda$ *and* $c_2, c_4, \ldots.$, *are numerical coefficients defined by*

$$c_\nu = 1 - \frac{1}{2^\nu} + \frac{1}{3^\nu} - \frac{1}{4^\nu} + \ldots.. \quad (180)$$

The asymptotic formula (179) is valid if we neglect quantities of the order $e^{-u_0} = \Lambda^{-1}$.

Proof: Split the range of integration at $u_0 = \log \Lambda$. We then have

$$\left.\begin{aligned} \int_0^\infty \frac{du}{\frac{1}{\Lambda} e^u + 1} \frac{d\varphi(u)}{du} &= \int_0^{u_0} \frac{d\varphi(u)}{du} du \\ &+ \int_0^{u_0} \left(\frac{1}{\frac{1}{\Lambda} e^u + 1} - 1\right) \frac{d\varphi(u)}{du} du + \int_{u_0}^\infty \frac{du}{\frac{1}{\Lambda} e^u + 1} \frac{d\varphi(u)}{du}, \end{aligned}\right\} \quad (181)$$

or

$$\left.\begin{aligned} \int_0^\infty \frac{du}{\frac{1}{\Lambda} e^u + 1} \frac{d\varphi(u)}{du} &= \varphi(u_0) - \int_0^{u_0} \frac{du}{1 + \Lambda e^{-u}} \frac{d\varphi(u)}{du} \\ &+ \int_{u_0}^\infty \frac{du}{\frac{1}{\Lambda} e^u + 1} \frac{d\varphi(u)}{du}. \end{aligned}\right\} \quad (182)$$

In the first integral occurring on the right-hand side of (182) put

$$u = u_0(1 - t), \quad (183)$$

and in the second integral occurring on the right-hand side put

$$u = u_0(1 + t) . \tag{184}$$

We now have (remembering that $u_0 = \log \Lambda$)

$$\left.\begin{aligned}\int_0^\infty \frac{du}{\frac{1}{\Lambda}e^u + 1}\frac{d\varphi(u)}{du} = \varphi(u_0) - u_0 \int_0^1 \frac{\varphi'[u_0(1-t)]}{1 + e^{u_0 t}}\,dt \\ + u_0 \int_0^\infty \frac{\varphi'[u_0(1+t)]}{1 + e^{u_0 t}}\,dt .\end{aligned}\right\} \tag{185}$$

In the first integral occurring on the right-hand side of (185) we can extend the range of integration to ∞; this will introduce an error of the order e^{-u_0}, which is beyond the range of accuracy of the asymptotic formula we are establishing. Hence, we have

$$\begin{aligned}\int_0^\infty \frac{du}{\frac{1}{\Lambda}e^u + 1}\frac{d\varphi(u)}{du} \\ \simeq \varphi(u_0) + u_0 \int_0^\infty \frac{\varphi'[u_0(1+t)] - \varphi'[u_0(1-t)]}{1 + e^{u_0 t}}\,dt \\ = \varphi(u_0) + 2 \sum_{\nu = 2, 4, 6, \dots} \frac{u_0^\nu \varphi^{(\nu)}(u_0)}{(\nu - 1)!}\int_0^\infty \frac{t^{\nu-1}}{1 + e^{u_0 t}}\,dt .\end{aligned} \tag{186}$$

On the other hand, we have

$$\left.\begin{aligned}\int_0^\infty \frac{t^{\nu-1}}{1 + e^{u_0 t}} = \int_0^\infty t^{\nu-1}(e^{-u_0 t} - e^{-2u_0 t} + e^{-3u_0 t} - \dots)dt \\ = \frac{(\nu - 1)!}{u_0^\nu}\left(1 - \frac{1}{2^\nu} + \frac{1}{3^\nu} - \dots\right) .\end{aligned}\right\} \tag{187}$$

Since the constants c_ν are defined according to (180), we have

$$\int_0^\infty \frac{du}{\frac{1}{\Lambda}e^u + 1}\frac{d\varphi(u)}{du} \simeq \varphi(u_0) + 2[c_2\varphi''(u_0) + c_4\varphi^{(iv)}(u_0) + \dots] , \tag{188}$$

which proves the lemma. We may note that

$$c_2 = \frac{\pi^2}{12}\,; \qquad c_4 = \frac{7\pi^4}{720}\,; \qquad c_6 = \frac{31\pi^6}{30{,}240}\,. \tag{189}$$

In order to apply (188) to an asymptotic evaluation of the integrals (170), (171), and (172), we must first transform them into suitable forms. Let

$$\vartheta mc^2 \cosh\theta = u\,. \tag{190}$$

First consider the integral for P. Then

$$P = \frac{8\pi m^4 c^5}{3h^3}\,\frac{1}{\vartheta mc^2}\int_0^\infty \frac{du}{\frac{1}{\Lambda}e^u + 1}\,\frac{d\varphi(u)}{du}\,, \tag{191}$$

where we have put

$$\frac{d\varphi(u)}{du} = \sinh^3\theta\,. \tag{192}$$

Hence,

$$\varphi(u_0) = \int_0^{u_0} \sinh^3\theta\,du = \vartheta mc^2 \int_0^{\theta_0} \sinh^4\theta\,d\theta\,, \tag{193}$$

which is an integral we have already computed (Eq. [16]). Introducing, as before, the variable x which is defined by (Eq. [18]),

$$x = \frac{p_0}{mc} = \sinh\theta_0\,, \tag{194}$$

we have

$$\varphi(u_0) = \frac{\vartheta mc^2}{8} f(x)\,, \tag{195}$$

where $f(x)$ is defined as in equation (20). From (192) we derive that

$$\left(\frac{d^2\varphi}{du^2}\right)_{u=u_0} = \frac{3}{\vartheta mc^2}\sinh\theta_0\cosh\theta_0 = \frac{3}{\vartheta mc^2}\,x(x^2+1)^{1/2} \tag{196}$$

and

$$\left.\begin{aligned}\left(\frac{d^4\varphi}{du^4}\right)_{u=u_0} &= \frac{3}{(\vartheta mc^2)^3}\,\frac{\cosh\theta_0}{\sinh^3\theta_0}\,(2\sinh^2\theta_0 - 1)\\ &= \frac{3}{(\vartheta mc^2)^3}\,\frac{(x^2+1)^{1/2}(2x^2-1)}{x^3}\,.\end{aligned}\right\} \tag{197}$$

By Sommerfeld's lemma it now follows that

$$P = \frac{\pi m^4 c^5}{3h^3} f(x) \Bigg[1 + \frac{4\pi^2}{(\vartheta mc^2)^2} \frac{x(x^2+1)^{1/2}}{f(x)} + \frac{7\pi^4}{15(\vartheta mc^2)^4} \frac{(x^2+1)^{1/2}(2x^2-1)}{x^3 f(x)} + \ldots \Bigg] . \tag{198}$$

To evaluate the integral for N we write it as

$$N = \frac{8\pi V m^3 c^3}{h^3} \frac{1}{\vartheta mc^2} \int_0^\infty \frac{du}{\frac{1}{\Lambda} e^u + 1} \frac{d\varphi(u)}{du} , \tag{199}$$

where $\varphi(u)$ is now defined by

$$\frac{d\varphi(u)}{du} = \tfrac{1}{2} \sinh 2\theta . \tag{200}$$

From (200) we derive that

$$\varphi(u_0) = \frac{\vartheta mc^2}{3} x^3 , \tag{201}$$

$$\left(\frac{d^2\varphi}{du^2}\right)_{u=u_0} = \frac{1}{\vartheta mc^2} \frac{2x^2+1}{x} , \tag{202}$$

and

$$\left(\frac{d^4\varphi}{du^4}\right)_{u=u_0} = \frac{3}{(\vartheta mc^2)^3} \frac{1}{x^5} . \tag{203}$$

We thus have

$$N = \frac{8\pi V m^3 c^3}{3h^3} x^3 \left[1 + \frac{\pi^2}{(\vartheta mc^2)^2} \frac{2x^2+1}{2x^4} + \frac{7\pi^4}{40(\vartheta mc^2)^4} \frac{1}{x^8} + \ldots \right] . \tag{204}$$

The integral (171) for U can be evaluated similarly. We find that

$$U = \frac{\pi V m^4 c^5}{3h^3} g(x) \left[1 + \frac{4\pi^2}{(\vartheta mc^2)^2} \frac{(3x^2+1)(x^2+1)^{1/2} - (2x^2+1)}{x g(x)} + \ldots \right] , \tag{205}$$

where $g(x)$ is defined as in equation (23).

On comparing (198), (204), and (205) with the corresponding expressions (19), (21), and (22), we see that the dominant terms in the present expansions agree with our earlier expressions for complete degeneracy. On the other hand, we now see that the necessary condition for the convergence of the foregoing expansions is that

$$\frac{4\pi^2}{(\vartheta mc^2)^2}\,\frac{x(x^2+1)^{1/2}}{f(x)} \ll 1\,, \tag{206}$$

$$\frac{\pi^2}{(\vartheta mc^2)^2}\,\frac{2x^2+1}{2x^4} \ll 1\,, \tag{207}$$

and

$$\frac{4\pi^2}{(\vartheta mc^2)^2}\,\frac{(3x^2+1)(x^2+1)^{1/2}-(2x^2+1)}{xg(x)} \ll 1\,. \tag{208}$$

As $x \to 0$, the foregoing inequalities take the limiting forms (cf. Eqs. [24] and [29])

$$\frac{\pi^2}{(\vartheta mc^2)^2}\,\frac{5}{2x^4} \ll 1\,; \qquad \frac{\pi^2}{(\vartheta mc^2)^2}\,\frac{1}{2x^4} \ll 1\,; \qquad \frac{\pi^2}{(\vartheta mc^2)^2}\,\frac{5}{2x^4} \ll 1\,. \tag{209}$$

Again, as $x \to \infty$, the inequalities (206), (207), and (208), take the limiting forms

$$\frac{\pi^2}{(\vartheta mc^2)^2}\,\frac{2}{x^2} \ll 1\,; \qquad \frac{\pi^2}{(\vartheta mc^2)^2}\,\frac{1}{x^2} \ll 1\,; \qquad \frac{\pi^2}{(\vartheta mc^2)^2}\,\frac{2}{x^2} \ll 1\,. \tag{210}$$

From (209) and (210) it is clear that *a necessary and sufficient condition for the setting-in of degeneracy is*

$$\frac{4\pi^2}{(\vartheta mc^2)^2}\,\frac{x(x^2+1)^{1/2}}{f(x)} \ll 1\,. \tag{211}$$

The inequality (211) implies the other two ([207] and [208]).

In using (211) as a criterion for degeneracy, it should be remembered that x is related (in a first approximation) to the mean concentration, n, of the electrons by

$$n = \frac{8\pi m^3c^3}{3h^3}\,x^3\,; \qquad x = \frac{h}{mc}\left(\frac{3n}{8\pi}\right)^{1/3}. \tag{212}$$

Further, ϑ is $1/kT$.

For astronomical applications, the criterion of degeneracy is stated more conveniently in a somewhat different form, which will be obtained in the next chapter.

We shall conclude the discussion of the case of degeneracy with a derivation of the specific heat, C_V, of the electrons at constant volume. To evaluate C_V we note that, by definition,

$$C_V = \left(\frac{\partial U}{\partial T}\right)_{N,\,V} . \tag{213}$$

According to (204), the condition of the constancy of N is equivalent to

$$\left(\frac{\partial x}{\partial T}\right)_V = -\frac{\pi^2}{T(\vartheta mc^2)^2}\,\frac{2x^2+1}{3x^3} . \tag{214}$$

By (205)

$$\left.\begin{aligned}\left(\frac{\partial U}{\partial T}\right)_V = \frac{\pi V m^4 c^5}{3h^3}\Bigg[&\frac{dg}{dx}\left(\frac{\partial x}{\partial T}\right)_V \\ &+ \frac{8\pi^2}{T(\vartheta mc^2)^2}\,\frac{(3x^2+1)(x^2+1)^{1/2}-(2x^2+1)}{x}\Bigg] .\end{aligned}\right\} \tag{215}$$

It is easily found that

$$\frac{dg}{dx} = 24x^2[(x^2+1)^{1/2}-1] . \tag{216}$$

By (214), (215), and (216) we have

$$C_V = \frac{8\pi^3 V m^4 c^5}{3h^3 T(\vartheta mc^2)^2}\, x(x^2+1)^{1/2} ; \tag{217}$$

or by (204), the specific heat per electron is given by

$$\frac{C_V}{N} = \frac{\pi^2 k^2}{mc^2}\,\frac{(x^2+1)^{1/2}}{x^2}\, T . \tag{218}$$

7. *The nondegenerate Maxwell-Juttner case.*—Let us now consider the other limiting case when Λ^{-1} is very large compared to unity. We can then neglect the unity occurring in the denominators of the integrands of (170), (171), and (172). The present case is therefore

the opposite extreme to the one considered in § 1. The integrals for N, P, and U can now be written as

$$N = \frac{8\pi V m^3 c^3}{h^3} \Lambda \int_0^\infty e^{-\vartheta mc^2 \cosh\theta} \sinh^2\theta \cosh\theta \, d\theta \,, \tag{219}$$

$$U = \frac{8\pi V m^4 c^5}{h^3} \Lambda \int_0^\infty e^{-\vartheta mc^2 \cosh\theta} \sinh^2\theta \cosh\theta \,(\cosh\theta - 1) d\theta \,, \tag{220}$$

and

$$PV = \frac{8\pi V m^4 c^5}{3h^3} \Lambda \int_0^\infty e^{-\vartheta mc^2 \cosh\theta} \sinh^4\theta \, d\theta \,. \tag{221}$$

The last integral can be simplified by an integration by parts. We find that

$$PV = \frac{8\pi V m^3 c^3}{h^3} \frac{1}{\vartheta} \Lambda \int_0^\infty e^{-\vartheta mc^2 \cosh\theta} \sinh^2\theta \cosh\theta \, d\theta \,. \tag{222}$$

Comparing (219) and (222), we see that

$$N = PV\vartheta \qquad \text{or} \qquad PV = NkT \,. \tag{223}$$

In other words, *Boyle's law is identically true for the nondegenerate case.* This important result was first established by Juttner, though it is implicit in some earlier work by Planck.

The integrals occurring in (219) and (220) can be evaluated explicitly in terms of Bessel functions. We use the formula[6]

$$\int_0^\infty e^{-z\cosh\theta} \cosh \nu\theta \, d\theta = K_\nu(z) \,, \tag{224}$$

where $K_\nu(z)$ is related to the Hankel function with imaginary argument as

$$K_\nu(z) = \tfrac{1}{2}\pi i \, e^{\frac{1}{2}\nu\pi i} H_\nu^{(1)}(iz) \,. \tag{225}$$

Since

$$\sinh^2\theta \cosh\theta = \tfrac{1}{4}(\cosh 3\theta - \cosh\theta) \tag{226}$$

and

$$\sinh^2\theta \cosh^2\theta = \tfrac{1}{8}(\cosh 4\theta - 1) \,, \tag{227}$$

[6] The formulae are contained in G. N. Watson's *Bessel Functions*, Cambridge; see pp. 79, 181, and 202 in Watson's book. Equation (224) is due to Schläfli.

we have, according to (219), (220), (224), (226), and (227),

$$N = \frac{8\pi V m^3 c^3}{h^3} \Lambda \, \tfrac{1}{4}[K_3(\vartheta mc^2) - K_1(\vartheta mc^2)] \,, \tag{228}$$

and

$$U = \frac{8\pi V m^4 c^5}{h^3} \Lambda \, \{\tfrac{1}{8}[K_4(\vartheta mc^2) - K_0(\vartheta mc^2)] - \tfrac{1}{4}[K_3(\vartheta mc^2) - K_1(\vartheta mc^2)]\} \,. \tag{229}$$

Using the recurrence formula,

$$K_{\nu-1}(z) - K_{\nu+1}(z) = -\frac{2\nu}{z} K_\nu(z) \,, \tag{230}$$

we find that

$$K_3(z) - K_1(z) = \frac{4}{z} K_2(z) \tag{231}$$

and

$$K_4(z) - K_0(z) = \frac{2}{z} [3K_3(z) + K_1(z)] \,. \tag{232}$$

Equations (228) and (229) can therefore be simplified to the forms

$$N = \frac{8\pi V m^3 c^3}{h^3} \Lambda \frac{1}{\vartheta mc^2} K_2(\vartheta mc^2) \tag{233}$$

and

$$U = \frac{8\pi V m^4 c^5}{h^3} \Lambda \frac{1}{\vartheta mc^2} [\tfrac{1}{4}\{3K_3(\vartheta mc^2) + K_1(\vartheta mc^2)\} - K_2(\vartheta mc^2)] \,. \tag{234}$$

From (233) and (234) we find that

$$U = N \left[\frac{3K_3(\vartheta mc^2) + K_1(\vartheta mc^2)}{4K_2(\vartheta mc^2)} - 1 \right] mc^2 \,. \tag{235}$$

By (223), equation (235) can also be written as

$$\frac{U}{PV} = \vartheta mc^2 \left[\frac{3K_3(\vartheta mc^2) + K_1(\vartheta mc^2)}{4K_2(\vartheta mc^2)} - 1 \right] . \tag{236}$$

If $\vartheta mc^2 \gg 1$ we can use the asymptotic formula

$$K_\nu(z) \sim \left(\frac{\pi}{2z}\right)^{1/2} e^{-z} \left[1 + \frac{4\nu^2 - 1^2}{1!\,8z} + \frac{(4\nu^2 - 1^2)(4\nu^2 - 3^2)}{2!\,(8z)^2} + \ldots . \right] . \tag{237}$$

Equation (236) now reduces to

$$\frac{U}{PV} \sim \frac{3}{2}\left(1 + \frac{5}{4\vartheta mc^2}\right) \qquad (\vartheta mc^2 \gg 1)\,. \quad (238)$$

The inequality $\vartheta mc^2 \gg 1$ is equivalent to the condition

$$kT \ll mc^2 \qquad \text{or} \qquad T \ll 5.90 \times 10^9 \text{ degrees Kelvin.} \quad (239)$$

On the other hand, if $\vartheta mc^2 \ll 1$ (i.e., $T \gg 6 \times 10^9$), we should use

$$K_\nu(z) \sim \frac{1}{2}\frac{(\nu - 1)!}{(\frac{1}{2}z)^\nu}\,. \quad (240)$$

From (236) and (240) we find that

$$\frac{U}{PV} \to 3\,, \qquad T \to \infty\,, \qquad \vartheta mc^2 \to 0\,. \quad (241)$$

We thus see that U/PV varies from 1.5 to 3 just as in the completely degenerate case. Here this variation is associated with increasing temperature, while there it arose because of increasing density. In either case the change of the ratio $(U:PV)$ from 1.5 to 3 is associated with an increasing number of electrons in the assembly with velocities approaching that of light. In Table 24, due to Chandrasekhar, the ratio $(U:PV)$ as a function of ϑmc^2 is shown.

TABLE 24

THE INTERNAL ENERGY AND THE SPECIFIC HEAT OF A PERFECT GAS

ϑmc^2	U/PV	C_V/Nk	ϑmc^2	U/PV	C_V/Nk
0	3.0	3.0	1.0	2.3704	2.7515
.1	2.9049	2.9952	1.5	2.2129	2.6031
.2	2.8193	2.9817	2.0	2.1024	2.4778
.3	2.7422	2.9617	2.5	2.0209	2.3743
.4	2.6726	2.9370	3.0	1.9586	2.2883
.5	2.6097	2.9089	3.5	1.9093	2.2173
.6	2.5527	2.8787	4.0	1.8695	2.1571
.7	2.5008	2.8473	4.5	1.8367	2.1062
.8	2.4535	2.8153	5.0	1.8093	2.0614
0.9	2.4102	2.7832	∞	1.5	1.5

From (235) we now derive that

$$\frac{1}{N}\left(\frac{\partial U}{\partial T}\right)_{T,\,V} = \frac{mc^2}{4}\,\frac{\partial}{\partial(\vartheta mc^2)}\left[\frac{3K_3(\vartheta mc^2) + K_1(\vartheta mc^2)}{K_2(\vartheta mc^2)}\right]\frac{\partial \vartheta mc^2}{\partial T}\,; \quad (242)$$

or, writing $z = \vartheta mc^2$, we find

$$\frac{1}{N}\left(\frac{\partial U}{\partial T}\right)_{T,\,V} = -kz^2 \frac{d}{dz}\left[\frac{3K_3(z)+K_1(z)}{4K_2(z)}\right]. \tag{243}$$

Thus the specific heat (per electron) at constant volume is given by

$$\frac{C_V}{Nk} = \frac{z^2}{4}\left[\frac{3K_3+K_1}{K_2^2}K_2' - \frac{3K_3'+K_1'}{K_2}\right]. \tag{244}$$

Using the formula

$$K_{\nu-1} + K_{\nu+1} = -2K_\nu' , \tag{245}$$

we can re-write (244) as

$$\frac{C_V}{Nk} = \frac{z^2}{8}\left[\frac{3(K_4+K_2)+(K_2+K_0)}{K_2} - \frac{(3K_3+K_1)(K_1+K_3)}{K_2^2}\right]. \tag{246}$$

By using (237) and (240), we can show that the quantity on the right-hand side of the foregoing equation varies from 1.5 to 3 as ϑmc^2 decreases from infinity to zero. More directly, from (223), (238), and (243) we derive that

$$\frac{C_V}{Nk} \sim -z^2 \frac{\partial}{\partial z}\left\{\frac{3}{2}\left(\frac{1}{z}+\frac{5}{4z^2}+\dots\right)\right\} \qquad (z\to\infty)\,, \tag{246'}$$

or

$$\frac{C_V}{Nk} = \frac{3}{2}\left(1+\frac{5}{2\vartheta mc^2}+\dots\right) \qquad (\vartheta mc^2 \to \infty)\,. \tag{247}$$

In Table 24 the quantity C_V/Nk is tabulated with ϑmc^2 as argument. Since $C_P - C_V = Nk$, it is clear that the ratio of the specific heats varies from $5/3$ for $T \ll mc^2/k$ to $4/3$ as $T \gg mc^2/k$.

8. *The nondegenerate case: a second approximation.*—If the exponential terms occurring in the denominators of the integrands in (170), (171), and (172) are large (but not infinitely large) compared to unity, we can expand

$$\left(\frac{1}{\Lambda}e^{\vartheta mc^2\cosh\theta}+1\right)^{-1}$$

as an infinite series and obtain

$$N = \frac{8\pi V m^3 c^3}{h^3} \sum_{n=1}^{\infty} (-)^{n+1} \Lambda^n \int_0^{\infty} e^{-n\vartheta mc^2 \cosh\theta} \sinh^2\theta \cosh\theta\, d\theta\,, \tag{248}$$

$$\left.\begin{aligned} U = \frac{8\pi V m^4 c^5}{h^3} \sum_{n=1}^{\infty} (-)^{n+1} \Lambda^n \int_0^{\infty} e^{-n\vartheta mc^2 \cosh\theta} \sinh^2\theta \\ \times \cosh\theta\,(\cosh\theta - 1)\, d\theta\,, \end{aligned}\right\} \tag{249}$$

and

$$PV \doteq \frac{8\pi V m^4 c^5}{3h^3} \sum_{n=1}^{\infty} (-)^{n+1} \Lambda^n \int_0^{\infty} e^{-n\vartheta mc^2 \cosh\theta} \sinh^4\theta\, d\theta\,. \tag{250}$$

Equation (250) can be transformed into

$$PV = \frac{8\pi V m^3 c^3}{h^3} \frac{1}{\vartheta} \sum_{n=1}^{\infty} \frac{(-)^{n+1}\Lambda^n}{n} \int_0^{\infty} e^{-n\vartheta mc^2 \cosh\theta} \sinh^2\theta \cosh\theta\, d\theta\,. \tag{251}$$

The integrals occurring in the foregoing equations are of the same form encountered in § 7. By (224), (226), and (227) we now have

$$N = \frac{8\pi V m^3 c^3}{h^3} \sum_{n=1}^{\infty} (-)^{n+1} \frac{\Lambda^n}{n\vartheta mc^2} K_2(n\vartheta mc^2)\,, \tag{252}$$

$$PV = \frac{8\pi V m^3 c^3}{h^3} \frac{1}{\vartheta} \sum_{n=1}^{\infty} (-)^{n+1} \frac{\Lambda^n}{n^2\vartheta mc^2} K_2(n\vartheta mc^2)\,, \tag{253}$$

and

$$\left.\begin{aligned} U = \frac{8\pi V m^4 c^5}{h^3} \sum_{n=1}^{\infty} (-1)^{n+1} \frac{\Lambda^n}{n\vartheta mc^2} \{\tfrac{1}{4}[3K_3(n\vartheta mc^2) + K_1(n\vartheta mc^2)] \\ - K_2(n\vartheta mc^2)\}\,. \end{aligned}\right\} \tag{254}$$

9. *The unrelativistic case.*—So far we have distinguished between degeneracy and nondegeneracy, but we have allowed in either case for the relativistic mass variation with velocity. However, in certain astronomical applications (as in most terrestrial applications of the

Fermi-gas laws) it is permissible to neglect the relativistic effects and write with sufficient accuracy

$$E = \frac{p^2}{2m}\,; \qquad mdE = pdp\,. \tag{255}$$

Inserting (255) in our general formulae (Eqs. [161] and [163]), we obtain

$$N = \frac{4\pi V}{h^3}(2m)^{3/2}\int_0^\infty \frac{E^{1/2}dE}{e^{\alpha+\vartheta E}+1}\,, \tag{256}$$

$$U = \frac{4\pi V}{h^3}(2m)^{3/2}\int_0^\infty \frac{E^{3/2}dE}{e^{\alpha+\vartheta E}+1}\,, \tag{257}$$

and

$$PV = \frac{2}{3}\,\frac{4\pi V}{h^3}(2m)^{3/2}\int_0^\infty \frac{E^{3/2}dE}{e^{\alpha+\vartheta E}+1}\,. \tag{258}$$

Comparing (257) and (258), we find that

$$U = \tfrac{3}{2}PV \tag{259}$$

is valid independent of degeneracy conditions—provided, of course, that the relativistic effects are neglected. This is a generalization of the result we have proved directly for the case of complete degeneracy (Eq. [31]) and for complete nondegeneracy (Eq. [238]). Put

$$\vartheta E = u\,, \qquad \alpha = -\log\Lambda\,, \tag{260}$$

and introduce the integral U_υ, defined by

$$U_\upsilon = \frac{1}{\Gamma(\upsilon+1)}\int_0^\infty \frac{u^\upsilon du}{\frac{1}{\Lambda}e^u+1}\,; \tag{261}$$

equations (256), (257), and (258) can now be written more conveniently in the forms

$$N = \frac{2V}{h^3}(2\pi mkT)^{3/2}U_{1/2} \tag{262}$$

and

$$PV = \tfrac{2}{3}U = \frac{2V}{h^3}(2\pi mkT)^{3/2}kTU_{3/2}\,. \tag{263}$$

a) Degenerate case.—For this case of Λ large we can obtain an asymptotic evaluation of the integral U_v by an application of Sommerfeld's lemma. We write

$$U_v = \frac{1}{\Gamma(v+2)}\int_0^\infty \frac{du}{\frac{1}{\Lambda}e^u + 1}\,\frac{d}{du}(u^{v+1})\,. \tag{264}$$

Then by the lemma we easily find that

$$U_v = \frac{(\log\Lambda)^{v+1}}{\Gamma(v+2)}\left[1 + 2\left\{c_2\frac{(v+1)v}{(\log\Lambda)^2} + c_4\frac{(v+1)v(v-1)(v-2)}{(\log\Lambda)^4} + \ldots\right\}\right]. \tag{265}$$

In particular,

$$U_{1/2} = \frac{4}{3\sqrt{\pi}}(\log\Lambda)^{3/2}\left[1 + \frac{\pi^2}{8(\log\Lambda)^2} + \ldots\right] \tag{266}$$

and

$$U_{3/2} = \frac{8}{15\sqrt{\pi}}(\log\Lambda)^{5/2}\left[1 + \frac{5\pi^2}{8(\log\Lambda)^2} + \ldots\right]. \tag{267}$$

Using (266) and (267), we find that our present expansions for N, PV, and U are equivalent to (204), (198), and (205) for the case where x is small. The connection between x and our present $\log\Lambda$ is readily seen to be (cf. Eqs. [173], [175], [194], and [260])

$$\log\Lambda = [(x^2+1)^{1/2} - 1]\vartheta mc^2\,. \tag{268}$$

b) Nondegenerate case.—If $\Lambda \ll 1$, we can expand $(1 + \Lambda e^{-u})$ in a series and evaluate U_v by integrating term by term. We find that

$$U_v = \frac{1}{\Gamma(v+1)}\int_0^\infty u^v du \sum_{n=1}^{\infty}(-)^{n+1}\Lambda^n e^{-nu} \tag{269}$$

$$= \Lambda - \frac{\Lambda^2}{2^{v+1}} + \frac{\Lambda^3}{3^{v+1}} - \frac{\Lambda^4}{4^{v+1}} + \ldots\,. \tag{270}$$

Equations (262), (263), and (270) now give deviations from Boyle's law, etc., owing to the exclusion principle even for "ordinary" densities.

This completes the analysis of the gas laws, which should be valid for an assembly of particles obeying the Pauli principle, though the discussion has been carried through explicitly only for the case of an electron gas.

10. *The vibrations of the normal modes of a radiation field.*—In order to consider a radiation field in a manner analogous to the treatment of an assembly of similar particles, it is first necessary to find suitable co-ordinates to describe its motion. We have to start, then, by an analysis of the number of possible modes of vibration in a given frequency interval; this stage of the analysis corresponds exactly to the discussion in § 2 of the number of independent eigenfunctions in a given energy interval.

Let ψ stand for any one of the components of the electric vector $\boldsymbol{E}$ or the magnetic vector $\boldsymbol{H}$. Then, according to the electromagnetic theory, we have

$$\frac{\partial^2\psi}{\partial t^2} = c^2\nabla^2\psi = c^2\left(\frac{\partial^2}{\partial x^2}+\frac{\partial^2}{\partial y^2}+\frac{\partial^2}{\partial z^2}\right)\psi\,. \tag{271}$$

Further, we have

$$\text{div}\,\boldsymbol{E} = \frac{\partial E_x}{\partial x}+\frac{\partial E_y}{\partial y}+\frac{\partial E_z}{\partial z} = 0\,. \tag{272}$$

Let us consider for simplicity an inclosure of the shape of a rectangular box,

$$0 \leqslant x \leqslant l_x\,; \qquad 0 \leqslant y \leqslant l_y\,; \qquad 0 \leqslant z \leqslant l_z\,. \tag{273}$$

Let ψ_0 be the value of ψ for a given $t = 0$. We shall assume that we can expand ψ_0 as a multiple Fourier series of the form

$$\psi_0 = \sum_{k_x=0}^{\infty}\sum_{k_y=0}^{\infty}\sum_{k_z=0}^{\infty} A_{k_xk_yk_z} \begin{matrix}\cos\\ \sin\end{matrix}\frac{k_x\pi x}{l_x}\begin{matrix}\cos\\ \sin\end{matrix}\frac{k_y\pi y}{l_y}\begin{matrix}\cos\\ \sin\end{matrix}\frac{k_z\pi z}{l_z}\,. \tag{274}$$

Similarly, if ψ'_0 is the value of $\partial\psi/\partial t$ at $t = 0$, we can write

$$\psi'_0 = \sum_{k_x=0}^{\infty}\sum_{k_y=0}^{\infty}\sum_{k_z=0}^{\infty} A'_{k_xk_yk_z} \begin{matrix}\cos\\ \sin\end{matrix}\frac{k_x\pi x}{l_x} \begin{matrix}\cos\\ \sin\end{matrix}\frac{k_y\pi y}{l_y} \begin{matrix}\cos\\ \sin\end{matrix}\frac{k_z\pi z}{l_z}. \tag{275}$$

From (274) and (275) it follows that the solution of (271) is

$$\left.\begin{aligned}\psi = \sum_{k_x=0}^{\infty}\sum_{k_y=0}^{\infty}\sum_{k_z=0}^{\infty} &\left\{A_{k_xk_yk_z}\cos 2\pi\nu t + \frac{A'_{k_xk_yk_z}}{2\pi\nu}\sin 2\pi\nu t\right\} \\ &\cdot \begin{matrix}\cos\\ \sin\end{matrix}\frac{k_x\pi x}{l_x} \begin{matrix}\cos\\ \sin\end{matrix}\frac{k_y\pi y}{l_y} \begin{matrix}\cos\\ \sin\end{matrix}\frac{k_z\pi z}{l_z},\end{aligned}\right\} \tag{276}$$

where

$$(2\pi\nu)^2 = \pi^2c^2\left\{\frac{k_x^2}{l_x^2} + \frac{k_y^2}{l_y^2} + \frac{k_z^2}{l_z^2}\right\}, \tag{277}$$

and ν is the frequency of the radiation considered.

It is clear that in each of the expressions (274), (275), and (276) there are eight possible terms and eight independent coefficients, A and A', for given k_x, k_y, and k_z.

We must now consider the boundary conditions more closely. We assume that the walls of the inclosure are perfect conductors, so that if $\psi = E_x$, then E_x should vanish on the two walls parallel to the (y, z) plane; that is, $E_x = 0$ at $y = 0$, $y = l_y$, $z = 0$, $z = l_z$, which leaves only two terms of the type

$$\begin{matrix}\cos\\ \sin\end{matrix}\frac{k_x\pi x}{l_x} \sin\frac{k_y\pi y}{l_y} \sin\frac{k_z\pi z}{l_z}. \tag{278}$$

Similarly, the Fourier expansions for E_y and E_z must contain respectively only terms of the types

$$\sin\frac{k_x\pi x}{l_x} \begin{matrix}\cos\\ \sin\end{matrix}\frac{k_y\pi y}{l_y} \sin\frac{k_z\pi z}{l_z} \tag{279}$$

and

$$\sin\frac{k_x\pi x}{l_x} \sin\frac{k_y\pi y}{l_y} \begin{matrix}\cos\\ \sin\end{matrix}\frac{k_z\pi z}{l_z}. \tag{280}$$

Since, however, the div $\boldsymbol{E}$ must vanish, the pure sine terms are impossible. Of the one remaining term in each of E_x, E_y, E_z, only two remain independent when (272) is satisfied. Thus there are two normal modes of vibrations for a given set of values k_x, k_y, and k_z which satisfy (277).

Let us return to equation (277), which we can write in the form

$$\frac{k_x^2}{a_x^2}+\frac{k_y^2}{a_y^2}+\frac{k_z^2}{a_z^2}=1\ , \tag{281}$$

where

$$a_x=\frac{2\nu}{c}\,l_x\ ; \qquad a_y=\frac{2\nu}{c}\,l_y\ ; \qquad a_z=\frac{2\nu}{c}\,l_z\ . \tag{282}$$

The number of normal modes of vibration with frequencies $\nu \leqslant \nu_0$ is equal to twice the number of points with integral co-ordinates inside an octant[7] of the ellipsoid (281) with $\nu = \nu_0$, which has the volume

$$\frac{4\pi}{3}\,a_x a_y a_z\ . \tag{283}$$

By (282) and (283) we thus have for the number of normal modes of vibrations in a radiation field with frequency $\leqslant \nu_0$

$$2\,\frac{1}{8}\,\frac{4\pi}{3}\,a_x a_y a_z=\frac{8\pi}{3c^3}\,\nu_0^3 l_x l_y l_z\ . \tag{284}$$

Hence, the number of normal modes of vibration with frequencies between ν and $\nu + d\nu$ is given by

$$\frac{8\pi V}{c^3}\,\nu^2 d\nu\ , \tag{285}$$

where we have replaced $l_x l_y l_z$ by V, the volume of the inclosure. Actually, the result has been obtained for an inclosure of a rectangular shape, but a somewhat more comprehensive analysis by Weyl shows that the result is completely general.

It is of interest to notice that if, in the expression (64), which gives

[7] Only an octant, since k_x, k_y, and k_z are by definition (Eq. [274]) nonnegative.

the number of quantum states for a particle of mass m and energy E in the range $E, E + \Delta E$, we put

$$E = h\nu \qquad \text{and} \qquad m = 0\,, \tag{286}$$

we obtain precisely the expression (285).[8]

11. *The statistics of light quanta.*—To be able to apply the laws of statistical mechanics to a field of radiation we first recall that, according to the quantum theory, each active mode of vibration with frequency ν is associated with an energy $h\nu$ of the field.

With some slight modifications the Pauli-Gibbs theory given in § 3 is capable of handling the present case.

Let us consider a field of radiation in a given volume V and with an internal energy U due to the active normal modes of vibration. A microscopic state of the radiation field will be completely determined by the specification of the number n_s of active modes of vibration with a frequency ν_s and energy $h\nu_s$. We then have

$$U = \sum_s n_s h\nu_s\,. \tag{287}$$

A possible sequence of numbers $n_1, n_2, \ldots, n_s, \ldots$, must satisfy (287). We shall write the different sequences of values for the n_s's which satisfy (287) in the form

$$\left.\begin{array}{l} n_0^{(1)},\ n_1^{(1)},\ldots,n_s^{(1)},\ldots, \\ n_0^{(2)},\ n_1^{(2)},\ldots,n_s^{(2)},\ldots, \\ \ldots\ldots\ldots\ldots\ldots, \\ n_0^{(i)},\ n_1^{(i)},\ldots,n_s^{(i)},\ldots, \\ \ldots\ldots\ldots\ldots\ldots, \\ n_0^{(W)},\ n_1^{(W)},\ldots,n_s^{(W)},\ldots, \end{array}\right\} \tag{288}$$

where W is the number of different solutions in integers of (287). The entropy, S, of the radiation field is now defined by (cf., Eq. [75])

$$e^{S/k} = W\,. \tag{289}$$

[8] This shows a certain formal equivalence (from the present point of view) of light quanta and a particle of zero rest mass which satisfies the Dirac equation.

As in the discussion in § 3 we now drop the restriction (287) on the n_s's by the passage to a canonical state where the energy U is no longer defined exactly but is distributed in such a way that U has a sharp maximum at a certain prescribed value say $\overline{U}$.

According to (287) and (289), we now have

$$e^{S/k-\vartheta U} = We^{-\vartheta \sum_s n_s^{(i)} h\nu_s} , \tag{290}$$

where ϑ is, for the present, an arbitrary constant. Equation (290) can be written more "symmetrically" in the form (cf. Eqs. [78] and [80])

$$e^{S/k-\vartheta U} = \sum_i e^{-\vartheta \sum_s n_s^{(i)} h\nu_s} . \tag{291}$$

We now drop the restriction (287) and write, instead of (291),

$$e^{S/k-\vartheta U} = \sum_{n_s} e^{-\vartheta \sum_s n_s h\nu_s} , \tag{292}$$

where the summation over the n_s's is taken over all the possible n_s's. But ϑ is now so chosen that

$$\frac{1}{k}\left(\frac{\partial S}{\partial U}\right)_V - \vartheta = 0 , \tag{293}$$

or, exactly as in § 3, $\vartheta = 1/kT$. If F is the free energy of the radiation field, equation (292) can now be written as

$$e^{-F/kT} = \prod_s \sum_{n_s} e^{-n_s h\nu_s/kT} . \tag{294}$$

We can show, exactly as in § 3, that (293) and (294) define for U an extremely sharp maximum at $U = \overline{U}$ (say), and that U is appreciably different from zero only in the range $\overline{U} \pm \Delta U$ where

$$\overline{(\Delta U)^2} = \frac{k}{\left|\left(\frac{\partial^2 S}{\partial U^2}\right)_{V,\, U=\overline{U}}\right|} , \tag{295}$$

where $\overline{(\Delta U)^2}$ is used to denote the "mean square deviation" from $\overline{U}$. In the same way, we can show that the entropy, defined according to (293) and (294), differs from that defined according to (289) only by a quantity of the order (cf. Eq. [105])

$$\frac{k}{S} \log \frac{\overline{N}}{\overline{U}} \sqrt{\frac{2\pi k}{\left|\left(\frac{\partial^2 S}{\partial U^2}\right)_{V,\, U=\overline{U}}\right|}} . \tag{295'}$$

Finally, we remark that (294) is now interpreted by the statement that the probability of a microscopic state defined by the sequence $(n_1, \ldots, n_s, \ldots)$ (which define the number of active modes with frequencies $[\nu_1, \ldots, \nu_s, \ldots])$ and an energy $U = \sum_s n_s h\nu_s$ is proportional to

$$e^{-\vartheta \sum_s n_s h\nu_s} = \prod_s e^{-n_s h\nu_s/kT} . \tag{296}$$

In order, then, to obtain the probability of a microscopic state with a definite total energy U which corresponds to the canonical distribution (296), we must sum over all the sequences $\{n_s\}$ which lead to the energy U.

For the internal energy U, we have, immediately,

$$U = \sum_s \frac{\sum_{n_s} n_s h\nu_s e^{-\vartheta n_s h\nu_s}}{\sum_{n_s} e^{-\vartheta n_s h\nu_s}} . \tag{297}$$

If, as in (142), we now define

$$\sigma_s = \sum_{n_s} e^{-\vartheta n_s h\nu_s} , \tag{298}$$

then we can write

$$U = -\sum_s \frac{\partial}{\partial \vartheta} \log \sigma_s . \tag{299}$$

This solves the statistical problem, and to obtain explicit formulae we have to evaluate σ_s. We shall assume that n_s can take all values from 0 to ∞. This means, according to the discussion of § 4, that the wave functions which describe the radiation field should be symmetrical in all the normal modes, each normal mode for this purpose being described as a simple harmonic oscillator. Hence,

$$\sigma_s = \frac{1}{1 - e^{-h\nu_s/kT}}, \tag{300}$$

or

$$-\frac{\partial}{\partial \vartheta} \log \sigma_s = \frac{h\nu_s}{e^{\vartheta h\nu_s} - 1}. \tag{301}$$

Therefore, by (299),

$$U = \sum_s \frac{h\nu_s}{e^{\vartheta h\nu_s} - 1}. \tag{302}$$

We can replace the sum by an integral and, weighting each frequency interval by the appropriate density of the normal modes specified by (285), we have

$$U = \frac{8\pi V}{c^3} \int_0^\infty \frac{h\nu^3 d\nu}{e^{h\nu/kT} - 1}. \tag{303}$$

It is clear that if we wish to find the energy in the radiation field in a given frequency interval, then we have a sum similar to (297), the summation now being extended only over the required frequency interval. We thus have

$$u_\nu d\nu = \frac{8\pi h\nu^3}{c^3} \frac{d\nu}{e^{h\nu/kT} - 1}, \tag{304}$$

which is Planck's law. Since the radiation is isotropic, the Planck intensity, B_ν, is related to u_ν by (Eq. [29], v)

$$B_\nu = \frac{c}{4\pi} u_\nu, \tag{305}$$

or

$$B_\nu = \frac{2h\nu^3}{c^2} \frac{1}{e^{h\nu/kT} - 1}. \tag{306}$$

This completes our discussion of the quantum statistics.

BIBLIOGRAPHICAL NOTES

The following general references may be noted.

1. W. Gibbs, *Elementary Principles of Statistical Mechanics*, Yale, 1902.

2. P. and T. Ehrenfest, *Encyklopädie der Mathematischen Wissenschaften*, **4**, Part 32, 1911.

3. P. Jordan, *Statistische Mechanik auf quantentheoretischer Grundlage*, Braunschweig, 1933.

4. R. H. Fowler, *Statistical Mechanics*, 2d ed., Cambridge, 1936.

5. H. A. Lorentz, *Lectures on Theoretical Physics*, **2**, 141–188. New York: Macmillan, 1927.

A general account of the physics of matter at high temperatures and densities is given by—

6. F. Hund, *Erg. exakt. Naturwiss.*, **15**, 189, 1936.

The following particular references may be given.

§ 1.—Ref. 4, chap. xvi; also,

7. R. H. Fowler, *M.N.*, **87**, 114, 1926, where the law $p = K_1\rho^{5/3}$ is derived and the first astronomical application given.

8. E. C. Stoner, *Phil. Mag.*, **7**, 63, 1929.

9. W. Anderson, *Zs. f. Phys.*, **54**, 433, 1929. Anderson was the first to recognize the importance of the relativistic effects in astronomical applications. See also

10. W. Anderson, *Tartu Pub.*, **29**, 1936.

11. E. C. Stoner, *Phil. Mag.*, **9**, 944, 1930, where the correct expression for the internal energy, U, for a completely degenerate electron gas is obtained for the first time.

12. S. Chandrasekhar, *M.N.*, **91**, 446, 1931.

13. S. Chandrasekhar, *Ap. J.*, **74**, 81, 1931. In references 12 and 13 the law $p = K_2\rho^{4/3}$ is used for the first time.

14. L. Landau, *Phys. Zs. d. Soviet Union*, **1**, 285, 1932.

15. T. E. Sterne, *M.N.*, **93**, 736, 1933.

The following reference (in which the law $p = K_2\rho^{4/3}$ is implicitly contained) may be noted.

16. J. Frenkel, *Zs. f. Phys.*, **50**, 234, 1928.

§ 2.—The number of normal modes of vibration, when relativity effects are taken into account, was first derived by—

17. P. A. M. Dirac, *Proc. R. Soc.*, A, **112**, 660, 1926 (the unnumbered equation on page 671 of this paper). For the corresponding unrelativistic treatment see Fowler, ref. 4, chap. ii. See also

18. E. Fermi, *Zs. f. Phys.*, **36**, 902, 1926.

19. J. von Neumann, *Zs. f. Phys.*, **48**, 868, 1928.

20. C. G. Darwin, *Proc. R. Soc.*, A, **118**, 654, 1928.

21. C. Møller and S. Chandrasekhar, *M.N.*, **95**, 673, 1935.

22. R. PEIERLS, *M.N.*, **96**, 780, 1936. This paper contains the most general derivation of the expression for the number of normal modes.

23. E. K. BROCK, *Phys. Rev.*, **51**, 586, 1937, where the Dirac equation in a spherical potential hole is solved exactly and the enumeration of states is rigorously carried out.

Also Jordan, ref. 3, chap. ii, §§ 1 and 2.

It has recently been contended by EDDINGTON (*M.N.*, **95**, 194, 1935) that the theory of a relativistic gas based on the expression for $Z(E)dE$ derived in this section (Eq. [66]) is incorrect. However, the investigations (refs. 21, 22, and 23), undertaken, incidentally, also with a view of examining Eddington's contention, have failed to support it. The general theory presented in this chapter is accepted by theoretical physicists (cf. Hund, ref. 6).

§ 3.—The account given here follows closely:

24. W. PAULI, *Zs. f. Phys.*, **41**, 81, 1926. Also ref. 1.

§ 4.—R. H. Fowler, ref. 4, chap. ii; also,

25. P. A. M. DIRAC, *The Principles of Quantum Mechanics*, 2d ed., chap. x, Oxford, 1935.

26. A. SOMMERFELD, *Zs. f. Phys.*, **47**, 1, 1928.

§ 5.—The fundamental equations in the forms given in equations (170), (171), and (172) are due to—

27. F. JUTTNER, *Zs. f. Phys.*, **47**, 542, 1928.

§ 6.—The lemma proved in this section is due to Sommerfeld (ref. 26). Generally, four cases have been distinguished: (*a*) unrelativistic nondegeneracy, (*b*) relativistic nondegeneracy, (*c*) unrelativistic degeneracy, and, finally, (*d*) relativistic degeneracy. However, in the discussion we have distinguished between degeneracy and nondegeneracy and taken account of the relativistic effects accurately in both cases. The presentation which results is more elegant.

Also,

28. R. C. MAJUMDAR, *Astr. Nachr.*, **247**, 217, 1932.

§ 7.—The nondegenerate case, including the relativistic effects, was first considered by—

29. F. JUTTNER, *Ann. d. Phys.*, **34**, 856, 1911.

30. F. JUTTNER, *Ann. d. Phys.*, **35**, 145, 1911.

Also,

31. M. PLANCK, *Ann. d. Phys.*, **26**, 1, 1908.

32. R. C. TOLMAN, *Phil. Mag.*, **28**, 583, 1914.

33. W. PAULI, *Zs. f. Phys.*, **18**, 272, 1923. The treatment given here of thermal equilibrium between radiation and free electrons requires the use of the relativistic statistics.

34. W. PAULI, "Relativitätstheorie," *Encyk. Math.* Wiss., **5**, Part 19, 641–674.

This section contains some unpublished investigations of the writer.

§ 8.—The analysis in this section is from Juttner, ref. 27.

§ 9.—Fermi, ref. 18; Sommerfeld, ref. 26.

§ 10.—Fowler, ref. 4, chap. iv; Pauli ref. 24; also,

35. H. WEYL, *Math. Ann.*, **71**, 441, 1911.

§ 11.—Fowler, ref. 4, chap. iv, § 431.

The following references may also be noted:

36. R. C. MAJUMDAR, *A.N.*, **243**, 5, 1931.
37. R. C. MAJUMDAR, *A.N.*, **247**, 217, 1932.
38. B. SWIRLES, *M.N.*, **91**, 861, 1931.
39. S. CHANDRASEKHAR, *Proc. R. Soc.*, A, **133**, 241, 1931.

The foregoing papers deal with the problem of opacity of degenerate matter. The following consider the various transport phenomena (conduction, viscosity, etc.).

40. D. S. KOTHARI, *Phil. Mag.*, **13**, 361, 1932.
41. D. S. KOTHARI, *M.N.*, **93**, 61, 1932.

Also

42. D. S. KOTHARI, *Proc. R. Soc.*, A, **165**, 486, 1938.

CHAPTER XI

DEGENERATE STELLAR CONFIGURATIONS AND THE THEORY OF WHITE DWARFS

The white dwarf stars differ from those we have considered so far in two fundamental respects. First, they are what might be called "highly underluminous"; that is, judged with reference to an "average" star of the same mass, the white dwarf is much fainter. Thus, the companion of Sirius, although it has a mass about equal to that of the sun, is yet characterized by a value of L which is only 0.003 times that of the sun. Second, the white dwarfs are characterized by exceedingly high values for the mean density; in fact, we encounter densities of the order of 10^6 and even 10^8 gm cm^{-3}. It is this second characteristic which is generally emphasized, though from a theoretical point of view the fact that $L/L_{\odot}$ is generally very small is of equal importance.

Since the radius of a white dwarf is very much smaller than that of a star on the main series, it follows that for a given effective temperature the white dwarf will be much fainter than the star on the main series. Similarly, for the same luminosity the white dwarf will be characterized by a very much higher effective temperature (i.e., much "whiter") than the main-series star; this, incidentally, explains the origin of the term "white dwarf."

We shall discuss the observational material in somewhat greater detail in § 3, but it should already appear plausible that the white dwarfs differ from other stars in some fundamental way. The clue to the understanding of the structure of these stars was discovered by R. H. Fowler, who pointed out that the electron gas in the interior of the white dwarfs must be highly degenerate in the sense made precise in the last chapter. We shall see that the white dwarfs can, in fact, be idealized to a high degree of approximation as completely degenerate configurations. In this chapter we shall be mainly concerned with the applications of the theory of degeneracy toward the elucidation of the structure of the white dwarfs.

1. *The gaseous fringe of the white dwarfs.*—It is clear that the extreme outer layers of a white dwarf must, in any case, be gaseous, i.e., nondegenerate, with the perfect gas law, $p \propto \rho T$, obeyed. The question then arises as to how far inward we can descend before degeneracy sets in. To answer this question we shall have to consider the criterion for degeneracy which was established in the last chapter (Eq. [211]) and which we shall now write in the form

$$\frac{(\vartheta mc^2)^2}{4\pi^2} \frac{f(x)}{x(1+x^2)^{1/2}} \gg 1 \, , \tag{1}$$

where

$$\vartheta = \frac{1}{kT} \, ; \qquad f(x) = x(2x^2-3)(x^2+1)^{1/2} + 3 \sinh^{-1} x \, . \tag{2}$$

Finally, x is related to the mean electron concentration, n, by (Eq. [212], x)

$$n = \frac{8\pi m^3 c^3}{3h^3} x^3 \, . \tag{3}$$

We shall write

$$\rho = \frac{8\pi m^3 c^3}{3h^3} \mu_e H \, x^3 = B x^3 \, , \tag{4}$$

where

$$B = \frac{8\pi m^3 c^3}{3h^3} \mu_e H = 9.82 \times 10^5 \mu_e \, . \tag{5}$$

Anticipating our result that the region of the white dwarf where the perfect gas law is valid is an outer fringe only, we can use for describing the structure of this gaseous fringe the theory of the stellar envelope given in chapter viii. On account of the very small values of L and R for the white dwarfs, the quantity α as defined in chapter viii (Eq. [54]) is very small indeed ($1-\beta \sim 10^{-4}$), so that we can use the analysis of § 3 of chapter viii. We then have

$$T = \frac{4}{17} \frac{\mu H}{k} \frac{GM}{R} \left(\frac{1}{\xi} - 1\right) \tag{6}$$

and

$$\rho = \frac{1}{30 f(0; w^*)} \bar{\rho} \left(\frac{1}{\xi} - 1\right)^{3.25} \, , \tag{7}$$

where $\bar{\rho}$ is the mean density, ξ is the radius vector expressed in terms of the radius of the star, and $f(0; w^*)$ is defined as in equation (55) of chapter viii. Inserting numerical values and expressing L, M, and R in solar units, we find that

$$T = 5.43 \times 10^6 \mu \frac{M}{R} \left(\frac{1}{\xi} - 1\right) \tag{8}$$

$$\rho = 0.761 \frac{\mu^{3.75} \bar{t}_e^{0.5}}{(1 - X_0^2)^{1/2}} \left(\frac{M^{7.5}}{LR^{6.5}}\right)^{1/2} \left(\frac{1}{\xi} - 1\right)^{3.25} . \tag{9}$$

By (4) and (9), we now find

$$x^3 = 7.75 \times 10^{-7} \frac{\mu^{3.75} \bar{t}_e^{0.5}}{\mu_e (1 - X_0^2)^{1/2}} \left(\frac{M^{7.5}}{LR^{6.5}}\right)^{1/2} \left(\frac{1}{\xi} - 1\right)^{3.25} . \tag{10}$$

By (1) and (8) we find that

$$3.04 \times 10^4 \frac{1}{\mu^2} \frac{R^2}{M^2} \left(\frac{1}{\xi} - 1\right)^{-2} \frac{f(x)}{x(x^2 + 1)^{1/2}} \gg 1 . \tag{11}$$

From (10) and (11) we can determine the point at which the right-hand side of (11) is unity; at this point we may say that "degeneracy sets in."

For most practical purposes it is found that it is sufficient to consider for $f(x)$ the limiting form which it takes for small values of x. By equation (24) of chapter x

$$f(x) \sim \tfrac{8}{5} x^5 \qquad (x \to 0) . \tag{12}$$

The inequality (11) now takes the simpler form,

$$4.86 \times 10^4 \frac{1}{\mu^2} \frac{R^2}{M^2} \left(\frac{1}{\xi} - 1\right)^{-2} x^4 \gg 1 . \tag{13}$$

Eliminating x between (10) and (13), we find that

$$2.54 \times 10^{-3} \frac{\mu^{2.25} \bar{t}_e^{0.5}}{\mu_e (1 - X_0^2)^{1/2}} \left(\frac{M^{4.5}}{LR^{3.5}}\right)^{1/2} \left(\frac{1}{\xi} - 1\right)^{1.75} \gg 1 . \tag{14}$$

Now, since for the white dwarfs L and R are quite small, it follows that for values of ξ appreciably different from unity the right-hand side is, in fact, much greater than unity. Thus, if we consider the case of the companion of Sirius, for which (according to Kuiper) Log $M = -0.01$, Log $L = -2.52$, and Log $R = -1.71$, equation (14) takes the form

$$43 \frac{\mu^{2.25}\bar{t}_e^{0.5}}{\mu_e(1 - X_0^2)^{1/2}} \left(\frac{1}{\xi} - 1\right)^{1.75} \gg 1 . \tag{15}$$

If we assume that $\mu \sim \mu_e = 1.0$, $X_0 = \frac{1}{3}$, $\bar{t}_e = 10$,[1] then the right-hand side of (15) is unity for $\xi = 0.94$. At this point, according to Table 17, the mass traversed from the boundary is only 0.23 per cent of the mass of the star; further, it is found that at this point $x = 0.12$, in agreement with our assumption that x is small. Finally, at $\xi = 0.94$, according to (8) and (9), ρ is found to be 1730 gm cm^{-3}, while T is 1.7×10^7 degrees. For some of the other white dwarfs the situation is even more "favorable," in the sense that the gaseous fringe is of even smaller extent. We thus see that the material of the white dwarf must be almost entirely degenerate; this result is implicitly contained in Fowler's work, but the arguments, essentially in the form we have given them, are due to Strömgren and Sidentopf.

2. *Completely degenerate configurations.*—We have seen in § 1 that the gaseous fringe of a white dwarf is of quite negligible extent, and that, further, the radiation is entirely negligible—indeed, in the gaseous fringe $1 - \beta \sim 10^{-4}$ or less. It is almost certain (cf. the discussion in § 6) that in the interior $1 - \beta$ does not exceed its value in the gaseous fringe, and we are thus led to consider equilibrium configurations which are completely degenerate and in which the radiation pressure is entirely neglected. The general theory given in this section is due to Chandrasekhar.

The equation of state can be written as (cf. Eqs. [19], [20], and [21] of the last chapter)

$$P = Af(x) ; \qquad \rho = n\mu_e H = Bx^3 , \tag{16}$$

[1] According to Strömgren, under the conditions of the gaseous fringe of a white dwarf, the guillotine factor $\bar{t}_e$ must be quite large.

where

$$A = \frac{\pi m^4 c^5}{3h^3} = 6.01 \times 10^{22} ; \qquad B = \frac{8\pi m^3 c^3 \mu_e H}{3h^3} = 9.82 \times 10^5 \mu_e \qquad (17)$$

and

$$f(x) = x(2x^2 - 3)(x^2 + 1)^{1/2} + 3 \sinh^{-1} x . \qquad (18)$$

The equation of equilibrium is (Eq. [6], iii)

$$\frac{1}{r^2} \frac{d}{dr} \left(\frac{r^2}{\rho} \frac{dP}{dr} \right) = - 4\pi G \rho . \qquad (19)$$

Substituting for P and ρ according to (16), we have

$$\frac{A}{B} \frac{1}{r^2} \frac{d}{dr} \left(\frac{r^2}{x^3} \frac{df(x)}{dr} \right) = -4\pi G B x^3 . \qquad (20)$$

From the definition of $f(x)$ we easily verify that

$$\frac{df(x)}{dr} = \frac{8x^4}{(x^2 + 1)^{1/2}} \frac{dx}{dr} , \qquad (21)$$

or

$$\frac{1}{x^3} \frac{df(x)}{dr} = \frac{8x}{(x^2 + 1)^{1/2}} \frac{dx}{dr} = 8 \frac{d\sqrt{x^2 + 1}}{dr} . \qquad (22)$$

Hence, equation (20) can be re-written as

$$\frac{1}{r^2} \frac{d}{dr} \left(r^2 \frac{d\sqrt{x^2 + 1}}{dr} \right) = - \frac{\pi G B^2}{2A} x^3 . \qquad (23)$$

Put

$$y^2 = x^2 + 1 . \qquad (24)$$

Then,

$$\frac{1}{r^2} \frac{d}{dr} \left(r^2 \frac{dy}{dr} \right) = - \frac{\pi G B^2}{2A} (y^2 - 1)^{3/2} . \qquad (25)$$

Let x take the value x_0 at the center. Further, let y_0 be the corresponding value of y at the center. Introduce the new variables η and ϕ, defined as follows:

$$r = \alpha \eta ; \qquad y = y_0 \phi , \qquad (26)$$

where

$$a = \left(\frac{2A}{\pi G}\right)^{1/2} \frac{1}{By_0}; \qquad y_0^2 = x_0^2 + 1 . \tag{27}$$

The differential equation finally takes the form

$$\frac{1}{\eta^2} \frac{d}{d\eta} \left(\eta^2 \frac{d\phi}{d\eta}\right) = -\left(\phi^2 - \frac{1}{y_0^2}\right)^{3/2} . \tag{28}$$

By (26) we have to seek a solution of (28) such that ϕ takes the value unity at the origin. Further, it is clear that the derivative of ϕ must vanish at the origin. The boundary is defined at the point where the density vanishes, and this by (24) means that if η_1 specifies the boundary, then

$$\phi(\eta_1) = \frac{1}{y_0} . \tag{29}$$

From our definitions of the various quantities it is easily seen that

$$\rho = \rho_0 \frac{y_0^3}{(y_0^2 - 1)^{3/2}} \left(\phi^2 - \frac{1}{y_0^2}\right)^{3/2} , \tag{30}$$

where

$$\rho_0 = Bx_0^3 = B(y_0^2 - 1)^{3/2} \tag{31}$$

specifies the central density. Also, we may notice that the scale of length, a, introduced in (27), has, in terms of the natural constants, the form

$$a = \frac{1}{4\pi m \mu_e H y_0} \left(\frac{3h^3}{2cG}\right)^{1/2} , \tag{32}$$

or, inserting numerical values,

$$a = \frac{7.71 \times 10^8}{\mu_e \, y_0} = l_1 y_0^{-1} \text{ cm} . \tag{33}$$

We shall now consider a little more closely the structure of the configurations governed by the differential equation (28).

a) The potential.—The function ϕ has a physical meaning. If V is the inner gravitational potential, then from the general theory (chap. iii, § 2)

$$\frac{dV}{dr} = -\frac{1}{\rho} \frac{dP}{dr} . \tag{34}$$

From (16), (18), and (22) we see that

$$\frac{dV}{dr} = -\frac{8A}{B} y_0 \frac{d\phi}{dr} ; \tag{35}$$

or, integrating, we find that

$$V = -\frac{8A}{B} y_0\phi + \text{constant} . \tag{36}$$

If we choose the zero of the potential at infinity, we have by (29) that the "constant" in (36) is $[(8A/B) - GM/R]$ (cf. Eq. [10], iii). Hence,

$$V = -\frac{8A}{B} y_0 \left(\phi - \frac{1}{y_0}\right) - \frac{GM}{R} \qquad (r \leqslant R) \tag{37}$$

b) The mass relation.—The mass, interior to a specified point η, is given by

$$M(\eta) = 4\pi \int_0^\eta \rho r^2 dr = 4\pi\alpha^3 \int_0^\eta \rho\eta^2 d\eta . \tag{38}$$

By (30),

$$M(\eta) = 4\pi\rho_0 \frac{\alpha^3 y_0^3}{(y_0^2 - 1)^{3/2}} \int_0^\eta \left(\phi^2 - \frac{1}{y_0^2}\right)^{3/2} \eta^2 d\eta ; \tag{39}$$

or, using the differential equation (28),

$$M(\eta) = -4\pi\rho_0 \frac{\alpha^3 y_0^3}{(y_0^2 - 1)^{3/2}} \eta^2 \frac{d\phi}{d\eta} . \tag{40}$$

Substituting for α and ρ_0 according to (27) and (31), we have

$$M(\eta) = -4\pi \left(\frac{2A}{\pi G}\right)^{3/2} \frac{1}{B^2} \eta^2 \frac{d\phi}{d\eta} . \tag{41}$$

The mass of the whole configuration is given by

$$M = -4\pi \left(\frac{2A}{\pi G}\right)^{3/2} \frac{1}{B^2} \left(\eta^2 \frac{d\phi}{d\eta}\right)_{\eta=\eta_1} . \tag{42}$$

We notice that in (41) and (42) y_0 does not occur explicitly. It is, of course, implicitly present inasmuch as y_0 occurs in the differential equation defining ϕ.

c) The relation between the mean and the central density.—Let $\bar{\rho}(\eta)$ be the mean density of the material inside η. Then

$$M(\eta) = \tfrac{4}{3}\pi a^3 \eta^3 \bar{\rho}(\eta) \ . \tag{43}$$

Comparing (40) and (43), we have

$$\frac{\bar{\rho}(\eta)}{\rho_0} = -3 \frac{y_0^3}{(y_0^2 - 1)^{3/2}} \frac{1}{\eta} \frac{d\phi}{d\eta} \ . \tag{44}$$

From (44) we deduce that the relation between the mean and the central density of the whole configuration is

$$\rho_0 = -\bar{\rho} \left(1 - \frac{1}{y_0^2}\right)^{3/2} \frac{\eta_1}{3\phi'(\eta_1)} \ , \tag{45}$$

where ϕ' denotes the derivative of ϕ. It is of interest to notice the similarity between the present relations (42) and (45) and the corresponding relations in the theory of polytropes (Eqs. [69] and [78] of chap. iv).

d) An approximation for configurations with small central densities.—By definition, $y_0^2 = x_0^2 + 1$, and we need a first-order approximation when x_0^2 is small. We shall neglect all quantities of order x_0^4 and higher. Then,

$$y_0 = 1 + \tfrac{1}{2}x_0^2 \ . \tag{46}$$

Put

$$\phi^2 - \frac{1}{y_0^2} = \theta \ . \tag{47}$$

In our present approximation we have

$$\phi = 1 - \tfrac{1}{2}(x_0^2 - \theta) \ . \tag{48}$$

At the origin, ϕ takes the value unity. Hence,

$$\theta(0) = x_0^2 \ . \tag{49}$$

From (28) we derive the following differential equation for θ.

$$\frac{1}{2} \frac{d^2\theta}{d\eta^2} + \frac{1}{\eta} \frac{d\theta}{d\eta} = - \theta^{3/2} \ . \tag{50}$$

Finally, introduce the variable ξ, according to

$$\xi = 2^{1/2}\eta . \tag{50'}$$

Equation (50) now reduces to

$$\frac{1}{\xi^2}\frac{d}{d\xi}\left(\xi^2\frac{d\theta}{d\xi}\right) = -\theta^{3/2} , \tag{51}$$

which is the Lane-Emden equation with index $n = 3/2$, but the solution we need is *not* the Lane-Emden function $\theta_{3/2}$. According to (49), we need a solution of (51) which takes the value x_0^2 at $\xi = 0$. Now, according to the homology theorem of chapter iv, § 8, as applied to the case $n = 3/2$, if $\theta(\xi)$ is a solution of (51), then $C^4\theta(C\xi)$ is also a solution, where C is an arbitrary real number. Hence, from $\theta_{3/2}$ we can derive a function satisfying (49) by a homologous transformation of $\theta_{3/2}$:

$$\theta = x_0^2\,\theta_{3/2}(\sqrt{x_0}\,\xi) . \tag{52}$$

Hence, by (48), (50), and (52)

$$\phi = 1 - \tfrac{1}{2}x_0^2\{1 - \theta_{3/2}(\sqrt{2x_0}\,\eta)\} + O(x_0^4) , \tag{53}$$

which relates ϕ with $\theta_{3/2}$. From (53) we see that for these configurations the boundary η_1 must be such that

$$\theta_{3/2}(\sqrt{2x_0}\,\eta_1) = 0 , \tag{54}$$

since, according to (29) and (46), $\phi(\eta_1) = y_0^{-1} = 1 - \frac{1}{2}x_0^2$. If $\xi_1(\theta_{3/2})$ is the boundary of the Lane-Emden function, then from (54) we deduce that

$$\eta_1 = \frac{\xi_1(\theta_{3/2})}{\sqrt{2x_0}} . \tag{55}$$

Again, from (53) we have

$$\frac{d\phi}{d\eta} = \tfrac{1}{2}x_0^2\sqrt{2x_0}\,\frac{d\theta_{3/2}(\xi)}{d\xi} . \tag{56}$$

Combining (55) and (56), we find that

$$\left(\eta^2\frac{d\phi}{d\eta}\right)_{\eta=\eta_1} = \left(\frac{x_0}{2}\right)^{3/2}\left(\xi^2\frac{d\theta_{3/2}}{d\xi}\right)_{\xi=\xi_1(\theta_{3/2})} . \tag{57}$$

Further,

$$\left(\frac{1}{\eta}\frac{d\phi}{d\eta}\right)_{\eta=\eta_1} = x_0^3\left(\frac{1}{\xi}\frac{d\theta_{3/2}}{d\xi}\right)_{\xi=\xi_1(\theta_{3/2})} \tag{58}$$

From (58) and (45) we have

$$\rho_0 = -\bar{\rho}\,\frac{\xi_1(\theta_{3/2})}{3\theta'_{3/2}(\xi_1)}\,, \tag{59}$$

which is precisely the relation between the mean and the central density for a Lane-Emden polytrope of index $n = 3/2$. Again, from (42) and (57),

$$M = -4\pi\left(\frac{2A}{\pi G}\right)^{3/2}\frac{1}{B^2}\left(\frac{x_0}{2}\right)^{3/2}\left(\xi^2\frac{d\theta_{3/2}}{d\xi}\right)_{\xi=\xi_1(\theta_{3/2})}. \tag{60}$$

On the other hand, if $x_0 \to 0$, we can write the equation of state (16) in the form

$$P = \frac{8A}{5}x^5\,; \qquad \rho = Bx^3\,, \tag{61}$$

or

$$P = K_1\rho^{5/3}\,, \tag{62}$$

where

$$K_1 = \frac{8A}{5B^{5/3}} = \frac{1}{20}\left(\frac{3}{\pi}\right)^{2/3}\frac{h^2}{m(\mu_e H)^{5/3}} = \frac{9.91\times 10^{12}}{\mu_e^{5/3}}. \tag{63}$$

Hence, configurations with small central densities (i.e., x_0 small) are Lane-Emden polytropes of index $n = 3/2$. The results based on (63) and the theory of polytropes, and the approximation derived from the exact differential equation (28) for $x_0 \to 0$ are easily seen to be equivalent. In particular, using (63), the mass relation (60) can be re-written in the form

$$M = -4\pi\left(\frac{5K_1}{8\pi G}\right)^{3/2}\rho_0^{1/2}\left(\xi^2\frac{d\theta_{3/2}}{d\xi}\right)_{\xi=\xi_1(\theta_{3/2})}, \tag{64}$$

which is identical with the mass relation for a polytrope of index $n = 3/2$ based on the law (62) (cf. Eq. [69], iv).

e) *The limiting mass.*—From the differential equation (28) we see that

$$\phi \to \theta_3 \qquad \text{as} \qquad y_0 \to \infty\,. \tag{65}$$

But from (33) it follows that at the same time the radius tends to zero. From the mass relation (42), on the other hand, we see that the mass tends to a finite limit:

$$M \to -4\pi \left(\frac{2A}{\pi G}\right)^{3/2} \frac{1}{B^2} \left(\xi^2 \frac{d\theta_3}{d\xi}\right)_{\xi=\xi_1(\theta_3)} . \tag{66}$$

The existence of this limiting mass was first isolated by Chandrasekhar, though its existence had been made apparent from earlier considerations by Anderson and Stoner, who, however, did not consider the problem from the point of view of hydrostatic equilibrium.

For $x_0 \to \infty$ we can write (16) in the form

$$P = 2Ax^4 ; \qquad \rho = Bx^3 , \tag{67}$$

or

$$P = K_2 \rho^{4/3} , \tag{68}$$

where

$$K_2 = \frac{2A}{B^{4/3}} = \left(\frac{3}{\pi}\right)^{1/3} \frac{hc}{8(\mu_e H)^{4/3}} = \frac{1.231 \times 10^{15}}{\mu_e^{4/3}} . \tag{69}$$

By equation (70) of chapter iv the mass of a Lane-Emden configuration based on (68) is given by

$$M = -4\pi \left(\frac{K_2}{\pi G}\right)^{3/2} \left(\xi^2 \frac{d\theta_3}{d\xi}\right)_{\xi=\xi_1} , \tag{70}$$

which is seen to be equivalent to (66) on substituting for K_2 according to (69).

We shall denote by M_3 the limiting mass (66).[2] The mass relation (42) can then be written in the form

$$M(y_0) = M_3 \frac{\Omega(y_0)}{{}_0\omega_3} , \tag{71}$$

where

$${}_0\omega_3 = -\left(\xi^2 \frac{d\theta_3}{d\xi}\right)_{\xi=\xi_1(\theta_3)} ; \qquad \Omega(y_0) = -\left(\eta^2 \frac{d\phi}{d\eta}\right)_{\eta=\eta_1} . \tag{72}$$

[2] We denote the limiting mass by M_3 since, as $x \to \infty$, $\phi \to \theta_3$, the Lane-Emden function of index 3.

As the mass of the configuration increases monotonically with increasing y_0, we have the useful inequality

$$\Omega(y_0) < {}_0\omega_3 \qquad (y_0 \text{ finite}) . \qquad (73)$$

Finally, we may note that the insertion of numerical values in the formula for M_3 yields

$$M_3 = 5.75\mu_e^{-2} \times \odot . \qquad (74)$$

f) The internal energy.—By equation (23) of chapter x, the internal energy U of the configuration is given by

$$U = A\int_0^R \{8x^3[(1 + x^2)^{1/2} - 1] - f(x)\}dV ; \qquad (75)$$

or, using equations (16) and (17) which express the equation of state, we can re-write the foregoing in the form

$$U = \frac{8A}{B}\int_0^R \rho[(1 + x^2)^{1/2} - 1]dV - \int_0^R PdV . \qquad (75')$$

But by equation (32) of chapter iii the second term on the right-hand side of (75′) is $-\Omega/3$ where Ω is the potential energy. Hence,

$$U = \frac{8A}{B}\int_0^R [(1 + x^2)^{1/2} - 1]dM(r) + \tfrac{1}{3}\Omega ; \qquad (76)$$

or, expressing x in terms of ϕ (cf. Eqs. [24] and [27]), we have

$$U = \frac{8Ay_0}{B}\int_0^R \left(\phi - \frac{1}{y_0}\right) dM(r) + \tfrac{1}{3}\Omega . \qquad (76')$$

Using (37) for expressing ϕ in terms of the potential V, we obtain

$$U = -\int_0^R \left(V + \frac{GM}{R}\right) dM(r) + \tfrac{1}{3}\Omega . \qquad (77)$$

Finally, using equation (16) of chapter iii, we find

$$U = -\tfrac{5}{3}\Omega - \frac{GM^2}{R} . \qquad (78)$$

For the case under consideration the internal energy is due entirely to the kinetic energies of the motions of the electrons; we can, therefore, write

$$T = U = -\tfrac{5}{3}\Omega - \frac{GM^2}{R} . \tag{79}$$

The total energy, E, of the configuration is

$$E = U + \Omega = -\tfrac{2}{3}\Omega - \frac{GM^2}{R} . \tag{79'}$$

For stars of small mass the configurations are (as we have shown in section d, above) polytropes of index $n = 3/2$, and by equation (90) of chapter iv,

$$\Omega = -\frac{6}{7}\frac{GM^2}{R} \qquad (M \ll M_3) . \tag{80}$$

By (79) and (80) we have

$$T = -\tfrac{1}{2}\Omega , \tag{80'}$$

which is the statement of the virial theorem (chap. ii, § 10) derived on the basis of Newtonian mechanics. On the other hand, if $M \to M_3'$, then (again by Eq. [90], iv),

$$\Omega = -\frac{3}{2}\frac{GM^2}{R} \qquad (M \to M_3) . \tag{81}$$

By (79) and (81) we now have

$$T = -\Omega , \tag{81'}$$

which must be the statement of the virial theorem for material particles moving with very nearly the velocity of light.

g) General results.—In sections d and e we have considered certain limiting cases. However, the exact treatment on the basis of the differential equation (28) will provide much more quantitative information. The boundary conditions,

$$\phi = 1 , \qquad \frac{d\phi}{d\eta} = 0 \qquad \text{at} \qquad \eta = 0 ,$$

combined with a particular value for y_0 will determine ϕ completely and therefore the mass of the configuration as well. Equation (28) does not admit of a homology constant, and hence *each mass has a density distribution characteristic of itself which cannot be inferred from the density distribution in a configuration of a different mass.* This is the most fundamental difference between our present configurations and the polytropes. We thus see that each specified value for y_0 determines uniquely the mass M, the radius R, the ratio of the mean to the central density, and the march of the density distribution. We have (collecting our results):

$$\left.\begin{aligned} \frac{M}{M_3} &= \frac{\Omega(y_0)}{{}_0\omega_3}\,, \\ \frac{R}{l_1} &= \frac{\eta_1}{y_0}\,, \\ \frac{\rho_0}{B} &= (y_0^2 - 1)^{3/2}\,, \\ \frac{\bar{\rho}}{\rho_0} &= -\frac{1}{\left(1 - \dfrac{1}{y_0^2}\right)^{3/2}} \frac{3}{\eta_1}\left(\frac{d\phi}{d\eta}\right)_{\eta=\eta_1}. \end{aligned}\right\} \qquad (82)$$

In (82) we have introduced the unit of length ($l_1 = \alpha y_0$),

$$l_1 = \frac{1}{4\pi m \mu_e H}\left(\frac{3h^3}{2cG}\right)^{1/2} = 7.71\mu_e^{-1} \times 10^8\ \text{cm}\,, \qquad (82')$$

which, therefore, does not involve the factor in y_0. Further, the physical variables determining the structure of the configurations are:

$$\left.\begin{aligned} \rho &= \rho_0 \frac{1}{\left(1 - \dfrac{1}{y_0^2}\right)^{3/2}} \left(\phi^2 - \frac{1}{y_0^2}\right)^{3/2}, \\ \bar{\rho}(\eta) &= -\rho_0 \frac{1}{\left(1 - \dfrac{1}{y_0^2}\right)^{3/2}} \frac{3}{\eta}\frac{d\phi}{d\eta}\,, \\ \frac{M(\eta)}{M_3} &= \frac{\left(\eta^2 \dfrac{d\phi}{d\eta}\right)}{\left(\xi^2 \dfrac{d\theta_3}{d\xi}\right)_{\xi=\xi_1(\theta_3)}}. \end{aligned}\right\} \qquad (83)$$

h) *Numerical results.*—In section *g* we reduced the problem of the structure of degenerate gas spheres to a study of the function ϕ for different initially prescribed values of the parameter y_0. The integration has been numerically effected by Chandrasekhar for ten different values of the parameter:

$$\frac{1}{y_0^2} = 0.8, 0.6, 0.5, 0.4, 0.3, 0.2, 0.1, 0.05, 0.02, 0.01 .$$

The integration is started at the origin by a series expansion and then continued by standard numerical methods. The following expansion for ϕ near the origin may be noted here:

$$\left.\begin{aligned}\phi = 1 - \frac{q^3}{6}\eta^2 + \frac{q^4}{40}\eta^4 - \frac{q^5(5q^2+14)}{7!}\eta^6 + \frac{q^6(339q^2+280)}{3\times 9!}\eta^8 \\ - \frac{q^7(1425q^4+11436q^2+4256)}{5\times 11!}\eta^{10} + \dots ,\end{aligned}\right\} \quad (84)$$

where $q^2 = (y_0^2 - 1)/y_0^2$. The important quantities of interest are the boundary quantities occurring in equation (82). These are tabulated in Table 25. From the figures in Table 25 it is easy to calculate the

TABLE 25

THE CONSTANTS OF THE WHITE-DWARF FUNCTIONS

$1/y_0^2$	η_1	$-\eta_1^2\phi'(\eta_1)$	$\rho_0/\bar{\rho}$
0	6.8968	2.0182	54.182
0.01	5.3571	1.9321	26.203
0.02	4.9857	1.8652	21.486
0.05	4.4601	1.7096	16.018
0.1	4.0690	1.5186	12.626
0.2	3.7271	1.2430	9.9348
0.3	3.5803	1.0337	8.6673
0.4	3.5245	0.8598	7.8886
0.5	3.5330	0.7070	7.3505
0.6	3.6038	0.5679	6.9504
0.8	4.0446	0.3091	6.3814
1.0	∞	0	5.9907

mass in units of M_3, the radius in units of l_1, and the central density in units of B ($= 9.82 \times 10^5\mu_e$ gm cm^{-3}). These express the chief physical characteristics in the "natural system" of units occurring in the theory of these configurations (see Table 26). In Table 27 they are converted into the more conventional system of units which express the radius and the density in c.g.s. units and the mass in units of the

TABLE 26

THE PHYSICAL CHARACTERISTICS OF COMPLETELY DEGENERATE CONFIGURATION IN THE "NATURAL" UNITS

$1/y_0^2$	M/M_3	R/l_1	ρ_0/B
0	1	0	∞
0.01	0.95733	0.53571	985.038
0.02	0.92419	0.70508	343.
0.05	0.84709	0.99732	82.8191
0.1	0.75243	1.28674	27.
0.2	0.61589	1.66682	8.
0.3	0.51218	1.96102	3.56423
0.4	0.42600	2.22908	1.83711
0.5	0.35033	2.49818	1.
0.6	0.28137	2.79148	0.54433
0.8	0.15316	3.61760	0.125
1.0	0	∞	0

TABLE 27*

THE PHYSICAL CHARACTERISTICS OF COMPLETELY DEGENERATE CONFIGURATIONS

$1/y_0^2$	$M/\odot$	ρ_0 in Grams per Cubic Centimeter	ρ_{mean} in Grams per Cubic Centimeter	Radius in Centimeters
0	5.75	∞	∞	0
0.01	5.51	9.67×10^8	3.70×10^7	4.13×10^8
0.02	5.32	3.37×10^8	1.57×10^7	5.44×10^8
0.05	4.87	8.13×10^7	5.08×10^6	7.69×10^8
0.1	4.33	2.65×10^7	2.10×10^6	9.92×10^8
0.2	3.54	7.85×10^6	7.9×10^5	1.29×10^9
0.3	2.95	3.50×10^6	4.04×10^5	1.51×10^9
0.4	2.45	1.80×10^6	2.29×10^5	1.72×10^9
0.5	2.02	9.82×10^5	1.34×10^5	1.93×10^9
0.6	1.62	5.34×10^5	7.7×10^4	2.15×10^9
0.8	0.88	1.23×10^5	1.92×10^4	2.79×10^9
1.0	0	0	0	∞

* The values given in this table differ slightly from the published values (S. Chandrasekhar *M.N.*, **95**, 208, 1935, Table III). The difference is due to the change in the accepted values of the fundamental physical constants.

The calculations are for $\mu_e = 1$. For the other values of μ_e, M should be multiplied by μ_e^{-2}, R by μ_e^{-1}, and ρ_0 by μ_e.

sun. To see the order of magnitude of the quantities involved, it is of interest to point out that the mass $4.87\odot\mu_e^{-2}$ has a radius only slightly larger than the radius of the earth, while the mass $0.957M_3$ has a radius considerably less than the radius of the earth. In Fig-

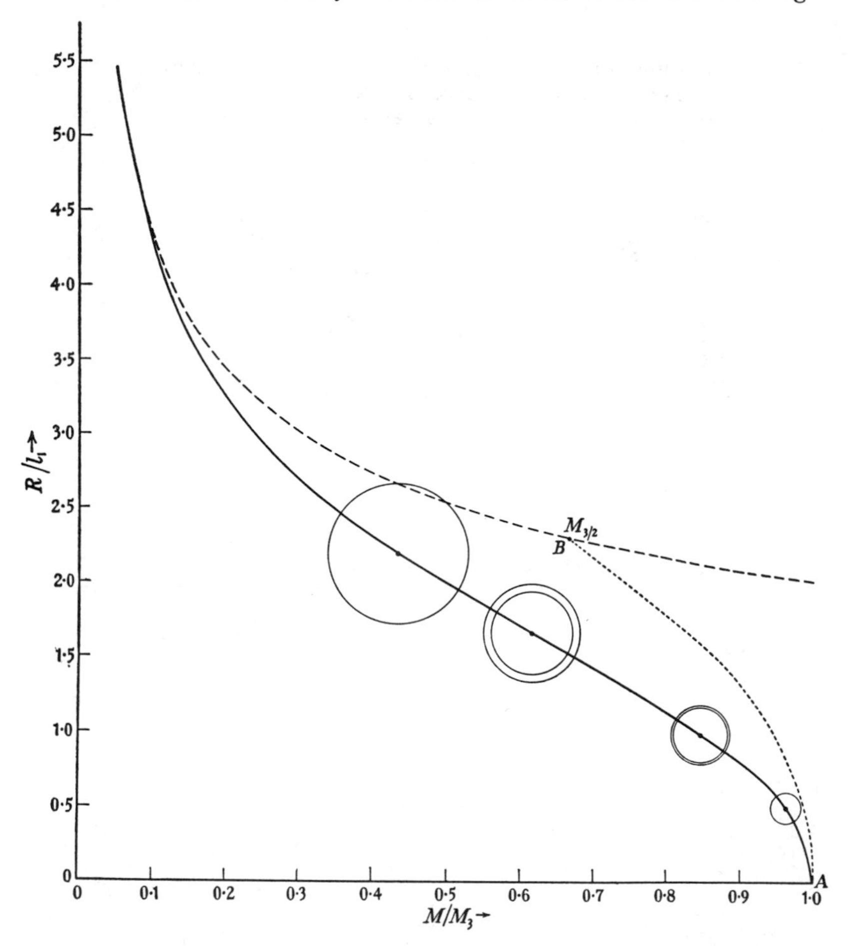

FIG. 31.—The solid-line curve represents the exact (mass, radius) relation for the completely degenerate configurations. This curve tends asymptotically to the dotted curve as $M \rightarrow 0$.

ures 31 and 32 we have illustrated the mass-radius and the mass-central density relationships. The dotted curves in the two cases are the corresponding relations based on the Lane-Emden polytrope of index $n = 3/2$ (the approximation considered in section *d*, above), and the exact curves tend toward these asymptotically for $M \to 0$. We notice from Figures 31 and 32 how marked the deviations of the dotted curves from the exact curves become even for quite small masses. Thus, for $M = 0.15M_3$ the central density predicted by the exact treatment is about 25 per cent greater and the radius about 5 per cent smaller. The relativistic effects on the equation of state

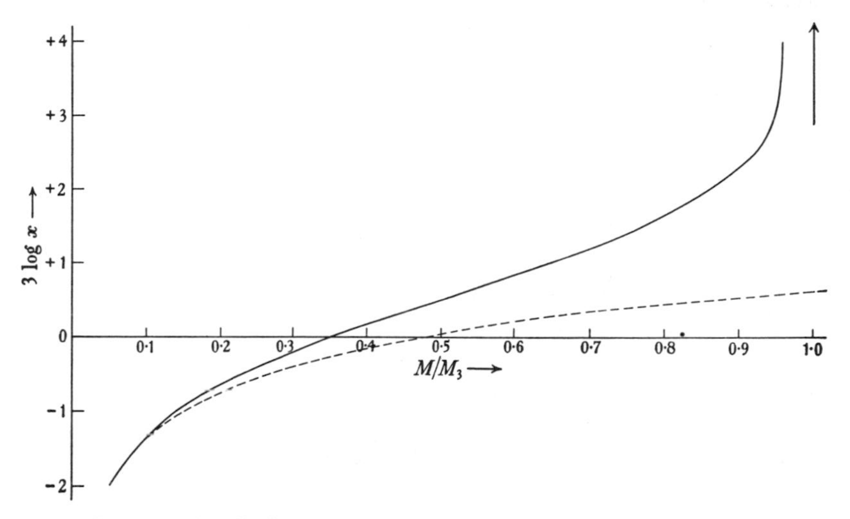

FIG. 32.—The solid-line curve represents the exact (mass, Log ρ_0) relation for the completely degenerate configurations. This curve tends asymptotically to the dotted curve as $M \to 0$.

are therefore quite significant even for small masses. They certainly cannot be ignored for masses greater than $0.2M_3$. Of course, the extrapolation of the $n = 3/2$ configurations for masses approaching M_3 is quite misleading. The completely degenerate configurations have a natural limit, and our discussion based on the differential equation shows how this limit is reached.

i) The relative density distributions in the different configurations.—Our main diagram (Fig. 33) now illustrates the relative density distributions in the configurations studied. Here we have plotted ρ/ρ_0 against η/η_1 for the different masses for which we have numerical results. The two limiting density-distributions specified by the Lane-

Emden functions $\theta_{3/2}$ and θ_3 are also shown (dotted) in the same diagram. The density distributions specified by the differential equa-

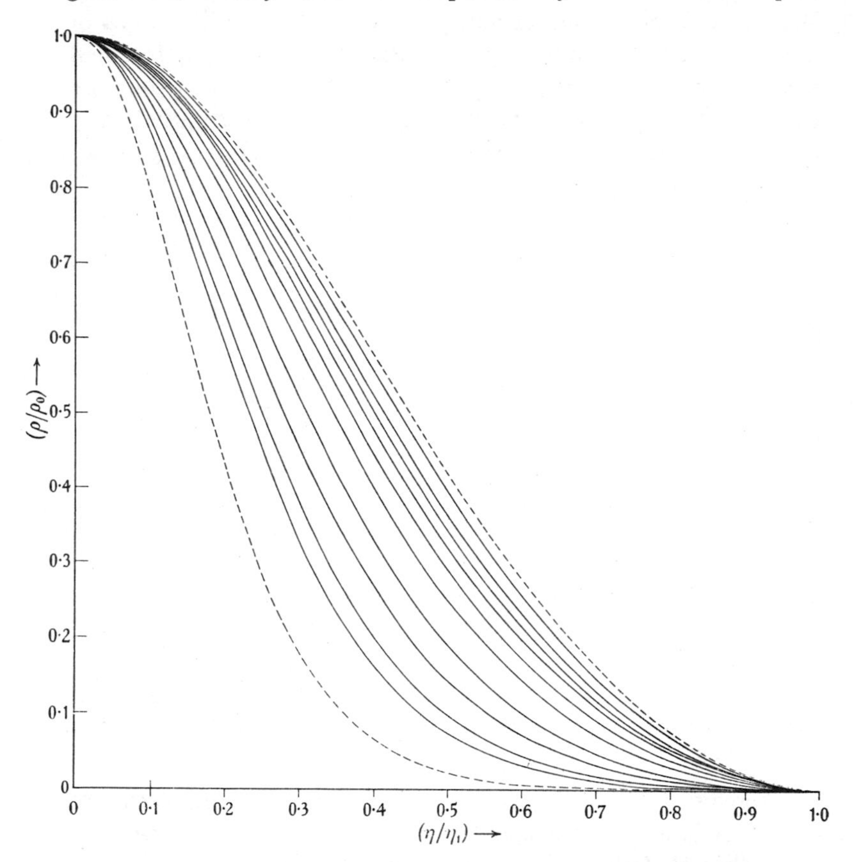

FIG. 33.—The relative density distributions in the completely degenerate configurations. The upper dotted curve corresponds to the polytropic distribution $n=\frac{3}{2}$, and the lower dotted curve to the polytropic distribution $n=3$. The inner curves represent the density distributions for $1/y_0^2=0.8$, 0.6, 0.5, 0.4, 0.3, 0.2, 0.1, 0.05, 0.02, and 0.01, respectively.

tion (28) thus form a continuous family which covers the range specified by the polytropic distributions of indices 3/2 and 3.

3. *The discussion of the observational material and of the theoretical mass-radius relation.*—We have already seen in § 1 that the gaseous fringe of the known white dwarfs can be neglected (in the first approximation) and that we can regard them (in the first approximation) as completely degenerate configurations. The theory developed in § 2 can therefore be applied, as it stands, to the known white dwarfs. A glance at Table 27 shows that the mean density, the mass, and the radius of these degenerate configurations are all of the right order of magnitude to provide the basis for the theoretical discussion of the structure of the white dwarfs. However, a really satisfactory test of the theory will consist in providing an observational basis for the existence of a mass such that as we approach it the mean density increases several times, even for a slight increase in the mass. At the present time there is just one case which seems to support this aspect of the theoretical prediction.

The case in question is Kuiper's white dwarf (AC 70°8247), which is, from several points of view, a most remarkable star; for instance—and this is very unfortunate—in this star no spectral lines have been detected so far and only a pure continuous spectrum has been observed. According to Kuiper, the most probable values of L and R are

$$\text{Log}\, L = -1.76\,, \qquad \text{Log}\, R = -2.38\,, \tag{85}$$

L and R being expressed in solar units. From (85) we derive that

$$\bar{\rho} = 19{,}600{,}000 \left(\frac{M}{\odot}\right) \text{gm cm}^{-3}\,. \tag{86}$$

It is seen that we have here an unusually dense star. If we assume that $\mu_e = 1.48$, then the mass-radius relation established in § 2 leads to a mass of $2.5\odot$, which would correspond to an actual mean density of 49,000,000 gm cm^{-3}. On the other hand, if we use the approximation $P = K\rho^{5/3}$ (Eqs. [62] and [63]), then from the mass-radius relation for the polytropes (Eq. [74], iv) we easily derive that

$$\text{Log}\, R = -\tfrac{1}{3}\, \text{Log}\, M - \tfrac{5}{3}\, \text{Log}\, \mu_e - 1.397\,, \tag{87}$$

where R and M are expressed in solar units. Assuming $\mu_e = 2.0$ (which is the maximum we can permit), we find that (87) leads to a mass of $28\odot$ for Kuiper's white dwarf; it should be noticed that this is the minimum mass predicted on the basis of (87). (If we assume for μ_e the more probable value of 1.5, then (87) leads to $M = 118\odot$.) Since the mass predicted on the model $P \propto \rho^{5/3}$ comes out unusually high, it seems likely that Kuiper's white dwarf does, in fact, provide a confirmation of the theory. In any case, it is clear that if spectral lines could be detected and identified in this star and the red shift measured, we might have a most valuable astronomical confirmation of the physical theory of degeneracy.[3]

However, since the theory is such a straightforward consequence of the quantum mechanics and, further, uses Dirac's theory of the electron only in that phase of its application which has been confirmed by laboratory experiments (Klein-Nishina formula, production of cosmic ray showers, etc.), there can be little doubt that it is essentially correct.

We have seen that the theory provides a unique mass-radius relation if the radius is measured in units of l_1 (Eq. [82]) and the mass in units of M_3. But these units involve the "molecular weight," μ_e, so that we can apply the theory to determine μ_e for white dwarfs for which both M and R are known, or to determine M for a white dwarf for which only the radius is known (assuming, however, a value for μ_e). It should be noticed that μ_e is not the same as the mean molecular weight μ used in the theory of gaseous stars. For, as the definition of μ_e we have used

$$\rho = n\mu_e H \, , \tag{88}$$

where n is the number of electrons per unit volume. For a mixture of elements which are all completely ionized we can write, in the notation of § 3 of chapter vii,

$$n = \frac{\rho}{H} \sum \frac{x_Z Z}{A_Z} \, , \tag{89}$$

where the element of atomic number Z and atomic weight A_Z is assumed to occur with an abundance x_Z by weight. The summa-

[3] There is a possibility that Wolf 219, another white dwarf discovered by Kuiper, for which Humason found recently a continuous spectrum, may be comparable to AC 70°8247. If confirmed, this star would be even more extraordinary than AC 70°8247, since it is of lower luminosity.

tion in (89) is extended over the elements present. Comparing (88) and (89), we derive

$$\mu_e = \frac{1}{\sum \left(x_Z \frac{Z}{A_Z} \right)} . \tag{90}$$

If X_0 is the abundance of hydrogen, we can re-write (90) as

$$\frac{1}{\mu_e} = X_0 + \sum_{Z \neq 1} \frac{x_Z Z}{A_Z} . \tag{91}$$

As a first approximation we can write $Z/A_Z = 1/2$ for all the metals and obtain

$$\mu_e = \frac{2}{1 + X_0} . \tag{92}$$

For the Russell mixture considered in chapter vii we find that

$$\mu_e = \frac{1}{0.492 + 0.508 X_0} . \tag{93}$$

We shall now consider briefly the other white dwarfs for which we have data.

a) *Sirius B.*—We have already considered this star in § 1. Using the data given there and using the theoretical mass radius relation, it is found that $\mu_e = 1.32$, $X_0 = 0.52$.

b) o_2 *Eridani B.*—According to Kuiper,

$$\text{Log } L = -2.26 , \qquad \text{Log } M = -0.35 , \qquad \text{Log } R = -1.74 . \tag{94}$$

The mean density is 91,000 gm cm^{-3}. The theoretical mass-radius relation leads to $X_0 = 0.15$.

c) *Van Maanen No. 2.*—From the reliably known parallax and spectral type, Kuiper derives for this star

$$\text{Log } L = -3.85 , \qquad \text{Log } R = -2.05 . \tag{95}$$

The radial velocity of this star has been determined and found to be +238 km/sec. According to Oort, most of this must be due to the Einstein gravitational red shift. Assuming that the full amount is due to the red shift (which will give the right order of magnitude), it is found, with the value of R given (Eq. [95]), that

$$\text{Log } M = 0.53 \, , \qquad \bar{\rho} = 6{,}800{,}000 \text{ gm cm}^{-3} \, . \tag{96}$$

The mass-radius relation now leads to $\mu_e = 1.206$, $X_0 = 0.66$.

4. *A stellar criterion for degeneracy.*—In the last chapter we showed that the criterion for the applicability of the degeneracy formulae is (Eq. [211], x),

$$\frac{4\pi^2}{(\vartheta mc^2)^2} \frac{x(1+x^2)^{1/2}}{f(x)} \ll 1 \, . \tag{97}$$

However, for applications to stellar problems it is more convenient to state the criterion for degeneracy in a rather different form.

Consider an assembly of N electrons contained in a volume V at temperature T. Then, on the basis of the perfect gas law, the electron pressure p_e would be given by

$$p_e = \left(\frac{N}{V}\right) kT \, . \tag{98}$$

At temperature T we also have radiation pressure of amount given by the Stefan-Boltzmann law

$$p_r = \tfrac{1}{3} a T^4 \, . \tag{98'}$$

Let us denote by P the total pressure $(= p_r + p_e)$ and introduce a parameter β_e, defined as follows:

$$P = p_r + p_e = \frac{1}{\beta_e} p_e = \frac{1}{1-\beta_e} p_r \, . \tag{99}$$

Eliminating T between the relations (99), we find

$$p_e = \left[k^4 \frac{3}{a} \frac{1-\beta_e}{\beta_e} \right]^{1/3} n^{4/3} \, , \tag{100}$$

where we have used n for (N/V). Let

$$n = \frac{8\pi m^3 c^3}{3h^3} x^3 \, , \tag{101}$$

as in equation (3). Then (100) can be transformed into

$$p_e = \frac{\pi m^4 c^5}{3h^3} \left(\frac{512\pi k^4}{h^3 c^3 a} \frac{1 - \beta_e}{\beta_e} \right)^{1/3} 2x^4 . \tag{102}$$

Since the radiation constant a can be expressed in terms of the other natural constants as (Eq. [107], v)

$$a = \frac{8}{15} \frac{\pi^5 k^4}{h^3 c^3} , \tag{103}$$

equation (102) can be simplified to

$$p_e = A \left(\frac{960}{\pi^4} \frac{1 - \beta_e}{\beta_e} \right)^{1/3} 2x^4 , \tag{104}$$

where A is defined as in (17). It must, of course, be understood that (104) is simply another form of (98).

Now for an assembly having the same number N of electrons in the volume V, we can formally calculate the electron pressure that would be given by the degenerate formula, namely,

$$p_{\text{deg}} = A f(x) . \tag{105}$$

We have already shown (Eq. [26], x) that for all finite values of x

$$\frac{f(x)}{2x^4} < 1 \qquad (x < \infty) . \tag{106}$$

Hence, comparing (104) and (105), we have the result that if for a prescribed N and T, the value of β_e, calculated on the basis of the perfect gas equation (98), be such that

$$\frac{960}{\pi^4} \frac{1 - \beta_e}{\beta_e} \geqslant 1 , \tag{107}$$

then the pressure given by the perfect gas formula is greater than that given by the degenerate formula—not only for the prescribed N and T, but for all values of N and T which specify the same β_e. Let β_ω be such that

$$\frac{960}{\pi^4} \frac{1 - \beta_\omega}{\beta_\omega} = 1 , \tag{108}$$

or

$$1 - \beta_\omega = 0.09212\ldots; \qquad \beta_\omega = 0.90788\ldots. \tag{109}$$

We can state the result just obtained in the following alternative form. *If for material at density* ρ *and temperature* T *the fraction* $(1 - \beta_e)$, *calculated according to (98), (98′), and (99), is greater than* $(1 - \beta_\omega)$, *then the system is definitely not degenerate.*

On the other hand, if

$$\frac{960}{\pi^4}\frac{1 - \beta_e}{\beta_e} < 1, \tag{110}$$

or

$$1 - \beta_e < 1 - \beta_\omega; \qquad \beta_e > \beta_\omega, \tag{111}$$

then for the specified β_e the electron assembly becomes degenerate for sufficiently high electron concentrations. The criterion for degeneracy under these circumstances would then be the following.

For the specified N and T, calculate β_e on the perfect gas law (i.e., $p_e = n_e kT$) and solve the equation

$$\left(\frac{960}{\pi^4}\frac{1 - \beta_e}{\beta_e}\right)^{1/3} = \frac{f(x)}{2x^4}. \tag{112}$$

(A solution exists, since [110] holds.) Denote the solution by x'. If x for the prescribed N (according to Eq. [101]) is much less than x', then the system is far removed from degeneracy, while if x is much greater than x' the system will be more or less completely degenerate.

Table 28 provides solutions of (112) for different values of $1 - \beta_e$.

If (110) holds, we can use the following approximation for the real equation of state:

$$\left.\begin{aligned} p_e &= Af(x) && (x \geqslant x') \\ &\text{and} \\ p_e &= 2A\left(\frac{960}{\pi^4}\frac{1 - \beta_e}{\beta_e}\right)^{1/3} x^4 && (x \leqslant x'), \end{aligned}\right\} \tag{113}$$

x' being such that

$$\left(\frac{960}{\pi^4}\frac{1 - \beta_e}{\beta_e}\right)^{1/3} = \frac{f(x')}{2x'^4}. \tag{114}$$

5. *The effect of radiation pressure. The mass* $\mathfrak{M} = M_3\beta_\omega^{-3/2}$.—In § 2 we considered the equilibrium of completely degenerate configurations, neglecting the radiation pressure entirely. This was justified in § 1, where it was shown that for the known white dwarfs these assumptions (of complete degeneracy and zero radiation pressure) were entirely justified and our object in the study of the completely degenerate configurations is primarily one of obtaining a satisfactory theory for the white dwarfs. It is, however, of some theoretical interest to consider the effect of "introducing" radiation pressure in these configurations.

Let us, in the first instance, consider a degenerate configuration which is built on the standard model. Then the total pressure, P, will be given by

$$P = \beta_e^{-1} p_e \, , \tag{115}$$

where p_e is the electron pressure and β_e is a constant. Then, according to equation (16),

$$P = \beta_e^{-1} A f(x) \; ; \qquad \rho = B x^3 \, . \tag{116}$$

It is clear that the analysis of § 2 applies to our present models if we replace A (wherever it occurs) by $\beta_e^{-1}A$. In particular, the mass relation (42) now takes the form

$$M(\beta_e; y_0) = -4\pi \left(\frac{2A}{\pi \beta_e G}\right)^{3/2} \frac{1}{B^2} \left(\eta^2 \frac{d\phi}{d\eta}\right)_{\eta=\eta_1} , \tag{117}$$

where ϕ is, as before, a solution of (28). We can also write (117) in the form

$$M(\beta_e; y_0) = M(1; y_0)\beta_e^{-3/2} \, , \tag{118}$$

in an obvious notation. In particular,

$$M(\beta_e; \infty) = M_3 \beta_e^{-3/2} \, . \tag{119}$$

From (118) and (119) it would at first sight appear that by allowing $\beta_e \to 0$ we can obtain degenerate configurations for any mass. This is, however, incorrect. For, according to the criterion of degeneracy established in § 4, β_e has to be greater than β_ω if the matter is to be regarded as degenerate, and we see that the maximum mass of the configurations which can be regarded as degenerate is therefore given by

$$\mathfrak{M} = M_3 \beta_\omega^{-3/2} . \tag{120}$$

The result just stated is extremely general and can be proved as follows: Consider a completely degenerate configuration of mass M, slightly less than M_3. The density will everywhere be so great that we can increase the radiation pressure from zero to a value only slightly less than $(1 - \beta_\omega)$ at each point of the configuration and still regard the matter as degenerate. According to (118), the mass of the new configuration so obtained will be approximately $M\beta_\omega^{-3/2}$. When $M \to M_3$, the result becomes exact. We have thus proved that the *maximum mass of a stellar configuration which, consistent with the physics of degenerate matter, can be regarded as wholly degenerate, is* $\mathfrak{M} = M_3\beta_\omega^{-3/2}$.

We may notice that

$$\mathfrak{M} = 1.156 M_3 = 6.65 \odot \mu_e^{-2} . \tag{121}$$

6. *Composite configurations.*—We shall now give some elementary considerations concerning stellar configurations with degenerate cores, a subject initiated by Milne. Milne, however, considered degenerate cores at such densities that the approximation $P = K\rho^{5/3}$ could be made. Since the exact treatment based on the differential equation (28) leads to the existence of the two masses M_3 and $\mathfrak{M}$, and since, further, there are no analogues to these on the approximate considerations, it is clear that very considerable care should be exercised in interpreting the results derived on the basis of the approximate considerations. In particular, the formal results which are derived for masses greater than $\mathfrak{M}$ have no physical meaning. On the other hand, it is possible to indicate the general characteristics of these composite configurations by allowing the degenerate core to be described by ϕ without any elaborate machinery.

First of all, it is important to bear in mind that, while in the degenerate regions the electrons contribute toward the pressure almost entirely, the situation is different in the gaseous region: depending on the abundance of hydrogen, the atomic nuclei would also contribute appreciably toward the gas pressure. The consideration of the composite configurations which allow for these factors is elementary but complicated. However, the essential features of the situation can be understood by considering the case where we can put $\mu_e = \mu$; this implies that $1 - \beta_e = 1 - \beta$.

According to (104), we have (for the case under consideration) in the gaseous region

$$P = \frac{1}{\beta} p_e = 2A \left(\frac{960}{\pi^4} \frac{1 - \beta}{\beta^4} \right)^{1/3} x^4 \tag{122}$$

and

$$\rho = Bx^3 . \tag{123}$$

Eliminating x between (122) and (123), we have

$$P = 2A \left(\frac{960}{\pi^4} \frac{1 - \beta}{\beta^4} \right)^{1/3} \frac{1}{B^{4/3}} \rho^{4/3} . \tag{124}$$

We shall assume that the gaseous region is governed by the standard-model equations, i.e., β is constant in (124). The gaseous region must then be governed by a solution $\theta(\xi)$ of the Lane-Emden equation of index 3—not necessarily θ_3. The mass relation (Eq. [70], iv) is now

$$M = -4\pi \left(\frac{2A}{\pi G} \right)^{3/2} \frac{1}{B^2} \left(\frac{960}{\pi^4} \frac{1 - \beta}{\beta^4} \right)^{1/2} \left(\xi^2 \frac{d\theta}{d\xi} \right)_{\xi = \xi_1(\theta)} , \tag{125}$$

which by (66) can be written as

$$M = M_3 \left(\frac{960}{\pi^4} \frac{1 - \beta}{\beta^4} \right)^{1/2} \frac{\omega_3}{{}_0\omega_3} , \tag{126}$$

where, in the notation of chapter iv,

$${}_0\omega_3 = -\left(\xi^2 \frac{d\theta_3}{d\xi} \right)_{\xi = \xi_1(\theta_3)} ; \qquad \omega_3 = -\left(\xi^2 \frac{d\theta}{d\xi} \right)_{\xi = \xi_1(\theta)} . \tag{127}$$

If the configuration is wholly gaseous, we have

$$M = M_3 \left(\frac{960}{\pi^4} \frac{1 - \beta}{\beta^4} \right)^{1/2} , \tag{128}$$

which is Eddington's quartic equation in a different form.

Now for a given mass M, equation (128) determines a $\beta = \beta(M)$. Start with this mass having an infinite radius and imagine it being slowly contracted. At first the configuration will be so rarefied that it will be wholly gaseous and the path of the "representative point" in the $(R, 1 - \beta)$ plane will be along the line parallel to the R-axis through $\beta = \beta(M)$. How far is this process of contraction possible? From our criterion of degeneracy we can now conclude that if $1 - \beta(M) > 1 - \beta_\omega$, then the process of contraction is theoretically possible to an unlimited extent. Since β_ω, according to definition, is given by

$$\frac{960}{\pi^4} \frac{1 - \beta_\omega}{\beta_\omega} = 1 , \tag{129}$$

it follows that a configuration for which $\beta(M) = \beta_\omega$ is, according to (128),

$$M_3 \beta_\omega^{-3/2} = \mathfrak{M} . \tag{130}$$

a) The domain of degeneracy.—For configurations of mass greater than $\mathfrak{M}$, the appropriate $1 - \beta(M)$ is greater than $(1 - \beta_\omega)$ and the representative point will travel down the straight line $\beta = \beta(M)$, however far the contraction may proceed. But the situation is different when the mass of the configuration is less than $\mathfrak{M}$. For such masses, $1 - \beta(M) < 1 - \beta_\omega$, and hence a stage must be reached when the configuration should begin to develop central regions of degeneracy. On the scheme of approximation (113) and (114), we can now easily see how far the process of contraction is possible before degeneracy sets in.

Let the central density be ρ_0. Then

$$\rho_0 = B x_0^3 . \tag{131}$$

Degeneracy would just begin to develop at the center for a value of $x = x_0$ such that

$$\frac{f(x_0)}{2x_0^4} = \left(\frac{960}{\pi^4} \frac{1 - \beta}{\beta}\right)^{1/3} . \tag{132}$$

For this configuration the mean density $\bar{\rho}$ is (according to Eq. [78], iv, which gives the ratio of the mean to the central density for a polytrope)

$$\bar{\rho} = -3 \left(\frac{1}{\xi} \frac{d\theta_3}{d\xi}\right)_{\xi = \xi_1(\theta_3)} B x_0^3 . \tag{133}$$

The radius R_0 of the configuration is, therefore, given by

$$\tfrac{4}{3}\pi R_0^3 = \frac{\text{Mass}}{\text{Mean density}} . \tag{134}$$

Substituting (125) and (133) in the foregoing expression, we obtain

$$R_0 = \left(\frac{2A}{\pi G}\right)^{1/2} \left(\frac{960}{\pi^4} \frac{1-\beta}{\beta^4}\right)^{1/6} \frac{1}{Bx_0} \xi_1(\theta_3) . \tag{135}$$

Define a unit of length by (cf. Eqs. [27] and [33])

$$l = \left(\frac{2A}{\pi G}\right)^{1/2} \frac{\xi_1(\theta_3)}{B} = \frac{7.71 \times 10^8 \times 6.897}{\mu} , \tag{136}$$

or, numerically,

$$l = 5.32 \times 10^9 \mu^{-1} \text{ cm} . \tag{137}$$

From (135), then,

$$\frac{R_0}{l} = \left(\frac{960}{\pi^4} \frac{1-\beta}{\beta^4}\right)^{1/6} \frac{1}{x_0} , \tag{138}$$

where x_0 is again determined from (132). By using (132), we can write (138) more conveniently as

$$\frac{R_0}{l} = \left(\frac{f(x_0)}{2x_0^4} \frac{1}{\beta}\right)^{1/2} \frac{1}{x_0} . \tag{139}$$

It is a fairly simple matter to calculate from (132) and (139) corresponding pairs of values for (R_0/l) and β. These are tabulated in Table 28. This $(R_0, 1-\beta)$ curve can therefore be drawn in the $(R, 1-\beta)$ plane (see Fig. 34). The region bounded by this curve and the two axes then defines the domain of degeneracy meaning that it is only in this region that the curves of constant mass are distorted from straight lines parallel to the R-axis.

From (132) and (139) we see that, as $\beta \rightarrow \beta_\omega$,

$$x_0 \rightarrow \infty , \quad R_0 \rightarrow 0 . \tag{140}$$

Hence, as we should expect, the $(R_0, 1-\beta)$ curve intersects the $(1-\beta)$ axis at a point where $\beta = \beta_\omega$. It can be proved easily that the $(R_0, 1-\beta)$ curve intersects the $(1-\beta)$ axis vertically.

b) The nature of the curves of constant mass for $M \leqslant M_3$ *in the domain of degeneracy.*—In (*a*), above, we have shown at what stage a configuration of mass less than $\mathfrak{M}$ (contracting from infinite ex-

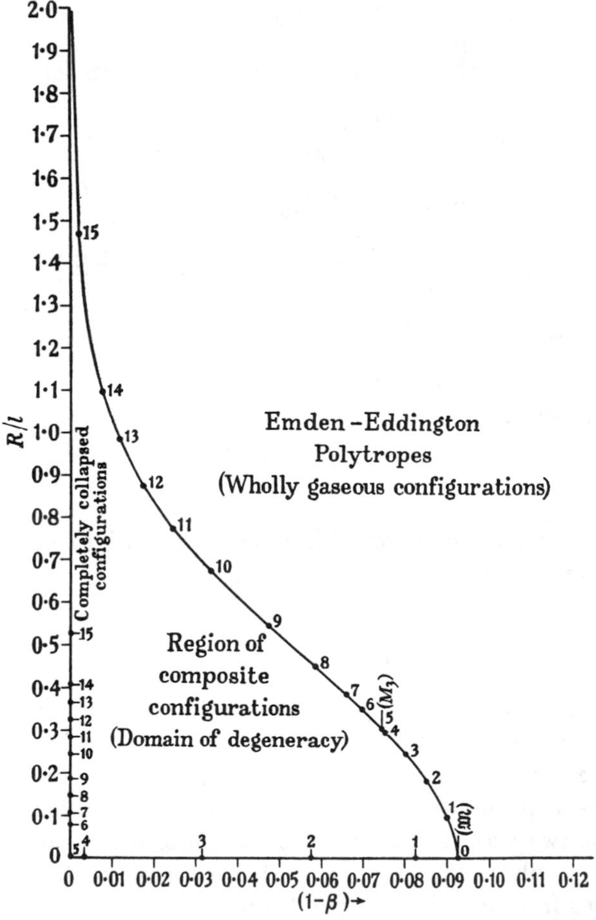

FIG. 34.—The curve running from $1-\beta=0.092\ldots$ to infinity along the R-axis is the $(R_0, 1-\beta)$ curve (see Eq. [139]). The points marked $(5, \ldots, 15)$ on the $(R_0, 1-\beta)$ curve and the R-axis are the end-points (in the domain of degeneracy) of the curves of constant mass for the values of M tabulated in Table 29. The points marked $(1, \ldots, 4)$ on the $(R_0, 1-\beta)$ curve and on the $(1-\beta)$ axis are the corresponding end-points for some curves of constant mass $(M_3 \leqslant M \leqslant \mathfrak{M})$ on the standard model (see Table 30).

tension) begins to develop degeneracy at the center. This happens when the appropriate line $(1-\beta) = 1-\beta(M)$ intersects the

(R_0, $1 - \beta$) curve. If the contraction continues further, the configuration will begin to develop finite degenerate cores, and our problem now is to examine how the curves of constant mass run inside the domain of degeneracy.

TABLE 28

THE STELLAR CRITERION FOR DEGENERACY AND THE (R_0, $1 - \beta$) CURVE

x	$1-\beta$	R_0/l	x	$1-\beta$	R_0/l
0	0	∞	2.8	0.06919	0.3515
0.2	0.00040	1.9868	3.0	.07149	.3304
0.4	.00282	1.3787	3.5	.07598	.2870
0.6	.00793	1.0956	4.0	.07920	.2535
0.8	.01505	0.9187	4.5	.08158	.2268
1.0	.02305	0.7934	5.0	.08337	.2051
1.2	.03101	0.6985	6.0	.08583	.1721
1.4	.03839	0.6235	7.0	.08739	.1481
1.6	.04495	0.5627	8.0	.08844	.1299
1.8	.05068	0.5123	9.0	.08918	.1157
2.0	.05561	0.4699	10.0	.08972	.1043
2.2	.05983	0.4337	20.0	.09150	.0524
2.4	.06344	0.4025	30.0	.09185	.0350
2.6	0.06653	0.3753	∞	0.09212	0

In § 2 we made an analysis of completely degenerate configurations. Each mass (less than M_3) has a certain uniquely determined radius. Thus, if the mass under consideration has a central density corresponding to $y = y_0$, then the radius, R, is given by

$$R = \alpha\eta_1 = \left(\frac{2A}{\pi G}\right)^{1/2} \frac{\eta_1}{By_0}, \tag{141}$$

where η_1 is the boundary of the corresponding function $\phi(y_0)$. In terms of the unit of length, l (Eq. [136]),

$$\frac{R}{l} = \frac{1}{y_0} \frac{\eta_1[\phi(y_0)]}{\xi_1(\theta_3)}. \tag{142}$$

These completely degenerate configurations correspond to $\beta = 1$. Hence, we know from (142) the point at which the curves of constant mass for $M < M_3$ must intersect the R-axis. Also, for any mass M we can calculate the value of β in the wholly gaseous state. Let $\beta\dagger$ be the value of β for a wholly gaseous configuration which in its completely degenerate state has a central density corresponding to $y = y_0$. Then, according to equations (71) and (128), we have the relation

$$\left(\frac{960}{\pi^4}\frac{1-\beta\dagger}{\beta\dagger^4}\right)^{1/2} = \frac{M}{M_3} = \frac{\Omega(y_0)}{{}_0\omega_3}, \tag{143}$$

where, as in equation (72),

$$\Omega(y_0) = -\eta_1^2\left(\frac{d\phi}{d\eta}\right)_{\eta=\eta_1(\phi(y_0))}. \tag{144}$$

Now the line through $\beta\dagger$ parallel to the R-axis will intersect the $(R_0, 1-\beta)$ curve at $[R_0(M(y_0)), 1-\beta\dagger]$. In the domain of degeneracy the continuation of the curve must in some way connect the point $[R_0(M(y_0)), 1-\beta\dagger]$ and the point R on the R-axis, where

$$\frac{R}{l} = \frac{1}{y_0(M)}\frac{\eta_1[\phi(y_0(M))]}{\xi_1(\theta_3)}. \tag{145}$$

From the numerical values for η_1, Ω, etc., for the ten different values of y_0 given in Table 25, the corresponding values of R/l (according to [145]) and $\beta\dagger$ can be evaluated. The results are given

TABLE 29

$1/y_0^2$	M/M_3	$1-\beta\dagger$	R/l
0	1.	0.07446	0
0.01	0.95733	.06966	0.07767
0.02	0.92419	.06596	0.10223
0.05	0.84709	.05746	0.14460
0.1	0.75243	.04732	0.18657
0.2	0.61589	.03358	0.24168
0.3	0.51218	.02414	0.28434
0.4	0.42600	.01718	0.32320
0.5	0.35033	.01187	0.36222
0.6	0.28137	.00779	0.40475
0.8	0.15316	.00236	0.52453
1.0	0	0	∞

in Table 29. We have thus fixed the "end-points" for the curves of constant mass for $M \leqslant M_3$ in the domain of degeneracy. The corresponding pairs of points on the $(R_0, 1-\beta)$ curve and the R-axis are shown in Figure 34.

It is clear that the curve for M_3 must pass through the origin of our system of co-ordinates. Further, if β_0 is the value of β for M_3 in the wholly gaseous state, then, according to (143),

$$\frac{960}{\pi^4}\frac{1-\beta_0}{\beta_0^4} = 1 \, , \tag{146}$$

or

$$1-\beta_0 = 0.07446 \, ; \qquad \beta_0 = 0.92554 \, . \tag{147}$$

c) The nature of the curves of constant mass for M > M$_3$ *in the domain of degeneracy.*—In (*b*), above, the end-points for the curves of constant mass (for configurations with mass less than, or equal to, M_3) have been fixed. We further saw that the curve for M_3 must pass through the origin. The question now arises: What happens for configurations with $\mathfrak{M} \geqslant M > M_3$? The answer to this question can be given quite simply if $(1-\beta)$ has the same value in the degenerate core as in the gaseous envelope. We have already shown (Eq. [119]) that the completely relativistic configuration has a mass

$$M = M_3\beta^{-3/2} \qquad (1 \geqslant \beta \geqslant \beta_\omega) \tag{148}$$

and is of zero radius. Hence, the curves of constant mass for $M > M_3$ must cross the $(1-\beta)$ axis at a point $(1-\beta^*)$, say, such that

$$M = M_3\beta^{*-3/2} \, . \tag{149}$$

Let us denote by $\beta\dagger$ the value of β in the wholly gaseous state. There is a simple relation between β^* and $\beta\dagger$. Comparing (143) and (149), we derive that

$$\beta^* = \left(\frac{\pi^4}{960}\frac{\beta\dagger^4}{1-\beta\dagger}\right)^{1/3} . \tag{150}$$

From (150) we see that $\beta^* = 1$, $\beta\dagger = \beta_0$ (Eq. [146]), is a solution; in other words, the appropriate curve for M_3 must pass through the origin which in fact it does. Again, $\beta^* = \beta\dagger = \beta_\omega$ is also a solution of (150); the appropriate curve for $\mathfrak{M}$ is therefore the full line through $(1-\beta_\omega)$ parallel to the R-axis, as we should have expected.

Table 30 gives a set of corresponding pairs of values for β^* and

β† (see also Fig. 34, where the corresponding pairs of points are marked [1, 2, 3, 4] on the [R_0, $1 - \beta$] curve and the [$1 - \beta$] axis).

The results described above (in [*b*]) are true for the usual standard model. If we consider as an another limiting case configurations in which $\beta = 1$ in the degenerate core and $\beta \leqslant 1$ in the gaseous envelope, then the discussion is similar but somewhat more complicated (cf. Chandrasekhar's papers quoted in the Bibliographical Notes).

TABLE 30

$1 - \beta$†	$1 - \beta^*$	$M/\mathfrak{M}$
0.09212	0.09212	1.
.090	.08220	0.9838
.085	.05768	0.9457
.080	.03143	0.9075
.075	.00319	0.8692
0.07446	0	0.8651

A more detailed discussion of composite configurations would consist in describing the mathematical methods for handling them precisely, i.e., by a consideration of the methods of fitting a solution of the Lane-Emden equation of index 3 to a solution of the differential equation for ϕ. Such discussions, however, are beyond the scope of the monograph. Reference may be made to the literature quoted in the Bibliographical Notes.

7. *Partially degenerate configurations.*—So far we have considered completely degenerate configurations and also stellar configurations with degenerate cores. For describing the degenerate state we have used the exact equation of state (allowing for relativistic effects) which should be valid if the degeneracy criterion is satisfied. In considering the composite configurations in § 6, we changed over from the perfect gas equation to the degenerate equation of state at a definite interface, the interface being defined in such a way that both the equations of state give the same numerical value for the pressure for the density and the temperature at the interface. We have seen that this approximation is quite good so long as we deal with configurations of not too small masses (in units of M_3). However, for stars of small mass ($\sim 0.1\ M_3$) the central density, even in the completely degenerate state, is not unduly high. Under these cir-

cumstances, we may expect that in actual stars (e.g., Krüger 60) the "transition zone" between the perfect gas region and the region of more or less complete degeneracy will be quite extensive. It is therefore a matter of some importance to allow for these incipiently degenerate regions in a satisfactory way.

We shall illustrate a method of approach to the problem just stated in one case, namely, that in which the configuration is so poor in hydrogen that we can put $\mu = \mu_e = 2$. Further, we shall assume that the star is of such small mass that relativistic effects can be neglected. Under these circumstances, the equation of state can be parametrically expressed as follows (cf. Eqs. [262] and [263], x):

$$p_{\text{gas}} = \frac{2}{h^3} (2\pi m)^{3/2} (kT)^{5/2} U_{3/2} , \tag{151}$$

$$\rho = \frac{2}{h^3} (2\pi m)^{3/2} (kT)^{3/2} \mu H \, U_{1/2} , \tag{152}$$

where U_v stands for the integral

$$U_v = \frac{1}{\Gamma(v+1)} \int_0^\infty \frac{u^v du}{\frac{1}{\Lambda} e^u + 1} . \tag{153}$$

We shall assume that $U_{1/2}$ and $U_{3/2}$ are known functions of Λ, so that the parametric representation of the equation of state is in terms of Λ.

We shall consider two classes of equilibrium configurations built on the equation of state (151) and (152) which allows for the transition between $p \propto \rho T$ to $p \propto \rho^{5/3}$ quite accurately.

a) The isothermal gas sphere.—In this case, T is assumed constant, and the equation of equilibrium,

$$\frac{1}{r^2} \frac{d}{dr} \left(\frac{r^2}{\rho} \frac{dp}{dr} \right) = -4\pi G \rho , \tag{154}$$

on inserting for p and ρ according to (151) and (152), becomes

$$\frac{1}{r^2} \frac{d}{dr} \left(\frac{r^2}{U_{1/2}} \frac{dU_{3/2}}{dr} \right) = -4\pi G \left(\frac{2}{h^3} (2\pi m)^{3/2} \right) (kT)^{1/2} (\mu H)^2 U_{1/2} . \tag{155}$$

Let

$$r = a\zeta \,, \tag{156}$$

where

$$a = \left(\frac{h^3}{8\pi G(2\pi m)^{3/2}(kT)^{1/2}(\mu H)^2} \right)^{1/2} . \tag{157}$$

Equation (154) now reduces to

$$\frac{1}{\zeta^2} \frac{d}{d\zeta} \left(\frac{\zeta^2}{U_{1/2}} \frac{d}{d\zeta} U_{3/2} \right) = -U_{1/2} \,. \tag{158}$$

Now it is easily seen that

$$\frac{d}{d\Lambda} U_v = \frac{1}{\Gamma(v+1)\Lambda^2} \int_0^\infty \frac{u^v e^u du}{\left(\frac{1}{\Lambda} e^u + 1 \right)^2} = \frac{1}{\Lambda} U_{v-1} \,. \tag{159}$$

Equation (158) can therefore be simplified to the form

$$\frac{1}{\zeta^2} \frac{d}{d\zeta} \left(\zeta^2 \frac{d \log \Lambda}{d\zeta} \right) = -U_{1/2}(\Lambda) \,. \tag{160}$$

If $\Lambda \ll 1$, we have (cf. Eq. [270], x)

$$U_v = \Lambda \,. \tag{161}$$

Hence, if we write

$$\Lambda = e^{-\psi} \,; \qquad \zeta = \xi \,, \tag{162}$$

equation (160) transforms to

$$\frac{1}{\xi^2} \frac{d}{d\xi} \left(\xi^2 \frac{d\psi}{d\xi} \right) = e^{-\psi} \,, \tag{163}$$

which is the isothermal equation of a classical perfect gas sphere (cf. § 22, iv).

On the other hand, if $\Lambda \gg 1$, then (Eq. [266], x)

$$U_{1/2} = \frac{4}{3\sqrt{\pi}} (\log \Lambda)^{3/2} \,. \tag{164}$$

Hence, if we make the substitutions

$$\log \Lambda = \theta \; ; \qquad \zeta = \left(\frac{3\sqrt{\pi}}{4}\right)^{1/2} \xi \, , \tag{165}$$

equation (160) reduces to

$$\frac{1}{\xi^2} \frac{d}{d\xi} \left(\xi^2 \frac{d\theta}{d\xi}\right) = -\theta^{3/2} \, , \tag{166}$$

which is the Lane-Emden equation of index $n = 3/2$. We thus see that, depending upon T, we obtain from (160) either the classical isothermal case ($T \rightarrow \infty$) or the polytrope, $n = 3/2$ ($T \rightarrow 0$). A closer study of the differential equation (160) than has yet been made will make it possible to study how the change from the classical isothermal gas sphere to the polytrope $n = 3/2$ takes place as Λ increases from very near 0 to ∞. The discussion of (160) may lead to results of cosmological importance.

b) The standard model.—We shall next consider the standard model built on the equation of state, (151) and (152). Quite generally, on the standard model, we have

$$P = \frac{1}{\beta} p_{\text{gas}} = \frac{1}{1 - \beta} \frac{a}{3} T^4 \, . \tag{167}$$

Let

$$Q_1 = \frac{2}{h^3} (2\pi m)^{3/2} \; ; \qquad Q_2 = k^4 \frac{3}{a} \frac{1 - \beta}{\beta} \, . \tag{168}$$

Equations (151), (152), and (167) can now be written as

$$p_{\text{gas}} = Q_1 (kT)^{5/2} U_{3/2} \, , \tag{169}$$

$$\rho = Q_1 (kT)^{3/2} \mu H \, U_{1/2} \, , \tag{170}$$

and

$$(kT)^4 = Q_2 p_{\text{gas}} \, . \tag{171}$$

From (169) and (171), we obtain

$$(kT)^{3/2} = Q_1 Q_2 U_{3/2} \, . \tag{172}$$

Substituting for kT according to (172) in (169) and (170), we obtain

$$\beta P = p_{\text{gas}} = Q_1^{8/3} Q_2^{5/3} U_{3/2}^{8/3} \tag{173}$$

and

$$\rho = Q_1^2 Q_2 \mu H \, U_{1/2} U_{3/2} \, . \tag{174}$$

Substituting (173) and (174) in the equation of equilibrium (154), we find

$$\frac{1}{r^2} \frac{d}{dr} \left(\frac{r^2}{U_{1/2} U_{3/2}} \frac{d}{dr} U_{3/2}^{8/3} \right) = -4\pi G \beta Q_1^{4/3} Q_2^{1/3} (\mu H)^2 U_{3/2} U_{1/2} \, . \tag{175}$$

By (159) we can simplify (175) somewhat into the form

$$\frac{1}{r^2} \frac{d}{dr} \left(r^2 U_{3/2}^{2/3} \frac{d \log \Lambda}{dr} \right) = -\frac{3\pi G \beta Q_1^{4/3} Q_2^{1/3} (\mu H)^2}{2} U_{3/2} U_{1/2} \, . \tag{176}$$

Let

$$r = \alpha \zeta = \left(\frac{2}{3\pi G \beta Q_1^{4/3} Q_2^{1/3} (\mu H)^2} \right)^{1/2} \zeta \, . \tag{177}$$

Equation (176) now reduces to

$$\frac{1}{\zeta^2} \frac{d}{d\zeta} \left(\zeta^2 U_{3/2}^{2/3} \frac{d \log \Lambda}{d\zeta} \right) = -U_{3/2} U_{1/2} \, . \tag{178}$$

If $\Lambda \ll 1$, we have $U_\nu = \Lambda$, and (178) can be written as

$$\frac{1}{\zeta^2} \frac{d}{d\zeta} \left(\zeta^2 \Lambda^{-1/3} \frac{d\Lambda}{d\zeta} \right) = -\Lambda^2 \, ; \tag{179}$$

or, if

$$\theta = \Lambda^{2/3} \, ; \qquad \zeta = \sqrt{\tfrac{3}{2}} \, \xi \, , \tag{180}$$

equation (179) reduces to the Lane-Emden equation of index $n = 3$, as would be expected.

On the other hand, if $\Lambda \gg 1$, then, according to equations (266) and (267) of chapter x,

$$U_{1/2} = \frac{4}{3\sqrt{\pi}} (\log \Lambda)^{3/2} \, ; \qquad U_{3/2} = \frac{15}{8\sqrt{\pi}} (\log \Lambda)^{5/2} \, . \tag{181}$$

If we now put

$$(\log \Lambda)^{8/3} = \theta\ ; \qquad \zeta = \sqrt{\frac{3\pi}{20}\left(\frac{15}{8\sqrt{\pi}}\right)^{2/3}}\ \xi\ , \tag{182}$$

equation (178) reduces to the Lane-Emden equation of index $n = 3/2$. We thus see that a detailed study of (178), which has not as yet been made, should give insight into the structure of partially degenerate stars. The numerical discussion of the models (Eqs. [160] and [178]) cannot be very difficult; a one-parametric series of integrations would be sufficient.

BIBLIOGRAPHICAL NOTES

No attempt at a complete bibliography is made.

1. R. H. Fowler, *M.N.*, **87**, 114, 1926. In this paper Fowler makes the fundamental discovery that the electron assembly in the white dwarfs must be degenerate in the sense of the Fermi-Dirac statistics.

2. E. C. Stoner, *Phil. Mag.*, **7**, 63, 1929.

3. E. C. Stoner, *Phil. Mag.*, **9**, 944, 1930.

In references 2 and 3 Stoner makes some further applications of Fowler's idea. Milne and Chandrasekhar independently applied the theory of polytropes to the case considered by Fowler:

4. E. A. Milne, *M.N.*, **90**, 769, 1930.

5. S. Chandrasekhar, *Phil. Mag.*, **11**, 592, 1931.

Relativistic effects were first considered by—

6. W. Anderson, *Zs. f. Phys.*, **54**, 433, 1929, but the correct formulation of Anderson's work is due to Stoner (ref. 3). Relativistic degeneracy in conjunction with polytropic theory was considered by Chandrasekhar:

7. S. Chandrasekhar, *M.N.*, **91**, 456, 1931.

8. S. Chandrasekhar, *Ap. J.*, **74**, 81, 1931.

The discussion of the theory of white dwarfs on the basis of the exact equation of state for a completely degenerate gas is due to—

9. S. Chandrasekhar, *M.N.*, **95**, 207, 1935.

Stellar configurations with degenerate cores are considered in—

10. S. Chandrasekhar, *M.N.*, **95**, 226, 1935.

11. S. Chandrasekhar, *M.N.*, **95**, 676, 1935.

Also,

12. S. Chandrasekhar, *Zs. f. Ap.*, **5**, 321, 1932.

13. P. ten Bruggencate, *Zs. f. Ap.*, **11**, 201, 1936.

§ 1.—The discussion given in this section is due to—

14. B. Strömgren, *Handb. d. Phys.*, **7**, 160–61, 1936.

15. H. Siedentopf, *A.N.*, **243**, 1, 1931.

§ 2.—In this section the analysis in reference 9 is reproduced.

§ 3.—The author is indebted to Dr. Kuiper for providing the observational material. Kuiper's discovery of his white dwarf is described in—

16. G. P Kuiper, *Pub. A.S.P.*, **47**, 307, 1935.

§§ 4, 5, 6.—The essential parts of the discussion in references 10, 11, and 12 are reproduced here.

§ 7.—This represents a hitherto unpublished investigation by the author. The functions $U_{1/2}$ and $U_{3/2}$ have recently been tabulated by E. C. Stoner:

17 E. C. Stoner, *Phil. Trans. Roy. Soc. A.*, **237**, 67, 1938.

CHAPTER XII

STELLAR ENERGY

In this chapter an attempt will be made to indicate some general trends in the current approach to the difficult problem concerning the origin of stellar energy. This subject is as yet in an early stage of development and the present brief account is intended primarily to indicate the directions in which the greatest progress is being, or is likely to be, made. This chapter, then, is on an entirely different level from the preceding ones, in which an attempt toward rigorous development has been made.[1]

1. *The Helmholtz-Kelvin time scale.*—We shall first examine the reason for postulating a source of stellar energy. To scc this, let us consider gaseous configurations in which the radiation pressure can be neglected; the majority of the normal stars (sun, Capella, etc.) are in this category. Then, as we have shown from the virial theorem in chapter ii, § 10,

$$E = -(3\gamma - 4)U = \frac{3\gamma - 4}{3(\gamma - 1)}\,\Omega\,, \tag{1}$$

where E, U, and Ω are the total, the internal, and the potential energies, respectively. If the configuration contracts and if, as a result of this, there is a change in the potential energy of amount $d\Omega$, then (as was shown) a fraction $(3\gamma - 4)/3(\gamma - 1)$ of the energy $|\Delta\Omega|$ "liberated" is radiated to the space outside, while the remaining fraction $[1 - (3\gamma - 4)/3(\gamma - 1)]$ is used in increasing the internal energy of the configuration. Hence, the amount of energy, $-\Delta E$, which is radiated to the space outside, consequent to a decrease in the potential energy of amount $|\Delta\Omega|$, is given by

$$-\Delta E = -\frac{3\gamma - 4}{3(\gamma - 1)}\,\Delta\Omega\,. \tag{2}$$

[1] This chapter was written in December, 1937, and consequently it has not been possible to include the more recent investigations of Gamow, Bethe, and others (see the Bibliographical Notes at the end of the chapter).

Now the contraction hypothesis of the origin of stellar energy postulates that the energy radiated by a star is due to a slow secular contraction. (The contraction hypothesis is also referred to as the "Helmholtz-Kelvin hypothesis.") Thus, if the potential energy alters by an amount $\Delta\Omega$ in time Δt, then the luminosity L is given by

$$L = -\frac{\Delta E}{\Delta t} = -\frac{3\gamma - 4}{3(\gamma - 1)}\frac{d\Omega}{dt}. \tag{3}$$

Now the contraction hypothesis allows an estimate to be made of the time during which the normal stars could have existed. To make this estimate, let us suppose that the configuration was initially in an infinitely extended state and that after a time t it has contracted to a radius R and a potential energy Ω. Then by (3) we should have

$$\int_0^t L dt = -\frac{3\gamma - 4}{3(\gamma - 1)}\Omega. \tag{4}$$

We can write

$$\Omega = -q\frac{GM^2}{R}, \tag{5}$$

where q is a numerical constant of order unity; if the configuration is a polytrope of index n, then (cf. Eq. [90], iv),

$$q = \frac{3}{5 - n}. \tag{6}$$

We can re-write (4) in the form

$$\overline{L}t = q\frac{3\gamma - 4}{3(\gamma - 1)}\frac{GM^2}{R}, \tag{7}$$

where $\overline{L}$ is the mean luminosity during the time t. Equation (7) allows us to establish the time scale of Helmholtz and Kelvin. For the time during which a star with an observed luminosity L could have existed while radiating at its present rate is given by

$$t_{\text{H.K.}} = q\frac{3\gamma - 4}{3(\gamma - 1)}\frac{GM^2}{LR}. \tag{8}$$

If we assume $\gamma = 5/3$, we find that the sun could have existed at its present rate of radiating energy to the space outside for a time not longer than

$$t_{\text{H.K.}}\text{ (sun)} = 1.59 \times 10^7 \times q \text{ years .} \tag{9}$$

If we assume that the sun has a polytropic density distribution of index 3, then $q = 1.5$ and we have

$$t_{\text{H.K.}}\text{ (sun)} = 2.4 \times 10^7 \text{ years .} \tag{10}$$

In the same way we find that

$$t_{\text{H.K.}}\text{ (Capella)} = 2.2 \times 10^5 \text{ years .} \tag{11}$$

It should be pointed out that (10) and (11) are not exact figures, but it is clear that no reasonable adjustment of the parameters can extend the time scale for the sun, for instance, to more than 10^8 years. The order of the "age" of the sun thus derived on the Helmholtz-Kelvin contraction hypothesis is found to conflict with other evidence which is essentially of a geological nature. Thus, the age of the terrestrial rocks as derived from the uranium-lead and helium-lead ratios of radioactive minerals is in the neighborhood of 1.6×10^9 years, and the sun must have existed in somewhat its present form for at least this length of time. Hence, the geological evidence completely disproves the contraction hypothesis for the sun, and therefore also for the normal stars. We are thus led to seek a different origin for the source of stellar energy.

2. *Transmutation of elements as the source of stellar energy.*—There is also evidence in addition to that of a geological nature, which points to an age for the sun (and therefore for the majority of the normal stars) of at least 1 or 2×10^9 years. Astronomical evidence, the discussion of which is beyond the scope of the present monograph, also points to a similar age (10^9–10^{10} years). It should, however, be mentioned that this does not necessarily mean that every single stellar object that is observed must have existed for this length of time; it only means that some aspects of the stellar system (e.g., the rotation of the galaxy) could not have existed for a time much longer than 10^{10} years.

Now a source of stellar energy which will allow for most stars a time scale of the order of 10^{10} years is the transmutation of elements —a suggestion which appears to have been first seriously considered by Harkins, Perrin, Eddington, and, more recently, by Atkinson and Houtermans. As we have seen in chapter vii, hydrogen is abundant in stellar interiors; and as we shall see in § 3, the probability of protons taking part in the transmutation processes is very much greater than for the nuclei of the higher atomic numbers. Consequently, most of the energy due to the transmutation processes will arise from the building-up of the elements of higher atomic numbers out of hydrogen. The mass of the hydrogen atom is 1.0081 in the scale $O = 16$ and it is seen that the energy available from the transmutation of a hydrogen atom is approximately 0.008 of its mass. In other words, the energy available from a gram of hydrogen is

$$0.008 \times c^2 = 7.2 \times 10^{18} \text{ ergs} . \tag{12}$$

Now each gram of material from the sun liberates, on the average, 2 ergs per second. Hence, the order of the length of time during which the sun (assumed to be initially a mass of pure hydrogen) can go on radiating at its present rate before all the available hydrogen is used up, is

$$\tfrac{1}{2} \times 7.2 \times 10^{18} \text{ sec} = 1.1 \times 10^{11} \text{ years} . \tag{13}$$

Thus, on the transmutation hypothesis, the maximum time scale for the sun is the "intermediate time scale." If we consider the more luminous stars, the time scale permitted will be very much more limited; and unless we are willing to accept the hypothesis of the annihilation of matter (for which hypothesis there is at the present time no physical basis), we simply have to accept the much shorter time scale for these stars. In any case, we shall restrict ourselves to a consideration of the transmutation of the elements as the source of stellar energy. Further, we shall see (§§ 3 and 4) that, for the order of the temperatures found for the stellar interiors (chaps. vi and vii), transmutations by proton capture of the lighter elements can take place at nonequilibrium rates. What is meant by a process occurring at a "nonequilibrium rate" will be made more precise presently—but it may be mentioned here that some investigators (Milne and Sterne) have considered the possibility of the transmutations occurring at equilibrium rates. We shall not, however, consider these theories, for, first, they require temperatures for the stellar interiors of an altogether different order ($\sim 10^9$ degrees or more) from those for which we have any evidence; second (as Strömgren has pointed out), if transmutations occurring at equilibrium rates are to be regarded as the source of stellar energy, then the Russell-Vogt theorem will not be applicable; but, as we have seen (chap. vii), the observational material strongly suggests the validity of the theorem; third, one of the main reasons for the consideration of transmutations occurring at equilibrium rates arose from the belief that stellar configurations built on the alternative hypothesis (transmutations occurring at nonequilibrium rates) are unstable, for which belief, however, there does not appear to be at present any convincing reason. It is beyond the scope of the monograph to go into greater details on these questions, but reference may be made to a general discussion of these matters by Strömgren in a recent article.[2]

3. *The transparency of potential barriers. The Gamow factor.*—We have seen in § 2 that the most profitable approach to the problem of the origin of stellar energy at the present time is made by examining the physical processes of the transmutation of the elements. For example, a process which we shall consider is the disintegration of lithium into two α-particles by the capture of a proton:

$$ {}^7_3Li + {}^1_1H \rightarrow 2\,{}^4_2He \, . \tag{14} $$

The foregoing disintegration has of course been carried out in the laboratory in the first experiments of Cockroft and Walton. We shall presently see that transmutations of the kind (14) can—and do, in fact—occur under the physical conditions which we have derived for stellar interiors (chaps. vi and vii); we shall also see that the "reaction" converse to (14), namely,

$$ 2\,{}^4_2He \rightarrow {}^7_3Li + {}^1_1H \, , \tag{15} $$

[2] B. Strömgren, *Erg. exakt. Naturwiss.*, **16**, 466, 1937—especially §§ 14 and 18.

does not occur. We are thus led to consider the transmutations of elements occurring at nonequilibrium rates as the source of stellar energy. This is different from assuming that reactions (14) and (15) both occur at almost equal but slightly different rates, and that it is the slight difference between the reaction rates in the two senses that is responsible for the generation of stellar energy. This assumption will have to be made if it is supposed that the transmutations occur at equilibrium rates. However, for the reasons stated toward the end of § 2 we shall not consider this alternative hypothesis.

If the transmutations occur at nonequilibrium rates, the problem which we have to consider is the evaluation of the probability of the penetration of potential barriers, which, on the basis of the classical mechanics, cannot be expected to happen. The physical situation can perhaps be understood by considering first the analogous problem of the α-decay of radioactive bodies.

Let us consider, for example, a uranium nucleus which is known to be α active. From the experiments of Rutherford on the scattering of α-particles by the uranium nucleus it has been inferred that the Coulomb law of attraction between the uranium nucleus and the α-particle is valid up to a distance at least as small as 3×10^{-12} cm; this means that we have

$$V(r) = \frac{2Ze^2}{r} \qquad (r > 3 \times 10^{-12} \text{ cm}) , \quad (16)$$

where we have written $V(r)$ to denote the potential energy between the two nuclei at the distance r. However, at much smaller distances we should expect deviations from the Coulomb law, since the stability of the uranium nucleus requires the existence of a "potential hole" at the center of the nucleus. The general nature of the function $V(r)$ must therefore be of the character shown in Figure 35, in which the dotted line represents the Coulomb potential and the solid line the actual potential. The inner part of the curve has been drawn arbitrarily, but for $r > 3 \times 10^{-12}$ cm the scattering experiments have shown that there is no appreciable deviation from the Coulomb law. At the same time, the uranium nucleus, being α active, emits α-particles which are found to have an energy of

6.6×10^{-6} ergs. It is, consequently, difficult to understand, on the classical picture, how particles contained inside the potential hole can go through a potential barrier which is at least twice as high as their total energy. According to classical mechanics, particles with energy 6.6×10^{-6} ergs could originate only from a point at a distance of 6×10^{-12} cm from the center, where the Coulomb potential

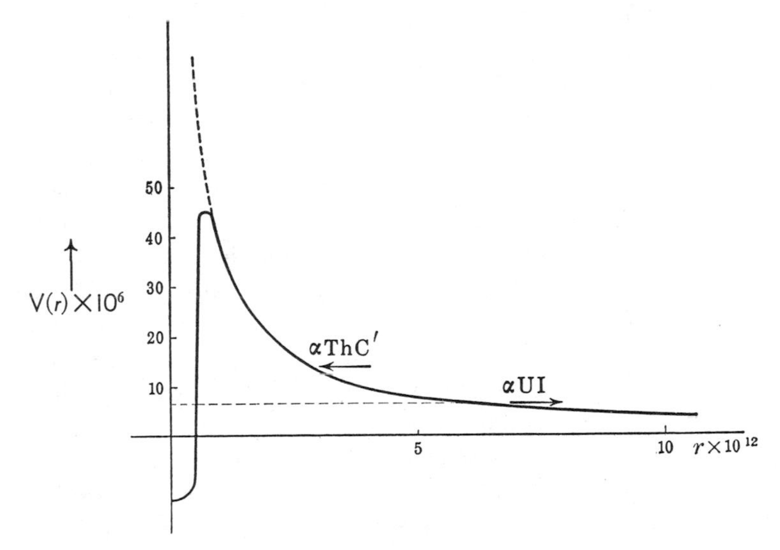

FIG. 35.—Potential energy of an α-particle in the field of a uranium nucleus

has the value 6.6×10^{-6} ergs. In this region, however, there can be no question of the α-particle being stably bound to the rest of the uranium nucleus. We thus see that apparently particles inside the potential hole with energies much less than that corresponding to the top of the potential barrier can, so to say, tunnel through. That this paradox of the classical picture does not exist in the quantum theory was, as is well known, shown to be the case by Gamow and by Condon and Gurney.

The possibility of a particle's going through a potential barrier is connected with the wave nature of the wave functions ψ and the interpretation of the square of the modulus of ψ, as the measure of the density of the probability of the particle being in a certain region. It is not true that the wave function vanishes in regions where the potential energy $V(r)$ is greater than the total energy E, and where on the classical mechanics the particle will have a negative kinetic energy. Actually, the wave function, although it decreases exponentially as we go out into the "forbidden region," is yet finite at great distances, thus giving a finite probability that a particle may appear in such regions, and, in particular, penetrate the potential barrier. An analogy (due to Gamow) from optics will illustrate the kind of phenomenon we are dealing with here. If a beam of light is incident on the boundary between two media with an angle of incidence greater than the critical angle, then, according to the concepts of geometrical optics, we will have a total reflection of the incident beam—the presumption being that all the light will be reflected at the surface separating the two media and that no disturbance in the second medium occurs. However, when this same problem is treated by the methods of the wave theory of light, it is found that there is, in fact, a finite disturbance in the second medium as well, which is appreciable for a distance of the order of a few wave lengths of light and falls off exponentially as we go farther out in the second medium. This disturbance in the second medium (which is predicted on the wave theory of light, and verified by experiment) for angles of incidence greater than the critical angle has no interpretation in the language of geometrical optics. In exactly the same manner the passage from classical mechanics to quantum mechanics allows the possibility of particles penetrating potential barriers, a feature which would be impossible to interpret in the language of classical mechanics.

For practical purposes it is sufficient to consider for the potential energy between two particles of charges $Z_1 e$ and $Z_2 e$ the form

$$V(r) = \frac{Z_1 Z_2 e^2}{r} \qquad (r \geqslant r^*) \qquad (17)$$

and

$$V(r) = V_0 \qquad (r < r^*) . \qquad (18)$$

$V(r)$, considered as a function of r, is shown in Figure 36. The potential energy is maximum at $r = r^*$, where it has the value

$$V_B = \frac{Z_1 Z_2 e^2}{r^*} \,. \tag{19}$$

V_B is sometimes called the top of the potential barrier. The energy of the α-particles ($Z_2 = 2$) emitted by radioactive bodies is much less than the appropriate V_B. As we have already indicated, quantum

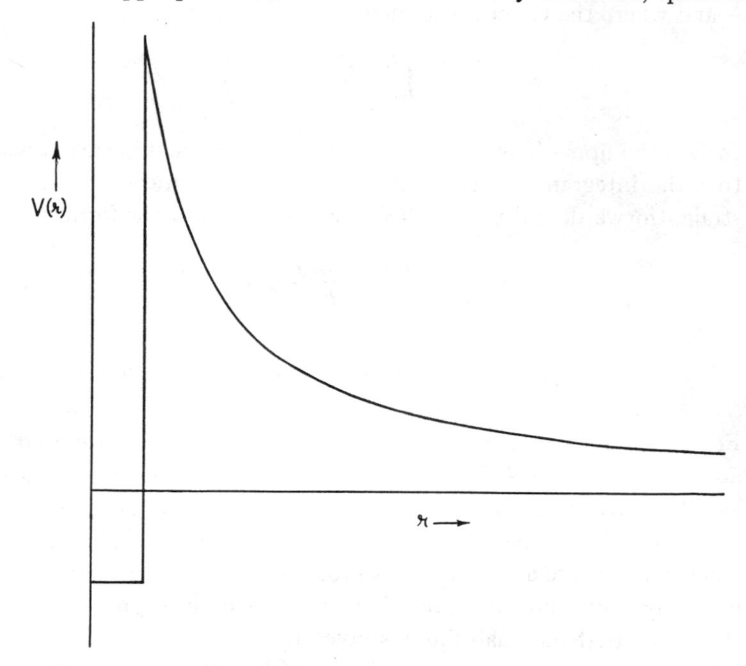

FIG. 36.—The nuclear model for the calculation of the transparency factor

mechanics allows us to calculate the probability of an α-particle initially inside $r = r^*$ being emitted. We can then evaluate the life-time of the radioactive nucleus. The calculation is a straightforward piece of analysis in quantum mechanics, and we shall not go into the details of the derivation (reference may be made to the literature quoted in the Bibliographical Notes at the end of the chapter). The result of such calculations is to give, for the number of particles

emitted per second by a radioactive nucleus of charge $Z_1 e$ (assumed to be at rest), the expression

$$\lambda' = \sqrt{2}\,\frac{\pi^2 \hbar^2 e^{-2G}}{M_2^{3/2} r^{*3}\left(\frac{Z_1 Z_2 e^2}{r^*} - E\right)^{1/2}}, \tag{20}$$

where M_2 is the mass and E the energy of the emitted particle—in the case of α-decay the emitted particle is, of course, an α-particle—and where the Gamow exponent, G, is given by

$$G = \frac{(2M_2)^{1/2}}{\hbar}\int_{r^*}^{r_E}\left(\frac{Z_1 Z_2 e^2}{r} - E\right)^{1/2} dr\,. \tag{21}$$

In (21) the upper bound of the integral, r_E, is defined in such a way that the integrand vanishes at $r = r_E$. The integration of (21) is straightforward, and the result can be expressed in the form

$$G = \frac{(2M_2)^{1/2}}{\hbar}\,\frac{Z_1 Z_2 e^2}{E^{1/2}}\,g(x)\,, \tag{22}$$

where

$$x = \frac{E}{V_B}\,; \qquad g(x) = \cos^{-1} x^{1/2} - x^{1/2}(1 - x)^{1/2}\,. \tag{23}$$

From (22) we see that G *increases with increasing nuclear charge, decreasing energy of the emitted particle, and decreasing nuclear radius* as defined by r^*. That this should be so is intuitively obvious, since increasing Z_1 and decreasing E and r^* imply that the particle will have to penetrate a higher and a broader potential barrier, which is naturally more difficult. Equation (20) can be interpreted by the statement that the "half-life"[3] is given by

$$\tau = \frac{\log 2}{\lambda'}\,. \tag{24}$$

Equation (24) can be alternatively expressed as

$$\tau = \frac{\tau_0}{W}\,, \tag{25}$$

[3] The "half-life" is defined as the time during which half the number of particles are emitted.

where W, the transparency of the potential barrier, is given by

$$W = e^{-2G} \tag{26}$$

and τ_0 is the lifetime without the potential barrier ($\tau_0 \sim 3.3 \times 10^{-21}$ sec. for radioactive nuclei); τ_0^{-1} will be, classically speaking, the number of times the α-particle will hit the inner wall of the potential hole per unit time. Equation (25) can, therefore, be interpreted as follows: W is the probability, per collision, for a particle to penetrate the potential barrier.

The expression for the transparency of the potential barrier which we have just given is, in fact, quite general and is precisely what is needed to calculate the probability of a particle penetrating into the potential hole from outside—more precisely, W is the probability, per collision, that a particle of charge Z_2e and energy E will penetrate into another nucleus of charge Z_1e, when the latter nucleus is assumed to be at rest. The modification of the transparency factor, W, when both the particles are in motion is obvious; we then regard E as the total kinetic energy of the two nuclei in a system of reference in which the center of mass of the two particles is at rest; further M is to be regarded as the reduced mass of the system,

$$M = \frac{M_1 M_2}{M_1 + M_2}, \tag{27}$$

where M_1 and M_2 are the masses of the two nuclei considered. The Gamow exponent can therefore be written as

$$G = \left(2 \frac{M_1 M_2}{M_1 + M_2}\right)^{1/2} \frac{Z_1 Z_2 e^2}{\hbar E^{1/2}} g(x) . \tag{28}$$

For most stellar applications a simplification of W is possible. As an empirical fact, it is found that

$$V_B \simeq 0.70 Z_1 Z_2 A_1^{-1/3} \text{ million electron volts} . \tag{29}$$

If we remember that a million electron volts corresponds to a temperature of the order of 10^9 degrees, it is clear that we can approximate $g(x)$, given by (23), by its value for $x \to 0$. Since

$$g(x) \to \frac{\pi}{2} \qquad \text{as} \qquad x \to 0 , \tag{30}$$

we have

$$G = \left(2\frac{M_1M_2}{M_1+M_2}\right)^{1/2}\frac{\pi^2 Z_1Z_2e^2}{hE^{1/2}}. \tag{31}$$

Finally, the transparency is given by

$$W = e^{-2G}. \tag{32}$$

4. *The penetration of nuclear barriers by charged particles with thermal velocities.*—We shall now calculate the number of successful captures of a nucleus of charge Z_2e by another nucleus of charge Z_1e which occur in a system at temperature T and in which the distribution of the velocities of the different particles is given by Maxwell's law (chap. x). It should be mentioned here that each successful capture does not necessarily imply a transmutation—it is possible that the captured particle may be re-emitted. We shall return to this question in § 5.

Now, the number of collisions (per unit volume and per unit time) in which the total kinetic energy of the relative motion of the colliding particles lies in the range E and $E + dE$ is given by

$$\frac{2N_1N_2\sigma_{12}^2}{(kT)^{3/2}}\left[\frac{2\pi(M_1+M_2)}{M_1M_2}\right]^{1/2}e^{-E/kT}\,EdE\,, \tag{33}$$ [4]

where $\pi\sigma_{12}^2$ is the effective cross-section for the collisions and N_1 and N_2 are the numbers of nuclei per unit volume of the two sorts in the system.

An approximate expression for σ_{12} can be given. If we represent the colliding particles by plane De Broglie waves, then for "head-on" collisions the collision cross-section is approximately given by the square of the De Broglie wave length which characterizes the incident particles in a system of reference in which the center of mass of the two particles is at rest. Thus,

$$\pi\sigma_{12}^2 \simeq \frac{\pi\hbar^2}{M^2v^2} = \frac{\pi\hbar^2}{2ME} \tag{33'}$$

where M is the reduced mass (cf. Eq. [27]), v is the relative velocity between the two particles, and E has the same meaning as in equa-

[4] R. H. Fowler, *Statistical Mechanics* (2d ed., Cambridge), p. 665, Eq. (1865).

tion (33). If we consider a non-head-on collision which is characterized by a relative angular momentum, then to obtain the appropriate cross-section we must multiply (33′) by a factor which is very small compared to unity. Hence, it is sufficient to restrict ourselves to head-on collisions only.

Now the probability that a collision will result in the capture (and thus lead to the possibility of a transmutation) is given by the factor W. Hence, according to (33) and (33′), the total number of penetrations occurring in the system per unit time is given by

$$\frac{N_1 N_2 \hbar^2}{2\pi (kT)^{3/2}} \left[\frac{2\pi (M_1 + M_2)}{M_1 M_2} \right]^{3/2} \int_0^\infty e^{-\frac{E}{kT} - 2G} \, dE \, . \tag{34}$$

Let

$$\frac{E}{kT} = y \tag{35}$$

and

$$Q = \left\{ \left(2 \frac{M_1 M_2}{M_1 + M_2} \right)^{1/2} \frac{\pi^2 Z_1 Z_2 e^2}{h (kT)^{1/2}} \right\}^{1/3} . \tag{36}$$

Equation (34) can then be written as

$$\frac{N_1 N_2 \hbar^2}{2\pi (kT)^{1/2}} \left[\frac{2\pi (M_1 + M_2)}{M_1 M_2} \right]^{3/2} \int_0^\infty e^{-y - 2Q^3 y^{-\frac{1}{2}}} \, dy \, . \tag{37}$$

To evaluate (37) we note that the exponential term in the integrand of the expression has a sharp maximum at

$$-1 + Q^3 y^{-3/2} = 0 \qquad \text{or} \qquad y = Q^2 \, . \tag{38}$$

Since the exponential term falls off very steeply on either side of $y = Q^2$, we write

$$y = u + Q^2 \tag{39}$$

and regard u as small, since the contribution to the integral arises essentially from the immediate neighborhood of $y = Q^2$. We then find

$$y + 2Q^3 y^{-1/2} = 3Q^2 + \frac{3}{4} \frac{u^2}{Q^2} + O(u^3) \, . \tag{40}$$

We can therefore write for the integral in (37)

$$e^{-3Q^2}\int_{-Q^2}^{\infty} e^{-\frac{3}{4}\frac{u^2}{Q^2}}\,du\,. \tag{41}$$

The quantity Q^2 is generally quite large compared to unity, and we can write with sufficient accuracy

$$2e^{-3Q^2}\int_{0}^{\infty} e^{-\frac{3}{4}\frac{u^2}{Q^2}}\,du = 2\sqrt{\frac{\pi}{3}}\,Qe^{-3Q^2}\,. \tag{42}$$

Hence, the number of encounters which result in successful captures that occur in the system per unit volume and per unit time is given by

$$\frac{N_1N_2\hbar^2}{(3\pi)^{1/2}(kT)^{1/2}}\left[\frac{2\pi(M_1+M_2)}{M_1M_2}\right]^{3/2} Qe^{-3Q^2}\,. \tag{43}$$

Substituting for Q according to (36), we have

$$N_1N_2\,\frac{2^{4/3}\pi^{4/3}}{3^{1/2}}\,\frac{\hbar^{5/3}(Z_1Z_2e^2)^{1/3}}{(kT)^{2/3}}\left(\frac{M_1+M_2}{M_1M_2}\right)^{4/3} e^{-3Q^2} \tag{44}$$

where we may notice that

$$3Q^2 = 3\left(2\,\frac{M_1M_2}{M_1+M_2}\right)^{1/3}\left(\frac{\pi Z_1Z_2e^2}{2\hbar}\right)^{2/3}\frac{1}{(kT)^{1/3}}\,. \tag{44'}$$

Inserting the numerical values of the atomic and other constants in (44) and (44′), we find that we can express the number of penetrations, $P(Z_1;\,Z_2)$, per gram of each of the two sorts of nuclei in the following form:

$$\left.\begin{aligned}\text{Log}\,P(Z_1;\,Z_2) = 39.480 + \text{Log}\left\{\frac{1}{A_1A_2}\left(\frac{A_1+A_2}{A_1A_2}\right)^{4/3}(Z_1Z_2)^{1/3}\right\}\\ -\left(\frac{A_1A_2Z_1^2Z_2^2}{A_1+A_2}\right)^{1/3}\frac{1.850\times 10^3}{T^{1/3}} - \frac{2\,\text{Log}\,T}{3}\,,\end{aligned}\right\} \tag{45}$$

where A_1 and A_2 are the atomic weights of the nuclei of charges Z_1 and Z_2, respectively. Since, according to our hypothesis concerning the origin of stellar energy, transmutations occurring at nonequilibrium rates are the fundamental physical processes, we have for the contribution to the rate of generation of energy by the transmutations of the kind we have been considering

$$\epsilon(Z_1 Z_2) = \text{constant } X_1 X_2 \rho T^{-2/3} e^{-\text{constant } [Z_1^2 Z_2^2/T]^{\frac{1}{3}}} , \tag{46}$$

where X_1 and X_2 are the abundances with which the nuclei of the two sorts occur. Before discussing the important formula (46), we shall first make a few remarks on its derivation. We assumed that the distribution of the velocities of the nuclei involved is according to Maxwell's law. As a criticism of this, it may be argued that transmutations generally lead to the emission of particles of very

TABLE 31

T (in Millions of Degrees)	Log P (1; 2) (See Eq [45])
1	7.85
2	13.27
3	15.89
4	17.54
5	18.70
10	21.78
20	24.20
30	25.35
40	26.07

considerably higher energies than would correspond to the temperature T. But this fact is not of great importance. These high energy particles which are emitted during some types of transmutations considered will be rapidly slowed down because of the very efficient stopping power of the stellar material. Thus, so long as we are not concerned with physical processes which are 10^{-4} or 10^{-5} times less frequent than those which are due to the particles with thermal energies, it is safe to assume a Maxwellian distribution of velocities for the nuclei taking part in the capture processes.

Returning to (44), we see that, according to this formula, the penetration of protons into the lighter nuclei is easily possible under the conditions which we have derived for the stellar interiors. Table 31 illustrates the point.

Another very important feature now becomes apparent. Transmutations by the capture of protons can occur only with elements of very low atomic number. Because of the occurrence of $(Z_1^2 Z_2^2)^{1/3}$ in the exponent in (46), the capture even of protons by the heavier nuclei becomes extremely unlikely.

If we turn to a different aspect of the situation, it may be argued that the rate at which captures occur for $T < 4 \times 10^6$ is exceedingly slow. But this question can only be settled by actual integrations for stellar models with an underlying law for the energy generation of the type (46). Integrations for one such set of configurations has been made by Steensholt, who finds that the process considered here is quite sufficient as a source of stellar energy. Finally, attention may be drawn to the extreme sensitiveness to temperature of the law (46); it is this circumstance which led to the belief,[5] to which we have referred at the end of § 2, that models built on the law (46) are likely to be unstable.

5. *Von Weizsäcker's theory.*—We have seen that the penetration of protons through the potential barriers of the lighter nuclei occurs in stellar interiors and that they will also suffice—with an adequate supply of the lighter elements—as a source of stellar energy. As a typical example we may consider the capture of protons by the lithium nucleus. Now this capture does not lead to the formation of a nucleus of a higher atomic number—instead, we have a disintegration process:

$$ {}^{7}_{3}Li + {}^{1}_{1}H = 2\,{}^{4}_{2}He \,. \qquad (47)[6] $$

We shall consider presently other examples of captures which result in similar disintegration processes, but it follows that we shall have an increasing proportion of the lighter nuclei. It is thus clear quite at the outset that we have to distinguish carefully between synthesis processes and disintegration processes—the German words *Aufbauprozesse* and *Abbauprozesse*, which von Weizsäcker has introduced, are very much more expressive. At this stage we should consider three possibilities:

a) That the heavy elements like lead, thorium, and uranium are now continually being formed in the stellar interiors and that all

[5] Not confirmed by Cowlings investigation (*M.N.*, **94**, 768, 1934) on the stability of such models.

[6] In writing equations representing nuclear reactions, we shall adopt the convention of prefixing the letter denoting the element on its upper left-hand corner by its atomic weight and on its lower left-hand corner by its atomic number.

the heavier elements now present have been synthesized in the stars during the past.

b) That a great (or an appreciable) fraction of the heavy elements now present in the stars have been formed at some earlier stage, and that at the present time, though we have a further synthesis of these elements, they do not occur at a sufficient rate (or have occurred for a sufficient length of time) to account for the actual abundances of the different elements.

c) That all the heavy elements now present in the stars have been formed at some earlier stage, and that at the present time we have only a further transformation of hydrogen (involving, principally, proton captures) into the lighter elements.

These three possibilities cannot of course be sharply distinguished. The difference between them is mainly one of degree, and we can easily conceive of a variety of other "intermediate" possibilities. However, as a working hypothesis, the second and the third have the disadvantage of not being capable of being made quite definite at present; in any case, need for the other possibilities can be felt only by attempting to follow the full consequences of the first hypothesis. This is the procedure von Weizsäcker has followed. The fundamental assumption, then, is the following:

Apart from secondary effects of minor importance, the transmutation of elements is the entire cause of the presence of all elements in the stars; they are all being synthesized continually in the stars which are assumed to have started as pure masses of hydrogen; further, transmutations are the only source of stellar energy.

The foregoing hypothesis, which we shall refer to as the "von Weizsäcker hypothesis," is made, it will be understood, entirely for the purpose of having a definite working basis, the partial failure or the complete success of which will indicate the necessity or otherwise of considering the other possibilities which we have mentioned.

From the von Weizsäcker hypothesis we can draw certain immediate inferences. First, it is clear that the lighter elements are formed by processes involving proton captures. These processes will be the most important among those in which the transmutations are caused by the capture of charged particles because, as we have seen in chapter vii, hydrogen is abundant in stellar interiors, and also because the occurrence of $\{\exp - (Z_1^2 Z_2^2/T^{\frac{1}{3}})\}$ in the formula giving the number of penetrations makes the capture even of α-particles very much less probable than the capture of protons. Second, the occurrence of $(Z_1^2 Z_2^2/T)^{1/3}$ in the exponential in (46) shows that even proton captures cannot be of any significance in the synthesis of the heavier nuclei. Von Weizsäcker's hypothesis therefore requires that some other physical process is fundamental for the synthesis of the heavier elements. Now the experiments of Fermi and others have shown that neutrons can be captured by the heaviest nuclei, so that it is plausible that this is the physical process responsible for the synthesis of the heavier elements in stellar interiors. We cannot, however, assume that there are free neutrons present in stellar interiors. The experiments of Fermi have again shown that the cross-sections for the capture of neutrons by the atomic nuclei are so large that, even if there were an appreciable amount of neutrons to start with, they would all have disappeared in a very short time. We thus infer that the only possibility consistent with von Weizsäcker's hypothesis is that there must be a source for a continuous supply of neutrons, and that the neutrons are formed as a by-product in such transmutations as do occur under the conditions of the stellar interiors. We shall now consider these questions in greater detail, following von Weizsäcker.

(1) *Transmutations due to proton captures.*—As we have already seen, among the transmutations arising from the capture of charged particles, those due to the capture of protons by the light nuclei are by far the most important. We shall now consider more closely the transmutations that can thus occur. In doing so we must distinguish between the synthesis and the disintegration processes. Now an empirically well-established rule which can be used for this purpose is the following.

The capture of a proton by a light nucleus can lead to a synthesis of a nucleus of a higher atomic number if, and only if, a disintegration process is not possible from pure energy considerations.

This rule can be understood by the use of the method of description of nuclear phenomena introduced by Bohr: When a proton penetrates through the potential barrier into an atomic nucleus, we have first the formation of an intermediate nucleus which in general will be in an excited state; this intermediate nucleus can follow one of three courses: it re-emits the captured proton, or it emits some other particle (generally an α-particle),[7] or, finally, it drops to the ground state with the emission of a γ-ray. If the first possibility occurs, we have a simple scattering phenomenon; if the second, a disintegration process; and if the third, a synthesis process. Now, since the lifetime of an excited nucleus with respect to γ-emission is long compared to the analogous "lifetime" τ_0 (introduced in § 3, Eq. [25]), it is clear that (unless the potential barrier is very high and broad and the energy of excitation of the intermediate nucleus is distributed among the nuclear constituents) we shall, in most cases, have the emission of a particle (if it is at all possible) before there has been time enough for a γ-emission. In the case of the lighter nuclei the potential barrier can be penetrated without undue difficulty; and since the number of nuclear constituents is small, the energy of excitation of the nucleus is not distributed quickly among the other particles. We thus see that for the lighter nuclei we have a pure energy criterion for distinguishing a disintegration from a synthesis process. Thus if we compare, for instance, two nuclei, one of which after the capture of a proton is able (from considerations of energy) to emit an α-particle and the other not, then in both cases a synthesis is a priori improbable. But while in the former case we almost always have a disintegration, in the latter case we will have the re-emission of the proton. In the second case, since the re-emission of the proton does not produce any effective change, it is clear that the occasional synthesis which can occur is the only one that matters. Thus, while in the first case every successful penetration of the nucleus by a proton will be quite invariably followed by a disintegration, in the second case we have a synthesis only in a small fraction of the total number of successful penetrations. Thus, in the latter case we shall have to multiply the expression (44) for the number of penetrations by another factor which gives the probability of a synthesis occurring.

Now the probability for the occurrence of a transmutation with the emission of a γ-ray after the successful penetration of a particle is given by the ratio of the probability of the emission of a γ-ray

[7] If this is possible from energy considerations.

$\Gamma_\gamma \simeq 10^{14}$ sec^{-1} (for the order of energies involved) and the proper frequency $\hbar/Mr^{*2}$ ($\simeq 10^{22}$ sec^{-1}) of the particle oscillating inside the nucleus. Hence, the factor by which we have to multiply (44) in order that we may obtain the number of transmutations is given by

$$\Gamma_\gamma \frac{M_1 M_2}{M_1 + M_2} \frac{r^{*2}}{\hbar} . \tag{48}$$

The total number of transmutations occurring in unit volume and in unit time is therefore given by[8]

$$N_1 N_2 \frac{2^{4/3}\pi^{4/3}}{3^{1/2}} \frac{\hbar^{2/3}(Z_1 Z_2 e^2)^{1/3} r^{*2} \Gamma_\gamma}{(kT)^{2/3}} \left(\frac{M_1 + M_2}{M_1 M_2}\right)^{1/3} e^{-3Q^2} . \tag{49}$$

Having thus settled as to when a synthesis (as distinguished from a disintegration) can occur, we shall next consider the stability of the nuclei synthesized by proton captures. We shall now have to distinguish between stable nuclei and those which are β active; β-decay of the unstable nucleus can consist either in the emission of electrons (β^--decay) or positrons (β^+-decay). Remembering that, according to current views, the nuclear constituents are protons and neutrons, we shall have Z-protons and $(A-Z)$ neutrons in a nucleus of charge Z and mass A. We shall now state the following rule:

Stable nuclei are those in which the number of protons in the nucleus is equal to, or 1 less than the number of neutrons, according as the mass-number A is even or odd. All other nuclei are unstable; nuclei with an excess of protons being β^+ active and those with an excess of neutrons being β^- active.

When some stable nuclei capture protons, they emit α-particles. Thus,

$$\left.\begin{aligned} {}^7_3Li + {}^1_1H &\rightarrow 2\,{}^4_2He , \\ {}^{11}_5B + {}^1_1H &\rightarrow {}^4_2He + {}^8_4Be , \\ {}^{15}_7N + {}^1_1H &\rightarrow {}^4_2He + {}^{12}_6C . \end{aligned}\right\} \tag{50}$$

[8] See a paper by G. Gamow and E. Teller (*Phys. Rev.*, **53**, 608, 1938) that has since appeared. Gamow and Teller have in addition considered "resonance penetrations" and indicate the importance of the consideration of such processes.

In the boron-proton reaction the final nucleus ${}^{8}_{4}Be$ (which is generally left in an excited state) disintegrates almost immediately into two α-particles.[9] It will be noted that in all the foregoing cases we have the formation of especially stable nuclei (${}^{4}_{2}He$, ${}^{12}_{6}C$), in which the number of neutrons and protons are equal. It will be found that by proton captures we cannot have a further synthesis of elements "over" these nuclei. On the other hand, if the capture of a proton by a stable nucleus leads to an intermediate nucleus which, according to our energy criterion, cannot emit an α-particle, then we will have the synthesis of a nucleus of a higher atomic number which will be β^+ active. We shall, however, see in (2) below that these β^+ active nuclei can be "stepping-stones" for the synthesis of still higher members of the periodic table.

It follows from what we have said that by successive captures of protons we cannot (under stellar conditions) have the synthesis of nuclei heavier than, say, oxygen. This is so for two reasons, both of which work in the same direction. First, by successive proton captures we are led (according to our stability criterion) to the synthesis of nuclei which will be more and more β^+ active. Second, the increasing nuclear charge will decrease exponentially the probability of a proton penetrating the nucleus (because of the factor $\{\exp - \text{constant } [Z_1^2 Z_2^2/T]^{1/3}\}$ in [44]). Since, for a further synthesis over a β^+ active nucleus, the proton will have to be captured before the β^+-decay, it is clear that we shall soon come to a stage where the lifetime of the nucleus for β^+- decay becomes less than the mean interval of time between two successive proton penetrations. This condition clearly sets an upper limit to the atomic number of the elements beyond which a further synthesis by proton captures cannot be possible. The actual point in the periodic table where further synthesis by proton captures in effect ceases will be, however, very much earlier, since a successful penetration does not (as we have seen) imply a successful synthesis.

Summarizing, we can say that nuclear reactions involving proton captures result essentially in an accumulation of the lighter nuclei which (as we shall see in greater detail below) in turn act as catalysts in the production of further α-particles. This, then, is the

[9] See N. Feather, *Nuclear Physics* (Cambridge, England, 1936), p. 191.

fundamental physical process which is effective as the source of stellar energy.

(2) *Nuclear transmutations by proton captures as an autocatalytic chain of cyclical reactions.*—We shall now examine more closely the actual nuclear reactions involving proton captures. Because of the incomplete nature of our information concerning the masses of some of the lighter nuclei, the following discussion (due to von Weizsäcker) should be regarded as partly tentative. However, the discussion discloses certain characteristic features which are likely to survive in the future discussions concerning stellar energy.

The natural starting-point is clearly the consideration of nuclear reactions in which both protons and α-particles are involved. At this point we encounter a difficulty; laboratory investigations have so far failed to disclose the existence of a nucleus of mass 5. Von Weizsäcker believed that the existence of ${}^5_3Li$ and ${}^5_2He$ could be conjectured. However, according to Bethe, the more recent experiments on artificial disintegrations exclude more or less definitely the possibility of a nucleus of mass 5.[10] In spite of this, we shall outline the nuclear reactions considered by von Weizsäcker as illustrative of the nature of such discussions.

Von Weizsäcker considers two possibilities:

$$ {}^5_3Li \quad \text{is} \quad \beta^+ \text{ active} \qquad (I) $$

and

$$ {}^5_3Li \quad \text{is} \quad \text{stable} . \qquad (II) $$

Let us first consider case I. The course of the reactions to be described can be followed by referring to Figure 37. The first nuclear reaction is (*1*) ${}^4_2He + {}^1_1H = {}^5_3Li$, which by hypothesis is β^+ active. We then have either (*1.1*) a β^+-decay of ${}^5_3Li$, in which case we would have ${}^5_3Li = {}^5_2He + \beta^+$, or (*1.2*) a further capture of a proton by ${}^5_3Li$ before it decays; in the latter case we have ${}^5_3Li + {}^1_1H = {}^6_4Be$, and this nucleus (according to our stability criterion) must be strongly β^+ active. In case (*1.1*) the most probable reaction is (*1.11*) ${}^5_2He + {}^1_1H = {}^4_2He + {}^2_1D$. We see that we have now completed a cycle. The α-particle with which we started has been recovered, and the whole cycle—(*1*), (*1.1*), (*1.11*)—can now be repeated; the net

[10] See, however, reference 20 in the Bibliographical Notes (p. 486).

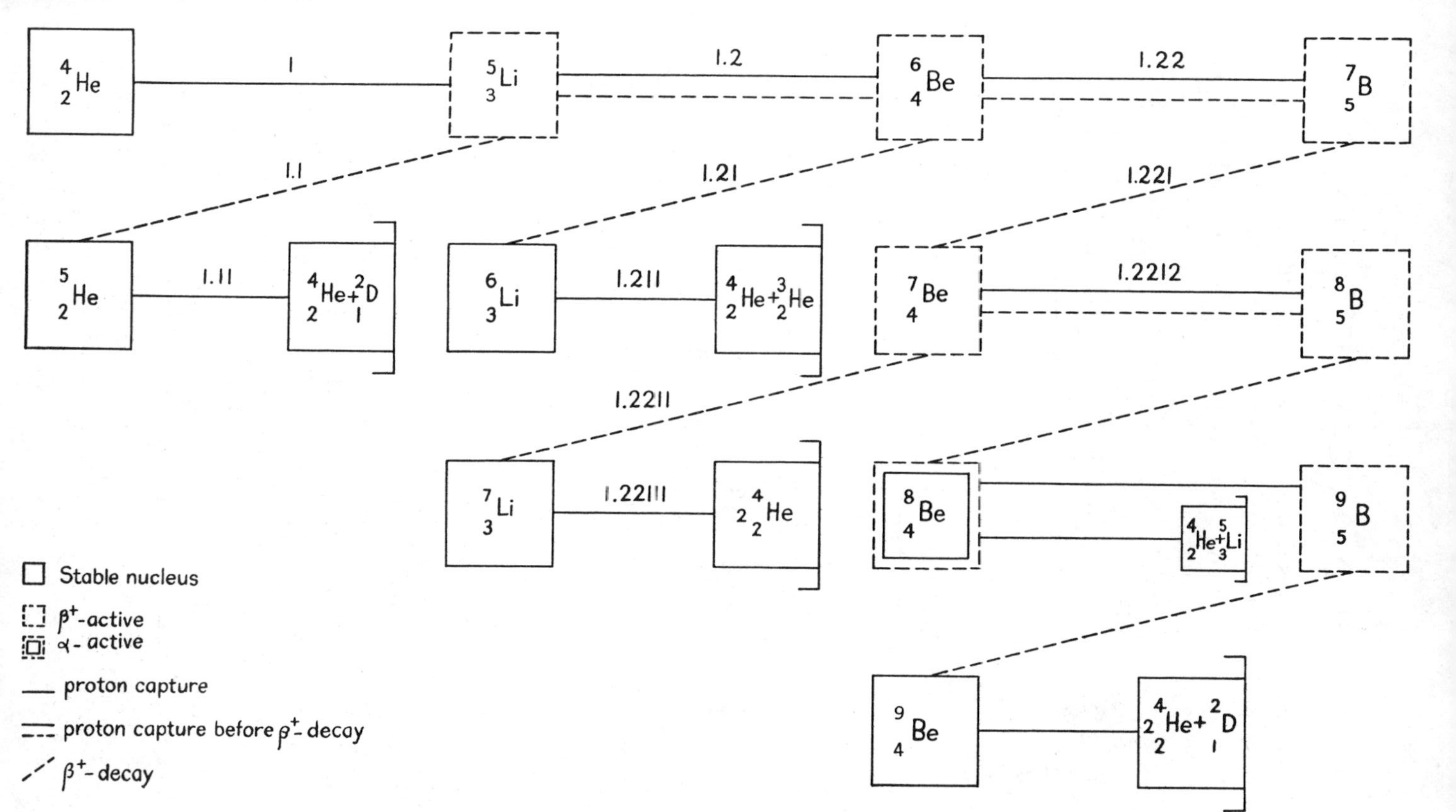

FIG. 37.—The nuclear reactions involving proton captures as autocatalytic chains

result of each cycle is that a deuteron and a positron[11] have replaced two protons. The helium thus acts as a catalyst in this cyclical chain of reactions. We shall postpone consideration of the part which ${}^2_1D$ plays in further reactions and return to the case (*1.2*). Here we again have two possibilities: either (*1.21*) the β^+-decay of ${}^6_4Be$ takes place before a proton capture (resulting in the formation of the stable isotope ${}^6_3Li$ of lithium) or (1.22) a proton is captured by ${}^6_4Be$ before the β^+-decay (in which case we have the synthesis of ${}^7_5B$, which must be strongly β^+ active). In case (*1.21*) the further capture of a proton by ${}^6_3Li$ will result in (*1.211*) ${}^6_3Li + {}^1_1H = {}^4_2He + {}^3_2He$. Again a cycle has been completed; the α-particle with which the chain of reactions—(*1*), (*1.2*), (*1.21*), (*1.211*)—started has been recovered, and the three protons which took part in the reaction chain have been replaced by ${}^3_2He$ and a positron. We shall consider the further reactions in which ${}^3_2He$ is involved a little later; but, returning to case (*1.22*), we can assume that ${}^7_5B$ is so strongly β^+ active that we effectively have only one possibility, namely, (*1.221*) (the β^+-decay of ${}^7_5B$ to ${}^7_4Be$, which must also be β^+ active). With ${}^7_4Be$ we again have the two possibilities: (*1.2211*) the β^+-decay of ${}^7_4Be$ to the stable ${}^7_3Li$, and (*1.2212*) the capture of a proton by ${}^7_4Be$ (before β^+-decay) to form the β^+ active ${}^8_5B$. In the case (*1.2211*) the further capture of a proton will result in (*1.22111*), the disintegration of ${}^7_3Li$ into two α-particles. At this point another cycle has been completed. We note, however, that at the end of this cycle we have two α-particles for every one with which the cycle—(*1*), (*1.2*), (*1.22*), (*1.221*), (*1.2211*), (*1.22111*)—started. In other words, the cycle which we have just considered can be called (in the language of chemistry) an "autocatalytic cycle," since the net result of the cycle is to increase the amount of the catalytic agent (${}^4_2He$) present. In the case (*1.2212*) the β^+-decay of ${}^8_5B$ will lead to the stable (or weakly α active) ${}^8_4Be$. The latter, on capturing a proton, will form the β^+ active ${}^9_5B$, which after its β^+-decay will result in the formation of the stable ${}^9_4Be$. Finally, the most probable reaction, which would result on ${}^9_4Be$ capturing a proton, will be its disintegration into two α-particles and a deuteron. Another auto-

[11] The positron would later combine with an electron and emit two γ-rays.

catalytic cycle has ended. It is important to notice that again we have the formation of deuterons.

We have yet to consider the further reactions in which ${}^3_2He$ and ${}^2_1D$ are involved; but before doing so, we may note that if we consider case II (where ${}^5_3Li$ is assumed to be stable), the whole sequence of the reaction chains is exactly the same as in case I, except that the first cycle—(*1*), (*1.1*), (*1.11*)—does not exist for case II. This results in one important difference between the two cases: if ${}^5_3Li$ is β^+ active, we have the formation of deuterons at an early stage in the reaction chains, whereas they appear at a relatively later stage in the sequence of the reaction chains if ${}^5_3Li$ is a stable nucleus.

We have considered the reaction chains among the lighter nuclei up to the synthesis of a stable isotope of beryllium mainly for the reason that, if ${}^5_3Li$ should be stable, then precisely at this point do we have the formation of deuterons; as we shall presently see, the production of deuterons is important in the further development of the von Weizsäcker theory. There is, however, not much point in continuing the reaction chains to include formally the synthesis of the higher members of the periodic table, as we have already explained (toward the end of [1], above).

We shall now consider the reactions in which ${}^3_2He$ and ${}^2_1D$ take part. If ${}^3_2He$ is β^+ active it would be transformed to ${}^3_1T$. It can also capture a proton and synthesize what must certainly be strongly β^+ active, namely, ${}^4_3Li$ (which, after its β^+-decay, will result in the formation of ${}^4_2He$).

As for the deuterons, since the probability of the capture process,

$$ {}^2_1D + {}^1_1H \rightarrow {}^3_2He \tag{A} $$

(with ${}^3_2He$ following the chain of reactions already described), is probably very small, we must also consider deuteron-deuteron reactions:

$$ {}^2_1D + {}^2_1D = {}^3_2He + {}^1_0n \tag{BI} $$

and

$$ {}^2_1D + {}^2_1D = {}^3_1T + {}^1_1H \,. \tag{BII} $$

The reactions BI and BII are the most efficient artificial transmutations that have so far been effected in the laboratory; we can estimate the capture cross-section for BI and BII to be about 10^5 times the capture cross-section for A. It is probable that, except in the earliest stages in the history of a star, deuterons are more than 10^{-5} times as abundant as protons, so that the reactions BI and BII become more important than A. We thus see that in the reaction cycles going on among the lighter nuclei we have found a process which will serve as a source of neutrons, the need for which we have already explained in § 5.

We thus see that the characteristic features of the nuclear reactions we have considered are (1) the nuclear reactions go in cycles; (2) in the cycles the α-particles play the part of catalytic agents, some cycles being even autocatalytic; (3) the reaction chains lead at some point to the production of deuterons; (4) the deuterons, if more than 10^{-5} times as abundant as hydrogen, will serve as a source of neutrons. These are the essential points in von Weizsäcker's discussion.

In order to simplify our discussion we shall, following von Weizsäcker, consider the following model reaction chain:

$$\left.\begin{aligned} {}^1_1H + {}^4_2He &= {}^5_3Li\,, \\ {}^5_3Li &= {}^5_2He + \beta^+\,, \\ {}^5_2He + {}^1_1H &= {}^4_2He + {}^2_1D\,, \end{aligned}\right\} \qquad (51)$$

and

$${}^2_1D + {}^2_1D = {}^3_2He + {}^1_0n\,; \qquad {}^2_1D + {}^2_1D = {}^3_1T + {}^1_1H\,. \qquad (52)$$

${}^3_2He$ and ${}^3_1T$ again lead to the formation of α-particles.

(3) *Synthesis of the heavy elements by neutron capture.*—Our discussion in (2) has brought out clearly the inadequacy of transmutations involving proton captures for the purpose of synthesizing the heavy nuclei. As we have seen, we must look for the synthesis of the heavier nuclei in transmutations involving neutron captures; for this purpose we need a continuous supply of neutrons. We have already shown that the reaction chains among the lighter nuclei do, in fact, include nuclear reactions (if the deuterons are more than 10^{-5} as abundant as hydrogen) which will serve as a source of neutrons. But before we can be sure that the neutrons do synthesize the heavy nuclei, we should make certain that the neutrons are, so

to say, not wasted by recombining with protons to form deuterons. The deuterons thus formed may again produce neutrons, but it is clear that for each such cycle there will be a reduction of the neutrons by a factor of 4. However, there is no very great danger of this happening, for the capture of neutrons by atomic nuclei takes place by what is called a "quantum mechanical resonance"; if the incident neutron has an energy nearly equal to an energy level for the neutron inside the nucleus, then we have a kind of "resonance" which makes the transition probability for the neutron penetrating the nucleus especially large; further, the energy levels for the neutrons inside the heavier nuclei lie very close together—it is this last circumstance which makes the capture of neutrons by the heavy nuclei relatively easy. However, the energy levels of the neutron inside a ${}^{2}_{1}D$ nucleus rapidly become spaced farther and farther apart; and (according to von Weizsäcker) in the energy range for the neutrons corresponding to thermal energies in stellar interiors there are very few resonance levels. This would make the capture cross-section for the neutron-proton reaction very small compared to what it is for the heavier nuclei. It thus seems safe to conclude that the neutrons produced in the deuteron-deuteron reaction will be available for the synthesis of the heavier elements.

There is one further point which should be mentioned, namely, that the liberated neutrons will soon attain thermal energies corresponding to the temperatures in the stellar interiors. The phenomenon we encounter here is essentially the same as that which has been found in laboratory experiments in which a block of paraffin slows down high-energy neutrons so quickly that they very soon attain the thermal energies corresponding to the room temperature; also stellar material with its abundance of hydrogen is, with respect to its stopping power of the neutrons, not very different from a paraffin block. To avoid misunderstanding, it should be stated that there is no contradiction here with our earlier remarks that the capture cross-section for the neutron-proton reaction is small compared to that for reactions with the other elements; for the probability of elastic scattering is more than one hundred times the corresponding probability for the synthesis of a deuteron.

The synthesis of the heavy elements being thus made plausible, we may note the physical factors which would govern such transmutations: (1) the rate at which the neutrons are liberated, and (2) the density of the resonance levels for the nucleus concerned in the energy range of the neutrons corresponding to the temperatures in stellar interiors.

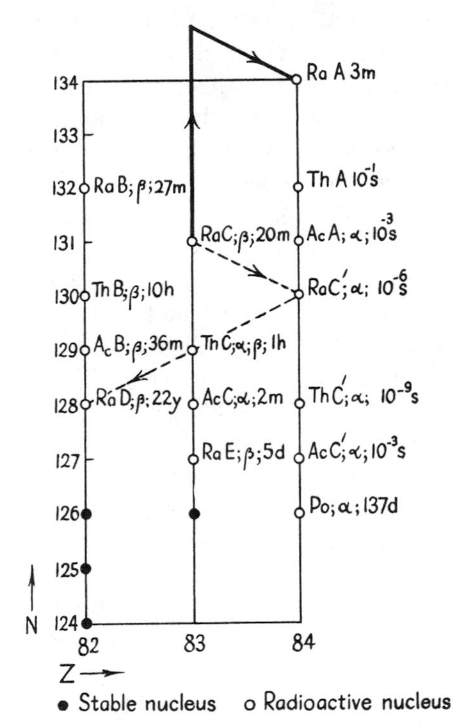

FIG 38—The synthesis of the heavy elements over lead ($Z=82$)

We now come to what is perhaps a difficulty in the von Weizsäcker theory in its present form. If we assume that all the heavy elements are synthesized at present in stellar interiors, we shall see that the rate at which neutrons should be captured by the heavier nuclei must be extremely rapid. Consider, for instance, the synthesis of the elements above lead and bismuth (see Fig. 38). In this region of the periodic table we have the elements of atomic number 84 (the C'-products) which are extremely α active and which on α-disintegration result in the isotopes of lead. These C'-products have half-lives with respect to α-decay which are extremely short (the half-lives are noted in Fig. 38). In order, then, to synthesize Th and U, it is clear that we must have neutron captures in such quick succession that we can go forward in the synthesis processes in spite of the decay which is taking place all the time. The synthesis will first continue along the sequence of the isotopes of bismuth ($Z = 83$, C-products). The most difficult "barrier" here is RaC, which is β active with a half-life of 20 minutes—RaC on β-decay goes over into RaC', which in 10^{-6} sec will go over into RaD (an isotope of lead) on α-disintegration. Further, the isotopes of bismuth with masses greater than RaC (i.e., greater than 214) must be all β active with much shorter lives than RaC. It is thus clear that RaC should capture four successive neutrons (all consecutively) before any β-decay occurs, in order that the β active isotope of bismuth with mass 218 may, on its β-decay, result in the synthesis of RaA, the lifetime of which for α-decay can be measured at least in minutes. We thus see that for the synthesis to go forward from lead it is necessary that the neutrons be captured at a very rapid rate; also, we cannot allow the mean interval between successive neutron captures to be less than, say, 1 minute. It is only then that we can have for instance the synthesis of uranium. Von Weizsäcker points out that this need for a very rapid succession of neutron captures cannot be an exception, in so far as uranium and thorium are not much less abundant than lead. It should be mentioned, however, that for the synthesis of the "moderately" heavy elements we should very probably not need so rapid a reaction-rate.

There are various problems suggested by the conclusions reached above. We shall consider some of them in (4) below, but we may note here that precisely the extreme α-activity of C'-products will allow a more or less straightforward explanation of the relatively large abundance of lead. Indeed, proceeding on similar lines, von Weizsäcker has examined the abundance of the different elements from the present point of view in some detail; he believes the possibility of the present scheme being sufficient (at least as a working basis) for a complete understanding of the general (Z, abundance) relation.[12] We shall not, however, go into these matters further here.

(4) *Astronomical implications.*—So far we have been concerned with general ideas; it now remains to examine the astronomical implications of the von Weizsäcker theory which we have described. There are a great many details that remain to be worked out, but we shall consider only two definite consequences of the theory.

i) *The helium content of the stars.*—If we consider the model process (51) and (52), we see that the mass of helium that is formed at the end of each cycle is much greater than the mass of the neutrons liberated. We can formally combine equations (51) and (52) and write

$$2\,{}_1^1H + {}_2^4He = {}_2^4He + {}_1^2D\,, \tag{53}$$

$${}_1^2D = \tfrac{1}{4}[{}_1^0n + {}_2^3He + {}_1^3T + {}_1^1H]\,, \tag{54}$$

and

$$\tfrac{1}{4}[{}_2^3He + {}_1^3T + 2\,{}_1^1H] = \tfrac{1}{2}[{}_2^4He]\,. \tag{55}$$

Combining the foregoing, we have

$$\tfrac{9}{4}\text{ proton} \rightarrow \tfrac{1}{2}\ \alpha\text{-particle} + \tfrac{1}{4}\text{ neutron}\,. \tag{56}$$

Thus, as the net result of a complete autocatalytic cycle corresponding to the model process, we have the liberation of one neutron for every two α-particles synthesized; in other words, the mass of helium synthesized is eight times the mass of the neutrons liberated. Now according to our discussion in (3) above, it is the neutrons which synthesize the heavier nuclei; we can therefore say quite roughly that the mass of all the elements (other than helium) synthesized will be equal to the mass of the neutrons liberated. We thus see that an immediate consequence of the theory is that the stars should contain a relatively high abundance of helium; indeed, according to the theory, we can expect helium to be as much as (at least) eight times as abundant as the "metals." This is an astronomical prediction which should be capable of verification. In

[12] Von Weizsäcker has since abandoned this hope. See the Bibliographical Notes (ref. 19) at the end of the chapter.

chapter vii, § 9, we have already referred to Strömgren's investigation of the helium content of the stars from the point of view of the mass-luminosity-radius relation. From the discussion (see especially Table 16), Strömgren concludes that the hydrogen-helium-Russell mixture hypothesis is "compatible with the observed masses, luminosities, and radii of the stars." This, then, can be regarded as supporting von Weizsäcker's theory in a general way.

ii) *The role of convection currents.*—According to the transmutation hypothesis of the origin of stellar energy, the rate of generation of energy will be given by a law of the type (46). Now a law of this type implies such a sensitive dependence of ϵ on T that we should expect the point-source model to be a suitable idealization for describing the structure of stars.

Since the nuclear transmutations among the lighter nuclei form an autocatalytic chain of reactions in which each cycle results in increasing the abundance of the lighter elements in some definite ratio, it is clear that the abundance of the lighter elements should increase exponentially with the number of cycles. If we consider the sun, the 10^{56} protons which it would contain if it were all hydrogen would all be used up after about 200 cycles, which then is the upper limit to the number of cycles that could have occurred in the sun. Now in each cycle the neutrons liberated have a mass of $\frac{1}{4}$ (see Eq. [56]), so that an average nucleus cannot have captured more than 50 neutrons. On the other hand, we have seen in (3), above, that for the synthesis of the heavy elements the neutrons should be captured at a very rapid rate—in fact, about once a minute. It is thus clear that the effective mass of the star which can, at a given instant of time, take part in the nuclear reactions must be extremely small. We have, therefore, to postulate that there exist convection currents and that, further, they succeed in effecting a continuous interchange of matter between those regions in which the nuclear reactions are taking place and the other parts of the star. There is, of course, no difficulty in admitting convection currents—indeed, from our discussion in chapter vi, § 2, it follows that in the point-source model the radiative gradient must necessarily become unstable in the central regions. But if there is convection, then, we should expect the currents to produce a more or less uniform condition over an appreciable volume in the neighborhood of the center, and it is difficult to see how we can succeed in confining the regions in which the nuclear reactions are supposed to take place to an extremely limited volume which the von Weizsäcker theory in its present form requires. It will have to be borne in mind in this connection, that we do not yet have a satisfactory method of dealing with problems involving convection currents. It is not suggested that these difficulties cannot be overcome, but they are problems for future investigations. This is perhaps an unsatisfactory state in which to leave the subject, but the object of devoting this amount of space to the von Weizsäcker theory is not because it is, as yet, a fully developed theory but because it introduces some ideas which are of general importance. There is still a variety of other problems (e.g., stability) that can be discussed in this connection, but they are all beyond the scope of this monograph.

BIBLIOGRAPHICAL NOTES

I. H. von Helmholtz announced his estimate of the time scale on the contraction hypothesis and suggested the meteoric theory of the origin of solar radiation at a popular lecture delivered at Königsberg on the occasion of the Kant commemoration in February, 1854. But Kelvin showed that "neither the meteoric theory of solar heat nor any other natural theory can account for the solar radiation continuing at anything like the present rate for many hundred millions of years."

Lord Kelvin's contributions are:

1. LORD KELVIN, *Brit. Assoc. Repts.*, *1861*, Part II, pp. 27–28 (reprinted in his *Collected Papers*, **5**, 141–144).

2. LORD KELVIN, *Les Mondes*, **3**, 473–480, 1863.

In a popular lecture delivered in 1897 Kelvin gives a most attractive account of his ideas on the time scale. See—

3. LORD KELVIN, *Collected Papers*, **5**, 205–230.

II. For general information on nuclear physics the following references may be given:

1. N. FEATHER, *Nuclear Physics*, Cambridge, 1936.

2. F. RASETTI, *Elements of Nuclear Physics*, New York: Prentice Hall, 1936.

3. G. GAMOW, *Structure of Atomic Nuclei and Nuclear Transformations*, Oxford, 1937.

4. C. F. VON WEIZSÄCKER, *Die Atomkerne*, Leipzig, 1937.

5. H. A. BETHE and OTHERS, *Revs. of Modern Phys.*, **8**, 82, 1936; **9**, 69, 245, 1937.

Transmutation of elements as the source of stellar energy was suggested by—

6. A. S Eddington, *Brit. Assoc. Repts.*, *1920*, p. 45.

7. J. Perrin, *Rev. du mois*, **21**, 113, 1920.

8. W. D. Harkins and E. D. Wilson, *Phil. Mag.*, **30**, 723, 1915.

After the discovery of the theory of α-decay by Gamow and by Condon and Gurney, the transmutation of elements arising from proton captures was considered by—

9. R. d'E. Atkinson and F. G Houtermans, *Zs. f. Phys.*, **54**, 656, 1929.

Further elaborations of the ideas contained in the paper of Atkinson and Houtermans are contained in—

10. A. H. Wilson, *M.N.*, **91**, 283, 1931.

11. G. Steensholt, *Zs. f. Ap.*, **5**, 140, 1932.

Further developments are contained in

12. R. d'E Atkinson, *Ap. J.*, **73**, 250, 308, 1931.

At the time of Atkinson's work (ref. 12), nuclear physics was in too rudimentary a stage, and consequently many of Atkinson's ideas have either been superseded or have to be reinterpreted.

The general theory described in § 5 is due to—

13. C. F. von Weizsäcker, *Phys. Zs.*, **38**, 176, 1937.

For a general account of von Weizsäcker's work see—

14. B. Strömgren, *Erg. exakt. Naturwiss.*, **16**, 465, 1935—particularly pp. 519–529.

The problem of the helium content is also examined in reference 14.

The following investigations have appeared since the writing of the monograph:

15. G Gamow and E. Teller, *Phys. Rev.*, **53**, 608, 1938.

16. G. Gamow, *Phys. Rev.*, **53**, 595, 1938.

In references 15 and 16 it is shown that resonance penetrations of charged particles can be of great importance if there are low-lying nuclear energy-levels (with excitation energies of the order of 10 kilovolts).

17. H. Bethe and C. H. Critchfield, *Phys. Rev.*, **54**, 248, 1938.

In reference 17 it is shown that the reaction ${}^1_1H + {}^1_1H = {}^2_1D + e^+$ can occur at quite appreciable rates under the conditions of the stellar interiors; indeed, the authors show that this reaction is sufficient to account for the energy generation in the sun.

18. G. Gamow, *Zs f. Ap.*, **16**, 113, 1938. This paper contains an attractive summary of the present state of the theories of the origin of stellar energy.

During the proof stage of this chapter another paper by von Weizsäcker has been received which carries somewhat further the discussion in reference 13.

19. C. F. von Weizsäcker, *Phys. Zs.*, **39**, 633, 1938.

In this paper von Weizsäcker believes that the difficulties mentioned in § 5 (pp. 481 and 483) are so serious that they require the abandoning of the hypothesis made at the beginning of § 5 (p. 469). Von Weizsäcker now proceeds on the basis of the alternative (*c*) mentioned on page 469.

Also, the possible nonexistence of an atomic nucleus of mass 5 requires the consideration of other types of nuclear reaction chains. Von Weizsäcker (and also Bethe) now suggest the following chain of reactions in which carbon plays the role of a catalyst:

$$
\begin{aligned}
&{}^{12}_{6}C + {}^{1}_{1}H = {}^{13}_{7}N\,; \qquad {}^{13}_{7}N = {}^{13}_{6}C + \beta^{+}\,,\\
&{}^{13}_{6}C + {}^{1}_{1}H = {}^{14}_{7}N\,,\\
&{}^{14}_{7}N + {}^{1}_{1}H = {}^{15}_{8}O\,; \qquad {}^{15}_{8}O = {}^{15}_{7}N + \beta^{+}\,,\\
&{}^{15}_{7}N + {}^{1}_{1}H = {}^{12}_{6}C + {}^{4}_{2}He\,.
\end{aligned}
$$

Gamow has since come to the conclusion that the foregoing chain of nuclear reactions represents the fundamental physical processes which serve as the primary source of stellar energy for the stars on the main series.

A still later paper by Joliot and Zlotowski appears to establish experimentally the existence and stability of ${}^{5}_{2}He$.

20. F. Joliot and I. Zlotowski, *Jour. d. Phys.* **9**, 403, 1938 (December).

APPENDIX I

PHYSICAL AND ASTRONOMICAL CONSTANTS

TABLE 32

	Number	Logarithm
Velocity of light (cm/sec) c	2.9978×10^{10}	10.4768
Electronic charge (e.s.u.) . . . . e	4.801×10^{-10}	$\overline{10}.6813$
Electronic mass (gm) . . . . m_e	9.105×10^{-28}	$\overline{28}.9593$
Mass of the proton (gm) . . . . H	1.672×10^{-24}	$\overline{24}.2232$
Mass of the hydrogen atom	1.673×10^{-24}	$\overline{24}.2235$
Planck's constant (erg sec) . . . . h	6.62×10^{-27}	$\overline{27}.8209$
Boltzmann's constant . . . . k	1.379×10^{-16}	$\overline{16}.1396$
The gas constant . . . . k/H	8.24×10^{7}	7.9161
The Stefan-Boltzmann constant . . . . $a = 8\pi^5 k^4/15c^3h^3$	7.55×10^{-15}	$\overline{15}.8779$
The unrelativistic degenerate constant $K_1\mu_e^{5/3} = \frac{1}{20}\left(\frac{3}{\pi}\right)^{2/3}\frac{h^2}{m_e H^{5/3}}$	9.91×10^{12}	12.996
The relativistic degenerate constant $K_2\mu_e^{4/3} = \frac{1}{8}\left(\frac{3}{\pi}\right)^{1/3}\frac{hc}{H^{4/3}}$	1.231×10^{15}	15.090
The constants of the equation of state of a degenerate gas: $A = \pi m_e^4 c^5/3h^3$	6.01×10^{22}	22.779
$B\mu_e^{-1} = 8\pi m_e^3 c^3 H/3h^3$	9.82×10^{5}	5.992
The Rydberg constant for infinite nuclear mass . . . R_∞	1.098×10^{5}	5.0406
The constant of gravitation (dynes cm²/gm²) . . . . . . G	6.67×10^{-8}	$\overline{8}.8241$
The mass of the sun (gm) . . . . $\odot$	1.985×10^{33}	33.2978
The radius of the sun (cm) . . . . $R_\odot$	6.951×10^{10}	10.8420
The luminosity of the sun (erg/sec) . . . . $L_\odot$	3.780×10^{33}	33.5775
The absolute bolometric magnitude of the sun	+4.63	. . . .
The mean density of the sun . . . . $\bar{\rho}_\odot$	1.4109	0.1495
The unit of length (cm) . . . . $l_1\mu_e = \left(\frac{2A}{\pi G}\right)^{1/2}\frac{1}{B}$	7.71×10^{8}	8.887
The limiting mass (gm) $M_3\mu_e^2 = -4\pi\left(\frac{2A}{\pi G}\right)^{3/2}\frac{1}{B^2}\left(\xi^2\frac{d\theta_3}{d\xi}\right)_{\xi_1}$	1.142×10^{34} $(=5.75\odot)$	34.058

APPENDIX II

THE MASSES OF THE LIGHT ATOMS

TABLE 33*

Atom	Mass†	Probable Error ×10⁵	Stability	Atom	Mass†	Probable Error ×10⁵	Stability
e	0.00055	0	....	$^{12}_{5}B$	12.019	70	β^- active
				$^{11}_{6}C$	11.01526	35	β^+ active
$^{1}_{0}n$	1.0c897	6	β^- active	$^{12}_{6}C$	12.00398	10	stable
$^{1}_{1}H$ ...	1.00813	2	stable	$^{13}_{6}C$	13.00761	15	stable
(D) $^{2}_{1}H$ ...	2.01473	2	stable	$^{14}_{6}C$	14.00767	12	β^- active
(T) $^{3}_{1}H$ ...	3.01705	7	stable	$^{13}_{7}N$	13.01004	13	β^+ active
$^{3}_{2}He$ ...	3.01707	12	stable‡	$^{14}_{7}N$	14.00750	8	stable
$^{4}_{2}He$ ...	4.00389	7	stable	$^{15}_{7}N$	15.00489	20	stable
$^{5}_{2}He$ ...	5.0137	40	n§ active	$^{16}_{7}N$	16.011	200	β^- active
$^{6}_{2}He$ ...	6.0208	50	β^- active	$^{15}_{8}O$	15.0078	40	β^+ active
$^{6}_{3}Li$ ...	6.01686	20	stable	$^{16}_{8}O$	16.00000	0	stable
$^{7}_{3}Li$ ...	7.01818	18	stable	$^{17}_{8}O$	17.00450	7	stable
$^{8}_{3}Li$ ...	8.025	100	β^- active	$^{18}_{8}O$	18.00369	20	stable
$^{8}_{4}Be$ ...	8.00792	28	α active(?)	$^{17}_{9}F$	17.0076	30	β^+ active
$^{9}_{4}Be$ ...	9.01504	25	stable	$^{18}_{9}F$	18.0056	40	β^+ active
$^{10}_{4}Be$ ...	10.01671	30	β^- active	$^{19}_{9}F$	19.00452	17	stable
$^{10}_{5}B$	10.01631	25	stable	$^{20}_{9}F$	20.007	250	β^- active
$^{11}_{5}B$	11.01292	17	stable				

* The values given in this table are taken from a paper by H. Bethe (*Rev. Mod. Phys.*, **9**, 373, 1937).

† To obtain the nuclear masses, the masses of the appropriate numbers of electrons should be subtracted from the values given.

‡ This nucleus will probably absorb a free electron and go over into $^{3}_{1}H$.

§ More recent experiments by Joliot and Zlotowski seem to indicate that $^{5}_{2}He$ is stable.

APPENDIX III[1]

THE MASSES, LUMINOSITIES, AND RADII OF THE STARS DERIVED HYDROGEN CONTENTS; CENTRAL DENSITIES; AND CENTRAL TEMPERATURES

TABLE 34*a**

VISUAL BINARIES

Star	Log M	Log L	Log R	X_0†	Log ρ_c	$1-\beta$	Log T_c
Sun.........	0.00	0.00	0.00	0.36	1.88	0.003	7.29
η Cas A.......	− .14	−0.09	−0.09	0.25	2.00	.003	7.32
B.......	− .33	−1.16	−0.25	[0.34]	[1.29]	[.001]	[7.25]
o_2 Eri C.......	− .70	−1.96	−0.37	....	....	....	....
α Aur A.......	+ .62	+2.08	+1.20	.30	−1.10	.051	6.71
B.......	+ .52	+1.90	+0.82	.30	−0.06	.033	6.99
α CMa A......	+ .37	+1.59	+0.25	.36	1.50	.013	7.39
α CMi A......	+ .17	+0.76	+0.23	.31	1.37	.007	7.25
ζ UMa $\overline{A}$......	+ .41	+1.48	+0.28	.45	1.45	.010	7.34
α Cen A.......	+ .04	+0.10	+0.09	.36	1.66	.003	7.23
B.......	− .06	−0.43	−0.06	[.43]	[2.00]	[.002]	[7.25]
ξ Boo A.......	− .06	−0.32	−0.06	[.39]	[1.98]	[.002]	[7.27]
B.......	− .12	−0.83	−0.10	[.48]	[2.07]	[.001]	[7.21]
ζ Her A.......	− .02	+0.59	+0.29	.12	0.99	.011	7.13
−8°4352 $\overline{AB}$.....	− .45	−1.40	−0.12	.20	1.79	.001	7.09
μ Her $\overline{BC}$.....	− .35	−1.37	−0.10	[.30]	[1.85]	[.001]	[7.10]
70 Oph A.......	− .05	−0.38	−0.03	[.42]	[1.93]	[.002]	[7.24]
B.......	− .13	−0.86	−0.16	[.50]	[2.22]	[.001]	[7.25]
Kr. 60 A........	−0.60	−1.77	−0.29	[0.19]	[2.15]	[0.009]	[7.15]

* In these tables L, M, and R are expressed in the corresponding solar units.

† These values were supplied by B. Strömgren. Those given in [ ] brackets are for stars too dense for the theory of chap. vii to be applicable with reasonable certainty.

[1] The data used in chaps. vii, viii, and xi will be found, occasionally, to differ slightly from the values given in this Appendix. The values given here correspond to Kuiper's final revision of the observational material and are taken from *Ap. J.*, **88**, 472, 1938. For the data for the stars of the Hyades cluster see Table 15, p. 287.

TABLE 34*b*

SPECTROSCOPIC BINARIES*

Star		Log *M*	Log *L*	Log *R*
Castor	C_1	−0.201	−1.16	−0.18
	C_2	−0.247	−1.24	−0.22
β Aur	A	+0.378	+1.83	+0.43
	B	+0.370	+1.83	+0.43
μ_1 Sco	$\overline{AB}$	+1.094	+3.35	+0.73
V Pup	$\overline{AB}$	+1.265	+3.86	+0.83
Y Cyg	$\overline{AB}$	+1.238	+4.51	+0.77
Ao Cas	A	+1.634	+5.97	+1.36
	B	+1.582	+5.58	+1.23
29 CMa	A	+1.66	+5.84	+1.31
	B	+1.53	+5.39	+1.13

* The values of X_0 for β Aur A and B are (according to Strömgren) 0.27 and 0.25, respectively. The corresponding values of $(1 - \beta)$ are 0.022 and 0.023. The central temperature for both stars is about 19 million degrees.

For the other stars in this table the theory of chaps. vi and vii cannot be applied with reasonable certainty (cf. chap. viii, §§ 6 and 7).

TABLE 34*c*

TRUMPLER'S STARS

Star	Log *M*	Log *L*	Log *R*	*Q**
NGC 2244, 15	1.76:	5.49	0.66	3
8	1.99	4.69	0.86	2
NGC 2264, 60	2.18	5.33	0.82	1
NGC 2362, 1	2.47	5.73	1.28	4
NGC 6871, 2	2.35	5.33	1.18	6
5	2.60	5.01	1.22	5
NGC 7380, 1	1.89	4.89	0.96	7

* The order of reliability of the measured red shifts according to a private communication from Dr. Trumpler.

TABLE 34*d*

WHITE DWARFS

Star	Log *M*	Log *L*	Log *R*
Sirius B	−0.01	−2.52	−1.71
o_2 Eri B	− .35	−2.25	−1.74
Van Maanen No. 2	+0.53:	−3.85	−2.05

APPENDIX IV

TABLES OF THE WHITE-DWARF FUNCTIONS

In the following tables (35–44) the solutions of the differential equation

$$\frac{1}{\eta^2}\frac{d}{d\eta}\left(\eta^2\frac{d\phi}{d\eta}\right) = -\left(\phi^2 - \frac{1}{y_0^2}\right)^{3/2}, \tag{1}$$

for different values of $1/y_0^2$ and satisfying the boundary conditions

$$\phi = 1\,, \quad \frac{d\phi}{d\eta} = 0 \qquad (\eta = 0)\,, \tag{2}$$

are given. In addition to the function ϕ and its derivative ϕ', certain other auxiliary functions are also tabulated. The quantities ρ/ρ_0, $\rho_0/\bar{\rho}(\eta)$ and $-\eta^2\phi'$ describe the physical structure of the completely degenerate configurations (see chap. xi, Eq. [83]).

Regarding the accuracy of the tables, it might be stated that errors exceeding three to four units in the last figures retained are not expected. The quantities ϕ and ϕ' have been checked by differencing. These tables of the white-dwarf functions (computed by Chandrasekhar) are published here for the first time.

TABLE 35

$$\frac{1}{y_0^2} = 0.01$$

η	ϕ	ρ/ρ_0	$-\phi'$	$\rho_0/\bar{\rho}(\eta)$	$-\eta^2\phi'$
0	1	1	0	1	0
0.1	0.998361	0.995041	0.032737	1.00299	0.00033
0.2	0.993472	0.980348	.064892	1.01197	0.00260
0.3	0.985420	0.956463	.095910	1.02704	0.00863
0.4	0.974345	0.924245	.125284	1.04832	0.02005
0.5	0.960433	0.884805	.152576	1.07600	0.03814
0.6	0.943911	0.839435	.177433	1.11032	0.06388
0.7	0.925036	0.789525	.199591	1.15156	0.09780
0.8	0.904088	0.736480	.218883	1.20008	0.14009
0.9	0.881358	0.681653	.235231	1.25626	0.19054
1.0	0.857140	0.626289	.248642	1.32055	0.24864
1.1	0.831725	0.571479	.259195	1.39347	0.31363
1.2	0.805392	0.518140	.267030	1.47555	0.38452
1.3	0.778403	0.467004	.272331	1.56739	0.46024
1.4	0.751003	0.418618	.275316	1.66966	0.53962
1.5	0.723410	0.373363	.276221	1.78306	0.62150
1.6	0.695820	0.331468	.275294	1.90833	0.70475
1.7	0.668404	0.293033	.272780	2.04629	0.78834
1.8	0.641308	0.258055	.268918	2.19778	0.87129
1.9	0.614657	0.226449	.263932	2.36370	0.95280
2.0	0.588552	0.198070	.258032	2.54500	1.0321
2.1	0.563075	0.172730	.251407	2.74267	1.1087
2.2	0.538289	0.150216	.244227	2.95774	1.1821
2.3	0.514243	0.130299	.236642	3.19130	1.2518
2.4	0.490970	0.112749	.228782	3.44446	1.3178
2.5	0.468492	0.097337	.220758	3.71838	1.3797
2.6	0.446821	0.083844	.212666	4.01428	1.4376
2.7	0.425959	0.072064	.204582	4.33338	1.4914
2.8	0.405902	0.061804	.196574	4.67697	1.5411
2.9	0.386640	0.052889	.188692	5.04634	1.5869
3.0	0.368158	0.045157	.180979	5.44283	1.6288
3.1	0.350437	0.038463	.173467	5.86781	1.6670
3.2	0.333457	0.032680	.166181	6.32268	1.7017
3.3	0.317193	0.027690	.159138	6.80884	1.7330
3.4	0.301621	0.023392	.152349	7.32773	1.7612
3.5	0.286714	0.019697	.145824	7.88079	1.7863
3.6	0.272447	0.016524	.139565	8.46950	1.8088
3.7	0.258793	0.013806	.133572	9.09532	1.8286
3.8	0.245724	0.011480	.127844	9.75970	1.8461
3.9	0.233215	0.009494	.122375	10.4642	1.8613
4.0	0.221240	0.007803	.117160	11.2101	1.8746
4.1	0.209775	0.006366	.112194	11.9991	1.8860
4.2	0.198794	0.005149	.107466	12.8324	1.8957
4.3	0.188274	0.004121	.102970	13.7116	1.9039
4.4	0.178192	0.003257	.098696	14.6380	1.9108
4.5	0.168527	0.002534	.094636	15.6130	1.9164
4.6	0.159258	0.001933	.090781	16.6378	1.9209
4.7	0.150365	0.001437	.087120	17.7137	1.9245
4.8	0.141828	0.001033	.083646	18.8419	1.9272
4.9	0.133630	0.000707	.080350	20.0235	1.9292
5.0	0.125752	0.000450	.077224	21.2594	1.9306
5.1	0.118179	0.000254	.074258	22.5505	1.9315
5.2	0.110895	0.000112	.071447	23.8874	1.9319
5.3	0.103885	0.000023	0.068782	25.3006	1.9321

$$\eta_1 = 5.3571 .$$

$\phi(\eta_1) = 0.1$; $\qquad -\eta_1^2\phi'(\eta_1) = 1.9321$,

$-\phi'(\eta_1) = 0.067325$; $\qquad \rho_0/\bar{\rho} = 26.203$.

TABLE 36

$$\frac{1}{y_0^2} = 0.02$$

η	ϕ	ρ/ρ_0	$-\phi'$	$\rho_0/\bar{\rho}(\eta)$	$-\eta^2\phi'$
0	1	1	0	1	0
0.1	0.998385	0.995066	0.032243	1.00297	0.00032
0.2	0.993571	0.980445	.063915	1.01191	0.00256
0.3	0.985640	0.956674	.094473	1.02691	0.00850
0.4	0.974730	0.924600	.123419	1.04808	0.01975
0.5	0.961024	0.885322	.150324	1.07562	0.03758
0.6	0.944745	0.840119	.174838	1.10977	0.06294
0.7	0.926145	0.790366	.196703	1.15081	0.09638
0.8	0.905498	0.737456	.215751	1.19910	0.13808
0.9	0.883091	0.682736	.231905	1.25502	0.18784
1.0	0.859214	0.627441	.245168	1.31903	0.24517
1.1	0.834151	0.572660	.255616	1.39163	0.30930
1.2	0.808180	0.519309	.263383	1.47337	0.37927
1.3	0.781558	0.468121	.268650	1.56486	0.45402
1.4	0.754526	0.419650	.271626	1.66676	0.53239
1.5	0.727302	0.374279	.272546	1.77979	0.61323
1.6	0.700078	0.332245	.271650	1.90471	0.69542
1.7	0.673024	0.293653	.269179	2.04233	0.77793
1.8	0.646287	0.258507	.265367	2.19353	0.85979
1.9	0.619988	0.226726	.260438	2.35921	0.94018
2.0	0.594229	0.198171	.254596	2.54037	1.0184
2.1	0.569093	0.172658	.248029	2.73801	1.0938
2.2	0.544642	0.149976	.240906	2.95321	1.1660
2.3	0.520925	0.129901	.233374	3.18708	1.2345
2.4	0.497977	0.112204	.225564	3.44081	1.2992
2.5	0.475818	0.096657	.217585	3.71560	1.3599
2.6	0.454462	0.083043	.209533	4.01272	1.4164
2.7	0.433911	0.071155	.201486	4.33348	1.4688
2.8	0.414162	0.060801	.193509	4.67923	1.5171
2.9	0.395206	0.051804	.185655	5.05138	1.5614
3.0	0.377026	0.044004	.177965	5.45134	1.6017
3.1	0.359606	0.037255	.170474	5.88061	1.6383
3.2	0.342924	0.031428	.163205	6.34067	1.6712
3.3	0.326957	0.026406	.156176	6.83309	1.7008
3.4	0.311680	0.022086	.149401	7.35941	1.7271
3.5	0.297068	0.018379	.142887	7.92125	1.7504
3.6	0.283094	0.015203	.136638	8.52020	1.7708
3.7	0.269732	0.012490	.130655	9.15788	1.7887
3.8	0.256954	0.010178	.124935	9.83594	1.8041
3.9	0.244736	0.008214	.119476	10.5560	1.8172
4.0	0.233050	0.006551	.114273	11.3197	1.8284
4.1	0.221873	0.005151	.109318	12.1286	1.8376
4.2	0.211179	0.003976	.104604	12.9842	1.8452
4.3	0.200944	0.002999	.100124	13.8882	1.8513
4.4	0.191146	0.002192	.095870	14.8419	1.8560
4.5	0.181763	0.001535	.091832	15.8467	1.8596
4.6	0.172773	0.001008	.088002	16.9038	1.8621
4.7	0.164156	0.000597	.084372	18.0143	1.8638
4.8	0.155892	0.000291	.080934	19.1791	1.8647
4.9	0.147963	0.000085	0.077681	20.3984	1.8651

$$\eta_1 = 4.9857 .$$

$$\phi(\eta_1) = 0.14142 ; \quad -\eta_1^2\phi'(\eta_1) = 1.8652 ,$$

$$-\phi'(\eta_1) = 0.07504 ; \quad \rho_0/\bar{\rho} = 21.486 .$$

TABLE 37

$$\frac{1}{y_0^2} = 0.05$$

η	ϕ	ρ/ρ_0	$-\phi'$	$\rho_0/\bar{\rho}(\eta)$	$-\eta^2\phi'$
0	1	1	0	1	0
0.1	0.998459	0.995141	0.030775	1.00293	0.00031
0.2	0.993863	0.980742	.061014	1.01173	0.00244
0.3	0.986291	0.957315	.090205	1.02649	0.00812
0.4	0.975873	0.925680	.117878	1.04735	0.01886
0.5	0.962780	0.886897	.143628	1.07447	0.03591
0.6	0.947222	0.842203	.167122	1.10811	0.06016
0.7	0.929439	0.792933	.188111	1.14854	0.09217
0.8	0.909689	0.740447	.206432	1.19613	0.13212
0.9	0.888244	0.686059	.222005	1.25125	0.17982
1.0	0.865380	0.630987	.234825	1.31437	0.23483
1.1	0.841369	0.576309	.244958	1.38600	0.29640
1.2	0.816474	0.522939	.252524	1.46670	0.36363
1.3	0.790944	0.471615	.257687	1.55709	0.43549
1.4	0.765010	0.422900	.260643	1.65785	0.51086
1.5	0.738882	0.377196	.261608	1.76972	0.58862
1.6	0.712747	0.334753	.260809	1.89348	0.66767
1.7	0.686771	0.295699	.258474	2.03000	0.74699
1.8	0.661096	0.260053	.254825	2.18019	0.82563
1.9	0.635843	0.227753	.250074	2.34504	0.90277
2.0	0.611111	0.198673	.244418	2.52557	0.97767
2.1	0.586983	0.172644	.238040	2.72291	1.0498
2.2	0.563522	0.149465	.231100	2.93823	1.1185
2.3	0.540777	0.128920	.223746	3.17276	1.1836
2.4	0.518783	0.110786	.216102	3.42781	1.2447
2.5	0.497563	0.094840	.208279	3.70475	1.3017
2.6	0.477130	0.080867	.200369	4.00503	1.3545
2.7	0.457489	0.068663	.192452	4.33017	1.4030
2.8	0.438637	0.058035	.184593	4.68173	1.4472
2.9	0.420567	0.048807	.176845	5.06137	1.4873
3.0	0.403263	0.040817	.169252	5.47081	1.5233
3.1	0.386710	0.033919	.161847	5.91181	1.5554
3.2	0.370886	0.027980	.154658	6.38621	1.5837
3.3	0.355770	0.022884	.147702	6.89590	1.6085
3.4	0.341338	0.018525	.140995	7.44284	1.6299
3.5	0.327563	0.014812	.134548	8.02889	1.6482
3.6	0.314419	0.011664	.128363	8.65616	1.6636
3.7	0.301881	0.009009	.122446	9.32659	1.6763
3.8	0.289921	0.006787	.116794	10.0421	1.6865
3.9	0.278513	0.004944	.111407	10.8048	1.6945
4.0	0.267631	0.003435	.106282	11.6162	1.7005
4.1	0.257248	0.002222	.101414	12.4782	1.7048
4.2	0.247340	0.001276	.096798	13.3920	1.7075
4.3	0.237881	0.000577	.092430	14.3588	1.7090
4.4	0.228846	0.000125	0.088306	15.3789	1.7096

$$\eta_1 = 4.4601\,.$$

$$\phi(\eta_1) = 0.22361\,; \qquad -\eta_1^2\phi'(\eta_1) = 1.7096\,,$$

$$-\phi'(\eta_1) = 0.08594\,; \qquad \rho_0/\bar{\rho} = 16.018\,.$$

TABLE 38

$$\frac{1}{y_0^2} = 0.1$$

η	ϕ	ρ/ρ_0	$-\phi'$	$\rho_0/\bar{\rho}(\eta)$	$-\eta^2\phi'$
0	1	1	0	1	0
0.1	0.998579	0.995270	0.028380	1.00285	0.00028
0.2	0.994340	0.978615	.056278	1.01142	0.00225
0.3	0.987355	0.958409	.083235	1.02579	0.00749
0.4	0.977739	0.927526	.108826	1.04609	0.01741
0.5	0.965647	0.889596	.132683	1.07250	0.03317
0.6	0.951270	0.845787	.154500	1.10526	0.05562
0.7	0.934823	0.797368	.174047	1.14465	0.08528
0.8	0.916542	0.745636	.191166	1.19102	0.12235
0.9	0.896674	0.691858	.205776	1.24478	0.16668
1.0	0.875471	0.637216	.217861	1.30636	0.21786
1.1	0.853184	0.582768	.227469	1.37630	0.27524
1.2	0.830056	0.529422	.234701	1.45516	0.33797
1.3	0.806319	0.477922	.239694	1.54358	0.40508
1.4	0.782186	0.428800	.242621	1.64226	0.47554
1.5	0.757857	0.382631	.243672	1.75198	0.54826
1.6	0.733508	0.339545	.243050	1.87356	0.62221
1.7	0.709296	0.299748	.240960	2.00792	0.69637
1.8	0.685358	0.263292	.237606	2.15605	0.76984
1.9	0.661810	0.230143	.233182	2.31900	0.84179
2.0	0.638751	0.200202	.227874	2.49792	0.91149
2.1	0.616260	0.173321	.221849	2.69405	0.97835
2.2	0.594400	0.149320	.215262	2.90869	1.0419
2.3	0.573221	0.127997	.208252	3.14327	1.1017
2.4	0.552760	0.109142	.200940	3.39929	1.1574
2.5	0.533040	0.092540	.193432	3.67836	1.2090
2.6	0.514077	0.077982	.185821	3.98218	1.2562
2.7	0.495876	0.065266	.178184	4.31257	1.2990
2.8	0.478438	0.054204	.170588	4.67144	1.3374
2.9	0.461756	0.044618	.163087	5.06081	1.3716
3.0	0.445816	0.036346	.155726	5.48280	1.4015
3.1	0.430604	0.029240	.148541	5.93960	1.4275
3.2	0.416101	0.023167	.141561	6.43352	1.4496
3.3	0.402285	0.018008	.134809	6.96689	1.4681
3.4	0.389131	0.013658	.128301	7.54210	1.4832
3.5	0.376616	0.010023	.122050	8.16154	1.4951
3.6	0.364712	0.007026	.116066	8.82752	1.5042
3.7	0.353394	0.004598	.110355	9.54224	1.5108
3.8	0.342632	0.002687	.104922	10.3076	1.5151
3.9	0.332400	0.001258	.099771	11.1251	1.5175
4.0	0.322668	0.000309	0.094906	11.9952	1.5185

$\eta_1 = 4.0690$.

$\phi(\eta_1) = 0.31623$; $\quad -\eta_1^2\phi'(\eta_1) = 1.5186$,

$-\phi'(\eta_1) = 0.09172$; $\quad \rho_0/\bar{\rho} = 12.626$.

TABLE 39

$$\frac{1}{y_0^2} = 0.2$$

η	ϕ	ρ/ρ_0	$-\phi'$	$\rho_0/\bar{\rho}(\eta)$	$-\eta^2\phi'$
0	1	1	0	1	0
0.1	0.998809	0.995540	0.023788	1.00269	0.00024
0.2	0.995255	0.982302	.047195	1.01077	0.00189
0.3	0.989395	0.960704	.069856	1.02432	0.00629
0.4	0.981320	0.931414	.091432	1.04346	0.01463
0.5	0.971155	0.895308	.111625	1.06837	0.02791
0.6	0.959050	0.853418	.130183	1.09928	0.04687
0.7	0.945179	0.806876	.146911	1.13647	0.07199
0.8	0.929733	0.756857	.161668	1.18026	0.10347
0.9	0.912914	0.704522	.174371	1.23107	0.14124
1.0	0.894929	0.650977	.184992	1.28932	0.18499
1.1	0.875985	0.597228	.193549	1.35555	0.23419
1.2	0.856286	0.544165	.200106	1.43033	0.28815
1.3	0.836027	0.492537	.204759	1.51430	0.34604
1.4	0.815393	0.442952	.207635	1.60821	0.40696
1.5	0.794554	0.395876	.208876	1.71284	0.46997
1.6	0.773667	0.351647	.208641	1.82908	0.53412
1.7	0.752870	0.310479	.207094	1.95792	0.59850
1.8	0.732286	0.272487	.204400	2.10041	0.66226
1.9	0.712023	0.237696	.200720	2.25775	0.72460
2.0	0.692170	0.206064	.196210	2.43121	0.78484
2.1	0.672804	0.177493	.191015	2.62220	0.84237
2.2	0.653985	0.151845	.185270	2.83225	0.89671
2.3	0.635764	0.128954	.179098	3.06302	0.94743
2.4	0.618176	0.108637	.172612	3.31631	0.99424
2.5	0.601249	0.090703	.165908	3.59407	1.0369
2.6	0.584999	0.074959	.159074	3.89842	1.0753
2.7	0.569436	0.061213	.152185	4.23160	1.1094
2.8	0.554561	0.049284	.145308	4.59603	1.1392
2.9	0.540372	0.039000	.138497	4.99426	1.1648
3.0	0.526858	0.030199	.131800	5.42898	1.1862
3.1	0.514007	0.022735	.125258	5.90298	1.2037
3.2	0.501800	0.016478	.118902	6.41911	1.2176
3.3	0.490219	0.011313	.112762	6.98018	1.2280
3.4	0.479240	0.007143	.106860	7.58886	1.2353
3.5	0.468838	0.003897	.101218	8.24749	1.2399
3.6	0.458987	0.001540	.095856	8.95768	1.2423
3.7	0.449657	0.000143	0.090796	9.71958	1.2430

$$\eta_1 = 3.7271 .$$

$$\phi(\eta_1) = 0.44721 ; \quad -\eta_1^2\phi'(\eta_1) = 1.2430 ,$$

$$-\phi'(\eta_1) = 0.08948 ; \quad \rho_0/\bar{\rho} = 9.9348 .$$

TABLE 40

$$\frac{1}{y_0^2} = 0.3$$

η	ϕ	ρ/ρ_0	$-\phi'$	$\rho_0/\bar{\rho}(\eta)$	$-\eta^2\phi'$
0	1	1	0	1	0
0.1	0.999025	0.995827	0.019473	1.00262	0.00019
0.2	0.996115	0.983429	.038655	1.01007	0.00155
0.3	0.991313	0.963163	.057264	1.03256	0.00515
0.4	0.984690	0.935601	.075038	1.04066	0.01201
0.5	0.976341	0.901498	.091742	1.06396	0.02294
0.6	0.966384	0.861752	.107178	1.09288	0.03858
0.7	0.954954	0.817358	.121183	1.12767	0.05938
0.8	0.942199	0.769361	.133637	1.16866	0.08553
0.9	0.928281	0.718812	.144462	1.21622	0.11701
1.0	0.913362	0.666724	.153622	1.27079	0.15362
1.1	0.897612	0.614046	.161115	1.33285	0.19495
1.2	0.881194	0.561631	.166976	1.40299	0.24044
1.3	0.864269	0.510234	.171267	1.48182	0.28944
1.4	0.846990	0.460436	.174074	1.57007	0.34119
1.5	0.829500	0.412780	.175501	1.66854	0.39488
1.6	0.811932	0.367636	.175664	1.77812	0.44970
1.7	0.794405	0.325276	.174689	1.89980	0.50485
1.8	0.777028	0.285874	.172705	2.03467	0.55956
1.9	0.759894	0.249518	.169839	2.18395	0.61312
2.0	0.743085	0.216226	.166217	2.34898	0.66487
2.1	0.726671	0.185952	.161961	2.53124	0.71425
2.2	0.710710	0.158610	.157185	2.73236	0.76078
2.3	0.695248	0.134074	.151994	2.95411	0.80405
2.4	0.680322	0.112198	.146487	3.19844	0.84376
2.5	0.665958	0.092818	.140750	3.46751	0.87969
2.6	0.652177	0.075763	.134864	3.76361	0.91168
2.7	0.638988	0.060860	.128898	4.08924	0.93967
2.8	0.626398	0.047938	.122916	4.44710	0.96366
2.9	0.614404	0.036833	.116971	4.84001	0.98372
3.0	0.603001	0.027393	.111111	5.27096	1.00000
3.1	0.592178	0.019478	.105377	5.74302	1.0127
3.2	0.581920	0.012964	.099806	6.25917	1.0220
3.3	0.572210	0.007754	.094430	6.82226	1.0283
3.4	0.563026	0.003784	.089278	7.43466	1.0321
3.5	0.554346	0.001065	0.084376	8.09790	1.0336

$$\eta_1 = 3.5803\,.$$

$$\phi(\eta_1) = 0.54772\,; \qquad -\eta_1^2\phi'(\eta_1) = 1.0337\,,$$

$$-\phi'(\eta_1) = 0.08064\,; \qquad \rho_0/\bar{\rho} = 8.6673\,.$$

TABLE 41

$$\frac{1}{y_0^2} = 0.4$$

η	ϕ	ρ/ρ_0	$-\phi'$	$\rho_0/\bar{\rho}(\eta)$	$-\eta^2\phi'$
0	1	1	0	1	0
0.1	0.999226	0.996135	0.015456	1.00232	0.00015
0.2	0.996916	0.984643	.030698	1.00932	.00123
0.3	0.993101	0.965821	.045517	1.02106	.00410
0.4	0.987833	0.940149	.059720	1.03763	.00956
0.5	0.981183	0.908264	.073130	1.05920	.01828
0.6	0.973239	0.877289	.085594	1.08596	.03081
0.7	0.964100	0.829008	.096985	1.11815	.04752
0.8	0.953881	0.783404	.107204	1.15607	.06861
0.9	0.942701	0.735049	.116184	1.20006	.09411
1.0	0.930687	0.684858	.123883	1.25053	.12388
1.1	0.917967	0.633701	.130290	1.30794	.15765
1.2	0.904671	0.582382	.135417	1.37282	.19500
1.3	0.890925	0.531620	.139299	1.44577	.23542
1.4	0.876851	0.482036	.141991	1.52747	.27830
1.5	0.862565	0.434155	.143562	1.61867	.32301
1.6	0.848174	0.388394	.144092	1.72023	.36888
1.7	0.833778	0.345077	.143671	1.83309	.41521
1.8	0.819468	0.304435	.142695	1.95832	.46136
1.9	0.805324	0.266616	.140358	2.09711	.50669
2.0	0.791418	0.231699	.137660	2.25075	.55064
2.1	0.777811	0.199698	.134394	2.42072	.59268
2.2	0.764555	0.170579	.130653	2.60861	.63236
2.3	0.751694	0.144268	.126523	2.81621	.66931
2.4	0.739261	0.120659	.122085	3.04547	.70321
2.5	0.727284	0.099625	.117415	3.29855	.73384
2.6	0.715783	0.081023	.112581	3.57778	.76105
2.7	0.704772	0.064704	.107646	3.88573	.78474
2.8	0.694256	0.050515	.102666	4.22517	.80490
2.9	0.684238	0.038307	.097690	4.59888	.82158
3.0	0.674716	0.027937	.092766	5.01002	.83489
3.1	0.665682	0.019274	.087931	5.46166	.84502
3.2	0.657126	0.012210	.083223	5.95681	.85220
3.3	0.649032	0.006662	.078674	6.49815	.85676
3.4	0.641384	0.002610	.074316	7.08766	.85909
3.5	0.634161	0.000216	0.070816	7.72543	0.85978

$\eta_1 = 3.5245$.

$\phi(\eta_1) = 0.63246$; $\quad -\eta_1^2\phi'(\eta_1) = 0.8598$,

$-\phi'(\eta_1) = 0.06922$; $\quad \rho_0/\bar{\rho} = 7.8886$.

TABLE 42

$$\frac{1}{y^2} = 0.5$$

η	ϕ	ρ/ρ_0	$-\phi'$	$\rho_0/\bar{\rho}(\eta)$	$-\eta^2\phi'$
0	1	1	0	1	0
0.1	0.999411	0.996471	0.011760	1.00212	0.000118
0.2	0.997653	0.985967	.023371	1.00851	.000935
0.3	0.994747	0.968730	.034689	1.01922	.003122
0.4	0.990730	0.945149	.045575	1.03434	.007292
0.5	0.985650	0.915750	.055906	1.05401	.013976
0.6	0.979571	0.881163	.065570	1.07841	.023605
0.7	0.972562	0.842106	.074472	1.10774	.036491
0.8	0.964704	0.799351	.082538	1.14227	.052824
0.9	0.956084	0.753699	.089710	1.18232	.072665
1.0	0.946793	0.705949	.095952	1.22822	.095952
1.1	0.936925	0.656883	.101245	1.28043	.122506
1.2	0.926576	0.607237	.105587	1.33938	.152045
1.3	0.915839	0.557680	.108995	1.40563	.184201
1.4	0.904807	0.508825	.111496	1.47979	.218533
1.5	0.893569	0.461195	.113133	1.56255	.254550
1.6	0.882208	0.415233	.113958	1.65467	.291732
1.7	0.870802	0.371298	.114027	1.75701	.329538
1.8	0.859425	0.329672	.113405	1.87057	.367432
1.9	0.848142	0.290560	.112160	1.99641	.404898
2.0	0.837012	0.254100	.110361	2.13573	.441445
2.1	0.826086	0.220367	.108078	2.28989	.476626
2.2	0.815410	0.189387	.105380	2.46036	.510038
2.3	0.805022	0.160974	.102332	2.64881	.541335
2.4	0.794953	0.135569	.098998	2.85706	.570227
2.5	0.785230	0.112593	.095438	3.08713	.596485
2.6	0.775871	0.092107	.091707	3.34121	.619941
2.7	0.766892	0.073992	.087858	3.62172	.640487
2.8	0.758302	0.058120	.083938	3.93126	.658076
2.9	0.750105	0.044362	.079991	4.27261	.672721
3.0	0.742303	0.032590	.076054	4.64868	.684490
3.1	0.734893	0.022685	.072166	5.06249	.693513
3.2	0.727868	0.014544	.068357	5.51695	.699979
3.3	0.721218	0.008093	.064659	6.01473	.704141
3.4	0.714931	0.003319	.061102	6.55774	.706344
3.5	0.708992	0.000390	0.057717	7.14658	0.707033

$$\eta_1 = 3.5330.$$

$$\phi(\eta_1) = 0.707107; \quad -\eta_1^2\phi'(\eta_1) = 0.70704,$$

$$-\phi'(\eta_1) = 0.056644; \quad \rho_0/\bar{\rho} = 7.3505.$$

TABLE 43

$$\frac{1}{y_0^2} = 0.6$$

η	ϕ	ρ/ρ_0	$-\phi'$	$\rho_0/\bar{\rho}(\eta)$	$-\eta^2\phi'$
0	1	1	0	1	0
0.1	0.9995788	0.996843	0.0084087	1.00285	0.000084
0.2	0.9983198	0.987436	.0167381	1.00761	.000670
0.3	0.9962374	0.971967	.0248708	1.01719	.002238
0.4	0.9933549	0.950741	.0327264	1.03070	.005236
0.5	0.9897042	0.924171	.0402225	1.04826	.010056
0.6	0.9853249	0.892758	.0472850	1.07003	.017023
0.7	0.9802638	0.857079	.0538498	1.09618	.026386
0.8	0.9745733	0.817762	.0598630	1.12694	.038312
0.9	0.9683110	0.775468	.0652821	1.16256	.052878
1.0	0.9615377	0.730873	.0700764	1.20336	.070076
1.1	0.9543172	0.684647	.0742266	1.24969	.089814
1.2	0.9467142	0.637436	.0777247	1.30194	.111924
1.3	0.9387939	0.589852	.0805729	1.36058	.136168
1.4	0.9306209	0.542460	.0827830	1.42612	.162255
1.5	0.9222579	0.495766	.0843752	1.49915	.189844
1.6	0.9137655	0.450217	.0853772	1.58033	.218566
1.7	0.9052010	0.406193	.0858223	1.67039	.248026
1.8	0.8966183	0.364011	.0857490	1.77016	.277827
1.9	0.8880672	0.323922	.0851990	1.88056	.307568
2.0	0.8795930	0.286120	.0842168	2.00263	.336867
2.1	0.8712367	0.250740	.0828478	2.13750	.365359
2.2	0.8630348	0.217868	.0811382	2.28647	.392709
2.3	0.8550189	0.187543	.0791338	2.45095	.418618
2.4	0.8472163	0.159727	.0768970	2.63191	.442927
2.5	0.8396500	0.134514	.0744175	2.83292	.465109
2.6	0.8323383	0.111723	.0717910	3.05402	.485307
2.7	0.8252960	0.091316	.0690381	3.29795	.503288
2.8	0.8185337	0.073204	.0661952	3.56698	.518970
2.9	0.8120588	0.057283	.0632962	3.86357	.532321
3.0	0.8058753	0.043447	.0603729	4.19033	.543356
3.1	0.7999840	0.031593	.0574542	4.54997	.552135
3.2	0.7943834	0.021622	.0545666	4.94529	.558762
3.3	0.7890688	0.013456	.0517354	5.37892	.563398
3.4	0.7840336	0.007051	.0489844	5.85315	.566260
3.5	0.7792685	0.002445	.0463373	6.36951	.567632
3.6	0.7747620	0.000016	0.0438166	6.92839	0.567863

$$\eta_1 = 3.6038 .$$

$$\phi(\eta_1) = 0.774597 ; \quad -\eta_1^2\phi'(\eta_1) = 0.56786 ,$$

$$-\phi'(\eta_1) = 0.043724 ; \quad \rho_0/\bar{\rho} = 6.9504 .$$

TABLE 44

$$\frac{1}{y_0^2} = 0.8$$

η	ϕ	ρ/ρ_0	$-\phi'$	$\rho_0/\bar{\rho}(\eta)$	$-\eta^2\phi'$
0	1	1	0	1	0
0.1	0.9998510	0.99777	0.0029774	1.00134	0.000030
0.2	0.9994053	0.99110	.0059309	1.00538	.000237
0.3	0.9986664	0.98008	.0088370	1.01217	.000795
0.4	0.9976402	0.96485	.0116730	1.02165	.001868
0.5	0.9963349	0.94563	.0144170	1.03400	.003604
0.6	0.9947606	0.92265	.0170489	1.04925	.006138
0.7	0.9929295	0.89620	.0195501	1.06751	.009580
0.8	0.9908555	0.86662	.0219039	1.08891	.014018
0.9	0.9885541	0.83425	.0240960	1.11358	.019518
1.0	0.9860421	0.79947	.0261144	1.14168	.026114
1.1	0.9833374	0.76268	.0279490	1.17341	.033818
1.2	0.9804587	0.72427	.0295923	1.20900	.042613
1.3	0.9774254	0.68465	.0310398	1.24867	.052457
1.4	0.9742574	0.64418	.0322885	1.29272	.063285
1.5	0.9709744	0.60326	.0333382	1.34144	.075011
1.6	0.9675963	0.56224	.0341907	1.39520	.087528
1.7	0.9641427	0.52146	.0348498	1.45436	.100716
1.8	0.9606326	0.48121	.0353210	1.51937	.114440
1.9	0.9570845	0.44178	.0356114	1.59070	.128557
2.0	0.9535161	0.40341	.0357296	1.66888	.142918
2.1	0.9499440	0.36632	.0356854	1.75450	.157373
2.2	0.9463841	0.33070	.0354891	1.84821	.171767
2.3	0.9428509	0.29669	.0351524	1.95073	.185956
2.4	0.9393579	0.26442	.0346870	2.06285	.199797
2.5	0.9359174	0.23398	.0341053	2.18546	.213158
2.6	0.9325403	0.20543	.0334198	2.31949	.225918
2.7	0.9292364	0.17882	.0326428	2.46604	.237966
2.8	0.9260144	0.15417	.0317867	2.62625	.249208
2.9	0.9228814	0.13147	.0308637	2.80139	.259564
3.0	0.9198435	0.11071	.0298853	2.99287	.268968
3.1	0.9169058	0.091856	.0288630	3.20217	.277373
3.2	0.9140720	0.074869	.0278077	3.43090	.284751
3.3	0.9113450	0.059699	.0267297	3.68081	.291086
3.4	0.9087265	0.046289	.0256389	3.95370	.296386
3.5	0.9062173	0.034584	.0245445	4.25146	.300670
3.6	0.9038174	0.024533	.0234553	4.57599	.303981
3.7	0.9015258	0.016094	.0223799	4.92910	.306381
3.8	0.8993407	0.009251	.0213264	5.31239	.307955
3.9	0.8972595	0.004042	.0203034	5.72690	.308815
4.0	0.8952787	0.000665	0.0193196	6.17285	0.309114

$$\eta_1 = 4.0446\,.$$

$$\phi(\eta_1) = 0.894427\,; \qquad -\eta_1^2\phi'(\eta_1) = 0.30912\,,$$

$$-\phi'(\eta_1) = 0.018896\,; \qquad \rho_0/\bar{\rho} = 6.3814\,.$$

GENERAL INDEX